TRANSGENIC PLANTS - ADVANCES AND LIMITATIONS

Transgenic Plants - Advances and Limitations

Editor:

Aleksei Anatoliyovych Stepanenko, Ph.D.

Delve Publishing

www.delvepublishing.com

Transgenic Plants - Advances and Limitations

Editor: **Aleksei Anatoliyovych Stepanenko, Ph.D.**

708 3rd Avenue, 6th Floor
New York, NY 10017
United States of America
www.delvepublishing.com

ISBN: 978-1-68095-768-6
Library of Congress Control Number: 2016951524

About the Editor

Aleksei Anatoliyovych Stepanenko, Ph.D.

Aleksei A. Stepanenko received his Masters's degree in Biochemistry from V. N. Karazin Kharkiv National University and PhD in Molecular biology from Institute of Molecular Biology and Genetics, where he is currently working as a Research Fellow. His scope of interests is genomic and phenotypic evolution of tumor cells under chemotherapeutic stresses and viro-immunological approaches as an alternative perspective in cancer therapy.

Contents

List of Contributors

Simon Schiml
Botanical Institute II, Karlsruhe Institute of Technology, 76049 Karlsruhe, Germany

Holger Puchta
Botanical Institute II, Karlsruhe Institute of Technology, 76049 Karlsruhe, Germany

Pit Sze Liew
Department of Veterinary Pathology and Microbiology, Faculty of Veterinary Medicine, Universiti Putra Malaysia, 43400 Serdang, Malaysia

Mohd Hair-Bejo
Department of Veterinary Pathology and Microbiology, Faculty of Veterinary Medicine, Universiti Putra Malaysia, 43400 Serdang, Malaysia

Edward P Rybicki
Biopharming Research Unit, Department of Molecular & Cell Biology and Institute of Infectious Disease and Molecular Medicine, University of Cape Town, Private Bag X3, Rondebosch, 7701 Cape Town, South Africa

Hikmet Budak
Plant Genomics Group, Molecular Biology, Genetics and Bioengineering Program, Faculty of Engineering and Natural Sciences, Sabanci University, Istanbul, Turkey

Babar Hussain
Plant Genomics Group, Molecular Biology, Genetics and Bioengineering Program, Faculty of Engineering and Natural Sciences, Sabanci University, Istanbul, Turkey

Zaeema Khan
Plant Genomics Group, Molecular Biology, Genetics and Bioengineering Program, Faculty of Engineering and Natural Sciences, Sabanci University, Istanbul, Turkey

Neslihan Z. Ozturk
Department of Agricultural Genetic Engineering, Faculty of Agricultural Sciences and Technologies, Nigde University, Ni ˇ gde, Turkey

Naimat Ullah
Plant Genomics Group, Molecular Biology, Genetics and Bioengineering Program, Faculty of Engineering and Natural Sciences, Sabanci University, Istanbul, Turkey

Yoshikazu Tanaka
Institute for Plant Science, Suntory Holdings Ltd., 1-1-1 Wakayamadai, Shimamoto, Mishima, Osaka 618-8503, Japan

Filippa Brugliera
Florigene Pty Ltd., 1 Park Drive, Bundoora, Victoria 3083, Australia

Steve Chandler
Florigene Pty Ltd., 1 Park Drive, Bundoora, Victoria 3083, Australia

Freddy Guihéneuf
Botany and Plant Science, School of Natural Sciences, Ryan Institute, National University of Ireland Galway, Galway, Ireland

Asif Khan
Research Group Germline Biology, Centre for Organismal Studies (COS), Heidelberg University, Heidelberg, Germany

Lam-Son P. Tran
Plant Abiotic Stress Research Group & Faculty of Applied Sciences, Ton Duc Thang University, Ho Chi Minh City, Vietnam
Signaling Pathway Research Unit, RIKEN Center for Sustainable Resource Science, Tsurumi, Japan

J. Schnell C. Girard P. Macdonald
Plant and Biotechnology Risk Assessment Unit, Canadian Food Inspection Agency, 1400 Merivale Road, Ottawa, ON K1A 0Y9, Canada

C. Girard
e-mail: cecile.girard@inspection.gc.ca

M. Steele A. Savoie
Animal Feed Division, Canadian Food Inspection Agency, 59 Camelot Drive, Ottawa, ON K1A 0Y9, Canada

A. Savoie
e-mail: annie.savoie@inspection.gc.ca

J. Bean L. Bourbonnie`re
Food Directorate, Health Canada, 251 Sir Frederick Banting Driveway, Ottawa, ON K1A 0K9, Canada

L. Bourbonnie`re
e-mail: luc.bourbonniere@hc-sc.gc.ca

M. Neuspiel
Office of the Chief Science Operating Officer, Canadian Food Inspection Agency, 1400 Merivale Road, Ottawa, ON K1A 0Y9, Canada

N. Dormann C. Pearson
Plant Biosafety Office, Canadian Food Inspection Agency, 59 Camelot Drive, Ottawa, ON K1A 0Y9, Canada

C. Pearson
e-mail: cindy.pearson@inspection.gc.ca

Richard Strasser
Department of Applied Genetics and Cell Biology, University of Natural Resources and Life Sciences, Vienna, Austria

Alexandre Martinière
Biochimie et Physiologie Moléculaire des Plantes, Institut de Biologie Intégrative des Plantes, UMR 5004 CNRS/UMR 0386 INRA/Montpellier SupAgro/Université Montpellier 2, Montpellier, France

Guilhem Desbrosses
Laboratory of Tropical and Mediterranean Symbioses (UMR113, Université Montpellier 2, Institut de Recherche pour le Développement,Cirad Montpellier SupAgro, Institut National de la Recherche Agronomique), Université Montpellier 2, Montpellier, France

Hervé Sentenac
Biochimie et Physiologie Moléculaire des Plantes, Institut de Biologie Intégrative des Plantes, UMR 5004 CNRS/UMR 0386 INRA/Montpellier SupAgro/Université Montpellier 2, Montpellier, France

Nadine Paris
Biochimie et Physiologie Moléculaire des Plantes, Institut de Biologie Intégrative des Plantes, UMR 5004 CNRS/UMR 0386 INRA/Montpellier SupAgro/Université Montpellier 2, Montpellier, France

Vadim Volkov
Faculty of Life Sciences and Computing, London Metropolitan University, London, UK

Yu-Chieh Kuo
Graduate School of Biotechnology and Bioengineering, Yuan Ze University, 135 Yuan-Tung Road, Taoyuan 32003, Taiwan

Chia-Chun Tan
Graduate School of Biotechnology and Bioengineering, Yuan Ze University, 135 Yuan-Tung Road, Taoyuan 32003, Taiwan
Department of Life Sciences, National Central University, 300, Jhongda Rd., Taoyuan 32001, Taiwan

Jung-Ting Ku
Graduate School of Biotechnology and Bioengineering, Yuan Ze University, 135 Yuan-Tung Road, Taoyuan 32003, Taiwan

Wei-Cho Hsu
Graduate School of Biotechnology and Bioengineering, Yuan Ze University, 135 Yuan-Tung Road, Taoyuan 32003, Taiwan

Sung-Chieh Su
Graduate School of Biotechnology and Bioengineering, Yuan Ze University, 135 Yuan-Tung Road, Taoyuan 32003, Taiwan

Chung-An Lu
Department of Life Sciences, National Central University, 300, Jhongda Rd., Taoyuan 32001, Taiwan

Li-Fen Huang
Graduate School of Biotechnology and Bioengineering, Yuan Ze University, 135 Yuan-Tung Road, Taoyuan 32003, Taiwan

Maria Lígia R. Macedo
Department of Food Technology and Public Health, University of Mato Grosso do Sul, Campo Grande, Mato Grosso do Sul 79070-900, Brazil

Caio F. R. Oliveira
Department of Food Technology and Public Health, University of Mato Grosso do Sul, Campo Grande, Mato Grosso do Sul 79070-900, Brazil
Department of Biochemistry and Tissue Biology, University of Campinas, Campinas, Sao Paulo 13083-862, Brazil

Carolina T. Oliveira
Department of Food Technology and Public Health, University of Mato Grosso do Sul, Campo Grande, Mato Grosso do Sul 79070-900, Brazil

Carolina Simó, Clara Ibáñez
Laboratory of Foodomics, Institute of Food Science Research (CIAL), Spanish National Research Council (CSIC), Nicolas Cabrera 9, Cantoblanco Campus, Madrid 28049, Spain

Alberto Valdés
Laboratory of Foodomics, Institute of Food Science Research (CIAL), Spanish National Research Council (CSIC), Nicolas Cabrera 9, Cantoblanco Campus, Madrid 28049, Spain

Alejandro Cifuentes
Laboratory of Foodomics, Institute of Food Science Research (CIAL), Spanish National Research Council (CSIC), Nicolas Cabrera 9, Cantoblanco Campus, Madrid 28049, Spain

Virginia García-Cañas
Laboratory of Foodomics, Institute of Food Science Research (CIAL), Spanish National Research Council (CSIC), Nicolas Cabrera 9, Cantoblanco Campus, Madrid 28049, Spain

Johnathan A. Napier
Department of Biological Chemistry and Crop Protection, Rothamsted Research, Harpenden, UK

Sarah Usher
Department of Biological Chemistry and Crop Protection, Rothamsted Research, Harpenden, UK

Richard P. Haslam
Department of Biological Chemistry and Crop Protection, Rothamsted Research, Harpenden, UK

Noemi Ruiz-Lopez
Instituto de la Grasa (CSIC), Seville, Spain

Olga Sayanova
Department of Biological Chemistry and Crop Protection, Rothamsted Research, Harpenden, UK

Marcelo Rogalski
Laboratório de Fisiologia Molecular de Plantas, Departamento de Biologia Vegetal, Universidade Federal de Viçosa, Viçosa, Brazil

Leila do Nascimento Vieira,
Laboratório de Fisiologia do Desenvolvimento e Genética Vegetal, Programa de Pós-graduação em Recursos Genéticos Vegetais, Centro de Ciências Agrárias, Universidade Federal de Santa Catarina, Florianópolis, Brazil

Hugo P. Fraga
Laboratório de Fisiologia do Desenvolvimento e Genética Vegetal, Programa de Pós-graduação em Recursos Genéticos Vegetais, Centro de Ciências Agrárias, Universidade Federal de Santa Catarina, Florianópolis, Brazil

Miguel P. Guerra
Laboratório de Fisiologia do Desenvolvimento e Genética Vegetal, Programa de Pós-graduação em Recursos Genéticos Vegetais, Centro de Ciências Agrárias, Universidade Federal de Santa Catarina, Florianópolis, Brazil

Susan Breen
Plant Sciences Division, Research School of Biology, The Australian National University, Canberra, ACT, Australia

Peter S. Solomon
Plant Sciences Division, Research School of Biology, The Australian National University, Canberra, ACT, Australia

Frank Bedon
Department of Economic Development, AgriBio, Bundoora, VIC, Australia
AgriBio, La Trobe University, Bundoora, VIC, Australia

Delphine Vincent
Department of Economic Development, AgriBio, Bundoora, VIC, Australia

Umesh P. Yadav
Department of Biological Sciences, University of North Texas, Denton, TX, USA

Brian G. Ayre
Department of Biological Sciences, University of North Texas, Denton, TX, USA

Daniel R. Bush
Department of Biology, Colorado State University, Fort Collins, CO, USA

Preface

The advances in transgenic techniques, genome editing systems, and functional genomics tools for identification and utility of signaling molecules for improvement in plant drought and saline tolerance as well as for directly manipulating key C and N transporters to improve crop yields and nutritional quality are reviewed. The use of plant insecticidal lectins and secreted peptides with antimicrobial activities as biotechnological tools for pest and pathogens control and crop protection in order to increase crop yields by minimizing quantity and quality losses is also described. Plants have been successfully used to produce recombinant cytokines, vaccines, antibodies, and other protein. Examples of plant-based vaccines for therapy of animal and even human viral diseases including the recent use of ZMapp for Ebolavirus infection are provided. There are important issues that have to be addressed in the future to overcome challenges in plant glycosylation engineering to produce active recombinant proteins. Several methods of improving protein yield and quality are proposed. This book also encompasses current progress in successful genetic manipulations of color in pot plants and cut-flower species, in development of transgenic plants as terrestrial sources of omega-3 fish oils, in development of genetically encoded plant pH-sensors, which can be used to measure pH in the microenvironment of living tissues. Finally, this book provides insight into the recent progress in genomics and metabolomics studies on transgenic crops focusing on the occurrence of unintended changes resulting from genetic manipulations as well as the pre-market biosafety assessments of genetically engineered plants and food and feeds derived from genetically engineered plants.

Editor
Aleksei Anatoliyovych Stepanenko, Ph.D.

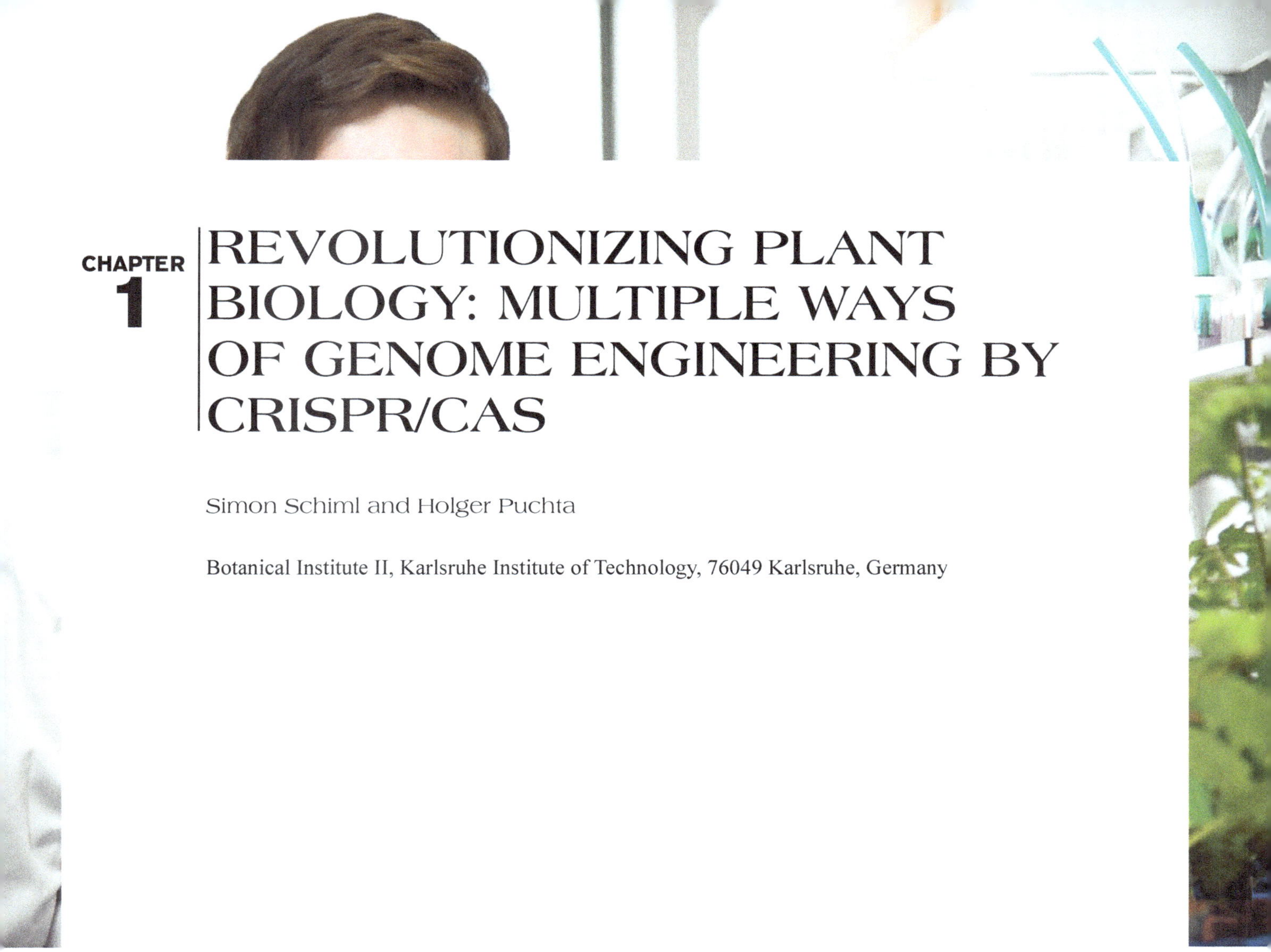

CHAPTER 1

REVOLUTIONIZING PLANT BIOLOGY: MULTIPLE WAYS OF GENOME ENGINEERING BY CRISPR/CAS

Simon Schiml and Holger Puchta

Botanical Institute II, Karlsruhe Institute of Technology, 76049 Karlsruhe, Germany

ABSTRACT

The precise manipulation of plant genomes relies on the induction of DNA double-strand breaks by site-specific nucleases to initiate DNA repair reactions that are either based on non-homologous end joining (NHEJ) or homologous recombination (HR). Recently, the CRISPR/Cas system emerged as the most important tool for genome engineering due to its simple structure and its applicability to a wide range of organisms. Here, we review the current status of its various applications in plants, where it is used for the successful generation of stable mutations in a steadily growing number of species through NHEJ. Furthermore, tremendous progress in plant genome engineering by HR was obtained by the setup of replicon mediated and in planta gene targeting techniques. Finally, other complex approaches that rely on the induction of more than one DNA lesion at a time such as paired nickases to avoid off-site effects or controlled genomic deletions are beginning to be applied routinely

BACKGROUND

Although with the rise of the CRISPR/Cas (clustered regularly interspaced short palindromic repeats and CRISPR-associated) technology double-strand break (DSB)-induced genome engineering moved into the centre of scientific interest in the last 2 years, the basic principles behind this approach were well known for decades. Previous experiments demonstrated that by induction of a unique DSB in plant genomes using a highly specific endonuclease, different types of genome manipulations could be achieved. On one hand, the frequency of gene targeting (GT), the precise integration of a T-DNA via HR with sequences identical to the genomic locus, can be increased by several orders of magnitude [1]. On the other hand, functional open reading frames can be destroyed by imprecise NHEJ [2]. Interestingly, even in the absence of homologies, integration of a T-DNA into a genomic locus by NHEJ can be enhanced by DSB induction [2]. Depending on the location, the induction of two DSBs can result in deletions [3] or reciprocal exchanges between chromosome arms [4].

For many years, these experiments were performed using mega nucleases such as I-*Sce*I [5, 6]. However, this method required the introduction of the respective target sequence into the plant genome prior to the genome engineering experiment itself. Therefore, targeting endogenous loci is excluded, as engineering the protein for new target sequences is strongly limited [7, 8]. A major improvement was marked by the introduction of zinc finger nucleases (ZFNs) [9, 10]: the DNA cleavage domain from the restriction enzyme *Fok*I was fused to the highly variable DNA binding domain (DBD) of a class

Citation: Simon Schiml and Holger Puchta, Revolutionizing plant biology: multiple ways of genome engineering by CRISPR/Cas, doi: 10.1186/s13007-016-0103-0.

of zinc finger transcription factors. By combining different zinc fingers in the DBD, different target sites in the DNA are recognized and are cleaved by the nuclease. Unfortunately there are drawbacks to this technique as not all combinations of zinc fingers function well, therefore every new nuclease has to be tested extensively. Furthermore, cloning of a new nuclease is quite time consuming. However, ZFNs are still in use today primarily because there are no open questions regarding their status as intellectual property. The discovery and molecular analysis of transcription activator-like effectors (TALEs) from the plant pathogen species *Xanthomonas* led to the third important class of engineered nucleases [11]. The TALE DBD consists of numerous repeats, varying only in two amino acid residues. In 2009, it was shown that each of the repeating sequences is able to recognize and bind exactly one nucleotide on the DNA [12, 13]. An engineered TALE nuclease was created by fusing the DBD once again to *Fok*I, in these experiments, cloning of a new nuclease was facilitated by adapting the GoldenGate cloning method [14].

The most recent, yet already the most important type of programmable nucleases utilizes the CRISPR/Cas system. It was originally discovered in the 1980s as a distinctive genomic locus in *E. coli* [15, 16] and was later characterized to serve as an adaptive immune system in many bacteria and archaea [17]. However, the molecular mechanism of the CRISPR/Cas system from *Streptococcus pyogenes* was not deciphered until 2012 [18]: foreign plasmid or viral DNA entering the bacterial cells are degraded by a single protein, the nuclease Cas9. The target specificity is governed by a short so-called CRISPR-RNA (crRNA), which is encoded in the CRISPR-locus and which is complementary to the invading DNA so that it can bind directly to the foreign DNA using a stretch of 20 nts. An additional short sequence motif next to the target sequence, termed protospacer adjacent motif (PAM), is needed for the correct recognition of the target site. A second short RNA, the trans-activating CRISPR-RNA (tracrRNA), binds to the crRNA, and a stable complex is formed with Cas9. The foreign DNA is then cleaved by two nuclease domains of Cas9. Furthermore, it was shown that the two RNAs can also be fused together to form a so-called single-guide RNA (sgRNA, Fig. 1). At the beginning of 2013, the groups of Feng Zhang at MIT and George Church at Harvard demonstrated that Cas9 can be used for genome engineering in human cell cultures, proving that it also works in eukaryotic systems [19, 20]. Since then, the CRISPR/Cas system has made unprecedented success as a tool for genome engineering due to its ease in cloning new sequence-specific nucleases and the fact that it works in almost any organism. Here, we review the major technical advances that have been made with Cas9 in plants.

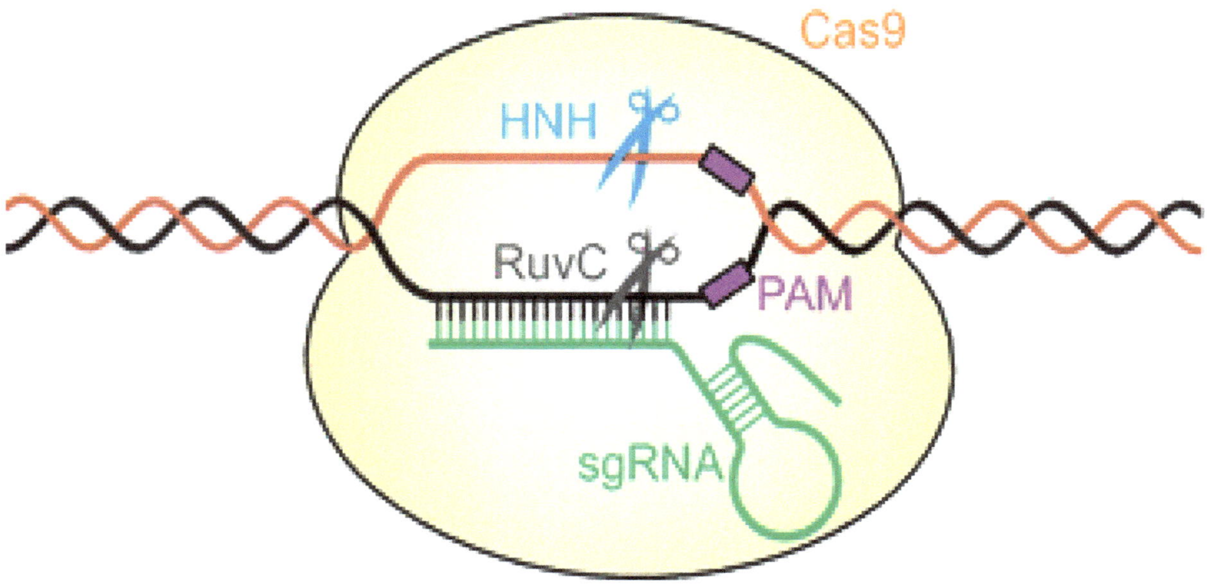

Figure. 1: Schematic representation of the Cas9 cleavage mechanism. The Cas9/sgRNA complex recognizes and binds the complementary sequence next to the PAM, which is highly specific for each Cas9 from different bacterial species. The RuvC domain and HNH motif of ...

Creating Heritable Mutations with RNA-Guided Cas9 in Plants

In August 2013, three studies were published in the same issue of nature biotechnology showing that Cas9 also works in plant cells. Scientists from the Chinese Academy of Sciences in Beijing demonstrated the disruption of endogenous genes in rice and wheat protoplasts as well as in rice calli [21] while researchers from Harvard reported cleavage activity of Cas9 in protoplasts of *Arabidopsis thaliana* and *Nicotiana benthamiana* [22]. Finally, the Sainsbury Lab in the UK used *N. benthamiana* leaf tissue for agroinfiltration, demonstrating Cas9-mediated disruption of the *PDS* gene (phytoene desaturase) [23].

However, stable inheritance of Cas9-mediated mutations in natural genes was not reported until 2014, when Feng et al. used the human codon-optimized Cas9 (hCas9) from F. Zhang [19] under the control of two CaMV 35S promoters and 12 different sgRNAs driven by the AtU6-26 promoter to target endogenous sites in *A. thaliana* [24]. U6 promoters control transcription by RNA polymerase III, which is specific for the production of short non-coding RNAs. The Cas9-harbouring constructs were stably integrated through *Agrobacterium tumefaciens*. T1 analyses revealed high mutation frequencies ranging from 30 % up to 92 % with 1-bp insertions being the dominant type of mutation. Investigation of T2 and T3 generations revealed Mendelian segregation and stable inheritance of mutations together with the rise of new somatic mutations, which are a result of ongoing DSB induction by the presence of the nuclease transgene. By using an interrupted GUS reporter construct [25], the ability of Cas9 to induce HR was also shown. By whole genome sequencing and Sanger sequencing of potential off-target sites, it was also demonstrated that the system is highly specific in plants as no significant off-target activity was detected. The same group also reported the generation of mutants for a set of genes in rice [26]. By transforming calli with subsequent regeneration of transgenic T0 plants, Cas9 activity could be observed early in plant development, leading to a high number of mutated plants. Additionally, heredity of some mutations to the T1 generation was demonstrated but not analysed in detail due to a high number of secondary mutations that created chimeric plants.

Soon after the reports mentioned above, the use of a distinct construct in *A. thaliana* was reported [27]. A codon optimized version of Cas9 is controlled by a plant Ubiquitin promoter, the sgRNA is expressed by the AtU6-26 promoter and both elements are located on the same T-DNA. Following stable transformation, homozygous mutations were induced at a frequency of up to 70 % in two endogenous marker genes of *A. thaliana* and in addition, inheritance of mutations into T2 and T3 generations as well as their segregation in a Mendelian fashion was demonstrated. Mutation patterns were analysed by next generation sequencing, which again exhibited small insertions of 1-bp to be the most dominant form of alteration. Furthermore, the first application of the Cas9-D10A nickase variant to plants was reported here. By inducing a point mutation to Cas9 that inactivates one of the nuclease domains, the nuclease is converted to a single-strand break (SSB) inducing nickase. NGS did not reveal a mutagenic potential of the nickase. However, when the constructs were applied to the HR reporter plant lines DGU.US and IU.GUS [28], the nickase was able to induce HR at least as efficiently as the Cas9 nuclease, rendering it a promising tool for HR-based genome engineering approaches.

Zhou et al. presented a study on the application of RNA-guided Cas9 in rice [29]. Here, a rice codon-optimized Cas9 was put under control of a maize ubiquitin promoter. Notably, two different sgRNAs, that have successfully been used in the mammalian system before (one with 48 nt tracr-tail [19] and one with 85 nt tracr tail [20]), as well as a dual guide system (dgRNA) with separate crRNA and tracrRNA, were used all under the control of a rice U6 promoter. When stably transformed to rice plants, the 48 nt sgRNA did not induce any detectable mutations, while the induction by the dgRNA was very low in the T0 generation (2 out of 16 transformants with a mono-allelic mutation). One might speculate, that these RNA constructs are not able to form a stable RNA-Cas9 complex under plant growth conditions. However, the 85 nt sgRNA did induce NHEJ events in the transformed generation with efficiencies ranging from 20 % up to 100 % of a small number of transformants. All of the mutations observed were already bi-allelic and the two mutated alleles showed Mendelian segregation in T1 and in transgene-free T2 plants. Furthermore, the induction of large deletions of up to 245 kb was observed in transgenic plants when two sgRNAs were applied, although inheritance was not investigated in this approach.

The van Eck group at the Cold Spring Harbor Laboratory was the first to demonstrate Cas9-mediated heritable mutations in tomato [30]. The hCas9 driven by the 35S promoter was combined with two sgRNAs under AtU6 control, intending to create a specific deletion in the*SlAGO7* gene for easy detection of mutations. Almost half of all transformants exhibited the recessive *wiry2-1* growth phenotype [31], indicating homozygous mutations early in development. Sanger sequencing confirmed the presence of respective mutations. However, in addition to the expected deletion, larger deletions that included the target size were observed, as well as small mutations affecting one or both of the target sequences. To show inheritance of mutations, wild-type flowers were pollinated by the mutant plants and the offspring were analysed. Heterozygous plants without the Cas9 transgene but with one wild-type and one mutant allele could be identified, thus confirming heredity of the mutations.

The group of Qi-Jun Chen from Beijing applied their RNA-guided Cas9 to *A. thaliana* and maize [32]. Using a set of different constructs (maize codon-optimised zCas9 or hCas9, Ubiquitin or 2 × 35S promoter, AtU6-26, OsU3 or TaU3 promoter for the sgRNA), targeted mutagenesis was demonstrated in both maize protoplasts and transgenic plants, with the combination of zCas9 and TaU3p exhibiting the highest efficiency. By using different Pol III promoters, it was also possible to assemble up to four different sgRNA expression cassettes on one vector for multiplex genome engineering. Heritability of mutations was confirmed in *A. thaliana* by checking for transgene-free T2 plants via PCR and subsequent sequencing of the Cas9 target sites. At the moment, the number of reports of new stably mutated plant species using CRISPR/Cas is rapidly growing [33, 34].

Harnessing different Cas9 Orthologues

A major step towards more complex applications of the CRISPR/Cas system was made by the adaption of additional Cas9 nucleases to genome engineering. The most widely used Cas9 nuclease originates from *Streptococcus pyogenes*. However, it was shown in mammalian cell culture that Cas9 orthologues from other species are also applicable for targeting unique genomic sites [35, 36]. Recently, it was shown that nucleases from *Streptococcus thermophilus* and *Staphylococcus aureus*also work efficiently in *A. thaliana* [37]. Stably transformed constructs contained a Cas9 expression system together with the species-specific sgRNA with distinct PAM-specificities. It was demonstrated that both nucleases led to highly efficient mutagenesis. For *S. aureus,* Cas9 targeting a specific PAM ('NNGGGT') increased the mutation frequency to almost 90 %, with the most dominant form of mutations being deletions. Mutations in the *ADH1* locus were shown to be heritable for both nucleases. Furthermore, it was demonstrated that cross interferences between Cas9 and sgRNA from different species do not occur. These findings provide the basis for more complex approaches, enabling the simultaneous control of different enzymatic activities in a single plant cell [38].

Inducing Genomic Change during Plant and Organ Development

Although the use of constitutive promoters exhibits high frequency mutagenesis for Cas9, other promoters are also desirable. This offers the possibility to achieve stable mutations more quickly, especially for plant species with long generation times. Additionally, conditional knockout or the mutagenesis of specific genes in unique organs is rendered possible by the use of respective promoters.

Together with scientists from Korea, George Coupland showed high-frequency mutagenesis in *A. thaliana* [39]. In their construct, instead of a constitutive promoter, the hCas9 was controlled by the *INCURVATA2*promoter, which is highly active in meristematic tissue. For the sgRNA, the AtU6-26 promoter was used. Mutagenesis frequencies for three endogenous targets ranged from 10 % to ~85 %, and T2 and T3 analysis revealed segregation of certain mutations as expected along with the induction of new mutations in the respective generation.

Qi-Jun Chen's group from Beijing reported a similar experiment [40]: the zCas9 was put under control of the promoter from the egg cell-specific *EC1.2* of *A. thaliana* to increase heritability by inducing mutations in egg cells. Together with respective sgRNAs, T1 double and triple mutant *Arabidopsis* plants could be obtained that also segregated in the T2 generation. Furthermore, different combinations out of eight promoters and two terminators were analysed for their mutagenesis efficiency. The highest number of T1 triple mutants (17 %) was observed using a combination of the *EC1.2* enhancer and *EC1.1*promoter together with the *rbcS E9* terminator. Although this approach does not seem to significantly increase heritability of mutations compared to using a somatic expression system, having the ability to control mutagenesis through different developmental stages can be of great importance for studying the function of certain genes.

Off-site Effects and How to avoid them

A major concern when using an RNA-guided Cas9 is off-site activity. Although extensive studies that address this issue have been performed in the bacterial and mammalian system, the exact extent of off-site activity is still not completely clear. For plants, there is little data currently available that addresses off-site activity. Sequencing of bioinformatically identified putative off-target sites showed no detectable events in *A. thaliana*, *N. benthamiana*, wheat, rice and sweet orange [21–24, 29, 41, 42]. Whole-genome sequencing of mutated *A. thaliana* plants also resulted in no off-target events [24]. In contrast to these results, a study using Cas9 in rice found a putative off-target site to be mutated in 1.6 % of the investigated plants, although this was still five times less frequent than the on-target site [43]. A study covering the application of Cas9 to soybean reported an off-target frequency of 13 % [44]. It has to be noted, however, that in this study, two paralogues of *DDM1* were targeted and therefore the high sequence similarity is likely to cause the high amount of off-target activity.

A solution to off-target activity proposed by experiments in human cell culture is the use of two Cas9 nickases [45]. By introducing a point mutation (D10A) into one of the nuclease domains of Cas9, the enzyme is converted into a SSB inducing nickase. The nickase can then be guided to two adjacent positions in the genome by two distinct sgRNAs resulting in SSBs on each of the two DNA strands (Fig. 2). The result is a DSB that can also lead to NHEJ-mediated mutagenesis. Hence, the specificity is increased, as independent off-site binding of the sgRNAs does not lead to mutations. It has already been demonstrated that this approach is also applicable to plants. By adding a second sgRNA to a Cas9 construct and subsequent stable transformation into *A. thaliana*, a mutagenesis rate comparable to that of the Cas9 single nuclease was achieved [46]. Notably, the mutation pattern shifted from small insertions for the nuclease to larger deletions being the most dominant mutagenesis outcome. Again, mutations were proven to be heritable by demonstrating the presence of mutations in transgene-free T2 plants.

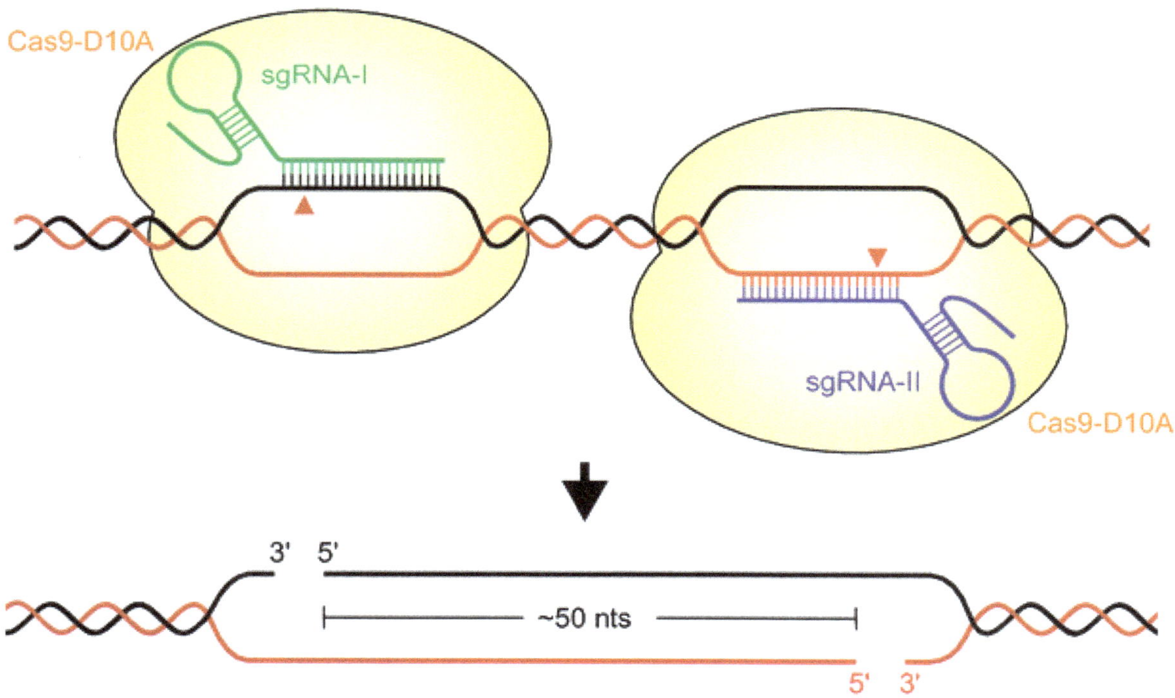

Figure. 2: Cas9 paired nickases approach. By using 2 sgRNAs, the D10A nickase variant can be guided to the two opposite DNA strands at adjacent positions. The resulting BSD exhibits long single-stranded 5′-overhangs

Utilizing Homologous Recombination

Site-specific integration of transgenes or precise genome alterations (referred to hereafter as gene targeting, GT) have always been major challenges in plant genome engineering. This is because NHEJ is by far the preferred mechanism to repair DNA breaks in somatic plant tissue. However, using the different classes of engineered nucleases, a wide variety of successful GT experiments have been performed [47]. A major step was taken with the development of the I-*Sce*I-based *in planta* GT system [48], which allowed for GT rates of more than 1 % in*A. thaliana* without having to rely on high transformation rates. Using Cas9, it was possible to improve this technique even further [46]: the number of T-DNAs needed for the system to work was reduced from three (donor sequence, artificial I-*Sce*I target sequence and I-*Sce*I expression system) to only one, harbouring both the donor sequence and the Cas9/sgRNA expression cassettes. Furthermore, the system's target flexibility allowed for the targeting of an endogenous locus in the*Arabidopsis* genome. The precise site-specific integration of the donor sequence into this locus and the possible inheritance of this manipulation into the next generation were demonstrated by using the experimental setup shown in Fig. 3.

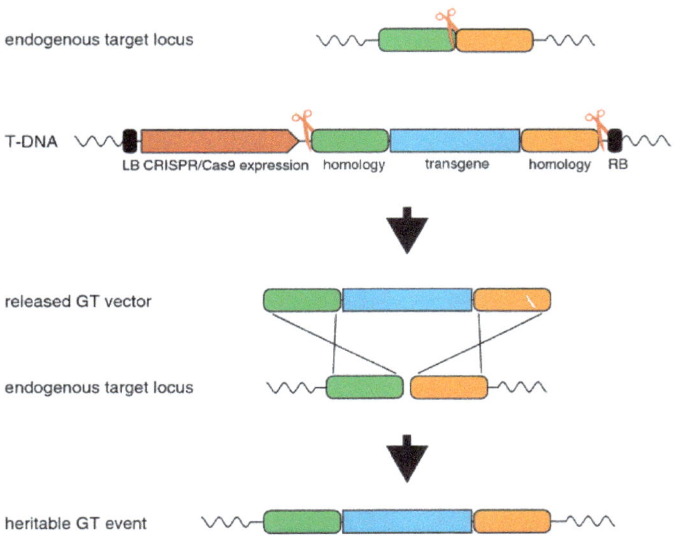

Figure. 3: Overview of the Cas9-mediated *in planta* GT system. The nuclease and the DNA donor sequence are located on one T-DNA that is stably transformed into the plant. The nuclease induces two DSBs that release the donor intermediate and a third DSB that activates .

In an innovative approach, the group of Dan Voytas combined Cas9-mediated mutagenesis and GT with geminivirus-based replicons [49].*Arabidopsis* plants were stably transformed with a specific T-DNA flanked by the viral large intergenic regions (LIRs). Upon co-transformation of the viral replication-initiation protein, replicational release, circularisation and rolling-circle replication of the replicon is initiated at the LIRs. The circularisation leads to two 35S promoters becoming correctly oriented in front of the desired gene, i.e., the nuclease ORF. With this strategy, NHEJ-mediated mutagenesis was demonstrated with ZFNs, TALENs and Cas9. Furthermore, it was shown that when a GT donor sequence was added to the replicon, ZFN-mediated GT could be achieved (Fig. 4). This strategy was also shown to be applicable using Cas9 in tomato [50].

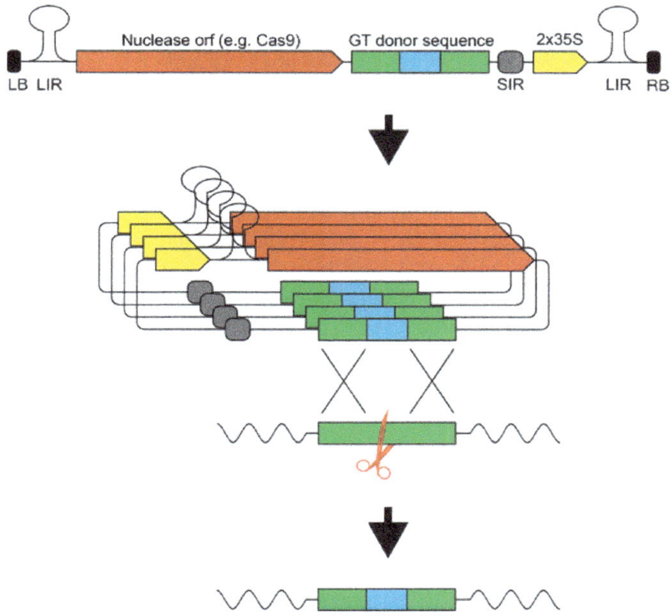

Figure. 4: Replicon-mediated GT. The replicon is released from the T-DNA, circularises at the LIRs and undergoes rolling circle replication. This leads to the promoter being positioned upstream of the nuclease ORF. Upon DSB-induction in the target locus, the integration .

Recently, scientists from DuPont Pioneer reported CRISPR/Cas-mediated GT in soybean [34]. Cas9-sgRNA constructs and a donor template were co-transformed into embryonic callus by particle bombardment. For two target sites on chromosome 4, the correct integration of the hygromycin resistance gene *HPT* was identified by PCR and confirmed by Southern blot in T0, and respective plants were regenerated. For one target in T1, three plants with the correct GT event and without any additional transgenes could be identified, confirming inheritance of the new allele. For the second target, only events with additional integration of either Cas9 at the target site or the donor template at random genomic sites were isolated. Notably, the amount of donor- and sgRNA-Cas9-DNA was not optimised to obtain true GT events. Furthermore, as this was a proof-of-principle study, only a small number of events was regenerated, and a true GT event for the second target is likely to be found when analysing a higher number of events.

Multiplex Genome Engineering

The architecture of the CRISPR/Cas system with the constant Cas9 protein and the sgRNA-derived target specificity provides the opportunity to target multiple sites at once, as is the case in the natural bacterial system. However, we are only beginning to exploit this possibility, and successful imitation of the bacterial system with polycistronic crRNAs and tracrRNAs has not yet been reported for plants. Therefore, the most common approach by some groups to achieve multiplex sgRNA expression is to simply assemble numerous sgRNA expression systems, each with its own promoter [32, 46, 51]. However, this method is limited as constructs become very large with a Pol III promoter for every sgRNA. Xie et al. developed a clever solution to overcome this problem [52]: tRNA sequences were put in between the sgRNA sequences on the construct to create a single polycistronic gene. Two host-endogenous nucleases cleave the expressed RNA at the borders between tRNA and sgRNA creating individual sgRNAs. Simultaneous multiplex mutagenesis was demonstrated in rice protoplasts and transgenic plants for different numbers of sgRNAs (up to eight) in a tRNA-sgRNA array.

Scientists from KAUST in Saudi Arabia demonstrated the potential of plant viruses for multiplex genome engineering [53]. In their study, the Tobacco rattle virus (TRV) was used to deliver sgRNAs to transgenic*N. benthamiana* that stably overexpressed Cas9. The two TRV RNAs were introduced into the plants by leaf agroinfiltration with a mixture of two bacterium cultures, one with a plasmid for RNA1, one for RNA2. The latter contained the sgRNA expression system. After reconstitution of the TRV in the infiltrated tissue, a systemic infection throughout the plant leads to sgRNA expression in all tissues and therefore efficient mutagenesis. By mixing bacterial cultures with different RNA2 vectors, simultaneous mutagenesis of two loci was achieved, indicating the potential for multiplex genome engineering by TRV-mediated sgRNA delivery.

Controlling Transcription with Cas9

Targeted genome manipulations with RNA-guided Cas9 have not only been achieved by altering the genome itself but also by influencing transcription. By guiding a catalytically inactive dead (d) Cas9 to a promoter or coding region, transcription can be efficiently blocked [54]. This effect can be enhanced by fusing a repressor protein such as the KRAB domain to dCas9. Likewise, an activator such as VP64 can also be fused to targeted dCas9 to activate transcription of a specific gene [55].

Magdy Mahfouz's group from KAUST was able to transfer this system to plants [56]. The C-terminus of the human codon-optimized dCas9 was fused to the EDLL domain [57] or to the TAL activation domain [58] to create artificial activators. A repressor was created by fusing the SRDX EAR motif [59]. Successful transcriptional activation or repression could be observed in infiltrated *N. benthamiana* leaves by measuring expression levels of a transient *GUS* gene or the endogenous*PDS*. Activation was highest when the activator was guided to the sense strand of the promoter near the transcriptional start site and both activator constructs performed at a comparable level. Repression of PDS was demonstrated for both dCas9 alone and for the dCas9: SRDX fusion construct and could be increased by guiding the complex to several target sites within the promoter and the first exon of the gene simultaneously.

Conclusions and Outlook

In the last 2 years, CRISPR/Cas has emerged as the most important tool for molecular biology due to its simplicity, versatility and efficiency. The immediate benefit for plant scientists is the possibility to rapidly create mutations in genes where no known T-DNA insertion or EMS mutant is available. Use of this method will therefore lead to a more complete understanding of gene function in plants. This approach can not only be applied to genes with unknown functions but also to genes for which we must revise our current knowledge due to the option to produce true knock-out mutants. For instance, scientists from UCSD recently demonstrated that a putative key player of auxin signalling in Arabidopsis (*ABP1*) does not possess the important function that was inferred from the analysis of plants obtained by more classical mutagenesis techniques [60]. However, NHEJ-mediated mutation is only a first, yet important step. Specific changes of single amino acids or integration of a larger piece of DNA in the plant genome can be achieved by using Cas9-based GT systems, while efficient multiplex systems will allow the complex rearrangement of chromosomes. Thus, as well as coming closer to developing synthetic plant genomes, we will be also able to obtain plants with a single engineered point mutation that cannot be discriminated from natural varieties. Such plants can even be obtained without the use of transgenic DNA [61], which will also help accelerate the acceptance of Cas9 mutagenized crop plants for agricultural use in the general public.

ACKNOWLEDGEMENTS

We apologize to all colleagues in the field for not being able to cite all of the rapidly growing number of reports published on the use of the RNA-guided Cas9 technology in plants due to space limitations. Furthermore, we want to thank our colleagues Friedrich Fauser and Jeannette Steinert for sharing projects and enthusiasm for Cas9-mediated genome engineering in our group.

REFERENCES

1. Puchta H, Dujon B, Hohn B. Two different but related mechanisms are used in plants for the repair of genomic double-strand breaks by homologous recombination. Proc Natl Acad Sci USA. 1996;93:5055–5060. doi: 10.1073/pnas.93.10.5055.
2. Salomon S, Puchta H. Capture of genomic and T-DNA sequences during double-strand break repair in somatic plant cells. EMBO J. 1998;17:6086–6095. doi: 10.1093/emboj/17.20.6086.
3. Siebert R, Puchta H. Efficient repair of genomic double-strand breaks by homologous recombination between directly repeated sequences in the plant genome. Plant Cell. 2002;14:1121–1131. doi: 10.1105/tpc.001727.
4. Pacher M, Schmidt-Puchta W, Puchta H. Two unlinked double-strand breaks can induce reciprocal exchanges in plant genomes via homologous recombination and nonhomologous end joining. Genetics. 2007;175:21–29. doi: 10.1534/genetics.106.065185.
5. Puchta H. The repair of double-strand breaks in plants: mechanisms and consequences for genome evolution. J Exp Bot. 2005;56:1–14. doi: 10.1093/jxb/eri123.
6. Jacquier A, Dujon B. An intron-encoded protein is active in a gene conversion process that spreads an intron into a mitochondrial gene. Cell. 1985;41:383–394. doi: 10.1016/S0092-8674(85)80011-8.
7. Steuer S, Pingoud V, Pingoud A, Wende W. Chimeras of the homing endonuclease PI-SceI and the homologous Candida tropicalis intein: a study to explore the possibility of exchanging DNA-binding modules to obtain highly specific endonucleases with altered specificity. Chem Biochem. 2004;5:206–213.
8. Gimble FS, Moure CM, Posey KL. Assessing the plasticity of DNA target site recognition of the PI-SceI homing endonuclease using a bacterial two-hybrid

selection system. J Mol Biol. 2003;334:993–1008. doi: 10.1016/j.jmb.2003.10.013.

9. Kim YG, Cha J, Chandrasegaran S. Hybrid restriction enzymes: zinc finger fusions to Fok I cleavage domain. Proc Natl Acad Sci USA. 1996;93:1156–1160. doi: 10.1073/pnas.93.3.1156.

10. Smith J, Bibikova M, Whitby FG, Reddy AR, Chandrasegaran S, Carroll D. Requirements for double-strand cleavage by chimeric restriction enzymes with zinc finger DNA-recognition domains. Nucleic Acids Res. 2000;28:3361–3369. doi: 10.1093/nar/28.17.3361.

11. Bonas U, Stall RE, Staskawicz B. Genetic and structural characterization of the avirulence gene avrBs3 from Xanthomonas campestris pv. vesicatoria. Mol Gen Genet. 1989;218:127–136. doi: 10.1007/BF00330575.

12. Boch J, Scholze H, Schornack S, Landgraf A, Hahn S, Kay S, et al. Breaking the code of DNA binding specificity of TAL-type III effectors. Science. 2009;326:1509–1512. doi: 10.1126/science.1178811.

13. Moscou MJ, Bogdanove AJ. A simple cipher governs DNA recognition by TAL effectors. Science. 2009;326:1501. doi: 10.1126/science.1178817.

14. Cermak T, Doyle EL, Christian M, Wang L, Zhang Y, Schmidt C, et al. Efficient design and assembly of custom TALEN and other TAL effector-based constructs for DNA targeting. Nucleic Acids Res. 2011;39:e82. doi: 10.1093/nar/gkr218.

15. Stern MJ, Ames GF, Smith NH, Robinson EC, Higgins CF. Repetitive extragenic palindromic sequences: a major component of the bacterial genome. Cell. 1984;37:1015–1026. doi: 10.1016/0092-8674(84)90436-7.

16. Ishino Y, Shinagawa H, Makino K, Amemura M, Nakata A. Nucleotide sequence of the iap gene, responsible for alkaline phosphatase isozyme conversion in Escherichia coli, and identification of the gene product. J Bacteriol. 1987;169:5429–5433.

17. Wiedenheft B, Sternberg SH, Doudna JA. RNA-guided genetic silencing systems in bacteria and archaea. Nature. 2012;482:331–338. doi: 10.1038/nature10886.

18. Jinek M, Chylinski K, Fonfara I, Hauer M, Doudna JA, Charpentier E. A programmable dual-RNA-guided DNA endonuclease in adaptive bacterial immunity. Science. 2012;337:816–821. doi: 10.1126/science.1225829.

19. Cong L, Ran FA, Cox D, Lin S, Barretto R, Habib N, et al. Multiplex genome engineering using CRISPR/Cas systems. Science. 2013;339:819–823. doi: 10.1126/science.1231143.

20. Mali P, Yang L, Esvelt KM, Aach J, Guell M, DiCarlo JE, et al. RNA-guided human genome engineering via Cas9. Science. 2013;339:823–826. doi: 10.1126/science.1232033.

21. Shan Q, Wang Y, Li J, Zhang Y, Chen K, Liang Z, et al. Targeted genome modification of crop plants using a CRISPR-Cas system. Nat Biotechnol. 2013;31:686–688. doi: 10.1038/nbt.2650.

22. Li J, Norville JE, Aach J, McCormack M, Zhang D, Bush J, et al. Multiplex and homologous recombination-mediated genome editing in Arabidopsis and Nicotiana benthamiana using guide RNA and Cas9. Nat Biotechnol. 2013;31:688–691. doi: 10.1038/nbt.2654.

23. Nekrasov V, Staskawicz B, Weigel D. Jones, Jonathan D G, Kamoun S. Targeted mutagenesis in the model plant Nicotiana benthamiana using Cas9 RNA-guided endonuclease. Nat Biotechnol. 2013;31:691–693. doi: 10.1038/nbt.2655.

24. Feng Z, Mao Y, Xu N, Zhang B, Wei P, Yang D, et al. Multigeneration analysis reveals the inheritance, specificity, and patterns of CRISPR/Cas-induced gene modifications in Arabidopsis. Proc Natl Acad Sci USA. 2014;111:4632–4637. doi: 10.1073/pnas.1400822111.

25. Mao Y, Zhang H, Xu N, Zhang B, Gou F, Zhu J. Application of the CRISPR-Cas system for efficient genome engineering in plants. Mol Plant. 2013;6:2008–2011. doi: 10.1093/mp/sst121.

26. Zhang H, Zhang J, Wei P, Zhang B, Gou F, Feng Z, et al. The CRISPR/Cas9 system produces specific and homozygous targeted gene editing in rice in one generation. Plant Biotechnol J. 2014;12:797–807. doi: 10.1111/pbi.12200.

27. Fauser F, Schiml S, Puchta H. Both CRISPR/Cas-based nucleases and nickases can be used efficiently for genome engineering in Arabidopsisthaliana. Plant J. 2014;79:348–359. doi: 10.1111/tpj.12554.

28. Orel N, Kyryk A, Puchta H. Different pathways of homologous recombination are used for the repair of double-strand breaks within tandemly arranged sequences in the plant genome. Plant J. 2003;35:604–612. doi: 10.1046/j.1365-313X.2003.01832.x.

29. Zhou H, Liu B, Weeks DP, Spalding MH, Yang B. Large chromosomal deletions and heritable small genetic changes induced by CRISPR/Cas9 in rice. Nucleic Acids Res. 2014;42:10903–10914. doi: 10.1093/nar/gku806.

30. Brooks C, Nekrasov V, Lippman ZB, van Eck J. Efficient gene editing in tomato in the first generation using the clustered regularly interspaced short palindromic repeats/CRISPR-associated9 system. Plant Physiol. 2014;166:1292–1297. doi: 10.1104/pp.114.247577.

31. Yifhar T, Pekker I, Peled D, Friedlander G, Pistunov A, Sabban M, et al. Failure of the tomato trans-acting short interfering RNA program to regulate AUXIN RESPONSE FACTOR3 and ARF4 underlies the wiry leaf syndrome. Plant Cell. 2012;24:3575–3589. doi: 10.1105/tpc.112.100222.

32. Xing H, Dong L, Wang Z, Zhang H, Han C, Liu B, et al. A CRISPR/Cas9 toolkit for multiplex genome editing in plants. BMC Plant Biol. 2014;14:327. doi: 10.1186/s12870-014-0327-y.

33. Schaeffer SM, Nakata PA. CRISPR/Cas9-mediated genome editing and gene replacement in plants: transitioning from lab to field. Plant Sci. 2015;240:130–142. doi: 10.1016/j.plantsci.2015.09.011.

34. Li Z, Liu Z, Xing A, Moon BP, Koellhoffer JP, Huang L, et al. Cas9-guide RNA directed genome editing in soybean. Plant Physiol. 2015

35. Ran FA, Cong L, Yan WX, Scott DA, Gootenberg JS, Kriz AJ, et al. In vivo genome editing using Staphylococcus

aureus Cas9. Nature. 2015;520:186–191. doi: 10.1038/nature14299.

36. Esvelt KM, Mali P, Braff JL, Moosburner M, Yaung SJ, Church GM. Orthogonal Cas9 proteins for RNA-guided gene regulation and editing. Nat Methods. 2013;10:1116–1121. doi: 10.1038/nmeth.2681.

37. Steinert J, Schiml S, Fauser F, Puchta H. Highly efficient heritable plant genome engineering using Cas9 orthologues from Streptococcus thermophilus and Staphylococcus aureus. Plant J. 2015.

38. Puchta H. Using CRISPR/Cas in three dimensions. Towards synthetic plant genomes, transcriptomes and epigenomes. Plant J. 2015.

39. Hyun Y, Kim J, Cho SW, Choi Y, Kim J, Coupland G. Site-directed mutagenesis in Arabidopsis thaliana using dividing tissue-targeted RGEN of the CRISPR/Cas system to generate heritable null alleles. Planta. 2015;241:271–284. doi: 10.1007/s00425-014-2180-5.

40. Wang Z, Xing H, Dong L, Zhang H, Han C, Wang X, et al. Egg cell-specific promoter-controlled CRISPR/Cas9 efficiently generates homozygous mutants for multiple target genes in Arabidopsis in a single generation. Genome Biol. 2015;16:144. doi: 10.1186/s13059-015-0715-0.

41. Jia H, Wang N. Targeted genome editing of sweet orange using Cas9/sgRNA. PLoS One. 2014;9:e93806. doi: 10.1371/journal.pone.0093806.

42. Upadhyay SK, Kumar J, Alok A, Tuli R. RNA-guided genome editing for target gene mutations in wheat. G3 (Bethesda). 2013;3:2233–8. doi:10.1534/g3.113.008847.

43. Xie K, Yang Y. RNA-guided genome editing in plants using a CRISPR-Cas system. Mol Plant. 2013;6:1975–1983. doi: 10.1093/mp/sst119.

44. Jacobs TB, LaFayette PR, Schmitz RJ, Parrott WA. Targeted genome modifications in soybean with CRISPR/Cas9. BMC Biotechnol. 2015;15:16. doi: 10.1186/s12896-015-0131-2.

45. Ran FA, Hsu PD, Lin C, Gootenberg JS, Konermann S, Trevino AE, et al. Double nicking by RNA-guided CRISPR Cas9 for enhanced genome editing specificity. Cell. 2013;154:1380–1389. doi: 10.1016/j.cell.2013.08.021.

46. Schiml S, Fauser F, Puchta H. The CRISPR/Cas system can be used as nuclease for in planta gene targeting and as paired nickases for directed mutagenesis in Arabidopsis resulting in heritable progeny. Plant J. 2014;80:1139–1150. doi: 10.1111/tpj.12704.

47. Puchta H, Fauser F. Gene targeting in plants: 25 years later. Int J Dev Biol. 2013;57:629–637. doi: 10.1387/ijdb.130194hp.

48. Fauser F, Roth N, Pacher M, Ilg G, Sánchez-Fernández R, Biesgen C, et al. In planta gene targeting. Proc Natl Acad Sci USA. 2012;109:7535–7540. doi: 10.1073/pnas.1202191109.

49. Baltes NJ, Gil-Humanes J, Cermak T, Atkins PA, Voytas DF. DNA replicons for plant genome engineering. Plant Cell. 2014;26:151–163. doi: 10.1105/tpc.113.119792.

50. Čermák T, Baltes NJ, Čegan R, Zhang Y, Voytas DF. High-frequency, precise modification of the tomato genome. Genome Biol. 2015;16:232. doi: 10.1186/s13059-015-0796-9.

51. Ma X, Zhang Q, Zhu Q, Liu W, Chen Y, Qiu R, et al. A Robust CRISPR/Cas9 system for convenient, high-efficiency multiplex genome editing in monocot and dicot plants. Mol Plant. 2015;8:1274–1284. doi: 10.1016/j.molp.2015.04.007.

52. Xie K, Minkenberg B, Yang Y. Boosting CRISPR/Cas9 multiplex editing capability with the endogenous tRNA-processing system. Proc Natl Acad Sci USA. 2015;112:3570–3575. doi: 10.1073/pnas.1420294112.

53. Ali Z, Abul-Faraj A, Li L, Ghosh N, Piatek M, Mahjoub A, et al. Efficient virus-mediated genome editing in plants using the CRISPR/Cas9 system. Mol Plant. 2015

54. Qi LS, Larson MH, Gilbert LA, Doudna JA, Weissman JS, Arkin AP, et al. Repurposing CRISPR as an RNA-guided platform for sequence-specific control of gene expression. Cell. 2013;152:1173–1183. doi: 10.1016/j.cell.2013.02.022.

55. Gilbert LA, Larson MH, Morsut L, Liu Z, Brar GA, Torres SE, et al. CRISPR-mediated modular RNA-guided regulation of transcription in eukaryotes. Cell. 2013;154:442–451. doi: 10.1016/j.cell.2013.06.044.

56. Piatek A, Ali Z, Baazim H, Li L, Abulfaraj A, Al-Shareef S, et al. RNA-guided transcriptional regulation in planta via synthetic dCas9-based transcription factors. Plant Biotechnol J. 2015;13:578–589. doi: 10.1111/pbi.12284.

57. Tiwari SB, Belachew A, Ma SF, Young M, Ade J, Shen Y, et al. The EDLL motif: a potent plant transcriptional activation domain from AP2/ERF transcription factors. Plant J. 2012;70:855–865. doi: 10.1111/j.1365-313X.2012.04935.x.

58. Morbitzer R, Römer P, Boch J, Lahaye T. Regulation of selected genome loci using de novo-engineered transcription activator-like effector (TALE)-type transcription factors. Proc Natl Acad Sci USA. 2010;107:21617–21622. doi: 10.1073/pnas.1013133107.

59. Hiratsu K, Matsui K, Koyama T, Ohme-Takagi M. Dominant repression of target genes by chimeric repressors that include the EAR motif, a repression domain, in Arabidopsis. Plant J. 2003;34:733–739. doi: 10.1046/j.1365-313X.2003.01759.x.

60. Gao Y, Zhang Y, Zhang D, Dai X, Estelle M, Zhao Y. Auxin binding protein 1 (ABP1) is not required for either auxin signaling or Arabidopsis development. Proc Natl Acad Sci USA. 2015;112:2275–2280. doi: 10.1073/pnas.1500365112.

61. Woo JW, Kim J, Kwon SI, Corvalán C, Cho SW, Kim H, et al. DNA-free genome editing in plants with preassembled CRISPR-Cas9 ribonucleoproteins. Nat Biotechnol. 2015

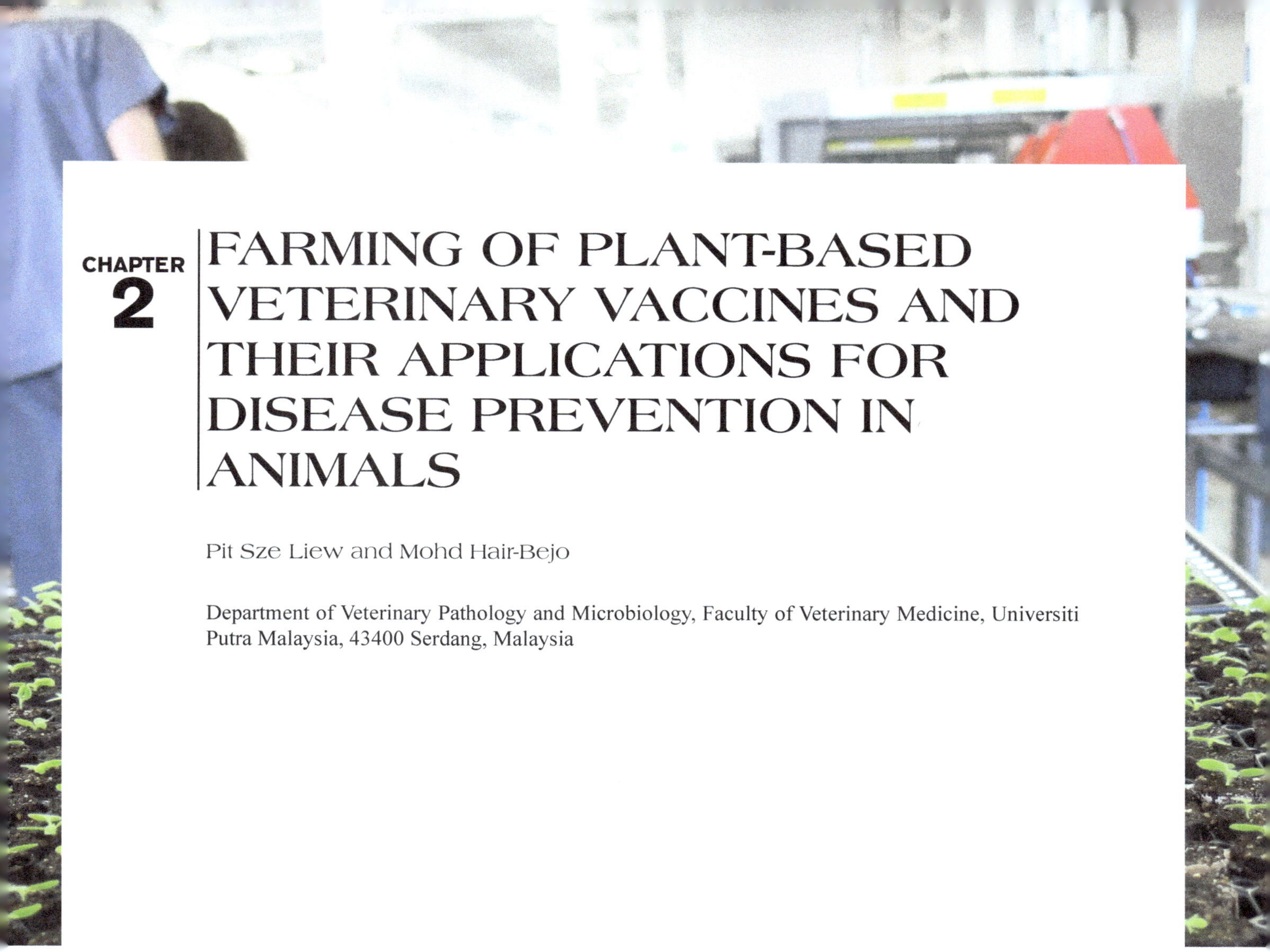

CHAPTER 2

FARMING OF PLANT-BASED VETERINARY VACCINES AND THEIR APPLICATIONS FOR DISEASE PREVENTION IN ANIMALS

Pit Sze Liew and Mohd Hair-Bejo

Department of Veterinary Pathology and Microbiology, Faculty of Veterinary Medicine, Universiti Putra Malaysia, 43400 Serdang, Malaysia

ABSTRACT

Plants have been studied for the production of pharmaceutical compounds for more than two decades now. Ever since the plantmade poultry vaccine against Newcastle disease virus made a breakthrough and went all the way to obtain regulatory approval, research to use plants for expression and delivery of vaccine proteins for animals was intensified. Indeed, in view of the high production costs of veterinary vaccines, plants represent attractive biofactories and offer many promising advantages in the production of recombinant vaccine proteins. Furthermore, the possibility of conducting immunogenicity and challenge studies in target animals has greatly exaggerated the progress. Although there are no edible plant-produced animal vaccines in the market, plant-based vaccine technology has great potentials. In this review, development, uses, and advantages of plant-based recombinant protein production in various expression platforms are discussed. In addition, examples of plant-based veterinary vaccines showing strong indication in terms of efficacy in animal disease prevention are also described.

INTRODUCTION

Plant molecular farming is a term used to describe the application of molecular biological techniques to the synthesis of commercial products in plants, which include a variety of carbohydrates, fats, and proteins, as well as secondary products [1]. The process of manufacturing a plant-based vaccine in the plant green factory begins with the selection of a target antigen of interest. The vaccine candidate is cloned into a plant expression cassette that is capable of promoting and terminating transgene expression. The expression cassette is then delivered into a plant for the production of a recombinant protein [2].

Citation: Pit Sze Liew and Mohd Hair-Bejo, Farming of Plant-Based Veterinary Vaccines and Their Applications for Disease Prevention in Animals, doi: 10.1155/2015/936940.

The delivery of an expression cassette carrying the gene of interest into the plant cells could be achieved by either stable or transient transformation. Transient gene expression represents a rapid and convenient system for verification and characterisation of the target gene product. Notwithstanding, it is now a routine practice in molecular farming for the production of foreign proteins [3]. The process circumvents the long development time and low protein accumulation levels associated with the use of transgenic plants. Besides, as the foreign protein production is only temporary, it requires no selection method to identify transformed plant cells. Compared to transient expression systems, the major advantage of stable transgenic lines is that the candidate gene sequence is incorporated into the plant genome and thus the acquired protein production trait is inherited. This allows the transfer of the desired character to the next and over multiple generations

[4]. Thus, seed stock could be established, which assures the continuing availability of the stock [5].

BRIEF HISTORY OF DEVELOPMENT OF PLANT-BASED RECOMBINANT PROTEIN PRODUCTION

The idea of plant molecular farming was burgeoning about 26 years ago when the proof of concept recombinant plant-derived pharmaceutical proteins was reported (Table 1; [1, 6]). Among these, biologically active human interferon alpha D was produced in turnip following inoculation with mutant cauliflower mosaic virus carrying human interferon alpha D at its ORF II [7]. In the same year, tobacco plants expressing either gamma or kappa immunoglobulin chains of mouse were crossed to generate progeny expressing both chains of immunoglobulin, showing the ability of plants to assemble heterologous biomolecules [8]. Then, the production of human serum albumin in tobacco and potato plants which was identical to the authentic human protein was reported [9].

Table 1: Various applications of plant-based expression system and key events related to their development

Product	Plant host(s)	Importance	References
Human pharmaceutical proteins			
Human interferon-α	Turnip	First recombinant plant-derived pharmaceutical proteins.	[7]
Human serum albumin	Tobacco, potato	Plant-derived proteins identical to the authentic human proteins.	[9]
Antibodies			
Mouse immunoglobulin	Tobacco	First plant-derived antibody showing the ability of plants to assemble heterologous biomolecules.	[8]
Antibody against hepatitis B	Tobacco	First commercialized plant-derived antibody used for vaccine purification in Cuba.	[10]
Antibody against *Streptococcus mutans* (CaroRx)	Tobacco	EU-approved medical device for prevention of tooth decay.	[6, 11]
Industrial and agricultural recombinant enzymes			
Avidin	Maize	Plant-derived products on the market.	[12]
β-glucuronidase	Maize		[13]
Trypsin	Maize		[14]
Vaccine antigens- human use			
Hepatitis B antigen	Tobacco	Plant-made HBsAg particles antigenically and physically similar to HBsAg particles derived from human serum and recombinant yeast.	[15]
	Lettuce, lupin	Specific antibody responses in human volunteers fed with transgenic lettuce.	[16]
	Potato	Increased specific serum immune response when given orally as a booster to vaccinated human volunteers.	[17]
Escherichia coli LT-B	Potato	First proof of concept for edible plant-derived vaccines that conferred protection to mice upon oral feeding.	[18]
	Maize	Specific serum and mucosal antibody responses in human volunteers ingested transgenic corn germ meal.	[19]
Norwalk virus	Tobacco, potato	Specific serum and mucosal antibody responses in mice fed orally with transgenic plants. Phase I clinical trials completed.	[20]
Vaccine antigens- veterinary use			
Newcastle disease virus HN proteins	Tobacco suspension cell culture	First USDA-approved plant-derived veterinary vaccine.	[21]

The idea has since expanded to the production of many industrial and agricultural recombinant enzymes. The leading examples were avidin [12], -glucuronidase [13], and trypsin [14]. Besides, the use of plants for the production of foreign proteins especially biomedically important materials is wide and varied. Several biopharmaceuticals like growth hormones, human blood components, and cytokines have been expressed in plants [6, 11]. Furthermore, many medically related proteins including antibodies or vaccines for human and veterinary use are continually being explored and developed. The monoclonal antibody against hepatitis B surface antigen (HBsAg), expressed in tobacco, was the first commercialized plant-derived antibody. Being marketed in Cuba, it has replaced the mouse derived monoclonal antibody for the routine purification of recombinant HBsAg for hepatitis B vaccine production [10]. Plant-made antibodies currently in the pipeline for commercialization include those against *Streptococcus mutans* and non-Hodgkin›s lymphoma [6, 11]. The monoclonal secretory antibody against *S. mutans* made in plant was shown to prevent colonisation of microbes in the human oral cavity [22]. It has since been approved as a medical device in 2003 by the European regulatory authority for human use and commercialized as CaroRx [6].

On the other hand, a variety of vaccine antigens have been well tested for expression in plants, be it for human or veterinary applications. Among the vaccine targets intended for human use, *S. mutans* surface protein A for dental caries and HBsAg for viral hepatitis B were the first to be explored [23]. It has been demonstrated that the plant-made 22 nm particles of HBsAg were similar to HBsAg particles derived from human serum and recombinant yeast, both antigenically and physically [15]. Subsequently, various vaccine candidates including *E. coli* heat-labile enterotoxin B subunit (LT-B), Norwalk virus particles, and cholera toxin B subunit have been expressed in plants [23, 24]. The bacterial antigen LT-B made the first proof of concept for edible plant-produced vaccines by showing that the orally delivered vaccine conferred protective efficacy to mice [18]. Moreover, the vaccine targeting antigens like viral hepatitis B, LT-B, and Norwalk virus are heading towards advanced stages of development and have completed the phase I clinical trials [6]. Notwithstanding, many vaccine antigens are making their way forward although they were in the early stage of development. These include and are not limited to vaccines for rotavirus infection [25], measles [26, 27], human immunodeficiency virus type 1 [28–30], human cytomegalovirus [31], respiratory syncytial virus [32, 33], *Staphylococcus aureus* [34], *Pseudomonas aeruginosa* [35, 36], and *Plasmodium falciparum* [37].

However, it was a plant-made poultry vaccine against Newcastle disease virus (NDV) that turned out to be the first plant-based vaccine to obtain regulatory approval from the United States Department of Agriculture, Center for Veterinary Biologics [21]. The NDV vaccine was shown to confer more than 90% protection rate in chicken upon challenge with NDV. Although the proof of concept poultry vaccine has not been commercialized and there are no other veterinary vaccines introduced into the market since then, animal pathogens have become the focus for expression in plants [38]. Indeed, the number and range of candidate antigens from animal microbes and viruses that have been expressed in plants are extensive [39]. The possibility of carrying out challenge experiments in specific animal species of interest has encouraged and resulted in enthusiastic development of plant-made veterinary vaccines.

MERITS OF PLANT PRODUCTION SYSTEM

Since the early 1990s, plants have gained an additional role of being bioreactors in molecular farming for new drugs and vaccines [38]. Whole plants, either by stable or transient transformation, were used to produce foreign target antigens of interest [40] that have the potential of being applied in routine vaccination strategies. Plants, therefore, represent attractive alternatives for vaccine production. Different parts of plants like the leaf and stem tissues, seeds, and fruits and root vegetables have been used for production of foreign proteins. Some of the commonest species of plants used include small flowering weed*Arabidopsis thaliana* that is widely used in plant science, tobaccos, alfalfa, spinach, potatoes, rice, beans, maize, tomatoes, strawberries, carrots, and many more [3]. Initially conceptualized as a platform for production of edible vaccines [41], expression of foreign proteins in plants aims to reduce the use of needles and cold chain for vaccine delivery especially in the developing countries. Besides, should the food plants be used for foreign protein expression especially for vaccines, they could be consumed directly or with only minimal processing. Nevertheless, plants offer many other advantages over other production systems.

The production cost in plants is only a fraction of that of mammalian cell systems and between 10 and 50 times lower than microbial system like *E. coli* fermentation for producing the same protein [42, 43]. The farming of transgenic plants requires relatively basic and economical propagation materials like sunlight, water, and nutrients. Furthermore, the harvest and downstream processing of transgenic plants require an uncomplicated technology and the scale-up of production is simple and rapid as they can be done by increasing the cultivation area [41].

Production of foreign proteins in plants is generally considered safe when compared to mammalian cell systems, as they are less likely to harbour microbes or prions that are pathogenic to humans or animals [41]. The conventional live veterinary vaccines especially of viral origins, intended for use in poultry, are typically propagated in chicken embryos or cell culture systems. These vaccines consist mainly of attenuated virus strains that have lost their virulent properties but are still replicative and demonstrate the desired antigenic features. Although the vaccines can mimic the course of naturally acquired immunity, they bear the risk of reversion to virulence, which could result in infection instead of protection [44]. Moreover, chicken embryos or animal cells propagation systems carry the inherent threat for unintentional contamination because they are rich in nutrients and thus susceptible to contamination. The avian leukosis virus has been found as a contaminant in commercial Marek›s disease vaccines of poultry [45]. Although the contaminated vaccines were promptly removed from the market, it was reported that routine quality assurance measures had failed to detect the contaminant virus. This has undeniably raised safety issues of vaccine production in animal-derived sources. Plants, when used as production system, have lower risks of contamination by extraneous infectious agents.

Other than the apparent advantages in terms of cost, scalability, and safety, plant expression platforms are able to carry out posttranslational modification of proteins like disulfide bonds formation and glycosylation [46]. The proteins could be targeted to and retained in the endoplasmic reticulum of the cell to allow N-glycosylation and avoid complex-type N-glycan modifications by the Golgi apparatus [43]. Being more structurally closer to those of insect cell expression system, the use of plant system, however, would require modification and harmonization should species-specific N-glycosylation be needed to produce therapeutic glycoproteins for veterinary or human purposes [47, 48]. While animal cell cultures are able to carry out posttranslational modification and reproduce glycoproteins with N-glycan structures specific to the animal species they are derived from, they are compensated by the apparent high production cost [49]. As with the insect cell and plant expression systems, the use of mammalian cell cultures from nonhuman origin for expression of glycoproteins for human use would need to be humanized [49]. Similarly, although the microbial production systems like *E. coli* and yeasts are relatively much cheaper than mammalian cell systems, they are not be able to synthesize glycosylated proteins with desired biological properties [48]. As bacteria do not glycosylate, while yeasts may hyperglycosylate, the immunogenicity of the proteins produced might be affected [46].

PLANT-BASED EXPRESSION PLATFORMS OF RECOMBINANT PROTEINS

Many plants have been explored and used for the production of recombinant proteins and vaccine antigens (Table 2; [41, 50]). Generally, these can be divided into leafy crops, fruits and root vegetables, and seeds. Soybean, alfalfa, and corn are among the most efficient plant systems for production of foreign proteins from an economic point of view [41]. Preliminary studies to generate stably transformed transgenic plants expressing proteins of interest have often seen the use of model plants such as *Arabidopsis thaliana* and tobacco [39]. With the completion of sequencing on the *Arabidopsis* nuclear genomes [51], transgenic plant research is blossoming. A variety of*Arabidopsis* lines and mutants with accessible genetics information are available. Hence, transformation protocols of *Arabidopsis* are established and could be performed successfully even by nonspecialists. However, the plant is not useful for commercial production as it has a low biomass [50].

Table 2: Comparison of different plant-based expression platforms for recombinant proteins production

Plant hosts	Advantages	Disadvantages
Model plants		
Arabidopsis thaliana	Often used for preliminary studies. A variety of *Arabidopsis* lines and mutants with accessible genetics information are available. Small genome size and short life cycle. Self-pollinating plant that could produce numerous seeds. Exceptional ease for transformation by *Agrobacterium*-mediated approach.	Low biomass.
Tobacco	Established transformation and expression protocols. High biomass yield. Nonfood and nonfeed crop. Less risk of feed and human food chains contamination. Low-alkaloid varieties are available, which requires less processing.	Risk of crossing with nontransgenic tobacco in the open field production. Contain high nicotine and other toxic alkaloids. Direct consumption not possible.
Leafy crops		
Alfalfa Clover	Established transformation protocols. Clonal propagation is possible. Direct consumption and useful for animal vaccines. High protein level. Large dry biomass yield. Many harvests per year. Homogenous N-linked glycan structures in alfalfa.	Risk of outcrossing with nontransgenic plants in the open field production. Low protein stability. Perishable tissues and must be processed soon after harvest. Deep root system in alfalfa makes them difficult for thorough cleaning from the production field.
Lettuce	Edible raw and useful for human vaccines. Fast growing.	
Fruits and root vegetables		
Potato	Established transformation protocols. Microtuber production is available for quick assay. Stable storage for longer periods in storage tissues without refrigeration. Clonal propagation, low risk of outcrossing in the open field production. Industrial processing of tuber is well established.	Low protein content. Raw tuber is not palatable, while cooking might cause denaturation of the foreign protein.
Tomato	Palatable in raw form. High biomass yield. Inherent high level of vitamin A may help in boosting immune responses. Industrial cultivation and processing are well established.	Low protein content. Acidic in nature and may be incompatible with some antigens or use for infants. Spoil readily.
Cereal and legume seeds		
Maize	Most widely used cereal crop for molecular farming. Large grain size and high per hectare biomass yield. In vitro manipulation and transformation of maize are well studied. Commercial production, processing, and scalability are established.	Cross-pollinating plant. Concerns for contamination of food maize crops.
Rice	High biomass yield. Commercial production, processing, and scalability are established. Self-pollinating, reduced risk of illegitimate gene flow due to pollen release.	Longer time-to-product period.
Barley	Self-pollinating.	Less widely grown. Inefficient transformation system.
Pea Soybean	Higher protein content than that of the cereals. Self-pollinating are risk of gene flow contamination are less.	Laborious and inefficient transformation procedures. Lower annual grain yield and higher production cost compared to maize and rice.

On the contrary, tobacco could achieve a relatively high biomass yield [52]. Besides, it is a nonfood and nonfeed crop; thus the risk of transgenic tobacco entering feed and human food chains is thus reduced [50]. However, the risk of crossing with nontransgenic tobacco in the open field production cannot be fully eliminated [41]. Moreover, transgenic tobacco cannot be consumed directly without downstream processing as it contains high amount of nicotine and other toxic alkaloids that must be removed completely before it could be delivered orally. Nevertheless, low-alkaloid varieties that require less processing are available. Besides, the suspension cultures of tobacco cells are devoid of these toxic metabolites and they can also be used to produce foreign proteins [53].

Other leafy crops that have been explored for molecular farming include alfalfa, clover, and lettuce [50]. Alfalfa and clover have a relatively established transformation protocols and they can be easily cultivated by clonal propagation [41]. The plant leaves can be consumed uncooked and this is particularly useful in the development of veterinary vaccines [50]. Moreover, these leafy crops contain a high protein level and they could achieve a large dry biomass yield per hectare of land [53]. Alfalfa could be harvested many times and up to nine times in a year. The N-glycosylation pattern in alfalfa is predominantly homogenous [50]. The consistency in the N-glycosylation process is of such importance particularly in the production of therapeutic proteins as the biological functions of these proteins are affected by the N-linked glycan structures [54, 55]. In contrast to the N-glycosylation in tobacco that showed a highly heterogeneous pattern, up to 75% of the monoclonal antibody expressed in alfalfa exhibited identical glycan structures suitable for downstream humanization processing [56]. However, they have the risk of outcrossing with nontransgenic plants in the open field production. Although having a deep root system reduces the need for chemical fertilizer, transgenic plants pose some difficulties for thorough cleaning from the production field [41].

Fruits and root vegetables like tomatoes and potatoes have also become preferred plant expression hosts. Potatoes have frequently been used because the transformation protocols to generate transgenic lines are established. Tuber extracts from the transgenic lines expressing the S1 glycoprotein gene of infectious bronchitis virus (IBV) have been shown to protect the chickens from clinical disease, as well as virus shedding upon challenge [57, 58]. Besides, microtuber production of potato is available for quick assay [41]. Foreign proteins produced are also stable and could be stored for longer periods in storage tissues without refrigeration [50]. The risk of outcrossing in the open field production is low as the plant could be clonally propagated. In addition, as the industrial processing of tuber is established, the cost of downstream processing can be greatly reduced. However, potato tuber contains a relatively low protein content [41] and it is not palatable although it can be eaten raw. While cooking can improve its palatability, it might lead to denaturation of the foreign protein, thus resulting in poor immunogenicity if it was used to produce vaccine antigens [59].

Therefore, tomatoes have become a more attractive alternative system [41], since they are palatable and can be eaten raw without cooking. Thus, vaccine antigens expressed in them do not risk to be denatured by heat treatments. The first vaccine candidate used for expression in tomatoes was the rabies virus glycoprotein [60]. Furthermore, they had been used to express the capsid proteins VP2 and VP6 of rotavirus, which were shown to be immunogenic to mice by intraperitoneal delivery [61]. The inherent high level of vitamin A in tomatoes may also help in boosting immune responses [59]. Tomatoes have a well-established industrial cultivation and processing system just like potatoes. However, the fruit are also relatively low in protein content and must be chilled after harvest in order to prevent spoilage [59]. Although freeze-drying technology is available to preserve the fruit, this adds an additional cost to the downstream processing.

Plant seeds represent another expression platform for synthesis of foreign protein and vaccine antigens, as well as their storage. In comparison to perishable plants like leafy crops and fruits, plant seeds enable long-term storage of the foreign protein produced [1]. The plant seeds are generally low in water content where most seeds have a water content of less than 10% of their total biomass, whereas in most cases the leaves contain more than 90% of water [62]. Besides, plant seeds are relatively high in protein content, which ranged from 10% to 40% of their wet weight, while in most leaves the protein percentage is less than 5% [62]. Thus, plant seeds represent a good vehicle to promote stable protein accumulation and storage. Moreover, protease activities in plant seeds are low and thus the risk of spoilage is greatly reduced as the foreign proteins produced are protected from proteolytic degradation [50]. It was demonstrated that antibodies and vaccine antigens accumulated in seeds remained stable without loss of activity for years at room temperature [63, 64]. Hence, seeds are suitable for direct consumption and useful especially for the development of animal vaccines. Industrial scale seed plantation and production are well established, beginning from cultivation, harvest, storage, distribution, and processing of the seeds [1].

Maize is the most widely used cereal crop for molecular farming [65]. The maize seeds or corn has a larger grain size and higher per hectare biomass yield compared to other cereals. The in vitro manipulation and transformation of maize have been widely studied, while its commercial production, processing, and scalability are also well established. The first commercialized corn-produced product was avidin [12] from the company ProdiGene and is available in the Sigma catalogue for diagnostic use [65]. In addition, corn has been used for the development of animal vaccines. The transmissible gastroenteritis virus (TGEV) envelope spike (S) protein expressed in corn was shown to induce protective antibodies in both piglets and gilts [66, 67]. Similarly, oral feeding of transgenic maize expressing the glycoprotein protein of NDV was shown to be immunogenic and conferred protective immunity to chicken [68].

Rice represents another leading platform for production of foreign proteins [1]. Like maize, rice has a high biomass yield, and its production, processing, and scalability have also been established. One apparent advantage of rice over maize is the reduced risk of illegitimate gene flow due to pollen release resulting from self-pollination [62]. In an oral feeding immunization trial of 2-week-old specific-pathogen-free (SPF) chickens with rice seeds expressing the VP2 protein of infectious bursal disease virus (IBDV), challenge and protection studies showed evidence of protective immunity in the chickens [69]. Barley is another commercial platform being studied other than maize and rice. The self-pollinating trait of barley is an important advantage to be considered in its development as a foreign protein production system. For example, subcutaneous injection of the F4 fimbriae adhesin protein of enterotoxigenic *E. coli* produced in barley grains was shown to induce neutralizing antibodies in mice [70].

Pea and soybean are the two commonest legume platforms being studied in plant molecular farming [1]. In one study, soybean seeds were used to express enterotoxigenic *E. coli* LT-B protein [71]. These transgenic seeds were able to induce both systemic and mucosal immunity in mice upon oral administration and conferred partial protection upon challenge. A major advantage of legumes over cereals is that the total protein content of legume grains is relatively much higher than that of cereals. Compared to total proteins (8 to 13%) from cereal grains, the total protein content in pea and soybean can be as high as 40% [72]. However, this apparent advantage is compensated by the laborious and inefficient transformation procedures of legumes in addition to their relatively lower annual grain yield and higher production cost when compared to maize and rice. Nevertheless, both pea and soybean are self-pollinating species; thus the risk of gene flow contamination is less.

Overall, one of the major disadvantages of seed-based expression is the relatively longer time-to-product period [62]. As the expression of protein is targeted to the seeds, the transgenic plants are grown through a flowering cycle to produce seeds. The assessment for the expression of foreign proteins could only begin when the seed is set. This also makes seed-based production systems become less appropriate for expression of certain foreign proteins like the influenza viral antigens [62]. Since the influenza vaccines are revised annually, the amount of seed produced in the given time may be insufficient to supply the population. Besides, seed-based production involves a flowering cycle that might increase the risk of pollen release and gene flow contamination by pollen transfer especially in the open field production system [53]. In contrast, transgenic leafy crops can be harvested before flowering and thus the risk of outcrossing is reduced.

PROOF OF CONCEPT PLANT-DERIVED VETERINARY VACCINES

The tremendous developments of plant-made veterinary vaccines have been mainly due to the ability to conduct challenge experiments in specific animal species of interest [4]. Key examples on the plant-produced immunogenic proteins tested against disease challenge in target animal species are shown in Table 3. One of the first demonstrations showing protective efficacy of the plant-derived vaccine was from mink enteritis virus (MEV), where a short, linear, and neutralizing epitope from the viral VP2 capsid protein was expressed in black-eyed bean [73]. Using a plant chimeric virus particles (CVPs) approach, the short epitope was inserted into the cowpea mosaic virus and displayed on the surface of CVPs upon infection in plants. Subcutaneous injection of 1 mg of the chimeric viral particles expressing MEV peptide on the CVPs surfaces protected mink against clinical disease and challenge from the virulent MEV. In yet another study, studies showed that the VP60 protein of rabbit haemorrhagic disease virus produced from the transgenic potatoes conferred protection to rabbits against infection upon parenteral immunization [74]. It was also immunogenic and induced partial protection to the rabbits upon oral delivery of the vaccine [75].

Table 3: Summary of plant-derived immunogenic veterinary viral antigens tested against disease challenge in target animals

Target animals	Disease antigens	Protein expressed	Plant host(s)	Expression approach	Findings	References
Mink	Mink enteritis virus	VP2 capsid protein	Black-eyed bean	CVPs using cowpea mosaic virus	Subcutaneous injection of 1 mg of the chimeric virus protected mink against clinical disease and challenge from the virulent virus.	[73]
Rabbit	Rabbit haemorrhagic disease virus	VP60 protein	Potato, leaf	Stable transformation	Animals primed subcutaneously with 12 μg of recombinant protein emulsified in oil adjuvant and boosted intramuscularly after 30, 60, and 90 days were protected from challenge with the virulent virus.	[74]
			Potato, tuber	Stable transformation	Rabbits fed with 500 μg of recombinant protein and boosted on 21, 42, and 63 days after primary vaccination were partially protected from challenge with the virulent virus.	[75]
Swine	Transmissible gastroenteritis virus	Envelope spike (S) protein	Maize	Stable transformation	Piglets fed with 2 mg S protein daily for 10 consecutive days prior to challenge showed fewer symptoms of infection than the control group vaccinated with commercial vaccine.	[66]
					Gilts previously primed with commercial vaccine and boosted orally two times, each with 26 mg S protein, showed a significant increase of TGEV neutralizing antibody titer in serum, colostrum, and milk.	[67]
Cloven-hoofed animals	Foot-and-mouth disease virus	VP1 capsid protein	*Chenopodium quinoa*	CVPs using bamboo mosaic virus	Two intramuscular injections with 5 mg of chimeric virus in SPF pigs at six weeks apart induced neutralizing antibodies and demonstrated a complete protection against challenge.	[76]
Poultry	Newcastle disease virus	Fusion protein	Maize	Stable transformation	Oral feeding of transgenic maize was shown to be immunogenic and conferred complete protection against challenge comparable to that by a commercial vaccine to the chicken.	[68]
	Infectious bronchitis virus	S1 glycoprotein	Potato, tuber	Stable transformation	Day-old chicks fed orally with 57.2 μg of recombinant protein and boosted at 7 and 14 days after were protected from clinical disease and virus shedding upon challenge.	[57]
	Infectious bursal disease virus	VP2 capsid protein	Rice seed	Stable transformation	Oral feeding with 10 mg of recombinant protein protected chicken from challenge with a highly virulent virus, showing a better bursal lesion score compared to chickens that received the live attenuated vaccine.	[69]
			Chenopodium quinoa	CVPs using bamboo mosaic virus	Intramuscular injection with 600 μg chimeric virus in oil adjuvant was shown to induce specific antibodies and protected SPF chickens upon challenge with a very virulent virus strain.	[77]
			Nicotiana benthamiana	Transient expression	Intramuscular injection with 12 μg of recombinant protein emulsified in oil adjuvant with booster doses given at 22 and 35 days later was shown to produce neutralizing antibodies in chickens, with reduced T-cell infiltration into the bursa of Fabricius upon challenge.	[78]

Indeed, the most promising proof of concept for an edible plant-based animal vaccine delivered orally was against TGEV of swine [67, 79,80]. The S protein of TGEV expressed in corn was mixed in medicated milk replacer and fed orally to 10-day-old piglets [66]. With a dose of 2 mg S protein in single feeding, the piglets were fed over a 10-day period before being challenged with a virulent TGEV orally. Compared to the control group vaccinated with the commercial vaccine, where 78% of the piglets developed diarrhoea, only 50% of the piglets fed with transgenic corn had diarrhoea. The study concluded that transgenic corn was able to confer partial protection to piglets against clinical disease and experimental challenge with virulent virus. In addition, further studies were conducted to examine vaccination with transgenic corn in gilts and the transfer of protective anti-S protein antibody to suckling piglets through colostrum [67]. In the study, all gilts were primed with a modified live TGEV vaccine orally at days 115 and 102 and intramuscularly at day 88 before farrowing. Following primary vaccination, the gilts were separated into groups and subjected to different types of booster treatments. When compared to the control group that did not receive any booster dose, gilts given a double oral booster of transgenic corn containing 26 mg of S protein at days 35 and 14 before farrowing showed a significant increase of TGEV neutralizing antibody titer in the serum, colostrum, and milk. Such responses were comparable to gilts that received modified live virus vaccine as a booster. The level of neutralizing antibody titer in milk was suggested to be adequately protective to the suckling piglets, although efficacy test was not performed in the piglets [67].

Furthermore, the possibility of conducting protective efficacy experiments in target animal has also allowed the development of plant-expressed foot-and-mouth disease virus (FMDV) vaccine for cloven-hoofed animals. The FMDV VP1 capsid protein carrying the virus neutralizing epitopes was the target of expression in various plants. Transgenic plants expressing either the complete protein or antigenic peptide of VP1 have been generated in plants like *Arabidopsis* [81], alfalfa [82, 83], and potato [84]. Earlier studies conducted in mice via intraperitoneal [81–84] and/or oral [82] delivery of the leaves extract showed the vaccine was immunogenic and protective. Moreover, the VP1 protein was also expressed with the use of plant viral display vector like the tobacco mosaic virus in tobacco leaf via the CVPs approach [85]. The entire VP1 protein expressed by the plant virus and the resulting CVPs injected intraperitoneally into mice conferred protection against viral challenge with live FMDV. Although these studies have shown an induction of protective immunity in mice, it was only later that the protective efficacy experiments was conducted in swine, one of the natural hosts for FMDV. In a related study, expression of the immunogenic VP1 peptide encompassing amino acids 128 to 164 via the CVPs approach was successfully carried out [76]. Using the bamboo mosaic virus (BaMV) as a plant viral display vector, the VP1 peptide was genetically fused to the modified coat protein gene of BaMV. Upon infection in the leaves of *Chenopodium quinoa*, the BaMV plant host, the VP1 peptide was displayed on the surface of CVPs. Two intramuscular injections with 5 mg of VP1-displaying CVPs in two-month-old SPF pigs at six weeks apart resulted in the induction of anti-FMDV neutralizing antibodies. The vaccine demonstrated a complete protection in pigs against FMDV challenge four weeks after the booster dose was administered.

In the poultry vaccine arena, several infectious pathogens of economic importance have been the attention of development of plant-made vaccines. The NDV is one of them, and the virus surface glycoprotein fusion and/or hemagglutinin neuraminidase are the targets of expression. In addition to the first approved plant-produced NDV vaccine [21] that was made in tobacco cell culture, NDV viral proteins had been expressed in other plant systems as well. Oral feeding of transgenic maize expressing the viral fusion protein was shown to be immunogenic and conferred protective immunity to chicken [68].

Besides, the IBV S1 glycoprotein contains virus neutralizing and hemagglutination-inhibiting epitopes have been the component of interest for vaccine development. By stably transforming the S1 glycoprotein gene into the potato plant, tuber extracts from the transgenic potatoes were used for vaccination and protective efficacy studies in chicken [57, 58]. Here, day-old chicks were fed orally with either 2.5 or 5 g of tuber extracts corresponding to 28.6 or 57.2 μg of S1 glycoprotein and feeding was repeated at days 7 and 14. Virus challenge performed via intranasal route using the virulent IBV seven days after the final vaccination showed all chicks fed with 5 g of tuber extracts were protected from clinical disease and virus shedding. This result was comparable to the control group vaccinated with commercial modified live vaccine.

The IBDV, being a highly contagious and deadly virus of young chickens, is another important pathogen of poultry. It is a double stranded RNA virus with two genome segments, termed A and B [86]. The VP2 capsid protein of IBDV segment A contains the virus neutralizing epitopes and was selected as the component for the development of plant-made vaccine [69, 87]. In one study, the VP2 gene of the variant IBDV strain E was expressed in *Arabidopsis thaliana* [87], while in another study the VP2 gene of a virulent IBDV strain with attenuated segment A was expressed in rice seeds [69]. In the oral feeding immunization trial with rice seeds, 2-week-old SPF chickens were fed with transgenic rice seeds at weekly interval for four times before being challenged with the virulent IBDV strain [69]. Chickens fed with 5 g of transgenic rice seed containing 40.21 μg of VP2 protein in a grain [69], amounting to approximately 10 mg dose of VP2 protein [88], gave the best result in challenge and protection studies. Evaluation based on lesion scoring of the bursa after challenge revealed that orally immunized chickens achieved better lesion score compared to chickens that received the live attenuated vaccine. The orally immunized chickens also contained less antigen present in the bursal tissue based on immunofluorescence assay. Furthermore, the full-length VP2 gene of a classical IBDV strain has been transiently expressed in *Nicotiana benthamiana* leaves and the recombinant protein was extracted for subunit vaccination [78]. Eighteen-day-old chicks injected intramuscularly with 12 μg of VP2 protein emulsified in oil adjuvant and boosted after 22 and 35 days were shown to produce anti-IBDV antibody with neutralizing ability. Apart from VP2 protein obtained by stable or transient transformation in plants, the CVPs approach was also used to generate viable chimeric BaMV virus carrying the VP2 P domain loop P_{BC} of a very virulent IBDV [77]. Intramuscular injection with 600 μg recombinant BaMV in oil adjuvant to

3-week-old SPF chickens was shown to induce IBDV-specific antibodies and protected the chickens upon challenge with a very virulent IBDV strain 28 days after vaccination. These studies concluded that plant-made VP2 protein represents a useful vaccination strategy against IBDV in chicken.

CONCLUSIONS

It is indeed surprising to see that 26 years down the road only two recombinant protein products from plants had made it through the regulatory processes to be licensed: monoclonal antibody against HBsAg and poultry vaccine against NDV. The idea of plant-made vaccines as edible vaccines has received much publicity and enthusiastic development since the first proof of concept recombinant plant-derived pharmaceutical proteins was reported. However, the progress made was not without hurdles [3]. Although the reports of successful expression of target antigens of interest were numerous, many of these failed to achieve expression levels suitable for commercialization [41]. Besides, the use of food plants for production of vaccine antigens has sprouted fears of contamination of the human food chain. Worries about regulatory issues have also deterred the development of plant-made vaccines. However, the regulatory pathway for plant molecular farming of vaccine antigens for veterinary use is far shorter when compared to products intended for human use. Therefore, this represents an opportunity that warrants the pursuit of plant-made vaccines for animals. The demonstration of safety and increase usage of plant-produced recombinant protein products in animals will lead to the acceptance and recognition of plant expression technology. This will in turn encourage their use for production of plant-based biopharmaceuticals for human use. Finally, with a better understanding of plant gene expression and molecular biology, the realisation of an ideal plant-made edible vaccine will not be far. Hence, the molecular farming of vaccines, be it for veterinary or human use, will be worth the exploration.

ACKNOWLEDGMENTS

The authors wish to appreciate Dr. Y. Abba for proofreading the paper draft. The study is funded by the Ministry of Higher Education, Malaysia.

REFERENCES

1. Lau O. S., Sun S. S. M. Plant seeds as bioreactors for recombinant protein production. *Biotechnology Advances*. 2009;27(6):1015–1022. doi: 10.1016/j.biotechadv.2009.05.005.
2. Walmsley A. M., Arntzen C. J. Plants for delivery of edible vaccines.*Current Opinion in Biotechnology*. 2000;11(2):126–129. doi: 10.1016/s0958-1669(00)00070-7.
3. Rybicki E. P. Plant-produced vaccines: promise and reality. *Drug Discovery Today*. 2009;14(1-2):16–24. doi: 10.1016/j.drudis.2008.10.002.
4. Santi L. Plant derived veterinary vaccines. *Veterinary ResearchCommunications*.2009;33(1,supplement):S61–S66. doi: 10.1007/s11259-009-9246-z.
5. Joensuu J. J., Niklander-Teeri V., Brandle J. E. Transgenic plants for animal health: plant-made vaccine antigens for animal infectious disease control. *Phytochemistry Reviews*. 2008;7(3):553–577. doi: 10.1007/s11101-008-9088-2.
6. Spök A., Twyman R. M., Fischer R., Ma J. K. C., Sparrow P. A. C. Evolution of a regulatory framework for pharmaceuticals derived from genetically modified plants. *Trends in Biotechnology*. 2008;26(9):506–517. doi: 10.1016/j.tibtech.2008.05.007.
7. De Zoeten G. A., Penswick J. R., Horisberger M. A., Ahl P., Schultze M., Hohn T. The expression, localization, and effect of a human interferon in plants. *Virology*. 1989;172(1):213–222. doi: 10.1016/0042-6822(89)90123-2.
8. HiattA.,CafferkeyR.,BowdishK.Productionofantibodies in transgenic plants. *Nature*. 1989;342(6245):76–78. doi: 10.1038/342076a0.
9. SijmonsP.C.,DekkerB.M.M.,SchrammeijerB.,Verwoerd T. C., Van Den Elzen P. J. M., Hoekema A. Production of correctly processed human serum albumin in transgenic plants. *Nature Biotechnology*.1990;8(3):217–221. doi: 10.1038/nbt0390-217.
10. Pujol M., Ramírez N. I., Ayala M., et al. An integral approach towards a practical application for a plant-made monoclonal antibody in vaccine purification. *Vaccine*. 2005;23(15):1833–1837. doi: 10.1016/j.vaccine.2004.11.023.
11. Ma J. K.-C., Barros E., Bock R., et al. Molecular farming for new drugs and vaccines. Current perspectives on the production of pharmaceuticals in transgenic plants. *EMBO Reports*. 2005;6(7):593–599. doi: 10.1038/sj.embor.7400470.
12. Hood E. E., Witcher D. R., Maddock S., et al. Commercial production of avidin from transgenic maize: characterization of transformant, production, processing, extraction and purification. *Molecular Breeding*. 1997;3(4):291–306. doi: 10.1023/a:1009676322162.
13. Witcher D. R., Hood E. E., Peterson D., et al. Commercial production of -glucuronidase (GUS): a model system for the production of proteins in plants. *Molecular Breeding*. 1998;4(4):301–312. doi: 10.1023/a:1009622429758.
14. Woodard S. L., Mayor J. M., Bailey M. R., et al. Maize (Zea mays)-derived bovine trypsin: characterization of the first large-scale, commercial protein product from transgenic plants. *Biotechnology and Applied Biochemistry*. 2003;38(2):123–130. doi: 10.1042/ba20030026.
15. Mason H. S., Lam D. M.-K., Arntzen C. J. Expression of hepatitis B surface antigen in transgenic plants.*Proceedings of the National Academy of Sciences of the United States of America*.1992;89(24):11745–11749. doi: 10.1073/pnas.89.24.11745
16. Kapusta J., Modelska A., Figlerowicz M., et al. A plant-derived edible vaccine against hepatitis B virus. *The FASEB Journal*.1999;13(13):1796–1799.
17. Thanavala Y., Mahoney M., Pal S., et al. Immunogenicity in humans of an edible vaccine for hepatitis B. *Proceedings of the National Academy of Sciences of*

the United States of America.2005;102(9):3378–3382. doi: 10.1073/pnas.0409899102

18. Mason H. S., Haq T. A., Clements J. D., Arntzen C. J. Edible vaccine protects mice against *Escherichia coli* heat-labile enterotoxin (LT): potatoes expressing a synthetic LT-B gene. *Vaccine*.1998;16(13):1336–1343. doi: 10.1016/s0264-410x(98)80020-0.
19. Tacket C. O., Pasetti M. F., Edelman R., Howard J. A., Streatfield S. Immunogenicity of recombinant LT-B delivered orally to humans in transgenic corn. *Vaccine*. 2004;22(31-32):4385–4389. doi: 10.1016/j.vaccine.2004.01.073.
20. Mason H. S., Ball J. M., Shi J.-J., Jiang X., Estes M. K., Arntzen C. J. Expression of Norwalk virus capsid protein in transgenic tobacco and potato and its oral immunogenicity in mice. *Proceedings of the National Academy of Sciences of the United States of America*.1996;93(11):5335–5340. doi: 10.1073/pnas.93.11.5335
21. Vermij P., Waltz E. USDA approves the first plant-based vaccine.*Nature Biotechnology*. 2006;24(3):233–234. doi: 10.1038/nbt0306-233.
22. Ma J. K. The caries vaccine: a growing prospect. *Dental Update*.1999;26(9):374–380.
23. Pascual D. W. Vaccines are for dinner. *Proceedings of the National Academy of Sciences of the United States of America*.2007;104(26):10757–10758. doi: 10.1073/pnas.0704516104
24. Tregoning J., Maliga P., Dougan G., Nixon P. J. New advances in the production of edible plant vaccines: chloroplast expression of a tetanus vaccine antigen, TetC. *Phytochemistry*. 2004;65(8):989–994. doi: 10.1016/j.phytochem.2004.03.004.
25. O'Brien G. J., Bryant C. J., Voogd C., Greenberg H. B., Gardner R. C., Bellamy A. R. Rotavirus VP6 expressed by PVX vectors in*Nicotiana benthamiana* coats PVX rods and also assembles into viruslike particles. *Virology*. 2000;270(2):444–453. doi: 10.1006/viro.2000.0314.
26. Huang Z., Dry I., Webster D., Strugnell R., Wesselingh S. Plant-derived measles virus hemagglutinin protein induces neutralizing antibodies in mice. *Vaccine*. 2001;19(15-16):2163–2171. doi: 10.1016/S0264-410X(00)00390-X.
27. Webster D. E., Cooney M. L., Huang Z., et al. Successful boosting of a DNA measles immunization with an oral plant-derived measles virus vaccine. *Journal of Virology*. 2002;76(15):7910–7912. doi: 10.1128/JVI.76.15.7910-7912.2002.
28. McLain L., Durrani Z., Wisniewski L. A., Porta C., Lomonossoff G. P., Dimmock N. J. Stimulation of neutralizing antibodies to human immunodeficiency virus type 1 in three strains of mice immunized with a 22 amino acid peptide of gp41 expressed on the surface of a plant virus. *Vaccine*. 1996;14(8):799–810. doi: 10.1016/0264-410x(95)00229-t.
29. Zhang G. G., Rodrigues L., Rovinski B., White K. A. Production of HIV-1 p24 protein in transgenic tobacco plants. *Applied Biochemistry and Biotechnology—Part B Molecular Biotechnology*. 2002;20(2):131–136. doi: 10.1385/mb:20:2:131.
30. Meyers A., Chakauya E., Shephard E., et al. Expression of HIV-1 antigens in plants as potential subunit vaccines. *BMC Biotechnology*.2008;8(1, article 53) doi: 10.1186/1472-6750-8-53.
31. Tackaberry E. S., Dudani A. K., Prior F., et al. Development of biopharmaceuticals in plant expression systems: cloning, expression and immunological reactivity of human cytomegalovirus glycoprotein B (UL55) in seeds of transgenic tobacco. *Vaccine*. 1999;17(23-24):3020–3029. doi: 10.1016/s0264-410x(99)00150-4.
32. Belanger H., Fleysh N., Shannon C. O. X., et al. Human respiratory syncytial virus vaccine antigen produced in plants. *The FASEB Journal*.2000;14(14):2323–2328. doi: 10.1096/fj.00-0144com.
33. Sandhu J. S., Krasnyanski S. F., Domier L. L., Korban S. S., Osadjan M. D., Buetow D. E. Oral immunization of mice with transgenic tomato fruit expressing respiratory syncytial virus-F protein induces a systemic immune response. *Transgenic Research*.2000;9(2):127–135. doi: 10.1023/a:1008979525909.
34. Brennan F. R., Jones T. D., Longstaff M., et al. Immunogenicity of peptides derived from a fibronectin-binding protein of *S. aureus*expressed on two different plant viruses. *Vaccine*. 1999;17(15-16):1846–1857. doi: 10.1016/s0264-410x(98)00485-x.
35. Brennan F. R., Jones T. D., Gilleland L. B., et al. Pseudomonas aeruginosa outer-membrane protein F epitopes are highly immunogenic in mice when expressed on a plant virus. *Microbiology*.1999;145(1):211–220. doi: 10.1099/13500872-145-1-211.
36. Staczek J., Bendahmane M., Gilleland L. B., Beachy R. N., Gilleland H. E., Jr. Immunization with a chimeric tobacco mosaic virus containing an epitope of outer membrane protein F of *Pseudomonas aeruginosa* provides protection against challenge with *P. aeruginosa*.*Vaccine*. 2000;18(21):2266–2274. doi: 10.1016/s0264-410x(99)00571-x.
37. Turpen T. H., Reinl S. J., Charoenvit Y., Hoffman S. L., Fallarme V., Grill L. K. Malaria epitopes expressed on the surface of recombinant tobacco mosaic virus. *Nature Biotechnology*.1995;13(1):53–57. doi: 10.1038/nbt0195-53.
38. Sala F., Rigano M. M., Barbante A., Basso B., Walmsley A. M., Castiglione S. Vaccine antigen production in transgenic plants: strategies, gene constructs and perspectives. *Vaccine*. 2003;21(7-8):803–808. doi: 10.1016/s0264-410x(02)00603-5.
39. Floss D. M., Falkenburg D., Conrad U. Production of vaccines and therapeutic antibodies for veterinary applications in transgenic plants: an overview. *Transgenic Research*. 2007;16(3):315–332. doi: 10.1007/s11248-007-9095-x.
40. Rigano M. M., Walmsley A. M. Expression systems and developments in plant-made vaccines. *Immunology and Cell Biology*.2005;83(3):271–277. doi: 10.1111/j.1440-1711.2005.01336.x.

41. Mason H. S., Warzecha H., Mor T., Arntzen C. J. Edible plant vaccines: applications for prophylactic and therapeutic molecular medicine. *Trends in Molecular Medicine*. 2002;8(7):324–329. doi: 10.1016/s1471-4914(02)02360-2.

42. Giddings G. Transgenic plants as protein factories. *Current Opinion in Biotechnology*. 2001;12(5):450–454. doi: 10.1016/s0958-1669(00)00244-5.

43. Mett V., Farrance C. E., Green B. J., Yusibov V. Plants as biofactories. *Biologicals*. 2008;36(6):354–358. doi: 10.1016/j.biologicals.2008.09.001.

44. Muskett J. C., Reed N. E., Thornton D. H. Increased virulence of an infectious bursal disease live virus vaccine after passage in chicks. *Vaccine*. 1985;3(4):309–312. doi: 10.1016/s0264-410x(85)90161-6.

45. Zavala G., Cheng S. Detection and characterization of avian leukosis virus in Marek's disease vaccines. *Avian Diseases*.2006;50(2):209–215. doi: 10.1637/7444-092405R.1.

46. Streatfield S. J., Howard J. A. Plant-based vaccines. *International Journal for Parasitology*. 2003;33(5-6):479–493. doi: 10.1016/s0020-7519(03)00052-3.

47. Wilson I. B. H. Glycosylation of proteins in plants and invertebrates. *Current Opinion in Structural Biology*. 2002;12(5):569–577. doi: 10.1016/S0959-440X(02)00367-6.

48. Brooks S. A. Appropriate glycosylation of recombinant proteins for human use. *Molecular Biotechnology*. 2004;28(3):241–255. doi: 10.1385/mb:28:3:241.

49. Saint-Jore-Dupas C., Faye L., Gomord V. From planta to pharma with glycosylation in the toolbox. *Trends in Biotechnology*.2007;25(7):317–323. doi: 10.1016/j.tibtech.2007.04.008.

50. Fischer R., Stoger E., Schillberg S., Christou P., Twyman R. M. Plant-based production of biopharmaceuticals. *Current Opinion in Plant Biology*. 2004;7(2):152–158. doi: 10.1016/j.pbi.2004.01.007.

51. Arabidopsis Genome Initiative. Analysis of the genome sequence of the flowering plant *Arabidopsis thaliana*. *Nature*. 2000;408:796–815.

52. Daniell H., Streatfield S. J., Wycoff K. Medical molecular farming: production of antibodies, biopharmaceuticals and edible vaccines in plants. *Trends in Plant Science*. 2001;6(5):219–226. doi: 10.1016/s1360-1385(01)01922-7.

53. Twyman R. M., Stoger E., Schillberg S., Christou P., Fischer R. Molecular farming in plants: host systems and expression technology. *Trends in Biotechnology*. 2003;21(12):570–578. doi: 10.1016/j.tibtech.2003.10.002.

54. Lerouge P., Cabanes-Macheteau M., Rayon C., Fischette-Lainé A.-C., Gomord V., Faye L. N-glycoprotein biosynthesis in plants: recent developments and future trends. *Plant Molecular Biology*. 1998;38(1-2):31–48. doi: 10.1023/a:1006012005654.

55. Gomord V., Fitchette A.-C., Menu-Bouaouiche L., et al. Plant-specific glycosylation patterns in the context of therapeutic protein production. *Plant Biotechnology Journal*. 2010;8(5):564–587. doi: 10.1111/j.1467-7652.2009.00497.x.

56. Bardor M., Loutelier-Bourhis C., Paccalet T., et al. Monoclonal C5-1 antibody produced in transgenic alfalfa plants exhibits a *N*-glycosylation that is homogenous and suitable for glyco-engineering into human-compatible structures. *Plant Biotechnology Journal*.2003;1(6):451–462. doi: 10.1046/j.1467-7652.2003.00041.x.

57. Zhou J.-Y., Wu J.-X., Cheng L.-Q., et al. Expression of immunogenic S1 glycoprotein of infectious bronchitis virus in transgenic potatoes. *Journal of Virology*. 2003;77(16):9090–9093. doi: 10.1128/JVI.77.16.9090-9093.2003.

58. Zhou J.-Y., Cheng L.-Q., Zheng X.-J., et al. Generation of the transgenic potato expressing full-length spike protein of infectious bronchitis virus. *Journal of Biotechnology*. 2004;111(2):121–130. doi: 10.1016/j.jbiotec.2004.03.012.

59. Gunn K. S., Singh N., Giambrone J., Wu H. Using transgenic plants as bioreactors to produce edible vaccines. *Journal of Biotech Research*.2012;4(1):92–99.

60. McGarvey P. B., Hammond J., Dienelt M. M., et al. Expression of the rabies virus glycoprotein in transgenic tomatoes. *Nature Biotechnology*. 1995;13(13):1484–1487. doi: 10.1038/nbt1295-1484.

61. Saldaña S., Guadarrama F. E., Olivera Flores T. D. J., et al. Production of rotavirus-like particles in tomato (*Lycopersicon esculentum* L.) fruit by expression of capsid proteins VP2 and VP6 and immunological studies. *Viral Immunology*. 2006;19(1):42–53. doi: 10.1089/vim.2006.19.42.

62. Boothe J., Nykiforuk C., Shen Y., et al. Seed-based expression systems for plant molecular farming. *Plant Biotechnology Journal*.2010;8(5):588–606. doi: 10.1111/j.1467-7652.2010.00511.x.

63. Stöger E., Vaquero C., Torres E., et al. Cereal crops as viable production and storage systems for pharmaceutical scFv antibodies. *Plant Molecular Biology*. 2000;42(4):583–590. doi: 10.1023/A:1006301519427.

64. Nochi T., Takagi H., Yuki Y., et al. Rice-based mucosal vaccine as a global strategy for cold-chain- and needle-free vaccination. *Proceedings of the National Academy of Sciences of the United States of America*. 2007;104(26):10986–10991. doi: 10.1073/pnas.0703766104.

65. Ramessar K., Sabalza M., Capell T., Christou P. Maize plants: an ideal production platform for effective and safe molecular pharming. *Plant Science*. 2008;174(4):409–419. doi: 10.1016/j.plantsci.2008.02.002.

66. Streatfield S. J., Jilka J. M., Hood E. E., et al. Plant-based vaccines: unique advantages. *Vaccine*. 2001;19(17–19):2742–2748. doi: 10.1016/s0264-410x(00)00512-0.

67. Lamphear B. J., Jilka J. M., Kesl L., Welter M., Howard J. A., Streatfield S. J. A corn-based delivery system for animal vaccines: an oral transmissible gastroenteritis

virus vaccine boosts lactogenic immunity in swine. *Vaccine*. 2004;22(19):2420–2424. doi: 10.1016/j.vaccine.2003.11.066.

68. Guerrero-Andrade O., Loza-Rubio E., Olivera-Flores T., Fehérvári-Bone T., Gómez-Lim M. A. Expression of the Newcastle disease virus fusion protein in transgenic maize and immunological studies.*Transgenic Research*. 2006;15(4):455–463. doi: 10.1007/s11248-006-0017-0.

69. Wu J., Yu L., Li L., Hu J., Zhou J., Zhou X. Oral immunization with transgenic rice seeds expressing VP2 protein of infectious bursal disease virus induces protective immune responses in chickens. *Plant Biotechnology Journal*. 2007;5(5):570–578. doi: 10.1111/j.1467-7652.2007.00270.x.

70. Joensuu J. J., Kotiaho M., Teeri T. H., et al. Glycosylated F4 (K88) fimbrial adhesin FaeG expressed in barley endosperm induces ETEC-neutralizing antibodies in mice. *Transgenic Research*. 2006;15(3):359–373. doi: 10.1007/s11248-006-0010-7.

71. Moravec T., Schmidt M. A., Herman E. M., Woodford-Thomas T. Production of *Escherichia coli* heat labile toxin (LT) B subunit in soybean seed and analysis of its immunogenicity as an oral vaccine. *Vaccine*. 2007;25(9):1647–1657. doi: 10.1016/j.vaccine.2006.11.010.

72. Stoger E., Ma J. K.-C., Fischer R., Christou P. Sowing the seeds of success: pharmaceutical proteins from plants. *Current Opinion in Biotechnology*. 2005;16(2):167–173. doi: 10.1016/j.copbio.2005.01.005.

73. Dalsgaard K., Uttentha Å., Jones T. D., et al. Plant-derived vaccine protects target animals against a viral disease. *Nature Biotechnology*.1997;15(3):248–252. doi: 10.1038/nbt0397-248.

74. Castañón S., Marín M. S., Martín-Alonso J. M., et al. Immunization with potato plants expressing VP60 protein protects against rabbit hemorrhagic disease virus. *Journal of Virology*. 1999;73(5):4452–4455.

75. Martín-Alonso J. M., Castañón S., Alonso P., Parra F., Ordás R. Oral immunization using tuber extracts from transgenic potato plants expressing rabbit hemorrhagic disease virus capsid protein. *Transgenic Research*. 2003;12(1):127–130. doi: 10.1023/A:1022112717331.

76. Yang C.-D., Liao J.-T., Lai C.-Y., et al. Induction of protective immunity in swine by recombinant bamboo mosaic virus expressing foot-and-mouth disease virus epitopes. *BMC Biotechnology*. 2007;7(1, article 62) doi: 10.1186/1472-6750-7-62.

77. Chen T.-H., Chen T.-H., Hu C.-C., et al. Induction of protective immunity in chickens immunized with plant-made chimeric bamboo mosaic virus particles expressing very virulent Infectious bursal disease virus antigen. *Virus Research*. 2012;166(1-2):109–115. doi: 10.1016/j.virusres.2012.02.021.

78. Gómez E., Lucero M. S., Chimeno Zoth S., Carballeda J. M., Gravisaco M. J., Berinstein A. Transient expression of VP2 in*Nicotiana benthamiana* and its use as a plant-based vaccine against infectious bursal disease virus. *Vaccine*. 2013;31(23):2623–2627. doi: 10.1016/j.vaccine.2013.03.064.

79. Lamphear B. J., Streatfield S. J., Jilka J. M., et al. Delivery of subunit vaccines in maize seed. *Journal of Controlled Release*.2002;85(1–3):169–180. doi: 10.1016/s0168-3659(02)00282-1.

80. Howard J. A. Commercialization of plant-based vaccines from research and development to manufacturing. *Animal Health Research Reviews*. 2004;5(2):243–245. doi: 10.1079/ahr200476.

81. Carrillo C., Wigdorovitz A., Oliveros J. C., et al. Protective immune response to foot-and-mouth disease virus with VP1 expressed in transgenic plants. *Journal of Virology*. 1998;72(2):1688–1690.

82. Wigdorovitz A., Carrillo C., Dus Santos M. J., et al. Induction of a protective antibody response to foot and mouth disease virus in mice following oral or parenteral immunization with alfalfa transgenic plants expressing the viral structural protein VP1. *Virology*. 1999;255(2):347–353. doi: 10.1006/viro.1998.9590.

83. Santos M. J. D., Wigdorovitz A., Trono K., et al. A novel methodology to develop a foot and mouth disease virus (FMDV) peptide-based vaccine in transgenic plants. *Vaccine*. 2002;20(7-8):1141–1147. doi: 10.1016/s0264-410x(01)00434-0.

84. Carrillo C., Wigdorovitz A., Trono K., et al. Induction of a virus-specific antibody response to foot and mouth disease virus using the structural protein VP1 expressed in transgenic potato plants. *Viral Immunology*. 2001;14(1):49–57. doi: 10.1089/08828240151061383.

85. Wigdorovitz A., Pérez Filgueira D. M., Robertson N., et al. Protection of mice against challenge with foot and mouth disease virus (FMDV) by immunization with foliar extracts from plants infected with recombinant tobacco mosaic virus expressing the FMDV structural protein VP1. *Virology*. 1999;264(1):85–91. doi: 10.1006/viro.1999.9923.

86. Nick H., Cursiefen D., Becht H. Structural and growth characteristics of infectious bursal disease virus. *Journal of Virology*.1976;18(1):227–234.

87. Wu H., Singh N. K., Locy R. D., Scissum-Gunn K., Giambrone J. J. Expression of immunogenic VP2 protein of infectious bursal disease virus in *Arabidopsis thaliana*. *Biotechnology Letters*. 2004;26(10):787–792. doi: 10.1023/b:bile.0000025878.30350.d5.

88. Mason H. S., Herbst-Kralovetz M. M. Plant-derived antigens as mucosal vaccines. *Current Topics in Microbiology and Immunology*.2012;354(1):101–120. doi: 10.1007/82_2011_158.

CHAPTER 3

PLANT-BASED VACCINES AGAINST VIRUSES

Edward P Rybicki

Biopharming Research Unit, Department of Molecular & Cell Biology and Institute of Infectious Disease and Molecular Medicine, University of Cape Town, Private Bag X3, Rondebosch, 7701 Cape Town, South Africa

ABSTRACT

Plant-made or "biofarmed" viral vaccines are some of the earliest products of the technology of plant molecular farming, and remain some of the brightest prospects for the success of this field. Proofs of principle and of efficacy exist for many candidate viral veterinary vaccines; the use of plant-made viral antigens and of monoclonal antibodies for therapy of animal and even human viral disease is also well established. This review explores some of the more prominent recent advances in the biofarming of viral vaccines and therapies, including the recent use of ZMapp for Ebolavirus infection, and explores some possible future applications of the technology.

INTRODUCTION

The science of "molecular farming", or the use of plants and plant cell cultures to produce high-value recombinant proteins, started with the production via transgenic tobacco and sunflower of chimaeric human growth hormone in 1986 [1], then of monoclonal antibodies in transgenic tobacco in 1989 [2], and human serum albumin in transgenic tobacco and cell cultures [3]. This was followed from 1992 onwards with the expression of the first candidate virus vaccine antigens: these were Hepatitis B virus (HBV) surface antigen (HBsAg) in 1992 [4], and a VP1 epitope of foot-and-mouth disease virus (FMDV) expressed on the surface of particles of recombinant Cowpea mosaic virus (CPMV) in 1993 ([5] see Table 1).

Citation: Edward P Rybicki, Plant-based vaccines against viruses, doi: 10.1186/s12985-014-0205-0.

Table 1: Highlights of historical and recent activity in virus vaccine biofarming

Virus and vaccine candidate	Reference
Hepatitis B virus	
First expression of HBV surface antigen in plants	[4]
Human clinical trial of plant-produced HBsAg	[16]
First production of HBsAg in transgenic banana	[17]
Use of rTMV-mediated transient expression to produce ~300 mg/kg HBsAg	[20]
Use of rBeYDV-mediated transient expression to produce 800 mg/kg HBc Ag	[24]
Tabletised lyophilised transgenic lettuce containing HBsAg VLPs is orally immunogenic in mice	[19]
Hepatitis C virus	
Use of rTMV to express R9 epitope of E2 protein as CTB fusion	[30]
Expression of chimaeric plant virus coat protein molecules containing R9	[31-33]
R9-CMV CP VLPs orally immunogenic via feeding of lettuce leaves in rabbits	[34]
Mixed Th1/Th2 response in mice to rodlike PMV-E2 epitope chimaeric VLPs	[36]
Co-expression of whole E2 and calnexin and calreticulin increases E2 accumulation in plants	[37]
Influenza viruses	
Evidence of high-yield expression of H5 haemagglutinin-derived VLPs via transient expression	[44]
10 million vaccine dose "rapid fire" milestone by Medicago Inc. in DARPA Blue Angel programme	[50]
Human clinical trial of plant-made H5N1 vaccine candidate	[47]
HA-only VLPs produced for H7N9 outbreak virus	[54]
Phase 1 trials of H1N1pdm and HPAI H5N1 HA-derived plant-made products	[48]
Testing of plant-made engineered soluble trimeric HA (H1N1pdm) in mice	[58]
Adjuvanting of monomeric H1N1pdm HA with SiO_2 and bis-(3',5')-cyclic dimeric guanosine monophosphate (c-di-GMP)	[59]
Emergency response influenza vaccine candidates made in South Africa	[43]
Conjugation of plant-made HA to TMV particles and successful testing in mice	[60]
Elicitation of neutralising Ab with elastin-like polypeptide fused with stabilised soluble trimer-forming H5N1 HA	[61]
Presentation of M2e epitope on surface of rTMV virions elicits protective immunity to homologous and heterologous challenge in mice	[64]
Papillomaviruses	
Proof of efficacy of a plant-made Cottontail rabbit and Rabbit oral papillomavirus vaccines	[74,75]
High yields of HPV-16 L1 and VLPs via agroinfiltration-mediated transient expression or via transplastomic expression	[76,77]
Transplastomic expression of capsomere-forming HPV-16 L1 fused with *Escherichia coli* LTB as a built-in adjuvant	[78]
Successful expression of HPV-8 and Bovine papillomavirus L1 VLPs in plants	[79,80]
Co-expression of HPV-16 L1 with *E coli* LTB and oral immunisation elicits increased intestinal mucosal IgA responses to L1	[90]
High-yield plant transient expression of chimaeric L1::L2 VLPs and proof of increased breadth of immune response	[91]
rPVX CP fusion with $L2_{108-120}$ epitope yields well and elicits high-titre anti-L2 protein sera in mice	[97]
Plant production, scale-up and protective efficacy in mouse model of therapeutic E7GGG-LicKM fusion protein vaccine	[85-87]
Plant expressed HPV-16 L1 with C-terminal string of E6 and E7 T-cell epitopes is viable prophylactic/therapeutic vaccine candidate	[98]
Production and proof of efficacy in mice of soluble E7GGG therapeutic vaccine in transplastomic *Chlamydomonas reinhardtii*	[100]
Proof of yield increase and efficacy in a mouse tumour model of shuffled E7 protein fused to Zera® peptide	[102]
Human immunodeficiency virus	
HIV-1 p24 capsid protein expressed successfully in transgenic tobacco	[107]
Transgenic maize as production platform for oral vaccine delivery tested using SIV major surface glycoprotein gp130	[108]
Transiently-expressed gp41-derived molecule fused to CTB elicits anti-membrane proximal region (MPR) antibodies in mice	[109,110]
Gag-derived antigens expressed transiently and transgenically as CTL-inducing booster immunogens	[113]
High-yield transplastomic expression of Gag VLPs	[117]
Phase I clinical trial of anti-HIV MAbs produced in transgenic tobacco	[116]
rTMV-based expression of Gag VLP-based deconstructed multiantigens incorporating gp 41 MPER region	[124]
First successful published expression of HIV Env in plants: 89.6.P gp140ΔCFI envelope protein expressed at high yield	[125]
Transient expression of HIV Env-H5 HA fusion molecule results in VLP formation	[126]
Veterinary and "One Health" vaccines	
Production in plants and proof of efficacy in sheep of intact VLPs of Bluetongue virus	[128]
Rabies virus vaccines produced in plants and shown to be protective including after oral administration	[138-141]
Crimean-Congo haemorrhagic fever virus Gc and Gn glycoproteins immunogenic in mice via oral or parenteral administration	[144]
Plant-produced Rift Valley fever virus Gn and N proteins immunogenic after feeding mice transgenic *Arabidopsis*	[145]
Expression via rBeYDV and successful immunogenicity trial in mice of Ebola GP1/anti-GP1 MAb HC	[158]
Anti-viral therapeutic antibodies	
Production and successful efficacy testing of anti-Rabies virus MAbs made in transgenic tobacco	[147]
High-yield production via rBeYDV of anti-Ebolavirus Zaire and West Nile virus MAbs	[149,150]
Transient expression via rTMV, and successful pre- and post-exposure efficacy trial in macaques, of anti-Ebola MAb cocktail	[151,152]
Successful efficacy trial in macaques of the ZMApp therapeutic MAb cocktail	[153]

Legend: rTMV = recombinant Tobacco mosaic virus, including ICON vectors. rBeYDV = recombinant Bean yellow dwarf mastrevirus.

Initially, the prevailing idea for the use of plant-produced vaccines was that these should be delivered in plant material, as edible vaccines – something that was already being questioned as early as 1996 [6]. However, this concept has now largely fallen out of favour, with the realisation that administration of vaccines to humans requires standardisation of dose and some measure of quality control, and the necessity for purification and formulation has largely been accepted [7]. Over the years since 1989, then, many proteins have been expressed as candidate vaccines or therapeutics, and many different plant-based expression systems have been tried, with a growing trend towards transient expression systems based on infiltration of whole plants with recombinant *Agrobacterium tumefaciens* (agroinfiltration) and the use of "deconstructed" plant viral vectors (reviewed in [8]).

Virus vaccines have been a large and exciting part of this field almost from its beginning, for disease agents ranging from Hepatitis B to C to Foot and mouth disease viruses, from Human papillomavirus and Human rotavirus to ovine Bluetongue and Rabbit haemorrhagic disease viruses, to mention just a few. Aspects of this history have been covered recently, and in particular for virus-like particle based vaccines including rotaviruses and Norwalk virus [9], Human papillomaviruses [10] and Hepatitis B virus [11], and so these will not be discussed in detail here except where there is new material to be covered.

This review will cover the relevant recent history of virus-specific candidate vaccines and virus-specific therapeutic antibodies made in plants, with a view to providing object examples of successful approaches and especially of dual human/animal use or "One Health" examples (http://www.onehealthinitiative.com/), in order to help inform future work.

RECENT CANDIDATE VIRAL VACCINES PRODUCED IN PLANTS

Hepatitis B Virus Vaccines

Hepatitis B virus (HBV) vaccines are one of the blockbuster vaccine success stories of modern times: since identification of the virus in the 1960s, it took less than 20 years for a subunit vaccine to get to market. However, this was in the form of 22 nm subviral particles purified from the serum of human carriers of HBV, and although highly effective, was expensive to produce and of limited supply – to say nothing of the ever-present risk associated with a blood product isolated from HBV carriers who may carry any number of other, as yet undetected viruses. It was a triumph of modern molecular biology, therefore, when a very similar virus-like particle (VLP) vaccine derived from expression of the HBV small surface antigen (S-HBsAg) was developed in 1984 [12]. While this was initially still expensive – US$40/dose, with three intramuscular doses being necessary – prices have come down very significantly, to the point that more than 110 countries now routinely immunise infants as part of the Extended Programme of Immunisation (EPI) [13]. The recombinant vaccines are highly effective and safe, and have helped set the standard for later introductions, such as of the recombinant VLP-based Human papillomavirus (HPV) vaccines.

However, there is still space for improvements in HBV vaccines, both in terms of cost of goods, and in specific antigen content. An increasing desire worldwide for "needle-free" vaccine delivery, for example, would require cheaper production of larger amounts of antigen for oral delivery, which current production modalities would not be able to meet. Problems with non-response of certain groups of people to the current vaccines have also necessitated the development of third generation products, containing the middle (M-HBsAg) and/or large (L-HBsAg) surface antigens, which contain the strongly immunogenic preS1 and/or preS2 domains. However, these vaccines are more expensive and less readily available [11]. Accordingly, plant production of HBsAg-based HBV vaccines has gone on for over 20 years, with a variety of products being made; pre-clinical testing of oral delivery of transgenic potato-delivered products for almost as long [14], with a preclinical trial of an orally-delivered product in 2001 [15], and human clinical trial in 2005 [16].

Oral delivery of HBsAg in transgenic plant material has not proved to be particularly effective, however, with immunogenicity generally being low [7]. This has effectively led to the general curtailment of the transgenic-plant-as vaccine efforts [9]. It is interesting in this regard that despite the concept of "vaccination via banana" having been hyped in the popular press since the 1990s (eg:http://www.theguardian.com/science/2000/sep/08/gm.infectiousdiseases), it was not until 2005 that HBsAg was first expressed in transgenic banana fruit in India, albeit at relatively low yield [17]. A recent review has proposed a combination of approaches, with parenteral vaccination with purified plant-produced HBsAg followed by oral boosting with less well purified antigen as tablets or capsules [18]: preliminary studies in mice using lyophilised HBsAg VLP-producing transgenic lettuce converted into tablets appear to add weight to the proposal [19].

The highest plant yield of conventional (=S, or small) HBsAg was achieved via use of deconstructed Tobacco mosaic virus-based cDNA MagnICON vectors (Icon Genetics, Halle, Germany): this was around 300 mg/kg wet weight in *Nicotiana benthamiana*, and the recombinant protein was full-length, formed disulphide-linked dimers, displayed the conformationally-determined 'a' antigenic determinant, and assembled correctly into VLPs. Interestingly, vacuum-mediated agroinfiltration of whole plants led to a better product, as determined by presence of the 'a' determinant [20].

Plant expression of HBV antigens other than the standard S HBsAg has been attempted in recent years by a number of groups. The Arizona Biodesign Institute group, who pioneered much of the biofarming of HBV vaccines, showed in 2004 that the S form of HBsAg could be used via transient agroinfiltration-mediated expression as a useful fusion partner, with an N-terminal fusions of up to 239 amino acids being tolerated without alteration of its particle-forming ability or major antigenic properties [21]. They followed this with the demonstration that the HBsAg middle protein (M protein) - with the highly immunogenic 55 amino acid pre-S2 region fused at N-terminus of the S protein – could be successfully produced in plants, and elicited stronger humoral immune responses than the S protein when injected into mice [22]. The same group used MagnICON vectors to express over 2 g/kg wet weight of plants of VLP-forming and highly immunogenic HBcAg, the "core" antigen [23]: HBc has considerable potential both as a therapeutic vaccine and as a display vehicle for other peptides, including the HBV preS region [11]. A subsequent paper from the Biodesign Institute group detailed the use of a ssDNA Bean yellow dwarf mastrevirus-based vector system (reviewed here [24]) to produce 800 mg/kg of HBcAg in *N benthamiana* [25].

It is safe to say that the potential for production of HBV vaccines in plants has been well realised: while wholly edible/orally-administered vaccines may remain a pipedream, levels of antigen production for both HBsAg and HBcAg have been achieved that are some of the highest for any plant-produced molecules, and the products have been shown to be antigenically appropriate and highly immunogenic.

Hepatitis C Vaccines

Between 130 and 150 million people worldwide are chronically infected with Hepatitis C virus (HCV), and around 350 000 people die annually from HCV-related liver disease [26]. While a significant proportion of those infected will clear their infections naturally, the majority will go on to develop chronic infections – with a 15-30% risk of cirrhosis of the liver within 20 years. The only treatment presently available is antivirals, with the most common regime being combination antiviral therapy with interferon and ribavirin, which are effective against all viral genotypes. Newer drugs are becoming available, such as the very promising sofosbuvir [27]; however, vaccines will be the only effective means of preventing infection from occurring.

Plant production of candidate HCV vaccines has a reasonably long history, given that the virus was only described in 1989 [28,29]. It is interesting that the first products have all been intended as therapeutic vaccines: in 2000, Nemchinov and colleagues described the use of a Tobacco mosaic virus (TMV)-derived vector in *N benthamiana* to express a synthetic hypervariable region 1 (HVR1)-derived peptide called R9, a potential neutralising epitope of HCV derived from the envelope protein E2, fused to the C-terminal of the B subunit of cholera toxin (CTB). The plant-derived HVR1/CTB reacted with immune sera from individuals infected with four of the major genotypes of HCV, and mice immunised intranasally with crude plant extract produced anti-HVR1 antibodies which specifically bound HCV VLPs [30]. This was followed by a succession of reports on plant expression of chimaeric plant virus coat protein molecules containing the same (R9) epitope: these included the use of Cucumber mosaic virus (CMV) CP, with immunoreactivity of sera from chronically infected patients with the recombinant CP [31] leading to use of purified R9-CMV to elicit in vivo responses in rabbits, and in vitro cellular responses as measured by interferon gamma production for lymphocytes from human patients [32,33]. Subsequently, the same group showed that purified R9-CMV particles were stable under simulated gastric and intestinal conditions, and could elicit a humoral immune response in rabbits fed with R9-CMV infected lettuce plants [34].

The R9 epitope has also been expressed as a fusion product with the Alfalfa mosaic virus (AMV) CP, via transfecting plants transgenic for AMV RNAs 1 and 2 with recombinant CP-encoding RNA3: the R9/ALMV-CP reacted with HVR1-specific monoclonal antibodies and immune sera from individuals infected with HCV [35].

In another approach, use of an E2-derived epitope as a plant-produced C-terminal fusion to Papaya mosaic virus (PMV) CP led to formation of flexible rodlike VLPs which were actively internalised in bone-marrow-derived antigen presenting cells (APCs). C3H/HeJ mice injected twice with the VLPs exhibited a humoral response lasting more than 120 days against both the CP and the E2 epitope, with the production of IgG1, IgG2a, IgG2b and IgG3 suggesting a Th1/Th2 mixed response [36].

Another interesting recent development was the filing of a USPTO patent application on plant production of the whole E2 protein, as apparently this has not been particularly successful because of misfolding. The patent covers agroinfecting a *N. benthamiana* cell expressing the recombinant E2 protein with one or both of the molecular chaperones calnexin and calreticulin, as this leads to an improvement in the yield of recombinant E2 protein [37]. This adds to what appears to be a successful expression, albeit at low level, of HCV E1 protein in plants [38].

It appears as though plant production of candidate therapeutic vaccines for HCV is quite well covered; however, the prophylactic vaccine space is far less well investigated: hopefully, this will change with optimisation of plant production of full-length E2 protein as detailed above.

Influenza Virus Vaccines

Human seasonal influenza is currently mainly caused by viruses from two distinct genera of *Influenzavirus*: these are three Influenza A viruses (H1N1, H1N1pdm09, and H3N2), and two lineages of Influenza B virus (Yamagata and Victoria). Annual attack rates globally are estimated at between 5–10% in adults and 20–30% in children, with about 3–5 million cases of severe illness, and about 250 000–500 000 deaths [39,40]. The conventional chicken egg-based inactivated whole-virus split vaccine technology has a production capacity as of 2011 of 1.42 billion doses of trivalent vaccine, and a production level of 620 million doses - albeit with a six-month lead-in period every year [41]. This is manifestly obviously not capable of dealing with pandemics, when

"…the potential vaccine supply would fall several billion doses short of the amount needed to provide protection to the global population" [42].

It is worth noting that this report was issued just three years before the 2009 pandemic – and was prophetic in its predictions of vaccine shortages, and of the overly long interval between identification of the pandemic agent, and availability of vaccines. The report went on to say:

A vaccine cannot be developed with certainty until after the pandemic virus emerges.

Current global capacity to manufacture influenza vaccines is limited.

Two doses may be needed for a pandemic vaccine because of the absence of pre-existing immunity. This could further delay time to achieve protection and add operational challenges to delivery.

High antigen content may be required; this would limit the total number of doses that can be made available with the current egg-based technology for inactivated vaccines. [my emphasis]

The target population for vaccination could potentially be the entire global population of over six billion; this would require comprehensive resources to support operational and logistic demands in many countries.

In respect of the last point, it is worth noting that in 2001, while northern hemisphere influenza vaccine production capacity was rated at 1,069 million doses, with actual production of 534 million doses, southern hemisphere capacity was 352 million doses, with a production of only 86 million [41]. Thus, provision of pandemic vaccines for the "Global South" would be - and was in fact, in 2009 [43] - gravely lacking with current technology. I am indebted to one of the referees of this review for the observation that "It is important that all the eggs are literally not placed in one basket – or all of the vaccines put in eggs".

Perhaps fortunately, then, influenza vaccines have been the major success story for plant-expressed antigens, largely due to two major factors: first, the haemagglutinin (HA) protein is the main determinant of neutralisation, and the only essential component of a vaccine; second, it appears to be well expressed in plants, and to fold properly. Perhaps the most significant factor of plant expression of HA, however, is the fact that expression of this alone is sufficient for the efficient formation at high yield of highly immunogenic VLPs, which bud from the plant cell outer membrane [44]. This is especially relevant to influenza vaccine development in light of the fact that HA-containing VLPs produced in insect cells elicit broader cross-neutralising immune responses than whole virion inactivated influenza virus or recombinant HA [45]. These properties have been exploited by a number of groups in work reported on in detail elsewhere [46-48]; accordingly, this review will be limited to recent developments of specific interest to pandemic and seasonal human vaccine development.

Plant-based technology lends itself very well to "orphan" or "niche" vaccines, because of what is in effect infinite scalability of production. It is also very well suited for manufacture of "rapid response" vaccines, such as those directed against pandemic influenza and bioterror agents: for this reason the US Defense Advanced Research Projects Agency (DARPA) in 2005 launched an Accelerated Manufacturing of Pharmaceuticals (AMP) programme that included an investigation of the suitability of plants as a manufacturing platform for the purpose. In 2009, as a response to the H1N1 "swine flu" virus pandemic, they initiated the "Blue Angel" effort, which was:

"…an accelerated and integrated effort to deliver effective interventions for pandemic influenza"

Part of this involved using US$100 million to fund a challenge for four companies to demonstrate a capability to use plants to produce 100 million doses of influenza vaccine a month. These were the Fraunhofer USA Center for Molecular Biotechnology in Delaware, Kentucky Bioprocessing in Owensboro, the Project GreenVax consortium with partners from Texas A&M University system and G-Con from Texas, and Medicago USA in North Carolina [49]. As of July 2012, Medicago Inc. had produced, as part of a "rapid fire" milestone, more than 10 million doses of an H1N1 VLP-based influenza vaccine candidate in one month, by Phase 1-appropriate current good manufacturing practices (cGMP) [50].

Latest developments from contenders in this challenge include preclinical and human clinical trials of a number of HA-based products. The first was of Medicago's H5N1 A/Indonesia/5/05 HA VLP vaccine candidate, which constituted "…the first ever report of administration of a plant-made VLP vaccine to humans" [47]. In preclinical work reported in this paper, two low doses of Alhydrogel® alum-adjuvanted plant-made VLPs (1.8 μg) in ferrets prevented pathology and reduced viral loads following heterotypic (A/Vietnam/1203/04 H5N1 clade 1 virus) lethal challenge. Interestingly, this protection occurred despite the fact that HAI titres to the A/Vietnam/1203/04 challenge virus were detectable in only 75–87.5% of the challenged ferrets: this prompted the comment that the correlates of protection for influenza virus infection are not fully understood, and that the VLPs are probably stimulating innate immune responses not seen for conventional vaccines.

The human trial of the H5 HA VLPs was performed in healthy adults 18–60 years of age who received 2 doses 21 days apart of 5, 10 or 20 mg of alum-adjuvanted H5 VLP vaccine or placebo (alum). Immunogenicity was evaluated using Haemagglutination-Inhibition (HAI), Single Radial Hemolysis (SRH) and MicroNeutralisation (MN) assays: results from all three assays were highly correlated, with clear dose-responses with all measures of immunogenicity. Almost 96% of those in the 2×10 or 2×20 μg dose groups mounted detectable MN responses, indicating promising immunogenicity. In their words, "These data are particularly encouraging in light of the fact that traditional, egg-based split H5N1 vaccines showed only modest immunogenicity in humans at doses as high as 45 mg with or without alum as an adjuvant".

One particularly encouraging aspect of this study was their testing of plant-specific glycan responses in the study group. One of the purported potential hazards of plant-made vaccines is the possibility that they might exacerbate pre-existing allergies to plant N-glycans, or even elicit such responses de novo [51]. In a clear negation of this potential for this vaccine, this study found no IgEs to plant-specific glycans in any of the 48 subjects, and no significant statistical difference in increase of IgGs to plant glycans between placebo and vaccine groups. That this may be important in influenza vaccines made in different expression systems is shown by evidence that glycosylation of recombinant A/California/04/09 HA made in insect cells and plants is different, with high mannose type glycans in plant-expressed HAs, and complex type glycoforms for the insect-expressed HA [52]. Further evidence that glycosylation of H5 HA may affect immunogenicity includes the fact that while antibody titres are higher with Sf9 insect cell-derived HA, neutralisation and HAI titres are much higher with CHO mammalian cell-produced HA [53] – although this does not seem to have been a problem with plant-produced H5 HA VLPs.

An illustration of the speed and scalability of transient expression in *N benthamiana* for emergency response to novel influenza virus outbreaks was shown recently by Medicago Inc., who produced grams of cGMP-grade plant-made H7N9 vaccine, as HA-only VLPs, in response to the outbreak in humans of that very severe influenza virus in China in 2013. The first vaccine lots were available only 19 days after the company accessed the H7 HA gene cDNA sequence, and as little as 3 μg in one dose of the H7 VLP vaccine administered with or without GLA (glucopyranosyl lipid A) adjuvant elicited high antibody titres in mice [54].

In 2012, accounts of Phase 1 trials of both H1N1pdm HA-derived (A/California/04/09) and HPAI H5N1 (A/ Indonesia/05/05) HA-derived products were published by researchers from what is now the Fraunhofer Center for Molecular Biotechnology in Delaware, USA (http://www.fraunhofer.org/MolecularBiotechnology), following proof that they could use a TMV-based transient expression system to produce both proteins [48]. Recombinant HA sequences both included KDEL ER retention and 6xHis affinity purification sequences at their C-termini and were purified using detergent-containing buffers, Ni-Sepharose and hydrophobic interaction and anion exchange chromatography columns, at scales up to 50 kg of *N benthamiana*biomass under cGMP. Mouse immunogenicity studies demonstrated that intramuscular (IM) injection of 1 μg of H1 HA or 5 μg of H5 HA protein adsorbed to 0.3% Alhydrogel as adjuvant elicited serum HAI antibody responses with titres ≥1:40 in at least 67% of vaccinated mice, with significant enhancement of immunogenicity and consequent dose-sparing afforded by use of the adjuvant. In rabbits, two IM doses of 45 μg of H1 HA plus Alhydrogel elicited HAI titres of ≥1:40 in 100% of animals, while two doses of 90 μg of H5 HA were required for the same titres in 80% of rabbits. Results in ferrets were similar, with lower responses to the H5 HA. Conclusions from this study were that both antigens were safe and immunogenic, and suited to human trial.

In contrast to the preclinical results, however, testing of the H5 HA vaccine in human volunteers showed no enhancement of immunogenicity by Alhydrogel: 15 and 45 μg doses with Alhydrogel adjuvant, and at 90 μg dose with and without Alhydrogel, were administered in a two-dose regimen three weeks apart; the highest responses were in the 90 μg unadjuvanted group [55].

The H1 HA vaccine antigen was tested with sera derived from human volunteers vaccinated with a conventional AS03 adjuvanted pdmH1N1 vaccine in order to assess its vaccine potential [56]. It was recognized strongly by serum antibodies and antibody-secreting cells from vaccines. Additionally, there was good correlation between results obtained using the plant-made antigen and the conventional vaccine antigen both by ELISPOT and by intracellular cytokine staining assays. The conclusion was that the candidate H1 HA had good vaccine potential, but needed a good adjuvant for use in a clinical trial. This has been done recently [57], in a first-in-human Phase 1 study with H1 HA given twice at dose levels of 15, 45 and 90 μg with and without Alhydrogel, in healthy adults 18–50 years of age. The highest seroconversion rates, measured by HAI (78%) and virus MN assays (100%), were in the 90 μg non-adjuvanted vaccine group after the second vaccine dose.

Thus, results from these two plant-made monomeric candidate HA vaccines are similar, with high doses being required to obtain suitable responses, and – unlike the Medicago trial - no obvious effect of using Alhydrogel adjuvant. It is also questionable whether vaccine antigens containing 6xHis tags would be licenced in a final product.

Another product from this group, however, has more promise: this is an engineered soluble trimeric HA (H1N1pdm) with a heterologous trimerisation motif, which induced serum antibodies active in HAI assays as well as protective immunity in mice given a lethal virus challenge. As may have been expected, the effective doses of the trimeric HA were much lower than previously required for the monomeric HA described above [58]. Further exploration of enhancing the immunogenicity of the H1N1pdm monomeric HA was also promising: this entailed doubly-adjuvanting the antigen with silica nanoparticles (SiO_2) and the mucosal adjuvant candidate bis-(3′,5′)-cyclic dimeric guanosine monophosphate (c-di-GMP) [59]. Mice were vaccinated intratracheally, which resulted in systemic humoral immune responses, and induction of a strong mucosal immune response with antigen-primed T-cells in the lungs.

While biofarming is often touted as being ideal for vaccine and other pharmaceutical production for the developing world, there is very little actually published from developing countries on use of the technology for their own purposes. One object example comes from my own lab in South Africa, where we explored the prospects of making emergency response pandemic influenza vaccines [43]. We were successful in making high-yield candidate subunit vaccines via transient agroinfiltration-mediated expression in *N benthamiana* from both a full-length (up to 130 mg/kg FW) and a soluble version (up to 675 mg/kg FW) of the H5N1 A/Viet Nam/1194/2004 HA, via a human codon-use optimization of the HA gene, and showed that while both proteins elicited usable titres of antibodies in mice and chickens by HAI assays, the full length protein was significantly better. Subsequent work showed that the full-length H5 HA could be extracted as semi-pure VLPs from the apoplastic space of unmacerated *N benthamiana* leaves via buffer infiltration and gentle centrifugation of cut leaves rolled into strips (H Inderthal, II Hitzeroth and EP Rybicki, unpublished) (Figure 1).

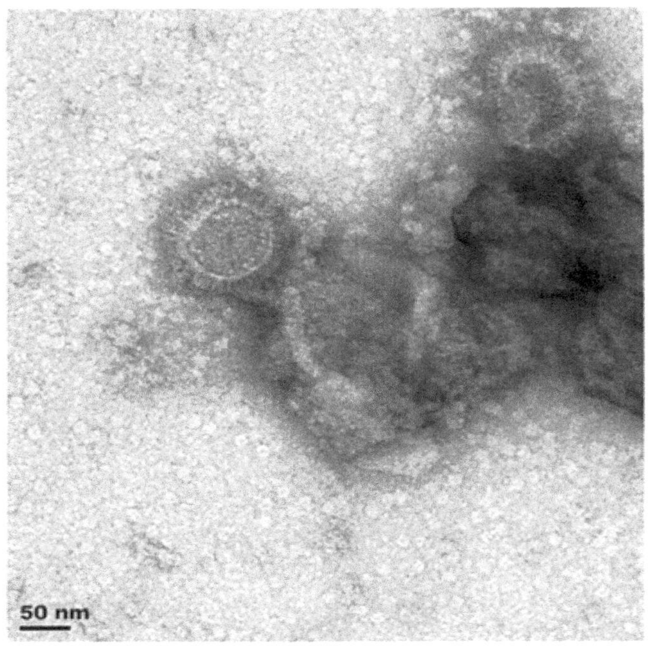

Figure 1: Transmission electron micrograph of virus-like particles composed of Influenzavirus A H5N1 haemagglutinin (HA) and plant lipids, produced by transient expression of the H5 HA via agroinfiltration of*Nicotiana benthamiana* leaves with recombinant *Agrobacterium ...*

A number of other different plant-produced influenza vaccine candidates are at earlier stages of development, but show promise: these include HA protein attached to a number of different partner molecules, as well as candidate "universal vaccines" based on the highly conserved M2e or M2 ion channel protein ectopic domain.

Chemical conjugation of plant-made monomeric H1N1 A/California/07/09 HA to the surface of Tobacco mosaic virus (TMV) virions produced a vaccine which elicited a potent antibody response as measured by HAI, and allowed dose-sparing in mice, with a single 15 µg IM dose being sufficient to protect all mice tested, if given with either alum (Alhydrogel) or a squalene oil-in-water based adjuvant (Addavax) adjuvant [60]. The TMV carrier is also unique in that it harnesses dendritic cell (DC) uptake and subsequent activation to stimulate effective antigen presentation, which influences the quality of the subsequent protective immune response. Using TMV as an antigen carrier allows repeated boosting without loss of immune activation to the partner antigen, even when anti-TMV antibodies were already present: in fact, prior exposure to TMV potentiated the response to HA. This group was able to produce 500 g of monomeric HA protein at pilot scale in less than a month – a good advertisement for the potential of this technology for an emergency response vaccine capability.

Another interesting approach was to make an elastin-like polypeptide fusion with a stabilised soluble trimer-forming H5N1 HA (ELPylated H5 HA, or H5-ELP): production was enhanced by ELPylation, and the molecules could be easily purified by the characteristic inverse transition cycling technique, and elicited neutralising antibodies in mice which bound plant-produced H5 HA VLPs and inactivated egg-produced virus [61].

Production of other influenza virus antigens has been limited, with the only other protein investigated to any extent being the M2e peptide, or ectopic domain of the M2 ion channel protein: this has considerable potential as a universal vaccine candidate, as it is highly conserved between all influenza A viruses. The insertion of a $M2e_{2-24}$ peptide in two different locations in the Human papillomavirus type 16 VLP-forming L1 protein has been reported [62], as has the use of surface display on a filamentous VLP-forming potexvirus CP [63], albeit without preclinical evaluation. However, a more recent investigation that used surface display on TMV virions was more noteworthy in that synthetic M2e peptides derived from a H1N1 virus inserted in between TMV U1 CP residues 155 and 156 were well displayed on virions. The candidate vaccines elicited high antibody titres in IM immunised mice, and were 100% protective against 5 LD50s of homologous challenge virus (A/PR/8/34), and 70% of heterologous (A/California/04/2009) challenge [64].

While plant-made influenza viruses in general, and HA-based vaccines in particular seem to be products with significant potential, there are still some avenues to explore – such as the expression and testing of neuraminidase (NA), matrix protein (M1) and nucleoprotein (NP), all of which can play a role in immunity to and recovery from influenza virus infections, and which could constitute valuable additions to the plant-made influenza vaccine armoury.

Papillomavirus Vaccines

Human papillomavirus (HPV) vaccines based on VLPs made by recombinant expression and assembly of the major capsid protein L1 are the latest blockbuster viral vaccines, with annual sales of both Merck's yeast-made Gardasil and GSK's insect cell-made Cervarix VLP-based vaccines close to or higher than US$1 billion as of 2012 [65]. Gardasil targets the two most prevalent HPVs found in 70% of cervical cancer cases – HPV-16 and HPV-18 – as well as the two most common genital wart-causing viruses, HPVs −6 and −11, while Cervarix contains only HPVs −16 and −18. While the global impact of the target disease is in the developing world – in 2001 women in developing countries accounted for ~85% of both annual cases of cervical cancer (~500 000 cases worldwide) and annual deaths from cervical cancer (~300 000 worldwide) [66] – the main market for the vaccines is in the developed world, largely due to their expense. While this is changing to some extent, with a number of governments negotiating favourable tenders for supply of HPV vaccines for EPI programmes, it remains a problem – and the imminent release of next-generation products with wider coverage – mixtures of more types of HPV L1, or chimaeric L1 vaccines including all or part of the minor capsid protein L2, or L2-based vaccines - will do nothing to fix this problem [67,68].

Another problem, as yet unaddressed by available vaccines, is therapeutic vaccination for those already infected with high-risk HPVs, or who have severe genital warts: neither Gardasil nor Cervarix has any effect on established infections, and neither will the new wider coverage VLP-based vaccines if they elicit the overwhelmingly humoral antibody responses seen so far [68]. Experimental vaccines based on the high-risk HPV oncogenes E6 and E7 have shown some promise in animal models; however, given that most of the burden of cervical disease as well as cancer is in developing countries [69], the problem of cost and small market size will again be a problem should any licenced vaccines be developed.

The plant-based production of Human papillomavirus vaccines has had quite a long history, that has been well reviewed up to 2010 for all HPV vaccines [10], and 2013 specifically for VLPs [9]. Briefly, the important early landmarks in farming of prophylactic papillomavirus vaccines are the proof that transgenically-expressed HPV-16 or HPV-11 L1 can assemble into immunogenic VLPS, albeit at low yield [70-73]; proofs of efficacy of a plant-made Cottontail rabbit papillomavirus (CRPV) L1-based vaccine [74] and Rabbit oral papillomavirus or CRPV L2 peptides surface-displayed on rTMV vaccines [75]; very high yields of HPV-16 L1 and VLPs via agroinfiltration-mediated transient expression [76] or via transplastomic (chloroplast) expression [77]. An interesting approach was shown in a study that used transplastomic plants to express a HPV-16 L1 mutated so as to form only pentameric capsomers rather than particles, fused to the *Escherichia coli* heat-labile enterotoxin subunit B (LTB) as a built-in adjuvant [78]. The resultant protein accumulated to 2% of total soluble protein (TSP), and had all the correct epitopes and biochemical activity of both parent molecules, indicating that it could be a viable vaccine candidate.

One of the few investigations to look at non-genital HPV L1s was a study of plant expression of the L1 of HPV-8, a high-risk cutaneous papillomavirus associated with epidermodysplasia verruciformis and non-melanoma skin cancer in immunocompromised people [79]. The expressed protein formed VLPs *in planta*, albeit at relatively low yield - and only if the 22aa C-terminal nuclear localisation signal was removed. This is similar to what was found for HPV-11 L1 [73], and in contrast to results for HPV-16, where removal lowered yield [76]. In only the third investigation of expression of the L1 of a non-human papillomavirus, reasonably high levels of protein (183 mg/ml biomass) were achieved for transient agroinfiltration-mediated expression of Bovine papillomavirus type 1 L1 [80]. The protein formed VLPs, albeit only T=1 30 nm particles rather than the 50 nm T=7 native structure, which were nevertheless highly immunogenic in rabbits.

The only documented expression of the minor papillomavirus capsid protein L2 was from our laboratory (reported in [81]): this was HPV-16 L2, human codon optimised, and expressed to levels of ~30 mg/kg – which we noted at the time could be useful, as the protein is present in virions at a maximum ratio of 1 L2 molecule per L1 capsomere, and given L1 yields that are ~10–20-fold higher, co-expression of the proteins could result in efficient L1+L2 VLP formation.

Candidate therapeutic vaccines based on the HPV-16 oncoprotein E7 have been investigated in some detail too: the first expression of the protein in plants – transient expression in *N benthamiana* via a recombinant Potato virus X-derived vector - was accompanied by a proof of efficacy in a mouse model, with protection from tumour development caused by the HPV-16 E7-expressing C3 cell line accompanied by strong cytotoxic T-cell responses [82]. The same group went on to show 5-fold enhanced expression of E7 by targeting of the protein to the secretory pathway [83]. The possibly immune-enhancing properties of *N benthamiana* extracts indicated in previous work were further explored, with proof of immunomodulatory activity on human monocyte derived dendritic cells (MDDCs) [84]. Possibly in response to concerns about using an unmodified oncoprotein as a vaccine, they engineered mutations that abolished interaction with cell cycle governing proteins (E7GGG), and showed that a plant-produced fusion product of this with the beta-1,3-1,4-glucanase (LicKM) of *Clostridium thermocellum* elicited the same protective response [85,86]. Experimental scale-up of production of this fusion protein as a therapeutic vaccine was achieved with 50% final yield (at 100 mg/kg biomass) of protein at 99% purity, by means of metal-ion affinity chromatography and gel filtration [87].

An interesting fusion of prophylactic and potentially therapeutic HPV vaccines was an investigation of the feasibility of producing chimaeric HPV-16 VLPs with L1 fused to a string of cytotoxic T-lymphocyte epitopes from HPV 16 E6 and E7 proteins, in transgenic tomato plants (*Lycopersicon esculentum*) [88]. While expression levels were low, VLPs were formed in planta and elicited both anti-L1 neutralising antibody and anti-E6/E7 CTL responses. A development from this work was the use of the chimaeric VLPs made in tomato plants for detection of HPV-16 specific antibodies in patients with grade 1 cervical intraepithelial lesions (CIN 1) [89], pointing up the potential use of plant-made products as inexpensive reagents.

New developments in the area of mainly prophylactic or VLP-based vaccines are limited to just a few studies, largely involving chimaeric L1 molecules or fusions of L1 to other partners. However, one recent interesting variant to the classic HPV L1 VLP approach was one in which HPV-16 L1 was co-expressed with *E coli* LTB in doubly transgenic tobacco (*N tabacum*) plants to reasonable yield (~0.3% TSP), and extracts were used to orally immunise mice (4x over 52 days) [90]. Combination with LTB dramatically increased intestinal mucosal IgA responses to L1 as well as L1-specific splenic cell proliferation and the expression of IFN-γ and IL-4 in CD4+ T cells and spleen cell supernatants.

A study of chimaeric HPV-16 L1 molecules from our group [91] investigated neutralising antibody epitopes derived from HPV-16 L2 comprising amino acid residues 108–120, 56–81 or 17–36 substituted into the C-terminal helix 4 (h4) region of L1, from amino acid 414, following evidence that these were the most effective in terms of eliciting a wider spectrum of cross-neutralising antibodies to high-risk HPVs [92-95]. Following previous experience with L1, we used a human codon-use optimised L1 gene with similarly-optimised inserts, transiently expressed in *N benthamiana* via agroinfiltration, and targeted for import into chloroplasts. All chimaeras were very highly expressed, with yields of up to 1.2 g/kg plant tissue. The L1 chimaera containing L2 amino acids 108–120 (L1:$L2_{108-120}$) was the most successful candidate in that it assembled into small VLPs (~30 nm), and elicited anti-L1 and anti-L2 responses in IM immunised mice, and immune sera neutralised homologous HPV-16 and heterologous HPV-52 pseudovirions. The other chimaeric L1s did not form distinct particles, and elicited significantly weaker humoral immune responses for the same dose of protein. The L1:$L2_{108-120}$ particle formation was in contrast to repeated previous expression of this and the other chimaeras in insect cells, where only loose aggregates of capsomers were formed for L1:$L2_{108-120}$ [95,96].

Another investigation of the potential of the $L2_{108-120}$ epitope was undertaken using a fusion of the peptide to the N-terminus of the Potato virus X coat Protein (PVX CP) for surface display, expressed via a recombinant PVX vector in *N benthamiana*: expression levels of up to 170 mg/kg biomass were obtained, and the chimaeric protein elicited anti-$L2_{108-120}$–reactive antibodies after SC injection or tattoo administration [97].

E7-based vaccines have continued to be investigated, with a number of approaches yielding promising results. A useful crossover approach mixing potentially prophylactic and therapeutic vaccines was that of the same group who developed the HPV-16 L1 chimaera with a string of T-cell epitopes from HPV 16 E6 and E7 fused to its C-terminus [98]. These workers immunised C57BL/6 mice with the antigen, and could demonstrate persistent anti-L1 IgG antibodies for over 12 months, with good neutralising activity. There was also efficient long-term protection from tumour growth induced by HPV-16 E6/E7-expressing TC-1 tumour cell challenges, and significant tumour reduction (57%) in animals with established tumours given therapeutic vaccination.

In another use of PVX CP fusions, this time with the E7GGG protein mentioned previously [99], it was shown that while both N- and C-terminal fusions with PVX CP formed long filamentous VLPs when expressed in *E coli*, and expression in *N benthamiana* was at high levels, the latter did not apparently form VLPs. Another production modality for E7GGG was investigated in the form of transplastomic*Chlamydomonas reinhardtii*, a well-characterised unicellular alga [100]. E7GGG was expressed alone or as a His6 or FLAG fusion protein, to levels of 0.12% of TSP. Affinity purification was performed for both tagged forms, and C57BL/6 mice were vaccinated SC with *C reinhardtii* extract or purified protein mixed with QuilA adjuvant. Both specific anti-E7 IgGs and E7-specific T-cell proliferation were shown in mice vaccinated with either inoculum, and tumour protection was shown after challenge with TC-1 tumour cell line expressing E7. The authors note that this was the first successful expression of a soluble E7 derivative in plants, as previously the protein had to be expressed as a fusion product in order to get a usable yield.

Another recent paper that explored the expression of E7 in transplastomic tobacco determined that targeting the protein by means of a transit peptide to the thylakoid lumen - a chloroplast inner compartment with different redox potential from the stroma, among other characteristics – increased production by more than 80-fold [101]. However, these authors did not inactivate the oncogenic potential of the protein, which means it presently has value only as an illustration of a useful yield enhancement technique.

Our group has recently explored the potential of a novel HPV-16 E7-derived gene construct – a synthetic shuffled HPV-16 E7 (16E7SH) that has lost its transforming properties, but retains all naturally-occurring CTL epitopes – for expression in plants as a candidate therapeutic vaccine [102]. The E7SH gene has previously successfully been used as a DNA vaccine in a mouse tumour model, eliciting potent cellular and humoral immune responses, including tumour protection and regression [103]. We fused the gene translationally to one encoding the Zera® peptide, a self-assembly domain of the maize gamma-zein seed storage protein that induces the accumulation of recombinant proteins into protein bodies (PBs) within the endoplasmic reticulum in a variety of eukaryotic expression systems [104]. This generally allows stabilisation of recombinant proteins, as well as enhanced accumulation and far easier purification. While E7SH alone was expressed at only a low level by agroinfiltration-mediated transient expression in *N benthamiana*, high-level expression of E7SH-Zera was achieved, with a maximum of 1.1 g/kg biomass, and the resultant protein bodies could be easily purified. Immune responses comparable to the E7SH DNA vaccine were demonstrated in mice, with specific humoral as well as cell-mediated immune responses, and significant tumour regression in vaccinated mice with pre-existing tumours. Interestingly, simply mixing Zera-only PBs and 16E7SH also enhanced immune responses, indicating an independent adjuvant activity for the Zera® component. Moreover, use of the E7SH-Zera gene as a DNA vaccine also resulted in increases in IFN-γ levels of mouse splenocytes compared to mice inoculated with the 16E7SH gene, a trend that was also observed for Granzyme B ELISPOT assays as well as in chromium release assays. We feel we have demonstrated proof of efficacy in a mouse tumour model of a novel HPV therapeutic vaccine candidate, which should be easy and cheap to produce and purify. For further development of this vaccine, a DNA vaccine prime followed by matched protein boost might be ideal in order to achieve further enhancements in immunogenicity.

The overall prospect for plant production of both prophylactic and therapeutic vaccines against HPV infections and disease appears bright: gold-standard VLP-based vaccine candidates can be made at high yield via transient expression, as can viable therapeutic vaccine candidates; efficacy has been demonstrated in animal models for both types of vaccine; early pipeline research appears to show the feasibility of making dual-purpose vaccines that may both prevent infection and hasten recovery from infection.

HIV Vaccines

There is little need to either reiterate the need for vaccines for Human immunodeficiency viruses (HIV) in general, or of the predominant HIV-1 in particular; there is also little need, in the face of a plethora of literature, to detail the approaches to, or problems inherent in, the development of said vaccines [105,106]. As could be expected, HIV vaccines were an early target for plant expression studies, with a variety of targets: indeed, HIV-1 p24 capsid protein was expressed successfully in transgenic tobacco, albeit at low yield (0.35% of TSP), as long ago as 2002 [107]; the suitability of transgenic maize as a production platform for oral delivery of HIV vaccines in seed extracts was tested using SIV major surface glycoprotein gp130 [108]; a novel gp41-derived molecule incorporating a CTB fusion as polymerising agent and adjuvant was produced via agroinfiltration-mediated expression in*N benthamiana* [109] and shown to produce mucosal and serum anti-membrane proximal region (MPR) antibodies in mice after mucosal prime-systemic boost immunisation [110]; a Tat monomer was produced in spinach via rTMV as a potential oral vaccine [111]; epitopes derived from HIV-1 Gag and Env were fused to HBsAg and expressed as VLPs in transgenic tomato fruit [112]; various Gag-derived antigens (p24, p41, p55) were produced in transgenic tobacco and via rTMV in *N benthamiana* for investigation of their utility as CTL-inducing immunogens by our group [113]; HIV-1 Nef was produced in *N benthamiana* using an agroinfiltration-mediated transient expression system [114]. The field was comprehensively reviewed recently [115], so this review will be limited discussion of major advances, and more recent work.

While many and varied antigens have been produced in various types of plant using a variety of expression systems, there have been few systematic investigations of antigens regarded as central to mainstream HIV vaccine production – that is, full-length Gag and Env – and no clinical trials of any candidate vaccine products, although plant-produced anti-HIV MAbs have gone to Phase 1 trial [116]. The most successful expression of full-length Pr^{55} Gag precursor polyprotein to date was done with subtype B HIV-1 Gag in transplastomic tobacco, with yields of enveloped ~100 nm diameter VLPs of up to 400 mg/kg biomass [117]. This was close to 10 000-fold more than a previous best by our group, with VLPs produced in transgenic tobacco [81], and represents a viable yield for a product with serious vaccine potential: recent evidence that lack of progression to AIDS is linked to strong CTL responses to Gag [118-120] reinforces the need for vaccines that can target such responses, and Pr^{55} Gag VLPs are potent elicitors of CTL, especially when used as a protein boost to a heterologous prime in primate models [121,122]. A recent paper details how HIV-1 p24 antigen expressed transgenically in either *Arabidopsis thaliana* or*Daucus carota* showed a priming effect in mice fed whole plant material, eliciting humoral immune responses detected as serum anti-p24-specific IgG after an intramuscular purified p24 protein boost [123]. It is interesting that dose-dependent antigen analyses using transgenic *A. thaliana* showed that low p24 antigen doses were superior to high doses.

A new plant-made HIV multiantigen that makes use of VLPs is "...enveloped particles... consisting of Gag and a deconstructed form of gp41 comprising the membrane proximal external, transmembrane and cytoplasmic domains (dgp41)" [124]. This combines the proven qualities of Gag VLPs as self-adjuvanting potent humoral and cellular response-inducing immunogen with an envelope protein comprising the membrane proximal external region (MPER) of HIV gp41, which is important in infection processes and in eliciting broadly neutralising anti-HIV antibodies. Both the *gag* and *gp41* genes were extensively deconstructed in terms of removal of potential methylation sites, and cryptic splice and polyadenylation sites, and plant codon-use optimised. Gag was expressed constitutively in transgenic *N benthamiana* plants under control of the CaMV 35S promoter, to levels of ~22 mg/kg biomass. Transgenic plants were then infiltrated with Agrobacterium transformed with a deconstructed MagnICON replicating vector expressing the dgp41 envelope protein that was targeted to the apoplast by means of a barley alpha-amylase signal peptide. Coexpression allowed >2-fold greater accumulation of both proteins, with dgp41 reaching ~9 mg/kg. Iodixanol density gradient centrifugation indicated that the coexpressed proteins cosedimented in a denser fraction than for pgp41 when expressed alone, and similarly to Gag-only fractions that were known to contain VLPs – and indeed, characteristic ~100 nm diameter enveloped VLPs were seen in extracts and *in situ* in plant leaf sections, and could be shown to bud into the medium from protoplasted co-transfected cells. I believe these authors are not exaggerating in their claim that "These findings provide further impetus for the journey towards a broadly efficacious and inexpensive subunit vaccine against HIV-1", given that their vaccine candidate includes both the whole of Gag, and a popular Env-derived target antigen.

The expression of full-length Env or even of gp120 in plants has been an elusive target: there is only a single published account of plant-based expression of HIV-1 Env, and despite this being highly successful, it was only as an adjunct to the demonstration of rapid and high-level transient production of functional HIV broadly neutralising monoclonal antibodies [125]. The Env was the HIV 89.6.P gp140ΔCFI envelope developed by the Vaccine Research Centre, NIH, and was expressed both in stable transgenic SR1 *N tabacum* and transiently via agroinfiltration in *N benthamiana*, as KDEL-tagged and native forms, for ER retention or secretion and consequent differential glycosylation. The gp140 was successfully expressed – to ~80 mg/kg biomass – and purified by leaf homogenization and clarification, followed by *Galanthus nivalis* agglutinin (GNA) lectin column affinity and DEAE ion exchange chromatography. As expected, ER-retained and secreted forms of gp140 were differentially glycosylated, with the former containing only OMT glycans (no complex glycans), and the latter with only low percentages of complex-type N-glycans. This low percentage of complex glycans is similar to the high mannose glycans on virion-associated HIV-1 Env, which differs from the higher percentage of complex glycans on Env produced in mammalian cell lines. The plant-made Env also bound HIV-1-specific MAbs produced either in CHO cells or also produced in plants, indicating no significant difference between it and mammalian cell-produced Env. This is the first convincing evidence for production of a HIV-vaccine-relevant Env molecule in plants, which, together with the evidence for Gag-based VLPs including Env derivatives from the previous work, indicates that true HIV VLPs could be produced in plant systems.

One other Env-related production modality in plants that could be of commercial interest for HIV as well as for other viruses of humans and animals is revealed in a patent filed by Medicago Inc.: this details the use of chimaeric constructs of "ectodomains" from enveloped virus trimeric surface proteins – such as retroviruses, rhabdoviruses, herpes-, corona-, paramyxo-, pox- and filoviruses - fused to an influenza virus HA protein transmembrane domain and cytoplasmic tail, in order to produce VLPs similar to the HA-only VLPs previously mentioned [126]. Their HIV Env construct was a fusion of various portions of the ConS ΔCFI gp145 - an engineered Env lacking the gp120-gp41 cleavage site, the fusion peptide, an immunodominant region in gp41 and the cytoplasmic tail (CT) domain [127] – and the transmembrane (TM) and CT domains of the HA2 portion of either the H3 or the H5 HA molecule. In the words of the patent, "Although native HIV Env protein poorly accumulates in plants, a chimeric HIV Env protein, fused to a transmembrane (TM) and cytoplasmic tail (CT) domains from influenza HA accumulates at high level, and buds into HIV VLPs in absence of core or matrix protein, in plants". This could provide a genuinely novel source of HIV Env antigens of enhanced immunogenicity, especially for use as a boost vaccine for heterologous prime-boost vaccination regimes.

Bluetongue Virus Vaccine

One of the more exciting success stories in veterinary biofarming in recent years is undoubtedly the proof of efficacy of a plant-made VLP-based vaccine for Bluetongue virus (BTV), a muticomponent dsRNA-containing orbivirus in the family *Reoviridae* [128]. This project was a part of the EU FP7 Plant Production of Vaccines (PlaProVa) initiative (http://www.plaprova.eu/), whose activities to do with VLP-based vaccines are partly reported on here [129]. Bluetongue disease is a relatively newly emerged problem in sheep and goats in northern Europe [130], and is almost certainly a result of climate change affecting the distribution of the Culicoides midges that transmit it [131]. Vaccines are seen as an essential part of disease control: however, unlike the case in endemic areas such as South Africa where attenuated live vaccines are used routinely, concerns about vaccine safety and the possible emergence of new strains of virus because of genomic reassortment with live vaccine strains, have resulted in a push for the development of recombinant protein-only vaccines [132].

It has been established that recombinant expression of BTV proteins in insect and other animal cell culture systems results in a variety of structures being formed: expression of VP3 alone produces subcore-like particles (SCLPs); VP3 and VP7 together produce core-like particles (CLPs); expression of these plus VP5 and VP2 produces authentic VLPs [133,134]. Building on evidence that these VLPs are protective in sheep challenged with live virus [135], the PlaProVa group investigated whether it was possible to do the same using plant-produced antigens. Gene constructs were based on the

Netherlands NET2006/04 strain of BTV-8, and sequences for VP2, VP3, VP5 and VP7 genes were *Nicotiana* codon-use optimised. The genes were cloned for expression into pEAQ-*HT*, an expression vector containing the Cowpea mosaic virus (CPMV)-derived HT or HyperTrans translational enhancer sequence from RNA2 [136]. Interestingly, infiltration of *N benthamiana* plants with either VP3 and VP5 alone produced necrotic symptoms; if these were combined with the other proteins, no necrosis was observed. Simple screening of clarified extracts by density gradient ultracentrifugation showed that expression of VP3 alone, or of VP3 + VP7, or of all four together, resulted in virion proteins co-sedimenting as high-MW aggregates. Some problems with stoichiometry resulting from relative overexpression of VP3 – an over-accumulation of SCLPs - were addressed by use of vectors expressing more than one protein: VP2 and VP5 were expressed via a dual HT vector, with VP3 and VP7 together on another vector with VP3 expression being detuned by use of a non-HT-containing 5′ UTR. This had the result of down-regulating the formation of CLPs, and a significant shift in the equilibrium towards VLP formation. Total BTV-8 protein yield by use of these vectors was >200 mg per kg biomass; the final yield of gradient-purified VLPs was ~70 mg per kg – gratifyingly high, given a vaccine dose (see below) of 50 µg VLPs/dose.

Testing of the plant-made antigens was done in sheep: immunogenicity was assessed by injecting two sheep with 20 µg of VLPs mixed 1:1 (v/v) with Freund›s incomplete adjuvant, and boosting at 21 and 42 days. Serum collected 18 days after first boost was positive for BTV antibodies using a commercial ELISA test kit, and serum from day 56 final bleed reacted with all four structural proteins in western blots, with strongest reactivity toward the major immunogenicity determinant (VP2) and the most abundant structural protein (VP7). Efficacy was tested by injecting four groups of five sheep with either 50 µg VLP or 200 µg CLP mixed 1:1 (v/v) with Montanide ISA70 VG adjuvant, or 5×10^4 $TCID_{50}$/mL commercial live attenuated BTV-8, and boosting on day 28. Animals were challenged with 1 mL infected sheep blood containing live BTV-8 on day 63, and clinical reactions monitored for 2 weeks. Plant-produced VLPs had an identical protective efficacy profile as assessed by clinical reaction index (CRI) as the live attenuated, BTV-8 vaccine, while plant-produced CLPs were poorly protective. Sera from both the VLP and the live attenuated vaccine group showed high serum neutralization titres after day 28; however, VLPs induced high antibody levels only after booster injection, whereas neutralising antibodies were elicited by the attenuated vaccine as soon as 7 days after vaccination. Plant-produced CLPs offered partial protection against live virus challenge, similar to insect cell–produced CLPs.

The results of this investigation are significant for biofarming and vaccinology in particular for a number of reasons. First, as a general finding, it showed it is possible to reasonably easily produce a multi-protein complex with varying stoichiometry: such an approach could also be applied to other complex viruses, such as the related African horsesickness virus, human rotavirus, or picornaviruses such as poliovirus or foot and mouth disease virus. Second, and for BTV in particular, the ability to produce high levels of properly-assembled VLPs that are protective in sheep is a valuable proof of concept and of efficacy for a vaccine against an important emerging disease. Additionally, the timescale of mere days for production following preparation of the agroinfiltration inoculum means that it is possible to respond quickly and potentially locally to outbreaks of emerging disease.

Dual-use or "One Health" Vaccines

The "One Health Initiative" (http://www.onehealthinitiative.com/) is a "…worldwide strategy for expanding interdisciplinary collaborations and communications in all aspects of health care for humans, animals and the environment", which it hopes to achieve by, *inter alia*, "Joint efforts in the development and evaluation of new diagnostic methods, medicines and vaccines for the prevention and control of diseases across species". There are a number of obvious viral vaccine targets for this initiative: these are all of the zoonotic viruses that affect domestic and farmed livestock for which there are either unsatisfactory human vaccines or therapeutics, or no human vaccines at all, and high-impact recently emerging viruses with no vaccines for animals or humans. A good example in the former category would be rabies virus; examples for the second would be agents such as Severe acute respiratory syndrome (SARS) and Middle Eastern respiratory syndrome (MERS) coronaviruses (CoVs), West Nile virus (WNV), and Ebola and Marburg filoviruses. Thomas Monath has also recently defined a "one health paradigm" [137] which specifies three frameworks for development and use of vaccines to control zoonoses:

Framework I vaccines target dead-end human and livestock hosts.

Framework II vaccines target [mainly arthropod-vectored] infections of domesticated animals as a means of preventing spread to humans.

Framework III vaccines target wild animal reservoirs.

His example for Framework 1 vaccines would be West Nile virus; for Framework II, Rabies virus, Rift Valley fever, Venezuelan equine encephalitis, and Hendra viruses; for Framework III, reservoir-targeted agents such as oral bait rabies.

Given the obviousness of the targets, it is surprising that very few of them have been explored for their potential as plant-made vaccines. Rabies is one of these, and relatively early on: a chimaera of Alfalfa mosaic virus CP and epitopes derived from glycoprotein G and the nucleoprotein was successfully expressed in plants via two different plant virus-based systems. Parenterally-injected extracts protected mice from challenge, and human volunteers who had ingested rCP-containing plant material produced rabies virus-neutralising antibodies [138]. A full-length synthetic G protein gene – with a plant secretion signal peptide, and ER retention signal – was found to express reasonably well (0.4% TSP) in transgenic tobacco, and purified protein elicited complete protection against virulent intracerebral challenge in intraperitoneally-immunised mice [139].

The development of an effective oral vaccine against rabies has also been reported, with single doses of 50 µg Vnukovo strain rabies glycoprotein G in transgenic maize seed containing the antigen at 1% of TSP, protecting all vaccinated mice from lethal vampire bat rabies virus challenge [140]. Further testing of this vaccine was done in sheep: maize kernels containing different doses of G protein (0.5, 1, 1.5 and 2 mg) were given in a single dose by the oral route, and 2 mg doses elicited a degree of protection comparable to that conferred by the injected commercial vaccine [141]. The authors state that "...this is the first study in which an orally administered edible vaccine showed efficacy in a polygastric model", which is an important landmark in veterinary and potentially One Health vaccinology.

Two important disease agents that are prime candidates for consideration as One Health vaccines and occur almost exclusively in developing countries, are the tick-borne Crimean-Congo haemorrhagic fever virus (CCHFV) and mosquito-borne Rift Valley fever virus (RVFV), both familial bunyaviruses. CCHFV has the wider distribution, including Africa, Asia, southern and eastern Europe and the Middle East [142], while RVFV is largely restricted to sub-Saharan Africa. However, both viruses are regarded as having emerging potential, with RVFV in particular regarded as having the potential to spread to Europe, Asia, and the Americas [143]. While attenuated live vaccines are available for RVFV, these are regarded as having significant side effects; there are no accepted vaccines for either virus for general use in humans.

A recent study investigated the expression of the two neutralizing epitope-rich CCHFV envelope glycoproteins Gc and Gn in hairy root cultures and in leaves derived from transgenic tobacco plants [144]. Proteins accumulated to levels of 1.8 mg/kg biomass in hairy roots and 1.4 mg/kg in leaves. Separate groups of mice were fed transgenic leaves or roots, or fed the plant material and injected SC with the plant-made proteins, or vaccinated with an attenuated CCHFV vaccine as a positive control. Mice in all the immunised groups had a consistent rise in anti- Gc and Gn IgG and IgA antibodies in serum and faeces, respectively. Mice in the group that was fed and parenterally boosted, however, exhibited a significant rise in anti-CCHFV IgG (titre of 1/32 000 compared to 1/256 for oral-only) after a single boost. Additionally, the plant-purified Gc and Gn proteins reacted with human immunoglobulins in serum from a patient who had recovered from CCHFV infection. The potential for recombinant protein production in plants as a CCHFV vaccine appears obvious, although yields seem low for routine production.

The production of Rift Valley fever virus antigens in plants is much less well studied: there is one PhD thesis that reports expression of a truncated Gn or soluble ectodomain construct and of the nucleoprotein (N) gene in transgenic *Arabidopsis thaliana* [145]. This study chose these two antigens because Gn is the more effective at eliciting neutralising antibodies of the two envelope glycoproteins and the recombinant Gn ectodomain (tGn) is known to be protective, and the N antigen elicits non-neutralising antibodies but also a strong cellular immune response that is partially protective. The N protein accumulated to high levels while truncated Gn protein did not accumulate to levels detectable by western blot. Feeding groups of mice transgenic plant material three times (0, 2 and 4 weeks) containing tGn or ~70 µg of N protein per 4 mice resulted in seroconversion after the second feeding for both antigens, with higher titres (10^3 – 10^4) for N compared to tGn (10^2 – 10^3). While these results are very preliminary as far as a vaccine goes, they indicate the feasibility of first, using transgenic or transiently-produced antigen in plant tissue as an oral vaccine in animals, and second, of producing antigen for complete or partial purification processes that could result in a human vaccine.

Anti-Viral Therapeutic Antibodies

A useful recent review details how plant-based antibody products may "...provide lower upfront cost, shorter time to clinical and market supply, and lower cost of goods (COGs)... [and] improvements in pharmacokinetics, safety and efficacy" [146]. As an object example, an important aspect of rabies disease prevention is therapy, given the many people in developing countries who get bitten by suspected rabid animals annually – and a major development in this area is the production of potent rabies-neutralising antibodies in plants, given the prevailing situation of mainly equine-produced sera being in short supply and of variable quality. A consortium of researchers that includes South Africans have recently described the engineering and production in transgenic *N tabacum* plants of both a humanised IgG version of the broadly neutralising murine MAb E559, and the murine version [147]. Purification via agarose protein A/G affinity chromatography yielded 1.8 mg/kg biomass (0.04% TSP) for the chimaeric Mab, and 1.2 mg/kg biomass (0.03% TSP) for the murine Ab. Both antibodies assembled properly, and were equivalent to hybridoma-produced MAbs in neutralising a panel of lyssaviruses that included all the phylogroup I viruses classical RABV, Duvenhage virus, European bat lyssavirus types 1 and 2, and Australian bat lyssavirus, although no neutralisation was seen for the phylogroup II viruses Lagos bat virus and Mokola virus. The efficacy in post-exposure prophylaxis of the humanised Ab was tested in hamsters injected with a lethal dose of CVS-11 strain of virus: the plant-produced antibody was apparently more effective than the commercial human rabies immunoglobulin (HRIG; Rabigam), in that survival for both treatment groups was >50% after 14 days, and zero and 11% for HRIG and plant Ab groups, respectively, after 28 days.

The production of anti-Ebola virus antibodies has also recently been explored in plants: this could yet become an important part of the arsenal to prevent disease in healthcare workers, given that at the time of writing an uncontrolled Ebola haemorrhagic fever outbreak was still raging in West Africa (http://www.who.int/csr/don/archive/disease/ebola/en/), had spread from Guinea to Sierra Leone, Liberia, Nigeria, Spain and to the USA (http://www.promedmail.org/direct.php?id=2823539), and the use of experimental solutions was not only being suggested [148], but was being put into practice.

As background, use of a high-yielding geminivirus-based transient expression system in *N benthamiana* that is particularly suited to simultaneous expression of several proteins had previously allowed expression of a MAb (6DB) known to protect animals from Ebola virus infection, at levels of 0.5 g/kg biomass [149]. The same group also used the same vector system (described in detail here [24]) in lettuce to produce potentially therapeutic MAbs against both Ebola and West Nile viruses [150].

A more comprehensive investigation was reported recently, of both plant production of Mabs and post-exposure prophylaxis of Ebola virus infection in rhesus macaques [151]. Three Ebola-specific mouse-human chimaeric MAbs (h-13 F6, c13C6, and c6D8; the latter two both neutralising) were produced in whole *N benthamiana* plants via agroinfilration of magnICON TMV-derived viral vectors. A mixture of the three MAbs – called MB-003 – given as a single dose of 16.7 mg/kg per Mab 1 hour post-infection followed by doses on days 4 and 8, protected 3 of 3 macaques from lethal challenge with 1 000 pfu of Ebola virus. The researchers subsequently showed significant protection with MB-003 treatment given 24 or 48 hours post-infection, with four of six monkeys testing surviving, compared to none in two controls. All surviving animals treated with MB-003 experienced insignificant if any viraemia, and negligible clinical symptoms compared to the control animals. A significant finding was that the plant-produced MAbs were three times as potent as the CHO cell-produced equivalents – a clear case of plant production leading to "biobetters". A follow-up of this work investigated efficacy of treatment with MB-003 after confirmation of infection in rhesus macaques, "according to a diagnostic protocol for U.S. Food and Drug Administration Emergency Use Authorization" [152]. In this experiment 43% of treated animals survived, whereas all controls tested here and previously with the same challenge protocol died from the infection.

An article published during the Ebola outbreak [153] detailed the therapeutic use in rhesus macaques of a cocktail of anti-Ebola Mabs called ZMapp. This is described as a successor to MB-003, incorporating components of the ZMAb cocktail developed by the National Microbiology Laboratory of the Public Health Agency of Canada and another antibody mixture, and was developed by Mapp Biopharmaceutical of San Diego with Defyrus of Toronto and manufactured by Kentucky BioProcessing. The cocktail proved able to rescue 100% of macaques when administered up to 5 days post live virus challenge. It was noteworthy that advanced disease in many animals - indicated by elevated liver enzymes, mucosal haemorrhages and generalised petechia - was reversed, leading to full recovery. ZMapp was also found to bind virions of the Guinean variant of Ebola implicated in the present outbreak. The authors claim that "ZMapp exceeds the efficacy of any other therapeutics described so far, and results warrant further development of this cocktail for clinical use".

In news received during consideration of this article that may vindicate this view, a report quoted as coming from the National Institute of Allergy and Infectious Diseases states that two US healthcare workers who contracted Ebola in Liberia were treated with ZMapp [154]. Despite being given up to nine days post-infection in one case, it appears to have been effective [155]. In later developments, the therapy was also given to another five people, two of whom died. The US government was negotiating at time of writing to produce large amounts of ZMapp as a first-line therapy [156,157]. As an illustration of the scale-up problem facing the manufacturers, in the macaque trial referred to above [151], animals were given three doses of ZMapp at 50 mg/kg intravenously at 3-day intervals. For an adult 70 kg human given the same dose, this would mean a total of 10.5 (3 × 3.5) grams of ZMapp: assuming optimal purified yield of each of the three MAbs individually at 100 mg/kg plant biomass, this would mean 105 kg of *N benthamiana* would be needed to dose just one person optimally. While biofarming may be the most scalable technology for producing MAbs and other therapeutics, this sort of scale requires resources far larger than currently exist.

A novel application of the same technology was also used to produce an Ebola immune complex (EIC) in *N benthamiana*, consisting of the Ebola envelope glycoprotein GP1 fused to the C-terminus of the heavy chain of the humanised 6D8 MAb, which binds a linear epitope on GP1. Geminivirus vector-mediated co-expression of the GP1-HC fusion and the 6D8 light chain produced assembled immunoglobulin, which was purified by protein G affinity chromatography. The resultant molecules bound the complement factor C1q, indicating immune complex formation. Subcutaneous immunisation of mice with purified EIC elicited high level anti-GP1 antibody production, comparable to use of GP1 VLPs [158]. This is the first published account of an Ebola virus candidate vaccine to be produced in plants.

Future Prospects for Plant-Produced Vaccines

The prospects for the increasing use of plants for the manufacture of pharmaceuticals in general look increasingly bright, especially with the recent approval and licensure of Protalix's carrot cell-produced Gaucher disease therapeutic enzyme glucocerebrosidase, traded as ELELYSO™ (taliglucerase alfa) (http://www.protalix.com/products/elelyso-taliglucerase-alfa.asp). This product was approved by the US Food and Drug Administration for injection in May 2012 for long-term replacement therapy for adults with a confirmed diagnosis of Type 1 Gaucher disease. While it is marketed as a biosimilar to the mammalian cell-produced enzyme, it is in fact a biobetter, given that it is naturally mannosylated – which aids uptake by macrophages, for example – unlike the conventional product, which has to be chemically altered. Other biologics such as therapeutic MAbs for use as anti-HIV agents in microbicides are also getting very close to registration – and the recent experience with anti-Ebola MAbs may well speed the process.

However, while the prospects for animal vaccines in general and possibly human therapeutic vaccines also seem favourable, given an increasing number of proofs of efficacy in this sphere, for prophylactic human vaccines specifically they are not as bright – in the short term, at least. The length and rigour of the human prophylactic vaccine developmental and clinical testing path compared to animal and human therapeutic vaccines are a huge obstacle to commercial production of novel biofarmed vaccines, as is the entrenched investment in conventional technology by Big Pharma. It may be that the technology will find a niche on the fringes of conventional vaccinology, where there is a need for small-scale production of vaccines for orphan diseases or for rapid responses to bioterror-related or emerging viral disease outbreaks. One area where biofarmed vaccines could break through soon could be rapid-response vaccines to novel influenza virus outbreaks: the capacity for such a response has already been demonstrated (see above); it remains to put it to the test in a real-life scenario. Once this happens, there is an increased likelihood of the technology being employed for other agents, such as MERS-CoV, RVFV and West Nile and Chikungunya viruses, where "One Health" principles are important.

CONCLUSIONS

The widespread application of biofarming – or more properly, plant molecular farming – that was anticipated in the 1990s has been long in coming, and there are still cynics that doubt that it will ever happen. However, it is almost certainly the case that there is now enough accumulated evidence of proofs of principle in humans and animals, and of efficacy mainly in animals but in humans too, that its great potential for the cheaper and more rapid and more widely scalable manufacture of high-value biologics and pharmaceuticals can no longer reasonably be denied. In the case of viral vaccines and therapeutics specifically, it has proved possible to routinely make fully-assembled complex VLPs and immunoglobulins that can either elicit potent protective immune responses, or disease-reducing therapeutic effects, as shown in animal models.

The recent history of plant-made HBV and HPV and HIV vaccine candidates in particular has shown huge yield increases for viral antigen production, largely as a result of application of sophisticated transient expression techniques that are rapidly becoming the industry norm. The surprising revelation that it was easy to make high-yield HA-only VLPs for influenza viruses in plants could transform that particular industry; so too the fact that plant-made BTV VLPs are probably as effective as insect- or mammalian cell-made VLPs and probably far cheaper to make, and far safer to make and use than attenuated BTV vaccines.

Thus, it seems a reasonably safe prediction that biofarmed viral vaccines will be approved more widely for at least animal use in the near future, and possibly also for use as therapy for existing infections and emergency response therapeutics and vaccines in humans in the slightly longer term – such as for rabies, and possibly for Ebola. Hopefully, approval for prophylactic vaccines for humans – such as for seasonal influenza – will follow in the slightly longer term.

ACKNOWLEDGEMENTS

EPR is grateful to Hedda Inderthal for the electron micrograph in Figure 1, and to Inga Hitzeroth and Ann Meyers and the rest of the team for their support in the Biopharming Research Unit. I acknowledge Kenneth Palmer and Tsafrir Mor and an unnamed referee for useful comments that improved this ms. I also acknowledge the National Research Foundation and Department of Science and Technology and the Poliomyelitis Research Foundation and the Medical Research Council and the University of Cape Town Research Committee in South Africa, and the PlaProVa project of the EU FP7 initiative, Era Biotech, and Medicago Inc. for funding.

REFERENCES

1. Barta A, Sommergruber K, Thompson D, Hartmuth K, Matzke MA, Matzke AJ. The expression of a nopaline synthase - human growth hormone chimaeric gene in transformed tobacco and sunflower callus tissue. Plant Mol Biol. 1986;6:347–357. doi: 10.1007/BF00034942.
2. Hiatt A, Cafferkey R, Bowdish K. Production of antibodies in transgenic plants. Nature. 1989;342:76–78. doi: 10.1038/342076a0.
3. Sijmons PC, Dekker BM, Schrammeijer B, Verwoerd TC, van den Elzen PJ, Hoekema A. Production of correctly processed human serum albumin in transgenic plants. Biotechnology (N Y) 1990;8:217–221. doi: 10.1038/nbt0390-217.
4. Mason HS, Lam DM, Arntzen CJ. Expression of hepatitis B surface antigen in transgenic plants. Proc Natl Acad Sci U S A.1992;89:11745–11749. doi: 10.1073/pnas.89.24.11745.
5. Usha R, Rohll JB, Spall VE, Shanks M, Maule AJ, Johnson JE, Lomonossoff GP. Expression of an animal virus antigenic site on the surface of a plant virus particle. Virology. 1993;197:366–374. doi: 10.1006/viro.1993.1598.
6. Plant vaccines: edible, but how credible?*CVI Forum* 1996,1996:10-13.
7. Rybicki EP. Plant-produced vaccines: promise and reality. Drug Discov Today. 2009;14:16–24. doi: 10.1016/j.drudis.2008.10.002.
8. Rybicki EP. Plant-made vaccines for humans and animals. Plant Biotechnol J. 2010;8:620–637. doi: 10.1111/j.1467-7652.2010.00507.x.
9. Scotti N, Rybicki EP. Virus-like particles produced in plants as potential vaccines. Expert Rev Vaccines. 2013;12:211–224. doi: 10.1586/erv.12.147.
10. Giorgi C, Franconi R, Rybicki EP. Human papillomavirus vaccines in plants. Expert Rev Vaccines. 2010;9:913–924. doi: 10.1586/erv.10.84.
11. Pniewski T. The twenty-year story of a plant-based vaccine against hepatitis B: stagnation or promising prospects? Int J Mol Sci.2013;14:1978–1998. doi: 10.3390/ijms14011978.
12. McAleer WJ, Buynak EB, Maigetter RZ, Wampler DE, Miller WJ, Hilleman MR. Human hepatitis B vaccine from recombinant yeast.Nature. 1984;307:178–180. doi: 10.1038/307178a0.
13. Hepatitis B.[http://www.who.int/csr/disease/hepatitis/whocdscsrlyo20022/en/index5.html]
14. Thanavala Y, Yang YF, Lyons P, Mason HS, Arntzen C. Immunogenicity of transgenic plant-derived hepatitis B surface antigen.Proc Natl Acad Sci U S A. 1995;92:3358–3361. doi: 10.1073/pnas.92.8.3358.
15. Kong Q, Richter L, Yang YF, Arntzen CJ, Mason HS, Thanavala Y. Oral immunization with hepatitis B surface antigen expressed in transgenic plants. Proc Natl Acad Sci U S A. 2001;98:11539–11544. doi: 10.1073/pnas.191617598.
16. Thanavala Y, Mahoney M, Pal S, Scott A, Richter L, Natarajan N, Goodwin P, Arntzen CJ, Mason HS. Immunogenicity in humans of an edible vaccine for hepatitis B. Proc Natl Acad Sci U S A.2005;102:3378–3382. doi: 10.1073/pnas.0409899102.
17. Kumar GB, Ganapathi TR, Revathi CJ, Srinivas L, Bapat VA. Expression of hepatitis B surface antigen in transgenic banana plants.Planta. 2005;222:484–493. doi: 10.1007/s00425-005-1556-y.
18. Pniewski T. Is an oral plant-based vaccine against hepatitis B virus possible? Curr Pharm Biotechnol. 2012;13:2692–2704. doi: 10.2174/138920112804724837.
19. Pniewski T, Kapusta J, Bociag P, Wojciechowicz J, Kostrzak A, Gdula M, Fedorowicz-Stronska O, Wojcik P, Otta H, Samardakiewicz S, Wolko B, Plucienniczak A. Low-dose oral immunization with lyophilized tissue of herbicide-resistant lettuce expressing hepatitis B surface antigen for prototype plant-derived vaccine

tablet formulation. J Appl Genet. 2011;52:125–136. doi: 10.1007/s13353-010-0001-5.

20. Huang Z, LePore K, Elkin G, Thanavala Y, Mason HS. High-yield rapid production of hepatitis B surface antigen in plant leaf by a viral expression system. Plant Biotechnol J. 2008;5:202–209. doi: 10.1111/j.1467-7652.2007.00316.x.

21. Huang Z, Mason HS. Conformational analysis of hepatitis B surface antigen fusions in an Agrobacterium-mediated transient expression system. Plant Biotechnol J. 2004;2:241–249. doi: 10.1111/j.1467-7652.2004.00068.x.

22. Huang Z, Elkin G, Maloney BJ, Beuhner N, Arntzen CJ, Thanavala Y, Mason HS. Virus-like particle expression and assembly in plants: hepatitis B and Norwalk viruses. Vaccine. 2005;23:1851–1858. doi: 10.1016/j.vaccine.2004.11.017.

23. Huang Z, Santi L, LePore K, Kilbourne J, Arntzen CJ, Mason HS. Rapid, high-level production of hepatitis B core antigen in plant leaf and its immunogenicity in mice. Vaccine. 2006;24:2506–2513. doi: 10.1016/j.vaccine.2005.12.024.

24. Rybicki EP, Martin DP. Virus-derived ssDNA vectors for the expression of foreign proteins in plants. Curr Top Microbiol Immunol.2014;375:19–45.

25. Huang Z, Chen Q, Hjelm B, Arntzen C, Mason H. A DNA replicon system for rapid high-level production of virus-like particles in plants.Biotechnol Bioeng. 2009;103:706–714. doi: 10.1002/bit.22299.

26. Hepatitis C: Fact Sheet No. 164.[http://www.who.int/mediacentre/factsheets/fs164/en/index.html]

27. FDA approves Sovaldi for chronic hepatitis C.[http://www.fda.gov/NewsEvents/Newsroom/PressAnnouncements/ucm377888.htm]

28. Alter HJ, Purcell RH, Shih JW, Melpolder JC, Houghton M, Choo QL, Kuo G. Detection of antibody to hepatitis C virus in prospectively followed transfusion recipients with acute and chronic non-A, non-B hepatitis. N Engl J Med. 1989;321:1494–1500. doi: 10.1056/NEJM198911303212202.

29. Choo QL, Kuo G, Weiner AJ, Overby LR, Bradley DW, Houghton M. Isolation of a cDNA clone derived from a blood-borne non-A, non-B viral hepatitis genome. Science. 1989;244:359–362. doi: 10.1126/science.2523562.

30. Nemchinov LG, Liang TJ, Rifaat MM, Mazyad HM, Hadidi A, Keith JM. Development of a plant-derived subunit vaccine candidate against hepatitis C virus. Arch Virol. 2000;145:2557–2573. doi: 10.1007/s007050070008.

31. Natilla A, Piazzolla G, Nuzzaci M, Saldarelli P, Tortorella C, Antonaci S, Piazzolla P. Cucumber mosaic virus as carrier of a hepatitis C virus-derived epitope. Arch Virol. 2004;149:137–154. doi: 10.1007/s00705-003-0190-x.

32. Piazzolla G, Nuzzaci M, Tortorella C, Panella E, Natilla A, Boscia D, De Stradis A, Piazzolla P, Antonaci S. Immunogenic properties of a chimeric plant virus expressing a hepatitis C virus (HCV)-derived epitope: new prospects for an HCV vaccine. J Clin Immunol.2005;25:142–152. doi: 10.1007/s10875-005-2820-4.

33. Nuzzaci M, Piazzolla G, Vitti A, Lapelosa M, Tortorella C, Stella I, Natilla A, Antonaci S, Piazzolla P. Cucumber mosaic virus as a presentation system for a double hepatitis C virus-derived epitope. Arch Virol. 2007;152:915–928. doi: 10.1007/s00705-006-0916-7.

34. Nuzzaci M, Vitti A, Condelli V, Lanorte MT, Tortorella C, Boscia D, Piazzolla P, Piazzolla G. In vitro stability of Cucumber mosaic virus nanoparticles carrying a Hepatitis C virus-derived epitope under simulated gastrointestinal conditions and in vivo efficacy of an edible vaccine. J Virol Methods. 2010;165:211–215. doi: 10.1016/j.jviromet.2010.01.021.

35. El Attar AK, Shamloul AM, Shalaby AA, Riad BY, Saad A, Mazyad HM, Keith JM. Expression of chimeric HCV peptide in transgenic tobacco plants infected with recombinant alfalfa mosaic virus for development of a plant-derived vaccine against HCV. Afr J Biotechnol. 2004;3:588–594.

36. Denis J, Majeau N, Acosta-Ramirez E, Savard C, Bedard MC, Simard S, Lecours K, Bolduc M, Pare C, Willems B, Shoukry N, Tessier P, Lacasse P, Lamarre A, Lapointe R, Lopez Macias C, Leclerc D. Immunogenicity of papaya mosaic virus-like particles fused to a hepatitis C virus epitope: evidence for the critical function of multimerization. Virology. 2007;363:59–68. doi: 10.1016/j.virol.2007.01.011.

37. Mason HS, Hong F, Bjorklund G: Methods of protein production and compositions thereof. US Patent Office Application20140127749, 2012.

38. Abdel-Mageed WS: Expression of HCV Glycoprotein E1 and its Evaluation for Vaccine Development.*PhD Dissertation.* Rheinisch-Westfälischen Technischen Hochschule, 2011.

39. Influenza (Seasonal) Fact Sheet No 211.[http://www.who.int/mediacentre/factsheets/fs211/en/]

40. Influenza vaccine viruses and reagents.[http://www.who.int/influenza/vaccines/virus/en/]

41. Partridge J, Kieny MP. Global production capacity of seasonal influenza vaccine in 2011. Vaccine. 2013;31:728–731. doi: 10.1016/j.vaccine.2012.10.111.

42. Global pandemic influenza action plan to increase vaccine supply.[http://www.who.int/csr/resources/publications/influenza/WHO_CDS_EPR_GIP_2006_1/en/]

43. Mortimer E, Maclean JM, Mbewana S, Buys A, Williamson AL, Hitzeroth II, Rybicki EP. Setting up a platform for plant-based influenza virus vaccine production in South Africa. BMC Biotechnol.2012;12:14. doi: 10.1186/1472-6750-12-14.

44. D'Aoust MA, Lavoie PO, Couture MM, Trepanier S, Guay JM, Dargis M, Mongrand S, Landry N, Ward BJ, Vezina LP. Influenza virus-like particles produced by transient expression in Nicotiana benthamiana induce a protective immune response against a lethal viral

challenge in mice. Plant Biotechnol J. 2008;6:930–940. doi: 10.1111/j.1467-7652.2008.00384.x.

45. Bright RA, Carter DM, Daniluk S, Toapanta FR, Ahmad A, Gavrilov V, Massare M, Pushko P, Mytle N, Rowe T, Smith G, Ross TM. Influenza virus-like particles elicit broader immune responses than whole virion inactivated influenza virus or recombinant hemagglutinin.Vaccine. 2007;25:3871–3878. doi: 10.1016/j.vaccine.2007.01.106.
46. D'Aoust MA, Couture MM, Charland N, Trepanier S, Landry N, Ors F, Vezina LP. The production of hemagglutinin-based virus-like particles in plants: a rapid, efficient and safe response to pandemic influenza. Plant Biotechnol J. 2010;8:607–619. doi: 10.1111/j.1467-7652.2009.00496.x.
47. Landry N, Ward BJ, Trepanier S, Montomoli E, Dargis M, Lapini G, Vezina LP. Preclinical and clinical development of plant-made virus-like particle vaccine against avian H5N1 influenza. PLoS One.2010;5:e15559. doi: 10.1371/journal.pone.0015559.
48. Shoji Y, Chichester JA, Jones M, Manceva SD, Damon E, Mett V, Musiychuk K, Bi H, Farrance C, Shamloul M, Kushnir N, Sharma S, Yusibov V. Plant-based rapid production of recombinant subunit hemagglutinin vaccines targeting H1N1 and H5N1 influenza. Hum Vaccin. 2011;7(Suppl):41–50. doi: 10.4161/hv.7.0.14561.
49. Pellerin C: DARPA Effort Speeds Biothreat Response. In*American Forces Press Service*: U.S. Department of Defense; 2010.
50. DARPA makes 10 million strides in the race to contain a hypothetical pandemic.[http://www.darpa.mil/NewsEvents/Releases/2012/07/25.aspx]
51. Gomord V, Fitchette AC, Menu-Bouaouiche L, Saint-Jore-Dupas C, Plasson C, Michaud D, Faye L. Plant-specific glycosylation patterns in the context of therapeutic protein production. Plant Biotechnol J.2010;8:564–587. doi: 10.1111/j.1467-7652.2009.00497.x.
52. Zhang S, Sherwood RW, Yang Y, Fish T, Chen W, McCardle JA, Jones RM, Yusibov V, May ER, Rose JK, Thannhauser TW. Comparative characterization of the glycosylation profiles of an influenza hemagglutinin produced in plant and insect hosts. Proteomics.2012;12:1269–1288. doi: 10.1002/pmic.201100474.
53. Lin SC, Jan JT, Dionne B, Butler M, Huang MH, Wu CY, Wong CH, Wu SC. Different immunity elicited by recombinant H5N1 hemagglutinin proteins containing pauci-mannose, high-mannose, or complex type N-glycans. PLoS One. 2013;8:e66719. doi: 10.1371/journal.pone.0066719.
54. Sippel K: Medicago reports positive results for H7N9 pre-clinical trial: First in the world. 2013.
55. Chichester JA, Jones RM, Green BJ, Stow M, Miao F, Moonsammy G, Streatfield SJ, Yusibov V. Safety and immunogenicity of a plant-produced recombinant hemagglutinin-based influenza vaccine (HAI-05) derived from A/Indonesia/05/2005 (H5N1) influenza virus: a phase 1 randomized, double-blind, placebo-controlled, dose-escalation study in healthy adults. Viruses. 2012;4:3227–3244. doi: 10.3390/v4113227.
56. Jul-Larsen A, Madhun AS, Brokstad KA, Montomoli E, Yusibov V, Cox RJ. The human potential of a recombinant pandemic influenza vaccine produced in tobacco plants. Hum Vaccin Immunother.2012;8:653–661. doi: 10.4161/hv.19503.
57. Cummings JF, Guerrero ML, Moon JE, Waterman P, Nielsen RK, Jefferson S, Gross FL, Hancock K, Katz JM, Yusibov V. Safety and immunogenicity of a plant-produced recombinant monomer hemagglutinin-based influenza vaccine derived from influenza A (H1N1) pdm09 virus: a Phase 1 dose-escalation study in healthy adults.Vaccine. 2014;32:2251–2259. doi: 10.1016/j.vaccine.2013.10.017.
58. Shoji Y, Jones RM, Mett V, Chichester JA, Musiychuk K, Sun X, Tumpey TM, Green BJ, Shamloul M, Norikane J, Bi H, Hartman CE, Bottone C, Stewart M, Streatfield SJ, Yusibov V. A plant-produced H1N1 trimeric hemagglutinin protects mice from a lethal influenza virus challenge. Hum Vaccin Immunother. 2013;9:553–560. doi: 10.4161/hv.23234.
59. Neuhaus V, Chichester JA, Ebensen T, Schwarz K, Hartman CE, Shoji Y, Guzman CA, Yusibov V, Sewald K, Braun A. A new adjuvanted nanoparticle-based H1N1 influenza vaccine induced antigen-specific local mucosal and systemic immune responses after administration into the lung. Vaccine. 2014;32:3216–3222. doi: 10.1016/j.vaccine.2014.04.011.
60. Mallajosyula JK, Hiatt E, Hume S, Johnson A, Jeevan T, Chikwamba R, Pogue GP, Bratcher B, Haydon H, Webby RJ, McCormick AA. Single-dose monomeric HA subunit vaccine generates full protection from influenza challenge. Hum Vaccin Immunother.2013;10:586–595. doi: 10.4161/hv.27567.
61. Phan HT, Pohl J, Floss DM, Rabenstein F, Veits J, Le BT, Chu HH, Hause G, Mettenleiter T, Conrad U. ELPylated haemagglutinins produced in tobacco plants induce potentially neutralizing antibodies against H5N1 viruses in mice. Plant Biotechnol J. 2013;11:582–593. doi: 10.1111/pbi.12049.
62. Matic S, Rinaldi R, Masenga V, Noris E. Efficient production of chimeric human papillomavirus 16 L1 protein bearing the M2e influenza epitope in Nicotiana benthamiana plants. BMC Biotechnol.2011;11:106. doi: 10.1186/1472-6750-11-106.
63. Tyulkina LG, Skurat EV, Frolova OY, Komarova TV, Karger EM, Atabekov IG. New viral vector for superproduction of epitopes of vaccine proteins in plants. Acta Nat. 2011;3:73–82.
64. Petukhova NV, Gasanova TV, Stepanova LA, Rusova OA, Potapchuk MV, Korotkov AV, Skurat EV, Tsybalova LM, Kiselev OI, Ivanov PA, Atabekov JG. Immunogenicity and protective efficacy of candidate universal influenza A nanovaccines produced in plants by Tobacco mosaic virus-based vectors. Curr Pharm Des. 2013;19:5587–5600. doi: 10.2174/13816128113199990337.
65. Edney A: Merck's HPV Vaccine Gardasil Shown Safe in Kaiser Study. vol. 2014: Bloomberg.com; 2012.

66. Parkin DM, Bray F, Ferlay J, Pisani P. Estimating the world cancer burden: Globocan 2000. Int J Cancer. 2001;94:153–156. doi: 10.1002/ijc.1440.

67. Dochez C, Bogers JJ, Verhelst R, Rees H. HPV vaccines to prevent cervical cancer and genital warts: an update. Vaccine. 2014;32:1595–1601. doi: 10.1016/j.vaccine.2013.10.081.

68. Tyler M, Tumban E, Chackerian B. Second-generation prophylactic HPV vaccines: successes and challenges. Expert Rev Vaccines.2014;13:247–255. doi: 10.1586/14760584.2014.865523.

69. Bosch FX, Broker TR, Forman D, Moscicki AB, Gillison ML, Doorbar J, Stern PL, Stanley M, Arbyn M, Poljak M, Cuzick J, Castle PE, Schiller JT, Markowitz LE, Fisher WA, Canfell K, Denny LA, Franco EL, Steben M, Kane MA, Schiffman M, Meijer CJ, Sankaranarayanan R, Castellsague X, Kim JJ, Brotons M, Alemany L, Albero G, Diaz M, de Sanjose S, et al. Comprehensive control of human papillomavirus infections and related diseases. Vaccine.2013;31(Suppl 7):H1–H31. doi: 10.1016/j.vaccine.2013.10.003.

70. Varsani A, Williamson AL, Rose RC, Jaffer M, Rybicki EP. Expression of Human papillomavirus type 16 major capsid protein in transgenic Nicotiana tabacum cv. Xanthi. Arch Virol. 2003;148:1771–1786. doi: 10.1007/s00705-003-0119-4.

71. Warzecha H, Mason HS, Lane C, Tryggvesson A, Rybicki E, Williamson AL, Clements JD, Rose RC. Oral immunogenicity of human papillomavirus-like particles expressed in potato. J Virol.2003;77:8702–8711. doi: 10.1128/JVI.77.16.8702-8711.2003.

72. Biemelt S, Sonnewald U, Galmbacher P, Willmitzer L, Muller M. Production of human papillomavirus type 16 virus-like particles in transgenic plants. J Virol. 2003;77:9211–9220. doi: 10.1128/JVI.77.17.9211-9220.2003.

73. Kohl TO, Hitzeroth II, Christensen ND, Rybicki EP. Expression of HPV-11 L1 protein in transgenic Arabidopsis thaliana and Nicotiana tabacum. BMC Biotechnol. 2007;7:56. doi: 10.1186/1472-6750-7-56.

74. Kohl T, Hitzeroth II, Stewart D, Varsani A, Govan VA, Christensen ND, Williamson AL, Rybicki EP. Plant-produced cottontail rabbit papillomavirus L1 protein protects against tumor challenge: a proof-of-concept study. Clin Vaccine Immunol. 2006;13:845–853. doi: 10.1128/CVI.00072-06.

75. Palmer KE, Benko A, Doucette SA, Cameron TI, Foster T, Hanley KM, McCormick AA, McCulloch M, Pogue GP, Smith ML, Christensen ND. Protection of rabbits against cutaneous papillomavirus infection using recombinant tobacco mosaic virus containing L2 capsid epitopes. Vaccine. 2006;24:5516–5525. doi: 10.1016/j.vaccine.2006.04.058.

76. Maclean J, Koekemoer M, Olivier AJ, Stewart D, Hitzeroth II, Rademacher T, Fischer R, Williamson AL, Rybicki EP. Optimization of human papillomavirus type 16 (HPV-16) L1 expression in plants: comparison of the suitability of different HPV-16 L1 gene variants and different cell-compartment localization. J Gen Virol. 2007;88:1460–1469. doi: 10.1099/vir.0.82718-0.

77. Fernandez-San Millan A, Ortigosa SM, Hervas-Stubbs S, Corral-Martinez P, Segui-Simarro JM, Gaetan J, Coursaget P, Veramendi J. Human papillomavirus L1 protein expressed in tobacco chloroplasts self-assembles into virus-like particles that are highly immunogenic. Plant Biotechnol J. 2008;6:427–441. doi: 10.1111/j.1467-7652.2008.00338.x.

78. Waheed MT, Thones N, Muller M, Hassan SW, Gottschamel J, Lossl E, Kaul HP, Lossl AG. Plastid expression of a double-pentameric vaccine candidate containing human papillomavirus-16 L1 antigen fused with LTB as adjuvant: transplastomic plants show pleiotropic phenotypes. Plant Biotechnol J. 2011;9:651–660. doi: 10.1111/j.1467-7652.2011.00612.x.

79. Matic S, Masenga V, Poli A, Rinaldi R, Milne RG, Vecchiati M, Noris E. Comparative analysis of recombinant Human Papillomavirus 8 L1 production in plants by a variety of expression systems and purification methods. Plant Biotechnol J. 2012;10:410–421. doi: 10.1111/j.1467-7652.2011.00671.x.

80. Love AJ, Chapman SN, Matic S, Noris E, Lomonossoff GP, Taliansky M. In planta production of a candidate vaccine against bovine papillomavirus type 1. Planta. 2012;236:1305–1313. doi: 10.1007/s00425-012-1692-0.

81. Rybicki EP, Williamson AL, Meyers A, Hitzeroth II. Vaccine farming in Cape Town. Hum Vaccin. 2011;7:339–348. doi: 10.4161/hv.7.3.14263.

82. Franconi R, Di Bonito P, Dibello F, Accardi L, Muller A, Cirilli A, Simeone P, Dona MG, Venuti A, Giorgi C. Plant-derived human papillomavirus 16 E7 oncoprotein induces immune response and specific tumor protection. Cancer Res. 2002;62:3654–3658.

83. Franconi R, Massa S, Illiano E, Mullar A, Cirilli A, Accardi L, Di Bonito P, Giorgi C, Venuti A. Exploiting the plant secretory pathway to improve the anticancer activity of a plant-derived HPV16 E7 vaccine.Int J Immunopathol Pharmacol. 2006;19:187–197.

84. Di Bonito P, Grasso F, Mangino G, Massa S, Illiano E, Franconi R, Fanales-Belasio E, Falchi M, Affabris E, Giorgi C. Immunomodulatory activity of a plant extract containing human papillomavirus 16-E7 protein in human monocyte-derived dendritic cells. Int J Immunopathol Pharmacol. 2009;22:967–978.

85. Massa S, Franconi R, Brandi R, Muller A, Mett V, Yusibov V, Venuti A. Anti-cancer activity of plant-produced HPV16 E7 vaccine.Vaccine. 2007;25:3018–3021. doi: 10.1016/j.vaccine.2007.01.018.

86. Venuti A, Massa S, Mett V, Vedova LD, Paolini F, Franconi R, Yusibov V. An E7-based therapeutic vaccine protects mice against HPV16 associated cancer. Vaccine. 2009;27:3395–3397. doi: 10.1016/j.vaccine.2009.01.068.

87. Buyel JF, Bautista JA, Fischer R, Yusibov VM. Extraction, purification and characterization of the plant-produced HPV16 subunit vaccine candidate E7 GGG. J Chromatogr B Analyt Technol Biomed Life Sci. 2012;880:19–26. doi: 10.1016/j.jchromb.2011.11.010.

88. Paz De la Rosa G, Monroy-Garcia A, Mora-Garcia Mde L, Pena CG, Hernandez-Montes J, Weiss-Steider B, Gomez-Lim MA. An HPV 16 L1-based chimeric human papilloma virus-like particles containing a string of epitopes produced in plants is able to elicit humoral and cytotoxic T-cell activity in mice. Virol J. 2009;6:2. doi: 10.1186/1743-422X-6-2.

89. Monroy-Garcia A, Gomez-Lim MA, Weiss-Steider B, la Rosa GP, Hernandez-Montes J, Perez-Saldana K, Tapia-Guerrero YS, Toledo-Guzman ME, Santiago-Osorio E, Sanchez-Pena HI, Mora-Garcia Mde L. A novel HPV 16 L1-based chimeric virus-like particle containing E6 and E7 seroreactive epitopes permits highly specific detection of antibodies in patients with CIN 1 and HPV-16 infection. Virol J.2011;8:59. doi: 10.1186/1743-422X-8-59.

90. Hongli L, Xukui L, Ting L, Wensheng L, Lusheng S, Jin Z. Transgenic tobacco expressed HPV16-L1 and LT-B combined immunization induces strong mucosal and systemic immune responses in mice. Hum Vaccin Immunother. 2013;9:83–89. doi: 10.4161/hv.22292.

91. Pineo CB, Hitzeroth II, Rybicki EP. Immunogenic assessment of plant-produced human papillomavirus type 16 L1/L2 chimaeras. Plant Biotechnol J. 2013;11:964–975. doi: 10.1111/pbi.12089.

92. Schellenbacher C, Roden R, Kirnbauer R. Chimeric L1-L2 virus-like particles as potential broad-spectrum human papillomavirus vaccines. J Virol. 2009;83:10085–10095. doi: 10.1128/JVI.01088-09.

93. Jagu S, Kwak K, Karanam B, Huh WK, Damotharan V, Chivukula SV, Roden RB. Optimization of multimeric human papillomavirus L2 vaccines. PLoS One. 2013;8:e55538. doi: 10.1371/journal.pone.0055538.

94. Wang JW, Roden RB. L2, the minor capsid protein of papillomavirus. Virology. 2013;445:175–186. doi: 10.1016/j.virol.2013.04.017.

95. Varsani A, Williamson AL, de Villiers D, Becker I, Christensen ND, Rybicki EP. Chimeric human papillomavirus type 16 (HPV-16) L1 particles presenting the common neutralizing epitope for the L2 minor capsid protein of HPV-6 and HPV-16. J Virol. 2003;77:8386–8393. doi: 10.1128/JVI.77.15.8386-8393.2003.

96. McGrath M, de Villiers GK, Shephard E, Hitzeroth II, Rybicki EP. Development of human papillomavirus chimaeric L1/L2 candidate vaccines. Arch Virol. 2013;158:2079–2088. doi: 10.1007/s00705-013-1713-8.

97. Cerovska N, Hoffmeisterova H, Moravec T, Plchova H, Folwarczna J, Synkova H, Ryslava H, Ludvikova V, Smahel M. Transient expression of Human papillomavirus type 16 L2 epitope fused to N- and C-terminus of coat protein of Potato virus X in plants. J Biosci.2012;37:125–133. doi: 10.1007/s12038-011-9177-z.

98. Monroy-Garcia A, Gomez-Lim MA, Weiss-Steider B, Hernandez-Montes J, Huerta-Yepez S, Rangel-Santiago JF, Santiago-Osorio E, Mora Garcia Mde L. Immunization with an HPV-16 L1-based chimeric virus-like particle containing HPV-16 E6 and E7 epitopes elicits long-lasting prophylactic and therapeutic efficacy in an HPV-16 tumor mice model. Arch Virol. 2014;159:291–305. doi: 10.1007/s00705-013-1819-z.

99. Plchova H, Moravec T, Hoffmeisterova H, Folwarczna J, Cerovska N. Expression of Human papillomavirus 16 E7ggg oncoprotein on N- and C-terminus of Potato virus X coat protein in bacterial and plant cells. Protein Expr Purif. 2011;77:146–152. doi: 10.1016/j.pep.2011.01.008.

100. Demurtas OC, Massa S, Ferrante P, Venuti A, Franconi R, Giuliano G. A Chlamydomonas-derived Human Papillomavirus 16 E7 vaccine induces specific tumor protection. PLoS One. 2013;8:e61473. doi: 10.1371/journal.pone.0061473.

101. Morgenfeld M, Lentz E, Segretin ME, Alfano EF, Bravo-Almonacid F. Translational fusion and redirection to thylakoid lumen as strategies to enhance accumulation of human papillomavirus E7 antigen in tobacco chloroplasts. Mol Biotechnol. 2014;56:1021–1031. doi: 10.1007/s12033-014-9781-x.

102. Whitehead M, Ohlschlager P, Almajhdi FN, Alloza L, Marzabal P, Meyers AE, Hitzeroth II, Rybicki EP. Human papillomavirus (HPV) type 16 E7 protein bodies cause tumour regression in mice. BMC Cancer. 2014;14:367. doi: 10.1186/1471-2407-14-367.

103. Ohlschlager P, Pes M, Osen W, Durst M, Schneider A, Gissmann L, Kaufmann AM. An improved rearranged Human Papillomavirus Type 16 E7 DNA vaccine candidate (HPV-16 E7SH) induces an E7 wildtype-specific T cell response. Vaccine. 2006;24:2880–2893. doi: 10.1016/j.vaccine.2005.12.061.

104. Torrent M, Llompart B, Lasserre-Ramassamy S, Llop-Tous I, Bastida M, Marzabal P, Westerholm-Parvinen A, Saloheimo M, Heifetz PB, Ludevid MD. Eukaryotic protein production in designed storage organelles. BMC Biol. 2009;7:5. doi: 10.1186/1741-7007-7-5.

105. Excler JL, Robb ML, Kim JH. HIV-1 vaccines: Challenges and new perspectives. Hum Vaccin Immunother. 2014;10:1734–1746. doi: 10.4161/hv.28462.

106. Chanzu N, Ondondo B. Induction of potent and long-lived antibody and cellular immune responses in the genitorectal mucosa could be the critical determinant of HIV vaccine efficacy. Front Immunol. 2014;5:202. doi: 10.3389/fimmu.2014.00202.

107. Zhang GG, Rodrigues L, Rovinski B, White KA. Production of HIV-1 p24 protein in transgenic tobacco plants. Mol Biotechnol.2002;20:131–136. doi: 10.1385/MB:20:2:131.

108. Horn ME, Pappu KM, Bailey MR, Clough RC, Barker M, Jilka JM, Howard JA, Streatfield SJ. Advantageous features of plant-based systems for the development of HIV vaccines. J Drug Target.2003;11:539–545. doi: 10.1080/10611860410001669992.

109. Matoba N, Magerus A, Geyer BC, Zhang Y, Muralidharan M, Alfsen A, Arntzen CJ, Bomsel M, Mor TS. A mucosally targeted subunit vaccine candidate eliciting HIV-1 transcytosis-blocking Abs.Proc Natl

Acad Sci U S A. 2004;101:13584–13589. doi: 10.1073/pnas.0405297101.

110. Matoba N, Kajiura H, Cherni I, Doran JD, Bomsel M, Fujiyama K, Mor TS. Biochemical and immunological characterization of the plant-derived candidate human immunodeficiency virus type 1 mucosal vaccine CTB-MPR. Plant Biotechnol J. 2009;7:129–145. doi: 10.1111/j.1467-7652.2008.00381.x.

111. Karasev AV, Foulke S, Wellens C, Rich A, Shon KJ, Zwierzynski I, Hone D, Koprowski H, Reitz M. Plant based HIV-1 vaccine candidate: Tat protein produced in spinach. Vaccine. 2005;23:1875–1880. doi: 10.1016/j.vaccine.2004.11.021.

112. Shchelkunov SN, Salyaev RK, Pozdnyakov SG, Rekoslavskaya NI, Nesterov AE, Ryzhova TS, Sumtsova VM, Pakova NV, Mishutina UO, Kopytina TV, Hammond RW. Immunogenicity of a novel, bivalent, plant-based oral vaccine against hepatitis B and human immunodeficiency viruses. Biotechnol Lett. 2006;28:959–967. doi: 10.1007/s10529-006-9028-4.

113. Meyers A, Chakauya E, Shephard E, Tanzer FL, Maclean J, Lynch A, Williamson AL, Rybicki EP. Expression of HIV-1 antigens in plants as potential subunit vaccines. BMC Biotechnol. 2008;8:53. doi: 10.1186/1472-6750-8-53.

114. Lombardi R, Circelli P, Villani ME, Buriani G, Nardi L, Coppola V, Bianco L, Benvenuto E, Donini M, Marusic C. High-level HIV-1 Nef transient expression in Nicotiana benthamiana using the P19 gene silencing suppressor protein of Artichoke Mottled Crinckle Virus. BMC Biotechnol. 2009;9:96. doi: 10.1186/1472-6750-9-96.

115. Scotti N, Buonaguro L, Tornesello ML, Cardi T, Buonaguro FM. Plant-based anti-HIV-1 strategies: vaccine molecules and antiviral approaches. Expert Rev Vaccines. 2010;9:925–936. doi: 10.1586/erv.10.79.

116. June 29 2011: Pharma-Planta launches a pivotal phase I clinical trial of a plant-derived microbicidal protein. [http://www.pharma-planta.net/index.php?pg=50]

117. Scotti N, Alagna F, Ferraiolo E, Formisano G, Sannino L, Buonaguro L, De Stradis A, Vitale A, Monti L, Grillo S, Buonaguro FM, Cardi T. High-level expression of the HIV-1 Pr55gag polyprotein in transgenic tobacco chloroplasts. Planta. 2009;229:1109–1122. doi: 10.1007/s00425-009-0898-2.

118. Berger CT, Frahm N, Price DA, Mothe B, Ghebremichael M, Hartman KL, Henry LM, Brenchley JM, Ruff LE, Venturi V, Pereyra F, Sidney J, Sette A, Douek DC, Walker BD, Kaufmann DE, Brander C. High-functional-avidity cytotoxic T lymphocyte responses to HLA-B-restricted Gag-derived epitopes associated with relative HIV control. J Virol. 2011;85:9334–9345. doi: 10.1128/JVI.00460-11.

119. Jia M, Hong K, Chen J, Ruan Y, Wang Z, Su B, Ren G, Zhang X, Liu Z, Zhao Q, Li D, Peng H, Altfeld M, Walker BD, Yu XG, Shao Y. Preferential CTL targeting of Gag is associated with relative viral control in long-term surviving HIV-1 infected former plasma donors from China. Cell Res. 2012;22:903–914. doi: 10.1038/cr.2012.19.

120. Mothe B, Llano A, Ibarrondo J, Daniels M, Miranda C, Zamarreno J, Bach V, Zuniga R, Perez-Alvarez S, Berger CT, Puertas MC, Martinez-Picado J, Rolland M, Farfan M, Szinger JJ, Hildebrand WH, Yang OO, Sanchez-Merino V, Brumme CJ, Brumme ZL, Heckerman D, Allen TM, Mullins JI, Gomez G, Goulder PJ, Walker BD, Gatell JM, Clotet B, Korber BT, Sanchez J, et al. Definition of the viral targets of protective HIV-1-specific T cell responses. J Transl Med. 2011;9:208. doi: 10.1186/1479-5876-9-208.

121. Chege GK, Burgers WA, Stutz H, Meyers AE, Chapman R, Kiravu A, Bunjun R, Shephard EG, Jacobs WR, Jr, Rybicki EP, Williamson AL. Robust immunity to an auxotrophic Mycobacterium bovis BCG-VLP prime-boost HIV vaccine candidate in a nonhuman primate model. J Virol. 2013;87:5151–5160. doi: 10.1128/JVI.03178-12.

122. Chege GK, Thomas R, Shephard EG, Meyers A, Bourn W, Williamson C, Maclean J, Gray CM, Rybicki EP, Williamson AL. A prime-boost immunisation regimen using recombinant BCG and Pr55(gag) virus-like particle vaccines based on HIV type 1 subtype C successfully elicits Gag-specific responses in baboons. Vaccine.2009;27:4857–4866. doi: 10.1016/j.vaccine.2009.05.064.

123. Lindh I, Brave A, Hallengard D, Hadad R, Kalbina I, Strid A, Andersson S. Oral delivery of plant-derived HIV-1 p24 antigen in low doses shows a superior priming effect in mice compared to high doses.Vaccine. 2014;32:2288–2293. doi: 10.1016/j.vaccine.2014.02.073.

124. Kessans SA, Linhart MD, Matoba N, Mor T. Biological and biochemical characterization of HIV-1 Gag/dgp41 virus-like particles expressed in Nicotiana benthamiana. Plant Biotechnol J. 2013;11:681–690. doi: 10.1111/pbi.12058.

125. Rosenberg Y, Sack M, Montefiori D, Forthal D, Mao L, Hernandez-Abanto S, Urban L, Landucci G, Fischer R, Jiang X. Rapid high-level production of functional HIV broadly neutralizing monoclonal antibodies in transient plant expression systems. PLoS One. 2013;8:e58724. doi: 10.1371/journal.pone.0058724.

126. D'Aoust MA, Couture MM, Lavoie PO, Vezina LP: Virus like particle production in plants. Organization EP ed., vol. WO2012083445 (A1); 2014.

127. Wang BZ, Liu W, Kang SM, Alam M, Huang C, Ye L, Sun Y, Li Y, Kothe DL, Pushko P, Dokland T, Haynes BF, Smith G, Hahn BH, Compans RW. Incorporation of high levels of chimeric human immunodeficiency virus envelope glycoproteins into virus-like particles. J Virol. 2007;81:10869–10878. doi: 10.1128/JVI.00542-07.

128. Thuenemann EC, Meyers AE, Verwey J, Rybicki EP, Lomonossoff GP. A method for rapid production of heteromultimeric protein complexes in plants: assembly of protective bluetongue virus-like particles. Plant Biotechnol J. 2013;11:839–846. doi: 10.1111/pbi.12076.

129. Thuenemann EC, Lenzi P, Love AJ, Taliansky M, Becares

M, Zuniga S, Enjuanes L, Zahmanova GG, Minkov IN, Matic S, Noris E, Meyers A, Hattingh A, Rybicki EP, Kiselev OI, Ravin NV, Eldarov MA, Skryabin KG, Lomonossoff GP. The use of transient expression systems for the rapid production of virus-like particles in plants. Curr Pharm Des. 2013;19:5564–5573. doi: 10.2174/1381612811319310011.

130. Zientara S, Sanchez-Vizcaino JM. Control of bluetongue in Europe. Vet Microbiol. 2013;165:33–37. doi: 10.1016/j.vetmic.2013.01.010.

131. Baylis M. Research gaps in understanding how climate change will affect arboviral diseases. Anim Health Res Rev. 2013;14:143–146. doi: 10.1017/S1466252313000145.

132. Calvo-Pinilla E, Castillo-Olivares J, Jabbar T, Ortego J, de la Poza F, Marin-Lopez A. Recombinant vaccines against bluetongue virus.Virus Res. 2014;182:78–86. doi: 10.1016/j.virusres.2013.11.013.

133. Roy P. Genetically engineered structure-based vaccine for bluetongue disease. Vet Ital. 2004;40:594–600.

134. Hewat EA, Booth TF, Roy P. Structure of correctly self-assembled bluetongue virus-like particles. J Struct Biol. 1994;112:183–191. doi: 10.1006/jsbi.1994.1019.

135. Stewart M, Dubois E, Sailleau C, Breard E, Viarouge C, Desprat A, Thiery R, Zientara S, Roy P. Bluetongue virus serotype 8 virus-like particles protect sheep against virulent virus infection as a single or multi-serotype cocktail immunogen. Vaccine. 2013;31:553–558. doi: 10.1016/j.vaccine.2012.11.016.

136. Sainsbury F, Thuenemann EC, Lomonossoff GP. pEAQ: versatile expression vectors for easy and quick transient expression of heterologous proteins in plants. Plant Biotechnol J. 2009;7:682–693. doi: 10.1111/j.1467-7652.2009.00434.x.

137. Monath TP. Vaccines against diseases transmitted from animals to humans: a one health paradigm. Vaccine. 2013;31:5321–5338. doi: 10.1016/j.vaccine.2013.09.029.

138. Yusibov V, Hooper DC, Spitsin SV, Fleysh N, Kean RB, Mikheeva T, Deka D, Karasev A, Cox S, Randall J, Koprowski H. Expression in plants and immunogenicity of plant virus-based experimental rabies vaccine. Vaccine. 2002;20:3155–3164. doi: 10.1016/S0264-410X(02)00260-8.

139. Ashraf S, Singh PK, Yadav DK, Shahnawaz M, Mishra S, Sawant SV, Tuli R. High level expression of surface glycoprotein of rabies virus in tobacco leaves and its immunoprotective activity in mice. J Biotechnol. 2005;119:1–14. doi: 10.1016/j.jbiotec.2005.06.009.

140. Loza-Rubio E, Rojas E, Gomez L, Olivera MT, Gomez-Lim MA. Development of an edible rabies vaccine in maize using the Vnukovo strain. Dev Biol (Basel) 2008;131:477–482.

141. Loza-Rubio E, Rojas-Anaya E, Lopez J, Olivera-Flores MT, Gomez-Lim M, Tapia-Perez G. Induction of a protective immune response to rabies virus in sheep after oral immunization with transgenic maize, expressing the rabies virus glycoprotein. Vaccine.2012;30:5551–5556. doi: 10.1016/j.vaccine.2012.06.039.

142. Oncu S. Crimean-Congo hemorrhagic fever: an overview. Virol Sin. 2013;28:193–201. doi: 10.1007/s12250-013-3327-4.

143. Boshra H, Lorenzo G, Busquets N, Brun A. Rift valley fever: recent insights into pathogenesis and prevention. J Virol.2011;85:6098–6105. doi: 10.1128/JVI.02641-10.

144. Ghiasi SM, Salmanian AH, Chinikar S, Zakeri S. Mice orally immunized with a transgenic plant expressing the glycoprotein of Crimean-Congo hemorrhagic fever virus. Clin Vaccine Immunol.2011;18:2031–2037. doi: 10.1128/CVI.05352-11.

145. Lagerqvist N: Rift Valley fever virus vaccine strategies. Karolinska Institutet, Department of Microbiology, Tumor and Cell Biology; 2013.

146. Whaley KJ, Morton J, Hume S, Hiatt E, Bratcher B, Klimyuk V, Hiatt A, Pauly M, Zeitlin L. Emerging antibody-based products. Curr Top Microbiol Immunol. 2014;375:107–126.

147. van Dolleweerd CJ, Teh AY, Banyard AC, Both L, Lotter-Stark HC, Tsekoa T, Phahladira B, Shumba W, Chakauya E, Sabeta CT, Gruber C, Fooks AR, Chikwamba RK, Ma JK. Engineering, expression in transgenic plants and characterisation of e559, a rabies virus-neutralising monoclonal antibody. J Infect Dis. 2014;210:200–208. doi: 10.1093/infdis/jiu085.

148. Senthilingam M: Ebola outbreak: Is it time to test experimental vaccines? , vol. 2014: CNN; 2014.

149. Chen Q, He J, Phoolcharoen W, Mason HS. Geminiviral vectors based on bean yellow dwarf virus for production of vaccine antigens and monoclonal antibodies in plants. Hum Vaccin. 2011;7:331–338. doi: 10.4161/hv.7.3.14262.

150. Lai H, He J, Engle M, Diamond MS, Chen Q. Robust production of virus-like particles and monoclonal antibodies with geminiviral replicon vectors in lettuce. Plant Biotechnol J. 2012;10:95–104. doi: 10.1111/j.1467-7652.2011.00649.x.

151. Olinger GG, Jr, Pettitt J, Kim D, Working C, Bohorov O, Bratcher B, Hiatt E, Hume SD, Johnson AK, Morton J, Pauly M, Whaley KJ, Lear CM, Biggins JE, Scully C, Hensley L, Zeitlin L. Delayed treatment of Ebola virus infection with plant-derived monoclonal antibodies provides protection in rhesus macaques. Proc Natl Acad Sci U S A. 2012;109:18030–18035. doi: 10.1073/pnas.1213709109.

152. Pettitt J, Zeitlin L, Kim do H, Working C, Johnson JC, Bohorov O, Bratcher B, Hiatt E, Hume SD, Johnson AK, Morton J, Pauly MH, Whaley KJ, Ingram MF, Zovanyi A, Heinrich M, Piper A, Zelko J, Olinger GG. Therapeutic intervention of Ebola virus infection in rhesus macaques with the MB-003 monoclonal antibody cocktail. Sci Transl Med. 2013;5:199ra113. doi: 10.1126/scitranslmed.3006608.

153. Qiu X, Wong G, Audet J, Bello A, Fernando L, Alimonti JB, Fausther-Bovendo H, Wei H, Aviles J, Hiatt E, Johnson A, Morton J, Swope K, Bohorov O, Bohorova N, Goodman C, Kim D, Pauly MH, Velasco

J, Pettitt J, Olinger GG, Whaley K, Xu B, Strong JE, Zeitlin L, Kobinger GP. Reversion of advanced Ebola virus disease in nonhuman primates with ZMapp. Nature. 2014;514:47–53.

154. Langreth R, Chen C, Nash J, Lauerman J: Ebola Drug Made From Tobacco Plant Saves U.S. Aid Workers. vol. 2014: Bloomberg.com; 2014.

155. Wilson J, Dellorto D: 9 questions about this new Ebola drug.vol. 2014: CNN; 2014.

156. Pollack A: U.S. Will Increase Production of the Ebola Drug ZMapp, but May Not Meet Demand. vol. 2014. New York: New York Times; 2014.

157. McCarthy M. US signs contract with ZMapp maker to accelerate development of the Ebola drug. BMJ. 2014;349:g5488. doi: 10.1136/bmj.g5488.

158. Phoolcharoen W, Bhoo SH, Lai H, Ma J, Arntzen CJ, Chen Q, Mason HS. Expression of an immunogenic Ebola immune complex in Nicotiana benthamiana. Plant Biotechnol J. 2011;9:807–816. doi: 10.1111/j.1467-7652.2011.00593.x.

CHAPTER 4

FROM GENETICS TO FUNCTIONAL GENOMICS: IMPROVEMENT IN DROUGHT SIGNALING AND TOLERANCE IN WHEAT

Hikmet Budak[1], Babar Hussain[1], Zaeema Khan[1], Neslihan Z. Ozturk[2] and Naimat Ullah[1]

[1] Plant Genomics Group, Molecular Biology, Genetics and Bioengineering Program, Faculty of Engineering and Natural Sciences, Sabanci University, Istanbul, Turkey

[2] Department of Agricultural Genetic Engineering, Faculty of Agricultural Sciences and Technologies, Nigde University, Ni ˇ gde, Turkey

ABSTRACT

Drought being a yield limiting factor has become a major threat to international food security. It is a complex trait and drought tolerance response is carried out by various genes, transcription factors (TFs), microRNAs (miRNAs), hormones, proteins, co-factors, ions, and metabolites. This complexity has limited the development of wheat cultivars for drought tolerance by classical breeding. However, attempts have been made to fill the lost genetic diversity by crossing wheat with wild wheat relatives. In recent years, several molecular markers including single nucleotide polymorphisms (SNPs) and quantitative trait loci (QTLs) associated with genes for drought signaling pathways have been reported. Screening of large wheat collections by marker assisted selection (MAS) and transformation of wheat with different genes/TFs has improved drought signaling pathways and tolerance. Several miRNAs also provide drought tolerance to wheat by regulating various TFs/genes. Emergence of OMICS techniques including transcriptomics, proteomics, metabolomics, and ionomics has helped to identify and characterize the genes, proteins, metabolites, and ions involved in drought signaling pathways. Together, all these efforts helped in understanding the complex drought tolerance mechanism. Here, we have reviewed the advances in wide hybridization, MAS, QTL mapping, miRNAs, transgenic technique, genome editing system, and above mentioned functional genomics tools for identification and utility of signaling molecules for improvement in wheat drought tolerance.

INTRODUCTION

Global warming has resulted in decreased precipitation and increased evaporation, causing more frequent drought spells worldwide. Drought reduces the plant yield up to 50% which is a great economic loss for the farming community (Akpinar et al., 2013). Consequently, development of drought tolerant wheat cultivars has become a serious challenge for the plant breeders to ensure the food security of the masses (Budak et al., 2013a). Drought is a multifaceted trait; plant responses to drought are affected by various factors including growth conditions, physiology, genotype, developmental stage, drought severity, and duration. Thus, drought tolerance mechanisms involve diverse gene expression patterns and as complex signaling pathways (Kantar et al., 2011a; Akpinar et al., 2012). Bread wheat is an important staple food worldwide, therefore efforts have been made to develop drought tolerant varieties (Budak et al., 2013b). Drought signaling pathways involve crosstalk among various biomolecules which makes breeding for drought tolerance an uphill task (Akpinar et al., 2012). In recent years, genomics knowledge based on Next Generation

Citation: Hikmet Budak, Babar Hussain, Zaeema Khan, Neslihan Z. Ozturk, and Naimat Ullah, From Genetics to Functional Genomics: Improvement in Drought Signaling and Tolerance in Wheat, doi: 10.3389/fpls.2015.01012.

Sequencing (NGS), gene editing systems (Shan et al., 2013), gene silencing (Yin et al., 2014), and over-expression methods (Saad et al., 2013) have increased our understanding about drought signaling pathways. At the transcriptome level, the RNA deep sequencing (Akpinar et al., 2015) and microarray analyses (Ergen et al., 2009) are employed to elucidate the differential expression of RNA transcripts involved in drought response. Moreover, microRNAs (miRNAs; Budak and Akpinar, 2015), hormones (Reddy et al., 2014), quantitative trait loci (QTLs; Barakat et al., 2015), metabolites (Xiao et al., 2012), transcription factors (TFs), and drought-related proteins (Lucas et al., 2011a; Alvarez et al., 2014) are key players in drought signaling. These factors regulate the gene expression in response to drought. TFs also interact with plant stress hormones, e.g., abscisic acid (ABA), jasmonic acid (JA), and salicylic acid (SA) in mediating drought response (Nakashima et al., 2014). To elucidate complex wheat drought signaling which will help in developing improved varieties, powerful tools are required for multiplexed or simultaneous detection of signaling molecules. Advances in functional genomics tools have provided us the opportunity to detect above mentioned molecules with ease, efficacy and accuracy thus opening a new era of crop improvement (Colmsee et al., 2012). In this review, we have summarized the advances in genetics, genomics, and functional genomics for identification of novel genes and their subsequent use in breeding programs for improved drought tolerance.

Signaling Pathways in Wheat for Drought Tolerance

Drought signaling is categorized into ABA-dependent and ABA-independent pathways as ABA is the first line of defense against drought. ABA-dependent signaling consists of two main gene clusters (regulons) regulated by ABA-responsive element-binding protein/ABA-binding factor (the AREB/ABF regulon) and the MYC/MYB regulon. Previous studies have shown that AP2/EREBP (ERF) TFs are engaged in both ABA-dependent and independent signaling pathways. Despite being two distinct and independent pathways, there is plausibly some crosstalk between both (Lata and Prasad, 2011). The AP2/ERF TFs family includes the ethylene-response factors, e.g., a TaERF promotes drought tolerance in wheat with increased proline and chlorophyll levels (Rong et al., 2014). The sucrose non-fermenting1-related protein kinase 2 family (SnRK2) consists of plant specific Ser/Thr kinases which are positive regulators of ABA signaling. SnRK2s were first reported to be involved in ABA signaling in wheat (PKABA1; Fujii and Zhu, 2012). The SnRK TFs are also involved in ABA independent pathway (Lata and Prasad, 2011). Although not specifically studied in wheat, the ABA-dependent pathways in rice and *Arabidopsis* have been extensively analyzed (Todaka et al., 2012). The ABA-independent regulons include the CBF/DREB (cold-binding factor/dehydration responsive element binding), NAC, and ZF-HD (zinc-finger homeodomain; Lata and Prasad, 2011). The transcriptional regulatory network based on DREBs is induced by dehydration in wheat. There are two known DREB regulons; DREB1/CBF and DREB2 (Edae et al., 2013). Above mentioned signaling pathways and their roles in drought tolerance have been extensively discussed in this review.

Genetics Based Improvement in Drought Signaling

The selection of an appropriate breeding strategy to develop drought tolerant cultivars is the key step for a successful breeding program. Therefore, strategies which can transfer specific genes, exploit wild relatives of crops, identify and transfer genes with ease, require less time, and labor to develop cultivars are of great value (Hussain, 2015). The advances in genetics, genomics, and functional genomics have enabled researchers to combine one or more advantages of different strategies to develop drought tolerant wheat. Here, we have described the advances in these methods.

Classical Breeding for Improving Drought Tolerance in Wheat

Classical breeding dates back to 5,000–10,000 years when man domesticated selective plant species based on their better taste. Domestication was followed by selection of high yielding genotypes and cross-hybridization to recombine tolerance genes from different sources (Hussain, 2015). The presence of genetic variation in wheat is the key to identify the contrasting parents for classical breeding (Shelden and Roessner, 2013). Significant genetic variation in wheat for drought tolerance has been identified for selection of diverse parents. Cross-hybridization of wheat diploid progenitors produce drought tolerant synthetic hexaploid (SHs) wheat (see the section Introgression of Drought Tolerance Genes from Wild Species) with introgression of several novel drought tolerance genes (Zhang et al., 2005). Most of plant breeders selected the drought and other stress tolerant wheat varieties on the basis of higher yields and ignored the physiological mechanism behind it. Therefore, few cultivars having drought and abiotic stress tolerance have been developed in comparison to the ones improved for high yield (Hussain et al., 2015). Under drought, plant machinery shifts its focus to ABA production for downstream activation of signaling and tolerance mechanisms which lowers the grain filling and yield. Therefore, the balance between yield and drought tolerance needs to be investigated (Alvarez et al., 2014). ABA content has been used as selection index for screening wheat under drought and contrasting parents for its production have been crossed. Several QTLs and genes involved in signaling pathways have been identified in subsequent segregating populations (Iehisa et al., 2014; see the section QTL Mapping for Drought Signaling Genes in Wheat). Longer time, intensive labor requirements, transfer of non-desirable genes, limited resistance resources, genetic barriers and limited understanding of tolerance mechanisms are problems associated with classical breeding (Hussain, 2015). Since breeding to date has not been focused on signaling pathways, there is a need to combine the available information collected by QTLs, MAS, and Omics tools with traditional traits for improved wheat drought tolerance.

Introgression of Drought Tolerance Genes from Wild Species

Significant loss of genetic diversity has occurred at three levels: (a) Species level (domestication), (b) Varietal level (green revolution), and (c) Gene level (breeding cycles). High yielding wheat developed through green revolution has less stress tolerance (Hussain, 2015). It is the time for plant breeders to look back and utilize this lost genetic diversity as some wild wheat relatives are potential sources of drought tolerance. For example, wild emmer wheat (*Triticum dicoccoides*) has inter and intra-varietal genetic diversity for water use efficiency (WUE), phenology, and contains several genes and QTLs for drought tolerance (Nevo and Chen, 2010). Gene expression studies in emmer wheat identified over 13,000 expressed sequence tags (ESTs) in response to drought (Ergen and Budak, 2009), and 33 outlier loci for drought tolerance were identified by single nucleotide polymorphisms (SNPs) markers (Ren et al., 2013). Transcriptomic analysis identified several genes and TFs involved in ethylene, IP3, and ABA dependent signaling pathways in wild emmer wheat (Ergen et al., 2009). SHs wheat constituted by crossing these wild relatives gained many novel QTLs and genes for ABA responsiveness and signaling (Iehisa et al., 2014). Role of *DREBs* in conferring drought tolerance to *T. dicoccoides*has also been established (Lucas et al., 2011b).

Introgression of drought tolerance to cultivated wheat from *Aegilops tauschii* was achieved by crossing it with durum wheat to make SHs wheat. DNA fingerprinting of SHs showed high genetic diversity with longer roots and higher soluble carbohydrates to resist water deficiency (Reynolds et al., 2007). D genome of *A. tauschii* contains several drought responsive genes potentiating the development of drought tolerant SHs wheat through crossing. *A. tauschii* and related SHs showed significant variation for ABA responsiveness when gene expression analyses were performed for ABA inducing *WABI5* and three downstream Cor/LEA protein coding genes (*Wrab18, Wrab17*, and *Wdhn13*) while the line with enhanced expression of *Wdhn13*showed salt and dehydration tolerance (Iehisa and Takumi, 2012). Proteomics approach has identified several proteins involved in ABA signaling (ABA 8′-hydroxylase, MPK6, dehydrin, 30S ribosomal protein S1, retrotransposon protein, a 70 kDa HSP) in the wild wheat relative, *Kengyilia thoroldiana* under drought. Proteins involved in antioxidative enzyme activity (thioredoxin peroxidase, ascorbate peroxidase, Cu/Zn superoxide dismutase) also showed increased expression levels (Yang et al., 2015). The value of wild wheat relatives as donors of drought genes has not only been established on morphological bases, but also validated with genomics (QTLs) and functional genomics (transcriptomics, proteomics, ESTs, SNPs) tools. Furthermore, their utility as drought gene donors has been confirmed in SHs. Therefore, we suggest that plant breeders should focus on wheat wild relatives to enhance the genetic diversity of wheat for drought tolerance.

Molecular Markers for Identification of Drought Signaling Genes

Selection in classical breeding is performed on the basis of morphological and economical traits which are highly influenced by the environment. Environmental influence on phenotypic expression creates confusion in selection of desirable traits. Discovery of DNA markers for economic and stress related crop traits have helped to select the desirable traits and parents with ease, efficacy and reliability in remarkably shorter time. Therefore markers, especially the SNPs have added more power to identify the genes linked to drought and other stresses (Budak et al., 2013b; Hussain, 2015). DNA markers for various genes involved in drought signaling have been reported, e.g., RAPD markers by using P21F/P21R and P25F/PR primers in A genome; and P18F/P18R primer in B genome mapped *DREB1* on 3A chromosome (Huseynova and Rustamova, 2010). In wheat, DREBs were tagged with five SNPs in A (P21F/P21R and P25F/PR primers), B (P18F/P18R primer), and D genome (P20F/P20R and P22F/PR primers). *DREB1*gene was tagged on chromosome 3A, 3B, and 3D. S646 and S770 SNPs were used and SNP S770 mapped *DREB-B1* between markers *Xfbb117*and *Xmwg818* on chromosome 3BL (Wei et al., 2009). SNPs identified the involvement of five signaling genes in yield and drought tolerance pathways. The *DREB1A* correlated with heading date, vegetation index and biomass, while flag leaf width, harvest index and leaf senescence were associated with *ERA1-B* and *ERA1-D* (enhanced response to ABA) genes. Other signaling genes, *1-FEH-A* and *1-FEH-B* (fructan-1-exohydrolase) were linked to yield and thousand kernel weight (Edae et al., 2013). Significant relationships between morpho-physiological traits and SNPs suggested key role of detected SNPs in drought tolerance.

High resolution melting (HRM) technology is the most powerful tool to identify the allelic variations. Use of HRM found that allelic variation in *DREB* TFs identified by SNPs led to variation in peptide sequences as well. The variation in peptide sequences was linked with differences in protein geometry and recognition of *cis*-elements involved in ABA signaling (Mondini et al., 2015). Two important TFs, i.e., *DREB1, WRKY1*, and a Na^{+} transporter, *HKT-1* conferring drought and salt tolerance were also mapped by SNPs (Mondini et al., 2012). SNPs were used to map *TaSnRK2.8* gene which plays important role in carbohydrate metabolism, protein–protein interaction, and ABA signaling (Zhang et al., 2013). Chromosome locations and primers for these markers are given in **Table** . It can be concluded that MAS has a lot of promise to identify the signaling genes. Strong association between signaling genes and drought related physiological traits suggest that MAS should be focused to identify the signaling genes (Edae et al., 2013). This can help to improve the drought tolerance with less effort, time and resources and can speed up the breeding programs in future.

Table 1: Molecular markers identified for wheat drought signaling genes

Marker type	Primer	Chromosome location	Target drought signaling gene	Reference
SNP	P21F/P21R and P25F/PR	3A	*DREB1*	Wei et al., 2009
	P18F/P18R	3BL	*DREB1*	Wei et al., 2009
SNP S770		*Xfbb117-Xmwg818*		
SNP	P20F/P20R and P22F/PR	3D	*DREB1*	Wei et al., 2009
RAPD	P25F/PR	3A	*DREB1*	Huseynova and Rustamova, 2010
RAPD	P18F/P18R	3A	*DREB1*	Huseynova and Rustamova, 2010
SNP	DREB1a and DREB1b	U16709.1*	*DREB1*	Mondini et al., 2012
SNP	WRKY1	DQ323885.1*	*WRKY1*	Mondini et al., 2012
SNP	HKT-1	AF303376.1*	*HKT-1*	Mondini et al., 2012
SNP	P21	3A	*DREB1A*	Edae et al., 2013
SNP	ERA1B	3A, 3B, 3D	*ERA1-B*	Edae et al., 2013
SNP	ERA1D	-	*ERA1-D*	Edae et al., 2013
SNP	W12	6A	*1-FEH-A*	Edae et al., 2013
SNP	W32	-	*1-FEH-B*	Edae et al., 2013
SNP	M13	5A	*TaSnRK2.8*	Zhang et al., 2013
SNP	DREB2a and DREB2b	-	*DREB2*	Mondini et al., 2015
SNP	DREB3a and DREB3b	-	*DREB3*	Mondini et al., 2015

**Ref gene.*

Transgenic Approaches for Improving Drought Signaling

Loss of tolerance genes by genetic erosion should be filled with such efficient and reliable methods that can transfer genes in a short time. Recombinant DNA technology has emerged as a powerful tool for the purpose. It provides the additional benefit of having no genetic barriers, thereby; so can transferring the genes from any wild relative, land race or other species (Hussain, 2015). Candidate genes for drought tolerance in wheat include TFs which regulate the signaling genes, genes encoding defense molecules [Reactive oxygen species (ROS), proline, JA, SA], and for production of defense proteins (Yang et al., 2010). Here, we have described the major achievements in wheat (see summary in **Table**).

Table 2: Improvement in wheat drought signaling and tolerance by transgenic approaches

Transformed gene	Improvement in signaling and tolerance	Reference
HVA1	Abscisic acid (ABA) signaling, produce Late embryogenesis abundant 3 (LEA3) for cell membrane integrity, higher biomass production and water use efficiency (WUE), drought, and salt tolerance	Sivamani et al., 2000
HVA1	ABA responsiveness, ABA signaling, higher WUE and relative water content (RWC), stable yield, drought tolerance	Bahieldin et al., 2005
GmDREB	ABA independent signaling (AIS), drought, and salt tolerance	Shiqing et al., 2005
GmDREB	AIS, 2-fold higher proline production, stay green, SURV, drought tolerance	Wang et al., 2006
GhDREB	AIS, higher soluble sugars and chlorophyll production, improved drought, salt, and freezing tolerance	Gao et al., 2009
DREB1A	Higher SURV, WUE, and yield under drought	Pierre et al., 2012
SNAC1	Activation of sucrose phosphate synthase, type 2C protein phosphatases and 1-phosphatidylinositol-3-phosphate-5-kinase genes for ABA signaling, high RWC, chlorophyll content and biomass, enhanced salinity and drought tolerance	Saad et al., 2013
Alfalfa aldose reductase	Antioxidant defense, ABA signaling, detoxification of aldehyde substrate, green biomass production, drought tolerance	Fehér-Juhász et al., 2014

Drought Signaling by Introducing *DREBs*

A soybean based DREB gene (*GmDREB*; Accession No. AF514908) was transformed to wheat by gene gun bombardment using ubiquitin and RD29A promoters, and transgenic plants with both promoters showed increased drought and salt tolerance (Shiqing et al., 2005). This increased tolerance of the crop was linked with to twofold higher proline production, stay green phenomenon under drought and survival and recovery on re-watering (SURV) after drought spell (Wang et al., 2006) suggesting a role of signaling pathway in downstream proline production. Transformation of wheat with a cotton originating DREB (*GhDREB*) improved drought, salt, and freezing tolerance due to higher production of soluble sugars and chlorophyll in leaves (Gao et al., 2009). Transgenic wheat with *DREB1A* was subjected to field screening on the basis of SURV and WUE. Although the event was selected in greenhouse, plants showed even higher yield under field drought (Pierre et al., 2012). However, there is dire need to find out activated genes or expressed proteins by these TFs to fully understand their role in signaling pathways.

Drought Signaling by Introducing *HVA1* Gene

Various studies aimed to find the function of ABA in regulating the expression of drought tolerance genes. A number of such genes code for the proteins involved in stomata closure to check the transpiration against the cell dehydration. Late embryogenesis abundant 3 (LEA3) in barley is one of such protein encoded by *Hordeum vulgare* abundant protein 1 (*HVA1*) gene. This gene is activated by ABA treatment and several crops transformed with the *HVA1* gene showed improved drought and salt tolerance (Nguyen and Sticklen, 2013). *HVA1* gene was transformed to spring wheat by gene gun and its over-expression under maize ubiq1 promoter resulted in improved biomass production, WUE, drought, and salt tolerance due to activation of ABA signaling (Sivamani et al., 2000). Field evaluation of transgenic wheat with *HVA1* for six cropping seasons at four locations (USA and Egypt, both irrigated and rainfed area) showed stable yield, higher WUE and relative water content (RWC). These tolerance traits were directly correlated with the expression of *HVA1* gene in transgenic plants (Bahieldin et al., 2005). It is important to note that to date, the research station has not released it as a variety (Hayes and Xue, 2014).

Drought Signaling by Introducing Other Genes

Transformation of wheat with *SNAC1* gene under the control of ubiquitin promoter showed enhanced salinity and drought tolerance at seedling stage in lab conditions. Transgenic plants showed higher biomass, RWC, and chlorophyll content. Expression studies of *SNAC1* by qPCR showed over-expression of sucrose phosphate synthase, type 2C protein phosphatases and 1-phosphatidylinositol-3-phosphate-5-kinase genes, which are involved in ABA signaling (Saad et al., 2013). Transgenic wheat with cDNA of alfalfa aldose reductase recipient gene, involved in antioxidant defense and exhibited 1.5–4.3 times more detoxification of aldehyde substrate and 12, 26, and 41% increase in green biomass production in three separate transgenic lines, resulting in enhanced drought tolerance (Fehér-Juhász et al., 2014). Although the transfer of*DREBs, HVA1*, and *NAC* have resulted in enhanced drought tolerance by improving signaling pathways, there are no studies to date that show whole downstream signaling cascade improved in transgenic plants.

GENOMICS BASED IMPROVEMENT OF DROUGHT SIGNALING

Although a reference whole genome sequence has not been reported for wheat to date, efforts have been put to identify potential genomic regions carrying genes for drought signaling pathways. QTLs, miRNAs, and genome editing systems (e.g., CRISPR/Cas system) are major genomics based methods applied to discover and manipulate related genomic regions. The role of these approaches in characterizing genes involved in drought signaling in wheat is discussed below.

QTL Mapping for Drought Signaling Genes in Wheat

Drought tolerance is a complex and quantitative trait encoded by many genes, and thus, the identification of genomic regions carrying genes for drought signaling is important. Doubled haploids (DHs), F_2populations (Ibrahim et al., 2012), recombinant inbred lines (RILS), near isogenic lines (NILS) are suitable populations for QTL mapping (Budak et al., 2013b). In wheat, the QTLs for yield and yield related traits in drought, and biomolecules involved in signaling pathways have been mapped. Inheritance of ABA accumulation and distribution in plants is not simple and several genes/QTLs are involved in it. A major QTL for ABA production was mapped on the long arm of the 5A chromosome between *Xpsr575* and *Xpsr426* loci (8 cM from *Xpsr426*) in single chromosome substitution line and their subsequent F_2 and DHs populations. Substitution lines were developed by from hybridization of low and high ABA producing Chinese Spring' and 'SQ1' genotypes (Quarrie et al., 1994). This QTL showed strong linkage with *Dhn1/Dhn2* locus depicting a direct association between ABA accumulation and early flowering based drought tolerance in wheat (Ibrahim et al., 2012). Nine QTLs were mapped in wheat in response to exogenously applied ABA, SA, JA, and ethylene, suggesting the presence of potential genes involved in signaling in these regions (Castro et al., 2008).

Several QTLs linked to ABA accumulation in leaves were mapped in an F_2 population developed from the cross of contrasting genotypes for ABA production. But one novel QTL was linked to both higher ABA content and smaller leaf size due to genetic linkage between the genome regions. The QTL location was a homoeolog of the major wheat gene *Vrn1* that code for number of tillers, ABA accumulation and leaf size (Quarrie et al., 1997). A major QTL for ABA production was mapped on chromosome 6D in an F_2 population developed from contrasting SHs. This QTL region has several genes for ABA responsiveness, seed dormancy, and regulation of LEA proteins that protect the cell machinery under dehydration stress (Iehisa et al., 2014). A major QTL for higher grain yield (21%), chlorophyll content and wider flag leaf on 7A chromosome was mapped in a DH wheat line by using psp3094 SSR. Exogenous ABA application activated this QTL suggesting that genes in this region might be involved in ABA signaling (Quarrie et al., 2007). Four homologs of *Arabidopsis* ABA signaling genes (*TmABF, TmVP1, TmERA1*, and *TmABI8*) were mapped in a wheat RILs population derived by crossing *T. boeoticum* and *T. monococcum*. The location of these QTLs was chromosome 3Am, 4Am, and 5Am (Nakamura et al., 2007).

Seven QTLs for ABA production in response to drought were identified on chromosomes 2A, 3A, 1B, 7B, and 5D in an F_4 population at 33% field capacity. The most important QTLs for ABA content were mapped on chromosomes 3B, 4A, and 5A on the marker location of*Wmc96, Trap9*, and *Barc164* (Barakat et al., 2015). In another study, five major QTLs for ABA responsiveness were identified on chromosomes 1B, 2A, 3A, 6D, and 7B in a wheat RILs population. A QTL located on chromosome 6D contributed 11.12% to variation for ABA against 5–8% contribution by other QTLs. Expression analysis showed allelic differences in QTLs for three ABA responsive Cor/LEA protein coding genes, *Wrab15, Wdhn13*, and *Wrab17*. The expression of these genes was influenced by QTLs present on chromosomes 2A, 7B, and 6D in ABA treated seedlings (Kobayashi et al., 2010). In conclusion, several QTLs for ABA production and downstream signaling pathways have been mapped in wheat (**Table Table33**) but most of studies have not focused further on this aspect. We recommend the use of functional genomics tools along with QTLs to identify the genes located in QTL regions.

Table 3: Quantitative trait loci (QTLs) mapped for drought signaling molecules in wheat

Chromosome location	Function	Reference
MAPMAKER QTL; 5A	Located between Xpsr575-Xpsr426, ABA production	Quarrie et al., 1994
2A	ABA production, smaller leaf size, homoeolog of wheat gene *Vm1* that controls number of tillers and ABA accumulation	Quarrie et al., 1997
7A	Activated on exogenous ABA application, involved in ABA signaling	Quarrie et al., 2007
3Am, 4Am, 5Am	QTLs for ABA signaling genes (*TmABF, TmVP1, TmERA1*, and *TmABI8*)	Nakamura et al., 2007
6A	Located at *Xgwm459* and *gwm334a region*, activated by exogenous ABA, salicylic acid (SA), jasmonic acid (JA), and ethylene application, contains signaling genes	Castro et al., 2008
1B, 2A, 3A, 6D, 7B	ABA responsiveness, regulate expression of ABA responsive LEA protein coding genes, i.e., *Wrab15, Wdhn13*, and *Wrab17*	Kobayashi et al., 2010
MAPMAKER QTL; 5A	Linked to *Dhn1/Dhn2* genes, early flowering based drought tolerance	Ibrahim et al., 2012
6D	ABA production and signaling, seed dormancy, regulation of LEA proteins	Iehisa et al., 2014
3A, 1B, 4A, 5A, 5D, 7B	Enhanced ABA production, located at marker locations of *Wmc96, Trap9*, and *Barc164*	Barakat et al., 2015

miRNAs Involved in Drought Signaling and Tolerance

Extensive application of NGS platforms has greatly contributed in identification of 20–22 nt short non-coding RNAs called as miRNAs that play regulatory roles in many processes (Budak et al., 2015a). miRNAs bind to their target transcripts through complementary base pairing, and either direct the cleavage of the target or repress its translation, leading to the decreased expression of the target transcript. Thus, miRNAs can act both at the transcriptional or post-transcriptional levels. miRNA mediated gene-silencing mechanism regulates the expression of plant hormones, TFs, and other developmental/stress signaling pathways (Curaba et al., 2014). Gene silencing involved in plant stress regulation is also mediated by naturally occurring small RNAs (siRNAs; Lu et al., 2012), and complementary double-stranded RNA (dsRNA) generated from inverted repeat (IR) transgenes (Frizzi and Huang, 2010). Post transcriptional gene silencing is carried out by miRNAs and virus-derived small interfering RNAs (vsiRNAs) which helps in discovering gene functions and developing crops with improved stress tolerance (Feng et al., 2013). miRNAs are important regulators in plant drought signaling because their target genes have key roles in metabolism and signal transduction (Yin et al., 2014).

The miRNA gene transcripts '*MIR*' are spatially and temporally influenced by cellular signaling factors, particularly plant hormones such as ABA under stresses (Jung et al., 2015). Some conserved plant miRNAs such as miR159 (*Triticum*, French bean, cotton, maize), miR164 (*Triticum, Brachypodium*, sugarcane), miR172 (*Triticum, Arabidopsis, Brachypodium, Oryza*, cotton) and miR393 (*Triticum, Oryza, Medicago, Pinguicula, Arabidopsis*) control the expression of key TFs which regulate development and signaling pathways (Gupta et al., 2014). miRNAs are involved in various drought related cellular pathways, including auxin signaling, ABA response, antioxidant defense, osmoprotection, cell growth, respiration, and photosynthesis, e.g., miR169 shows bread wheat specific differential expression under drought (Ding et al., 2013). Several signaling genes (ARF, *MYB33, MYB101, TIR1, AGO1*,) and growth regulation factors (GRF) are targeted and regulated by drought responsive miRNAs and *DREBs*(Covarrubias and Reyes, 2010). Most of the miRNAs have their specific putative targets which lead to regulation of specific genes/TF involved in signaling/tolerance mechanisms. In such a study, various bread wheat based miRNAs and their targets (shown in parentheses) were identified as tae-miR159a,b (MYB3), tae-miR159c-5p (Dihydro-flavonoid reductase-like protein), tae-miR171f (sensor histidine kinase), tae-miR395i (ATP sulfurylase), tae-miR156k (SBP), tae-miR166l-5p (FAM10 family protein), tae-miR168b (dehydrogenase/reductase), tae-miR444c.1 (MADS-box TF), tae-miR1432 (mitochondrial phosphate transporter), tae-miR160a (ARF), tae-miR164b (NAC), tae-miR166h (HD-ZIP4), tae-miR169d (CCAAT-box TF), tae-miR319c (Acyl-CoA synthetase), tae-miR393b,i (TIR1), tae-miR396a,c,g (GRF), tae-miR444d (IF3), tae-miR827-5p (finger-like protein). The above mentioned miRNAs regulated the expression of their targeted TFs/genes thus playing key roles in drought tolerance mechanism (Ma et al., 2015).

Similarly, increased expression of miR156 in *T. dicoccoides* targets the SBP TFs and promoted flowering while miR398 targets copper superoxide dismutases, cytochrome C oxidase, and regulates ROS production under drought stress. Increased expression level of miR1432 targets calcium-binding EF which activates signal transduction pathways. Other important drought responsive miRNAs in *T. aestivum*and *T. dicoccoides* are miR396, miR528, miR6248 (Kantar et al., 2011b; Budak et al., 2015b), miR1435, miR5024, and miR7714 (Akpinar et al., 2015). On the other hand, miR166 exhibits decreased expression in *T. dicoccoides* under drought which targets HD-ZIP3 TF and plays role in developmental while miR171 targets GRAS TF and is significant in abiotic stress responses (Kantar et al., 2011b). A summary of miRNAs involved in drought response and signaling is given in**Table Table44**. For a further comprehensive reading on drought responsive miRNAs, we would recommend our reader to go through other articles from our group (Lucas et al., 2011b; Budak et al., 2015a,b). From above discussion, it is evident that miRNAs have a key role in regulating drought tolerance pathways and should be exploited for wheat improvement.

Table 4: Wheat miRNAs involved in drought stress signaling

miRNAs	Differential expression under drought	Target genes related to signaling	Function	Reference
miR159a-5p	↑	MYB TF, GAMYB1 GAMYB2, genes for oligopeptide transport	Auxin Signaling, Oligopeptide transporter	Liu et al., 2013; Gupta et al., 2014
miR159a,b	↓	WRKY TF; MYB3; alkaline phosphatase, cytochrome P450, Mob1-like and TLD protein	Signaling	Ma et al., 2015
miR160a	↑	HSP 70; ARF; tetratrico peptide repeat (TPR)	Stress adaptation	Liu et al., 2015
miR164	↑	NAC domain TF, phytosulfokines, sHSPs 17; NAC TF; genes involved in MAPK signaling pathways	Signaling pathway, oxidative stress response	Gupta et al., 2014
miR166h	↑	Class III HD-ZIP protein 4	Stress response, phytohormones	Ma et al., 2015
miR167	↑	Dnaj heat shock n-terminal domain-containing protein	Auxin signaling pathway, developmental response	Liu et al., 2015
miR168	↑	Argonaute	Signal transduction, stress response	Gupta et al., 2014
miR169d	↓	CCAAT-box TF	ABA-responsive transcription, drought tolerance	Ma et al., 2015
miR171f	↓	Sensor histidine kinase	Stress response	Li et al., 2013
miR172	↑	Apetala2-like TF, ARF, helix-loop-helix DNA-binding protein	Signaling Pathway, stress response, development	Gupta et al., 2014
miR172a,b	↓	Floral homeotic protein, AP2 TFs	Flower development signaling pathway	
miR319	↓	MYB3, Acyl-CoA synthetase	Abiotic stress tolerance	Zhou et al., 2010
miR393	↑	bHLH TF, transport inhibitor response 1/auxin F-box	Signaling pathway, genes in auxin signaling, basal defense	Gupta et al., 2014
miR395i	↓	ATP sulfur lyasas and sulfur transporters	Abiotic stress	Sunkar et al., 2005
miR397	↑	Ice1 (inducer of CBF expression 1) TF, laccase	Response to water deprivation	Gupta et al., 2014
miR398	↑	COX, Superoxide dismutase (SOD) gene family	Respiration pathway	Kantar et al., 2011b
miR474	↓	PPR, protein kinase, kinesin, Leucine-rich repeat	Unknown	Kantar et al., 2011b
miR1029	↑	Apetala2-like TF, DREB TF	Signaling pathway abiotic stress	Gupta et al., 2014
miR1432	↑	Mitochondrial phoshphate transporter	-	Ma et al., 2015

CRISPR/Cas Genome Editing System

In addition to ZEN and TALEN, an efficient bacterium based genome editing system called the Clustered Regulatory Interspaced Short Palindromic Repeats (CRISPR), with associated protein Cas (CRISPR/Cas system) has emerged. The CRISPR are loci with variable short spacers interspersed by short repeats, later transcribed into non-coding RNAs (ncRNA). This ncRNA then forms a complex with the Cas and guides the complex to slice complementary target DNA. The development of single guide RNAs (sgRNAs) which are fusions of essential parts of *trans*-activating crRNA (tracrRNA) and the sgRNA of CRISPR RNA (crRNAs) proved to be an essential improvement in adopting the CRISPR-Cas system for targeted editing of complex eukaryotic genomes (Jiang et al., 2014). Following *Arabidopsis*, the system has also been demonstrated in rice and other crop plants. In protoplasts of bread wheat cultivar Kenong199, an ortholog of the barley MLO protein, *TaMLO* gene was targeted and showed high INDEL frequencies of 26.5–38.0%. The number of unique sgRNA target candidates generated on average were 21 per cDNA of either A or D genomes. The mean mutagenesis frequency in protoplasts was 28.5% with the transformation efficiency of 70–80% (Shan et al., 2013). The ability of this system to delete large DNA segments stably is valuable in wheat genomics given its large genome size and complexity. Development of transgenic wheat cultivars with stable drought tolerance through targeted genome editing will potentially revolutionize crop breeding. To date, the use of this system in engineering abiotic/drought stress tolerance or signaling has not been reported; however, in the future, it may prove to be a valuable tool in discovering the functions of signaling pathway components.

FUNCTIONAL GENOMICS FOR DISCOVERING DROUGHT SIGNALING MOLECULES

In recent years, the use of functional genomics tools has considerably increased in elucidating abiotic stress tolerance in plants. These methods include transcriptomics, metabolomics, proteomics, and ionomics and are capable of discovering and characterizing the expression of genes or other molecules in drought with high accuracy and efficiency. These highly sensitive tools can perform the spatial analysis of tissues which helps to understand the drought tolerance mechanisms, see **Figure** (Ergen et al., 2009). We have summarized the advances in OMICS for discovery of drought signaling molecules below

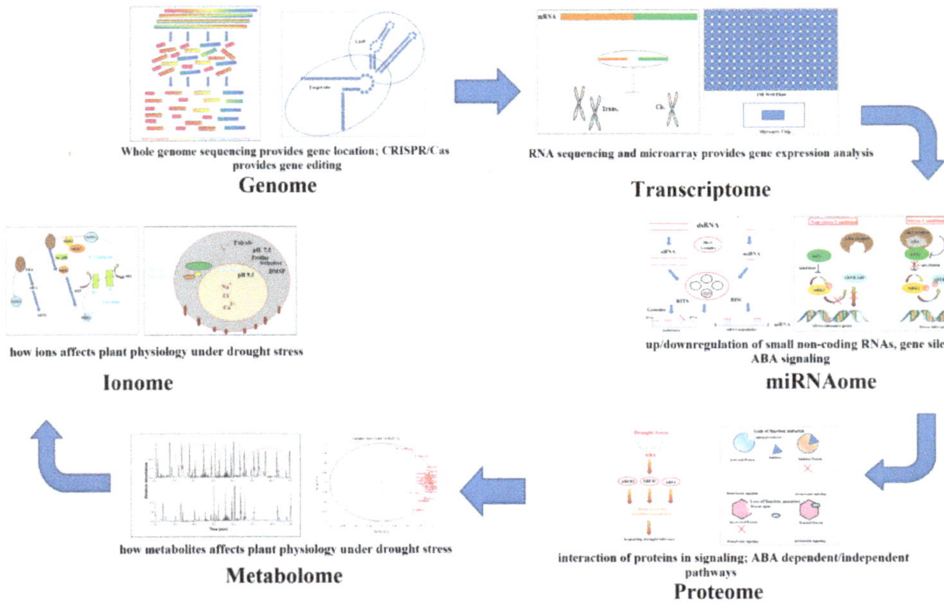

Figure 1: Schematic diagram showing flow of information between functional genomics techniques for studying signaling pathways in plants.

Transcriptomics for Identification of Drought Signaling Pathways

Transcriptomics is the study of whole set of RNA transcripts (transcriptome) produced by an organism under certain conditions, like drought. Microarrays were used first for profiling transcripts in response to various stresses, but polyploidy has limited the number of studies in wheat. Microarray based analysis became more common after the commercial release of wheat Affymetrix Gene Chip® (Santa Clara, CA, USA) which contained over 55,000 probes for wheat transcripts (Nair et al., 2012). Ergen et al. (2009) identified several novel genes and TFs involved in ABA, ethylene, and IP3 dependent signaling by Affymetrix Gene Chip® based spatial profiling of root and leaf tissues in wild emmer wheat. In leaf tissue, glutamine-dependent asparagine synthetase, a putative ATP-binding protein, homeo-domain TF Hox22 homolog (linked to LEA3 proteins), carotenoid producing 9-*cis*-epoxycarotenoid dioxygenase, proline-rich protein precursor and many proteinase inhibitors of Bowman Birk type showed highest increased expression levels under drought. While, germin-like protein, protein degrading cysteine proteinase precursors, bZIP TFs, cytochrome P450, OsRR9 homolog (a signal receiver domain), *LOL1*protein, and G–C content rich motif binding MYB like TF RADIALIS exhibited decreased expression levels. In root tissues, several dehydrins, LEA/COR protein *WRAB1*, putative lipases, a Rab GTPase homolog, several cold regulated proteins, 12-oxophytodienoic acid reductase, a cold shock protein A-2, MYB-domain Hv1 TF, glutathione transferase (ROS scavenger), PRP homologs and *WCOR719* (actin depolymerization factor) showed the highest increased expression levels. In contrast, many HSPs, RmlC-type cupin domain, B12D proteins, two nodulin 93 proteins, RING-H2 finger protein and GTP-binding EF exhibited decreased expression levels (Ergen et al., 2009).

Microarray gene chip based transcriptomic analysis of *TAM111* and*TAM112* wheat genotypes identified 123 genes for production of ABA, JA, auxin, cytokinin, brassinosteroid, gibberellins, and ethylene, and signaling pathways involving these hormones. These transcripts showed differential expression at grain filling stage and transcripts for ABA biosynthesis exhibited increased expression levels in both genotypes. Two transcripts similar to *PDR12* (coding for ABA transporter), transcripts for auxin, LEAs, dehydrins, HSPs, aquaporins, and redox homeostasis showed decreased expression levels. Transcript analysis revealed a key role of ABA in regulation of transcripts and physiological changes linked with drought adaptation. Higher leaf ABA production showed strong association with higher yield and biomass under field drought which reduced the stomatal conductance and was linked to elevated transcript changes in flag leaf (Reddy et al., 2014). Low chip-to-chip variation, high reproducibility and probe density of Affymetrix Gene Chip® are its advantages but a major disadvantage of high relative cost has limited large scale studies. Therefore, there is dire need for more reliable and cost effective methods to study the complex drought signaling pathways.

Deep sequencing (RNA-Seq) based transcriptomics has made it a cost effective and powerful tool. There were unique transcripts found in *T. dicoccoides*, and *T. durum* under drought that they were engaged in drought signaling pathways (Akpinar et al., 2015). It can be concluded that RNA-Seq based transcriptomic analyses identified various drought signaling genes such as bZIP, TdNAC, and MYB genes which can be exploited in future studies.

Proteomics for Identification of Drought Signaling Pathways

Though transcriptomics provides the gene expression profiles, these profiles do not necessarily reflect protein levels as some transcripts may not be translated. Proteomics analysis provides clues into the actual fluctuations of the protein levels involved in signaling, regulatory, and enzymatic functions encoded by genome through transcripts. Advanced bioinformatics tools have helped to identify, characterize, and annotate novel proteins (Colmsee et al., 2012). The proteomic analysis of two wheat genotypes having contrasting drought tolerances showed that among differentially expressed proteins, 26% were involved in carbohydrate metabolism, 23% in detoxification and defense, and 17% in storage proteins. In drought, WD40 repeat protein, catalase isozyme 1, LEAs, Triticin precursor, sucrose synthase, and alpha amylase inhibitors exhibited increased expression levels in tolerant and decreased expression level in sensitive cultivar. On the other hand, ascorbate peroxidase, small and large subunit ADP glucose pyrophosphorylase and G-beta like protein showed decreased expression levels in sensitive cultivar (Jiang et al., 2012). Our group performed proteomics analysis using two wild emmer wheat varieties (TR39477 and TTD22), and one durum wheat cv. (Kızıltan). After 9 days of drought exposure, 75 differentially expressed proteins were detected with many being common to all three wheat genotypes, e.g., manganese superoxide dismutase (MnSOD), a glutathione transferase showed increased expression in the durum wheat (Budak et al., 2013a).

Alvarez et al. (2014) identified 1656 proteins and two unique peptides in wheat through proteomics analysis by using roots of from drought tolerant (Nesser) and sensitive (Opata) varieties for ABA-responsiveness. Important signaling proteins including monomeric G-proteins and their regulators, two lipoxygenases, K channel β subunits, a plasma membrane proton ATPase, calnexin, and an elicitor-induced protein showed an increased response to ABA in drought tolerant cultivar and *vice versa* in the sensitive. Though signaling protein 14-3-3 homologs exhibited increased expression in the drought tolerant cultivar, they remained unchanged in the sensitive one. Out of 151 ABA-responsive proteins, 100 showed increased expression levels but the rest showed decreased expression levels. An interesting finding was the abundance of multiple porin proteins and β-expansin precursor in the sensitive cultivar suggesting that the cell wall structure and membrane permeability might have influenced different adaptation to drought in both cultivars. Furthermore, six LEA proteins and several phosphatases were also identified in both cultivars (Alvarez et al., 2014).

Metabolomics for Identification of Drought Signaling Pathways

Metabolites are considered as signaling molecules as they are associated with physiological processes and are exported from each organelle to cytoplasm in the form of retrograde signals. Gas chromatography mass-spectrometry (GC–MS), liquid chromatography mass-spectrometry (LC–MS), capillary electrophoresis mass-spectrometry (CE–MS), and nuclear magnetic resonance (NMR) are the major analytical tools in metabolomics to detect, identify, and analyze small molecules. Initially, the analysis of metabolites was limited to a few compounds having major roles in drought tolerance. But, advances in these methods have enabled us to identify a wider range of metabolites produced under a specific condition. Metabolite profiling is a powerful tool to characterize genotype or phenotype of an organism for dissecting novel signaling pathways (Xiao et al., 2012). Plants accumulate compatible solutes to protect them from drought and oxidative stress for survival. GC–MS based metabolite profiling in moss *Physcomitrella patens* showed accumulation of compatible solutes in response to drought stress (Zhan et al., 2014). In wheat, levels of proline, tryptophan, and the branched chain amino acids leucine, isoleucine, and valine increased under drought in tolerant cultivars but organic acid levels decreased (Krugman et al., 2011). There are various examples where metabolites act as intracellular signals, e.g., in response to cytosolic sugar levels, trehalose 6-phosphate (T6P) enhances the redox transfer to AGPase, mediated by thioredoxin, mainly depending on metabolite balance between the chloroplast and cytosol. Metabolite profiling for elucidating signals in plants has been characterization in form of methyl-erythritol-cyclo-diphosphate (MECD; Xiao et al., 2012).

Ionomics for Identification of Drought Signaling Pathways

Ionomics is a high throughput analysis of the ion composition in an organism under a certain condition. It has immense applications in forward and reverse genetics, screening of mutants, finding mechanisms of ion uptake, compartmentalization, transport, and exclusion, thus helps to understand the mechanisms of drought and other abiotic stresses in plants (Shelden and Roessner, 2013). In wheat, ionomics studies under drought stress have not been reported yet. Together, ionomics and other genomics data can provide a complete picture of cellular changes against drought, enabling a thorough understanding of the underlying mechanisms of tolerance (Colmsee et al., 2012). Furthermore, ionomics can help to identify novel genes coding for ions by utilizing phenotypic and genotypic data obtained from mapping populations. It can thus help in understanding the gene networks controlling the ion accumulation at different growth stages under drought stress (Satismruti et al., 2013). It should be noted that ionomics is a relatively new functional genomics tool with limited number of studies available, but spatial and highly sophisticated ion profiling will be a key in the future to understand the signaling pathways for drought tolerance.

SYSTEMS BIOLOGY APPROACHES TO DISCOVER SIGNALING PATHWAYS

System biology is a recent, fast growing and comprehensive analytical approach in life sciences to discover the control and regulation of intracellular biological systems, biochemical cycles and pathways in plants under different environmental stresses. Several computational studies have been involved, from few decades, to find the biochemical function of metabolites and small biomolecules responsible for biochemical pathways under drought stress (Gutiérrez et al., 2005). Most of the experimental methods are being used as the part of system biology including various target and untargeted metabolite analysis approaches used to identify the drought specific metabolites in different species of plants. GC–MS is one of the several approaches, was used to find metabolic compounds differentially expressed in tolerant (Excalibur and RAC875) and sensitive (Kukri) bread wheat cultivars under drought stress environment (Bowne et al., 2012). Another finding was aspartate-derived synthesis of few amino acids including lysine, methionine, and threonine in *Arabidopsis thaliana* done also explains the advantageous use of system biology approach, model based on measured kinetic parameters (Curien et al., 2009). Thus, system biology helps to understand the relationship and inter-connection among various types of bio-molecules involved in a certain tolerance mechanism.

CONCLUSION

Significant wheat yield losses due to drought lower farmers' income, food availability and ultimately affect the economy of various countries. Few drought tolerant wheat varieties have been developed as most of breeders have selected plants on the basis of morphological traits and ignored physiological basis of drought tolerance. Lost genetic variation for drought tolerance has been patched up by using wild wheat *T. dicoccoides* and *A. tauschii* in crossing and synthetic wheat showed improved drought tolerance. The improvement has been measured in terms of novel genes, QTLs, ESTs, and SNPs, e.g., three LEA protein coding genes (*Wrab18, Wrab17, Wdhn13*) involved in ABA signaling pathway were identified in SHs. Proteomics analysis identified ABA 8′-hydroxylase, MPK6, dehydrin, 30S ribosomal protein S1, and a 70 kDa HSP involved in ABA signaling in wild emmer wheat. In recent years, MAS have been used to select the plants which make the cultivar development process less time consuming. SNP markers have identified *DREB-B1, DREB1A, ERA1-B, ERA1-D, 1-FEH-A, 1-FEH-B, WRKY1, TaSnRK2.8*, and *HKT-1* genes in wheat. Many QTLs for ABA production, SA, JA, ethylene, and ABA based signaling, and QTL having genes for regulating other signaling genes (*TmABF, TmVP1, TmERA1* and *TmABI8, Wrab15, Wdhn13*, and*Wrab17)* have also been mapped. Transgenic approach provides the benefit of speedy gene transfer without any genetic barriers. Transformation of wheat with *GmDREB, GhDREB, DREB1A, HVA1, SNAC1*, and aldose reductase genes improved drought signaling and tolerance. Similarly, various miRNAs showed differential expression in drought and enhanced or silenced the expression of genes involved in drought signaling. However, the signaling genes, miRNAs, TF, etc., don't not express in isolation but interact with each other in signaling pathways (**Figure Figure22**).

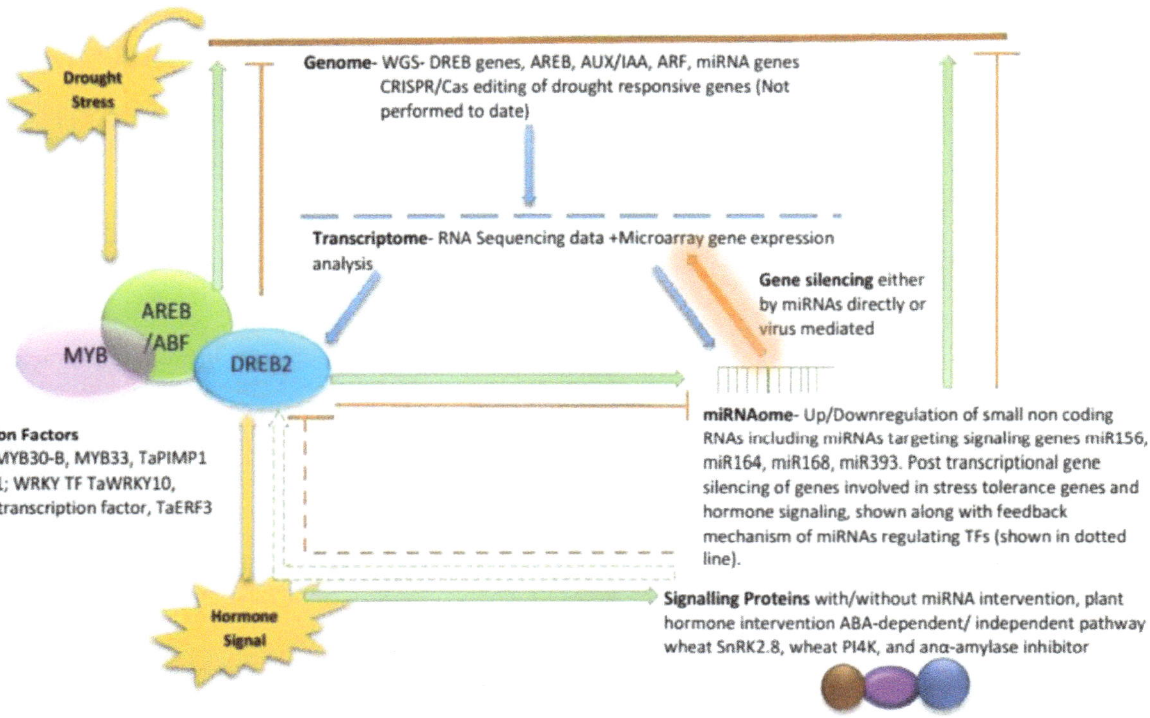

Figure 2: Putative stress signaling pathway in wheat where functional genomics contribute.

In recent years, functional genomics has emerged as a power tool to identify the molecules involved in drought signaling pathways. Transcriptomics analyses identified various genes including Hox22, bZIP TF, dehydrins, *WRAB1, WCOR719, HSPs*, LEA, *TaWRKY17, TaWRKY16, TaWRKY24, TaWRKY19-C, TaWRKY59, TaWRKY82, TaWRKY61, TaWLIP19, TaWRKY10, TaNAC69*, and *TaMYB33* for drought signaling pathways. Proteomics and metabolomics have identified several proteins (Monomeric G-proteins, lipoxygenases, potassium channel β subunits, calnexin, LEAs, phosphatases) and metabolites (Proline, tryptophan, leucine, isoleucine, valine) involved in drought signaling as summarized in **Table Table55**. In this way, highly efficient functional genomics tools have helped in identifying several important genes which can be exploited by breeders to develop drought tolerant wheat cultivars in futures. Similarly, genome-editing system CRSPR/Cas will be valuable in future for better understanding of drought tolerance mechanisms due to its ability to modify the genome. Study of miRNAs in future is also important in future as they are key regulators of signaling pathways.

Table 5: Functional genomics studies for identification of drought signaling molecule in recent years

Functional genomics tool	Drought signaling genes/molecules/mechanism identified	Reference
Transcriptomics	Hox22 TF linked to LEA3 proteins, ABA inducible LEA *WRAB1*, Dehydrins, Proline-rich protein precursor, Asparagine synthetase, 9-*cis*-epoxycarotenoid dioxygenase, Rab GTPase homolog, A-2 cold shock protein, MYB-domain Hv1 TF, glutathione transferase, *WCOR719*	Ergen et al., 2009
Transcriptomics	Genes for ABA, JA, auxin, cytokinin, brassinosteroid, gibberellins, and ethylene production and drought signaling based on these hormones. Transcripts for *PDR12*, auxin, LEAs, dehydrins, HSPs, aquaporins, redox homeostasis, and reduced stomatal conductance	Reddy et al., 2014
Proteomics	WD40 protein, catalase isozyme 1, LEA and alpha amylase inhibitors, ascorbate peroxidase, G-beta like protein, triticin precursor, sucrose synthase	Jiang et al., 2012
Proteomics	Ribulose-1,5-bisphosphate carboxylase large subunit, OsI_16800 protein, SORBIDRAFT_09g029170 protein, polyamine oxidase, Os02g0101500, Os03g0786100, Ferredoxin-NADP(H) oxidoreductase, Os03g0786100, Glutathione transferase, Mn superoxide dismutase, Cold regulated proteins	Budak et al., 2013a
Proteomics	Monomeric G-proteins and their regulators, lipoxygenases, K channel β subunits, plasma membrane proton ATPase, calnexin, an elicitor-induced protein, porin proteins, β-expansin precursor, LEAs, phosphatases	Alvarez et al., 2014
Proteomics	8′-hydroxylase, MPK6, dehydrin, 30S ribosomal protein S1, retrotransposon protein, 70 kDa HSP, thioredoxin peroxidase, ascorbate peroxidase, Cu/Zn superoxide dismutase	Yang et al., 2015
Metabolomics	Proline, tryptophan, leucine, isoleucine, and valine, organic acids	Krugman et al., 2011
Metabolomics	Trehalose 6-phosphate (T6P) promotes thioredoxin-mediated redox transfer to AGPase and MECD helped in elucidating signaling in plants	Xiao et al., 2012

REFERENCES

1. Akpinar B. A., Avsar B., Lucas S. J., Budak H. (2012). Plant abiotic stress signaling. *Plant Signal. Behav.* 7 1450–1455. 10.4161/psb.21894
2. Akpinar B. A., Kantar M., Budak H. (2015). Root precursors of microRNAs in wild emmer and modern wheats show major differences in response to drought stress. *Funct. Integr. Genomics*15 587–598. 10.1007/s10142-015-0453-0
3. Akpinar B. A., Lucas S. J., Budak H. (2013). Genomics approaches for crop improvement against abiotic stress. *Sci. World J.* 2013 361921 10.1155/2013/361921
4. Alvarez S., Roy Choudhury S., Pandey S. (2014). Comparative quantitative proteomics analysis of the ABA response of roots of drought-sensitive and drought-tolerant wheat varieties identifies proteomic signatures of drought adaptability. *J. Proteome Res.* 131688–1701. 10.1021/pr401165b
5. Bahieldin A., Mahfouz H. T., Eissa H. F., Saleh O. M., Ramadan A. M., Ahmed I. A., et al. (2005). Field evaluation of transgenic wheat plants stably expressing the HVA1 gene for drought tolerance. *Physiol. Plant.* 123 421–427. 10.1111/j.1399-3054.2005.00470.x
6. Barakat M. N., Saleh M. S., Al-Doss A. A., Moustafa K. A., Elshafei A. A., Zakri A. M., et al. (2015). Mapping of QTLs associated with abscisic acid and water stress in wheat. *Biol. Plant.* 59 291–297. 10.1007/s10535-015-0499-9
7. Bowne J. B., Erwin T. A., Juttner J., Schnurbusch T., Langridge P., Bacic A., et al. (2012). Drought responses of leaf tissues from wheat cultivars of differing drought tolerance at the metabolite level. *Mol. Plant* 5 418–429. 10.1093/mp/ssr114
8. Budak H., Akpinar B. A. (2015). Plant miRNAs: biogenesis, organization and origins. *Funct. Integr. Genomics* 15 523–531. 10.1007/s10142-015-0451-2
9. Budak H., Akpinar B. A., Unver T., Turktas M. (2013a). Proteome changes in wild and modern wheat leaves upon drought stress by two-dimensional electrophoresis and nanoLC-ESI-MS/MS. *Plant Mol. Biol.* 83 89–103. 10.1007/s11103-013-0024-5
10. Budak H., Kantar M., Kurtoglu K. Y. (2013b). Drought tolerance in modern and wild wheat. *ScientificWorldJournal* 2013 54824610.1155/2013/548246
11. Budak H., Kantar M., Bulut R., Akpinar B. A. (2015a). Stress responsive miRNAs and isomiRs in cereals. *Plant Sci.* 235 1–13. 10.1016/j.plantsci.2015.02.008
12. Budak H., Khan Z., Kantar M. (2015b). History and current status of wheat miRNAs using next-generation sequencing and their roles in development and stress. *Brief. Funct. Genomics* 14189–198. 10.1093/bfgp/elu021
13. Castro A. M., Tacaliti M. S., Giménez D., Tocho E., Dobrovolskaya O., Vasicek A., et al. (2008). Mapping quantitative trait loci for growth responses to exogenously applied stress induced hormones in wheat. *Euphytica* 164 719–727. 10.1007/s10681-008-9694-5
14. Colmsee C., Mascher M., Czauderna T., Hartmann A., Schlüter U., Zellerhoff N., et al. (2012). OPTIMAS-DW: a comprehensive transcriptomics, metabolomics, ionomics, proteomics and phenomics data resource for maize. *BMC Plant Biol.* 12:24510.1186/1471-2229-12-245
15. Covarrubias A. A., Reyes J. L. (2010). Post-transcriptional gene regulation of salinity and drought responses by plant microRNAs.*Plant Cell Environ.* 33 481–489. 10.1111/j.1365-3040.2009.02048.x
16. Curaba J., Singh M. B., Bhalla P. L. (2014). miRNAs in the crosstalk between phytohormone signalling pathways. *J. Exp. Bot.* 65 1425–1438. 10.1093/jxb/eru002
17. Curien G., Bastien O., Robert-Genthon M., Cornish-Bowden A., Cárdenas M. L., Dumas R. (2009). Understanding the regulation of aspartate metabolism using a model based on measured kinetic parameters. *Mol. Syst. Biol.* 5 271 10.1038/msb.2009.29
18. Ding Y., Tao Y., Zhu C. (2013). Emerging roles of microRNAs in the mediation of drought stress response in plants. *J. Exp. Bot.* 643077–3086. 10.1093/jxb/ert164
19. Edae E. A., Byrne P. F., Manmathan H., Haley S. D., Moragues M., Lopes M. S., et al. (2013). Association

mapping and nucleotide sequence variation in five drought tolerance candidate genes in spring wheat. *Plant Genome* 6 1310.3835/plantgenome2013.04.0010

20. Ergen N. Z., Budak H. (2009). Sequencing over 13 000 expressed sequence tags from six subtractive cDNA libraries of wild and modern wheats following slow drought stress. *Plant Cell Environ.* 32 220–236. 10.1111/j.1365-3040.2008.01915.x

21. Ergen N. Z., Thimmapuram J., Bohnert H. J., Budak H. (2009).Transcriptome pathways unique to dehydration tolerant relatives of modern wheat. *Funct. Integr. Genomics* 9 377–396. 10.1007/s10142-009-0123-1

22. Fehér-Juhász E., Majer P., Sass L., Lantos C., Csiszár J., Turóczy Z., et al. (2014). Phenotyping shows improved physiological traits and seed yield of transgenic wheat plants expressing the alfalfa aldose reductase under permanent drought stress. *Acta Physiol. Plant.* 36 663–673. 10.1007/s11738-013-1445-0

23. Feng H., Zhang Q., Li H., Wang X., Wang X., Duan X., et al. (2013). vsiRNAs derived from the miRNA-generating sites of pri-tae-miR159a based on the BSMV system play positive roles in the wheat response to *Puccinia striiformis* f. sp. tritici through the regulation of taMyb3 expression. *Plant Physiol. Biochem.* 6890–95. 10.1016/j.plaphy.2013.04.008

24. Frizzi A., Huang S. (2010). Tapping RNA silencing pathways for plant biotechnology. *Plant Biotechnol. J.* 8 655–677. 10.1111/j.1467-7652.2010.00505.x

25. Fujii H., Zhu J.-K. (2012). Osmotic stress signaling via protein kinases. *Cell. Mol. Life Sci.* 69 3165–3173. 10.1007/s00018-012-1087-1

26. Gao S. Q., Chen M., Xia L. Q., Xiu H. J., Xu Z. S., Li L. C., et al. (2009). A cotton (*Gossypium hirsutum*) DRE-binding transcription factor gene, GhDREB, confers enhanced tolerance to drought, high salt, and freezing stresses in transgenic wheat.*Plant Cell Rep.* 28 301–311. 10.1007/s00299-008-0623-9

27. Gupta O. P., Meena N. L., Sharma I., Sharma P. (2014). Differential regulation of microRNAs in response to osmotic, salt and cold stresses in wheat. *Mol. Biol. Rep.* 41 4623–4629. 10.1007/s11033-014-3333-0

28. Gutiérrez R. A., Shasha D. E., Coruzzi G. M. (2005). Systems biology for the virtual plant. *Plant Physiol.* 138 550–554. 10.1104/pp.104.900150

29. Hayes P., Xue S. (2014). Beating the heat. *Nat. Biotechnol.* 32610–613. 10.1038/nbt.2948

30. Huseynova I., Rustamova S. (2010). Screening for drought stress tolerance in wheat genotypes using molecular markers. *Biol. Sci.*65 132–139.

31. Hussain B. (2015). Modernization in plant breeding approaches for improvingbiotic stress resistance in crop plants. *Turk. J. Agric. For.* 39 515–530. 10.3906/tar-1406-176

32. Hussain B., Khan A. S., Ali Z. (2015). Genetic variation in wheat germplasm for salinity tolerance atseedling stage: improved statistical inference. *Turk. J. Agric. For.* 39 182–192. 10.3906/tar-1404-114

33. Ibrahim S. E., Schubert A., Pillen K., Léon J. (2012). Comparison of QTLs for drought tolerance traits between two advanced backcross populations of spring wheat. *Int. Acad. J.* 2216–227.

34. Iehisa J. C. M., Matsuura T., Mori I. C., Yokota H., Kobayashi F., Takumi S. (2014). Identification of quantitative trait loci for abscisic acid responsiveness in the D-genome of hexaploid wheat. *J. Plant Physiol.* 171 830–841. 10.1016/j.jplph.2014.02.003

35. Iehisa J. C. M., Takumi S. (2012). Variation in abscisic acid responsiveness of *Aegilops tauschii* and hexaploid wheat synthetics due to the D-genome diversity. *Genes Genet. Syst.* 879–18. 10.1266/ggs.87.9

36. Jiang S. S., Liang X. N., Li X. H. X., Wang S. L., Lv D. W., Ma C. Y., et al. (2012). Wheat drought-responsive grain proteome analysis by linear and nonlinear 2-DE and MALDI-TOF mass spectrometry. *Int. J. Mol. Sci.* 13 16065–16083. 10.3390/ijms131216065

37. Jiang W., Yang B., Weeks D. P. (2014). Efficient CRISPR/Cas9-mediated gene editing in *Arabidopsis thaliana* and inheritance of modified genes in the T2 and T3 generations. *PLoS ONE*9:e99225 10.1371/journal.pone.0099225

38. Jung I. L., Ryu M., Cho S. K., Shah P., Lee J. H., Bae H., et al. (2015). Cesium toxicity alters MicroRNA processing and AGO1 expressions in *Arabidopsis thaliana*. *PLoS ONE* 10:e012551410.1371/journal.pone.0125514

39. Kantar M., Lucas S. J., Budak H. (2011a). "Drought stress: molecular genetics and genomics approaches," in *Advances in Botanical Research Advances in Botanical Research* ed. Turkan I., editor. (Burlington, MA: Elsevier;) 445–493.

40. Kantar M., Lucas S. J., Budak H. (2011b). miRNA expression patterns of *Triticum dicoccoides* in response to shock drought stress. *Planta* 233 471–484. 10.1007/s00425-010-1309-4

41. Kobayashi F., Takumi S., Handa H. (2010). Identification of quantitative trait loci for ABA responsiveness at the seedling stage associated with ABA-regulated gene expression in common wheat. *Theor. Appl. Genet.* 121 629–641. 10.1007/s00122-010-1335-4

42. Krugman T., Peleg Z., Quansah L., Chagué V., Korol A. B., Nevo E., et al. (2011). Alteration in expression of hormone-related genes in wild emmer wheat roots associated with drought adaptation mechanisms. *Funct. Integr. Genomics* 11 565–583. 10.1007/s10142-011-0231-6

43. Lata C., Prasad M. (2011). Role of DREBs in regulation of abiotic stress responses in plants. *J. Exp. Bot.* 62 4731–4748. 10.1093/jxb/err210

44. Li Y.-F., Zheng Y., Jagadeeswaran G., Sunkar R. (2013). Characterization of small RNAs and their target genes in wheat seedlings using sequencing-based approaches. *Plant Sci.* 20 17–24. 10.1016/j.plantsci.2012.12.014

45. Liu J., Feng L., Li J., He Z. (2015). Genetic and epigenetic control of plant heat responses. *Front. Plant Sci.* 6:26710.3389/fpls.2015.00267

46. Liu S., Wang N., Zhang P., Cong B., Lin X., Wang S., et al. (2013). Next-generation sequencing-based transcriptome profiling analysis of *Pohlia nutans* reveals insight into the stress-relevant genes in Antarctic moss. *Extremophiles* 17 391–403. 10.1007/s00792-013-0528-6

47. Lu T., Zhu C., Lu G., Guo Y., Zhou Y., Zhang Z., et al. (2012).Strand-specific RNA-seq reveals widespread occurrence of novel cis-natural antisense transcripts in rice. *BMC Genomics* 13:72110.1186/1471-2164-13-721

48. Lucas S., Dogan E., Budak H. (2011a). TMPIT1 from wild emmer wheat: first characterisation of a stress-inducible integral membrane protein. *Gene* 483 22–28. 10.1016/j.gene.2011.05.003

49. Lucas S., Durmaz E., Akpnar B. A., Budak H. (2011b). The drought response displayed by a DRE-binding protein from*Triticum dicoccoides*. *Plant Physiol. Biochem.* 49 346–351. 10.1016/j.plaphy.2011.01.016

50. Ma X., Xin Z., Wang Z., Yang Q., Guo S., Guo X., et al. (2015).Identification and comparative analysis of differentially expressed miRNAs in leaves of two wheat (*Triticum aestivum* L.) genotypes during dehydration stress. *BMC Plant Biol.* 15:2110.1186/s12870-015-0413-9

51. Mondini L., Nachit M. M., Pagnotta M. A. (2015). Allelic variants in durum wheat (*Triticum turgidum* L. var. durum) DREB genes conferring tolerance to abiotic stresses. *Mol. Genet. Genomics* 290 531–544. 10.1007/s00438-014-0933-2

52. Mondini L., Nachit M., Porceddu E., Pagnotta M. A. (2012).Identification of SNP mutations in DREB1, HKT1, and WRKY1 genes involved in drought and salt stress tolerance in durum wheat (*Triticum turgidum* L. var durum). *OMICS* 16 178–187. 10.1089/omi.2011.0081

53. Nair P., Kandasamy S., Zhang J., Ji X., Kirby C., Benkel B., et al. (2012). Transcriptional and metabolomic analysis of*Ascophyllum nodosum* mediated freezing tolerance in*Arabidopsis thaliana*. *BMC Genomics* 13:643 10.1186/1471-2164-13-643

54. Nakamura S., Komatsuda T., Miura H. (2007). Mapping diploid wheat homologues of *Arabidopsis* seed ABA signaling genes and QTLs for seed dormancy. *Theor. Appl. Genet.* 114 1129–1139. 10.1007/s00122-007-0502-8

55. Nakashima K., Yamaguchi-Shinozaki K., Shinozaki K. (2014).The transcriptional regulatory network in the drought response and its crosstalk in abiotic stress responses including drought, cold, and heat. *Front. Plant Sci.* 5:170 10.3389/fpls.2014.00170

56. Nevo E., Chen G. (2010). Drought and salt tolerances in wild relatives for wheat and barley improvement. *Plant Cell Environ.*33 670–685. 10.1111/j.1365-3040.2009.02107.x

57. Nguyen T. Z., Sticklen M. (2013). Barley HVA1 gene confers drought and salt tolerance in transgenic maize (*Zea mays* L.).*Adv. Crop Sci. Technol.* 1 1–8. 10.4172/acst.1000105

58. Pierre C. S., Crossa J. L., Bonnett D., Yamaguchi-Shinozaki K., Reynolds M. P. (2012). Phenotyping transgenic wheat for drought resistance. *J. Exp. Bot.* 63 1799–1808. 10.1093/jxb/err385

59. Quarrie S., Kaminska A., Barnes J., Dodig D., Gennaro A. (2007). A QTL for grain yield on 7AL of wheat is activated by ABA and low nutrient treatments during flag leaf ontogeny.*Comp. Biochem. Physiol. A Mol. Integr. Physiol.* 146 S25310.1016/j.cbpa.2007.01.593

60. Quarrie S. A., Gulli M., Calestani C., Steed A., Marmiroli N. (1994). Location of a gene regulating drought-induced abscisic acid production on the long arm of chromosome 5A of wheat.*Theor. Appl. Genet.* 89 794–800. 10.1007/BF00223721

61. Quarrie S. A., Laurie D. A., Zhu J., Lebreton C., Semikhodskii A., Steed A., et al. (1997). QTL analysis to study the association between leaf size and abscisic acid accumulation in droughted rice leaves and comparisons across cereals. *Plant Mol. Biol.* 35155–165. 10.1023/A:1005864202924

62. Reddy S. K., Liu S., Rudd J. C., Xue Q., Payton P., Finlayson S. A., et al. (2014). Physiology and transcriptomics of water-deficit stress responses in wheat cultivars TAM 111 and TAM 112. *J. Plant Physiol.* 171 1289–1298. 10.1016/j.jplph.2014.05.005

63. Ren J., Chen L., Sun D., You F. M., Wang J., Peng Y., et al. (2013). SNP-revealed genetic diversity in wild emmer wheat correlates with ecological factors. *BMC Evol. Biol.* 13:16910.1186/1471-2148-13-169

64. Reynolds M., Dreccer F., Trethowan R. (2007). Drought-adaptive traits derived from wheat wild relatives and landraces. *J. Exp. Bot.* 58 177–186. 10.1093/jxb/erl250

65. Rong W., Qi L., Wang A., Ye X., Du L., Liang H., et al. (2014).The ERF transcription factor TaERF3 promotes tolerance to salt and drought stresses in wheat. *Plant Biotechnol. J.* 12 468–479. 10.1111/pbi.12153

66. Saad A. S. I., Li X., Li H. P., Huang T., Gao C. S., Guo M. W., et al. (2013). A rice stress-responsive NAC gene enhances tolerance of transgenic wheat to drought and salt stresses. *Plant Sci.* 20 33–40. 10.1016/j.plantsci.2012.12.016

67. Satismruti K., Senthil N., Vellaikumar S., Ranjani R. V., Raveendran M. (2013). Plant ionomics: a platform for identifying novel gene regulating plant mineral nutrition. *Am. J. Plant Sci.* 041309–1315. 10.4236/ajps.2013.47162

68. Shan Q., Wang Y., Li J., Zhang Y., Chen K., Liang Z., et al. (2013). Targeted genome modification of crop plants using a CRISPR-Cas system. *Nat. Biotechnol.* 31 686–688. 10.1038/nbt.2650

69. Shelden M. C., Roessner U. (2013). Advances in functional genomics for investigating salinity stress tolerance mechanisms in cereals. *Front. Plant Sci.* 4:123 10.3389/fpls.2013.00123

70. Shiqing G., Huijun X., Xianguo C., Ming C., Zhaoshi X., Liancheng L., et al. (2005). Improvement of wheat drought and salt tolerance by expression of a stress-inducible transcription factorGmDREB of soybean

(*Glycine max*). *Chin. Sci. Bull.* 502714–2723. 10.1007/BF02899641

71. Sivamani E., Bahieldin A., Wraith J. M., Al-Niemi T., Dyer W. E., Ho T. H. D., et al. (2000). Improved biomass productivity and water use efficiency under water deficit conditions in transgenic wheat constitutively expressing the barley HVA1 gene. *Plant Sci.*155 1–9. 10.1016/S0168-9452(99)00247-2

72. Sunkar R., Girke T., Jain P. K., Zhu J.-K. (2005). Cloning and characterization of microRNAs from rice. *Plant Cell* 17 1397–1411. 10.1105/tpc.105.031682

73. Todaka D., Nakashima K., Shinozaki K., Yamaguchi-Shinozaki K. (2012). Toward understanding transcriptional regulatory networks in abiotic stress responses and tolerance in rice. *Rice* 56 10.1186/1939-8433-5-6

74. Wang J.-W., Yang F.-P., Chen X.-Q., Liang R.-Q., Zhang L.-Q., Geng D.-M., et al. (2006). Induced expression of DREB transcriptional factor and study on its physiological effects of drought tolerance in transgenic wheat. *Acta Genet. Sin.* 33 468–476. 10.1016/S0379-4172(06)60074-7

75. Wei B., Jing R., Wang C., Chen J., Mao X., Chang X., et al. (2009). Dreb1 genes in wheat (*Triticum aestivum* L.): development of functional markers and gene mapping based on SNPs. *Mol. Breed.* 23 13–22. 10.1007/s11032-008-9209-z

76. Xiao Y., Savchenko T., Baidoo E. E. K., Chehab W. E., Hayden D. M., Tolstikov V., et al. (2012). Retrograde signaling by the plastidial metabolite MEcPP regulates expression of nuclear stress-response genes. *Cell* 149 1525–1535. 10.1016/j.cell.2012.04.038

77. Yang S., Li X., Ma Y., Sun X., Yang Y., Yang Y. (2015). Proteome response of wild wheat relative *Kengyilia thoroldianato* drought stress. *Can. J. Plant Sci.* 95 237–249. 10.4141/cjps-2014-294

78. Yang S., Vanderbeld B., Wan J., Huang Y. (2010). Narrowing down the targets: towards successful genetic engineering of drought-tolerant crops. *Mol. Plant* 3 469–490. 10.1093/mp/ssq016

79. Yin F., Gao J., Liu M., Qin C., Zhang W., Yang A., et al. (2014).Genome-wide analysis of water-stress-responsive microRNA expression profile in tobacco roots. *Funct. Integr. Genomics* 14319–332. 10.1007/s10142-014-0365-4

80. Zhan X., Zhang Y.-H., Chen D.-F., Simonsen H. T. (2014). Metabolic engineering of the moss *Physcomitrella patens* to produce the sesquiterpenoids patchoulol and Î ± /Î2-santalene.*Front. Plant Sci.* 5:636 10.3389/fpls.2014.00636

81. Zhang H., Mao X., Zhang J., Chang X., Jing R. (2013). Single-nucleotide polymorphisms and association analysis of drought-resistance gene TaSnRK2.8 in common wheat. *Plant Physiol. Biochem.* 70 174–181. 10.1016/j.plaphy.2013.04.010

82. Zhang P., Dreisigacker S., Melchinger A. E., Reif J. C., Mujeeb Kazi A., Van Ginkel M., et al. (2005). Quantifying novel sequence variation and selective advantage in synthetic hexaploid wheats and their backcross-derived lines using SSR markers.*Mol. Breed.* 15 1–10. 10.1007/s11032-004-1167-5

83. Zhou L., Liu Y., Liu Z., Kong D., Duan M., Luo L. (2010).Genome-wide identification and analysis of drought-responsive microRNAs in *Oryza sativa*. *J. Exp. Bot.* 61 4157–4168. 10.1093/jxb/erq237

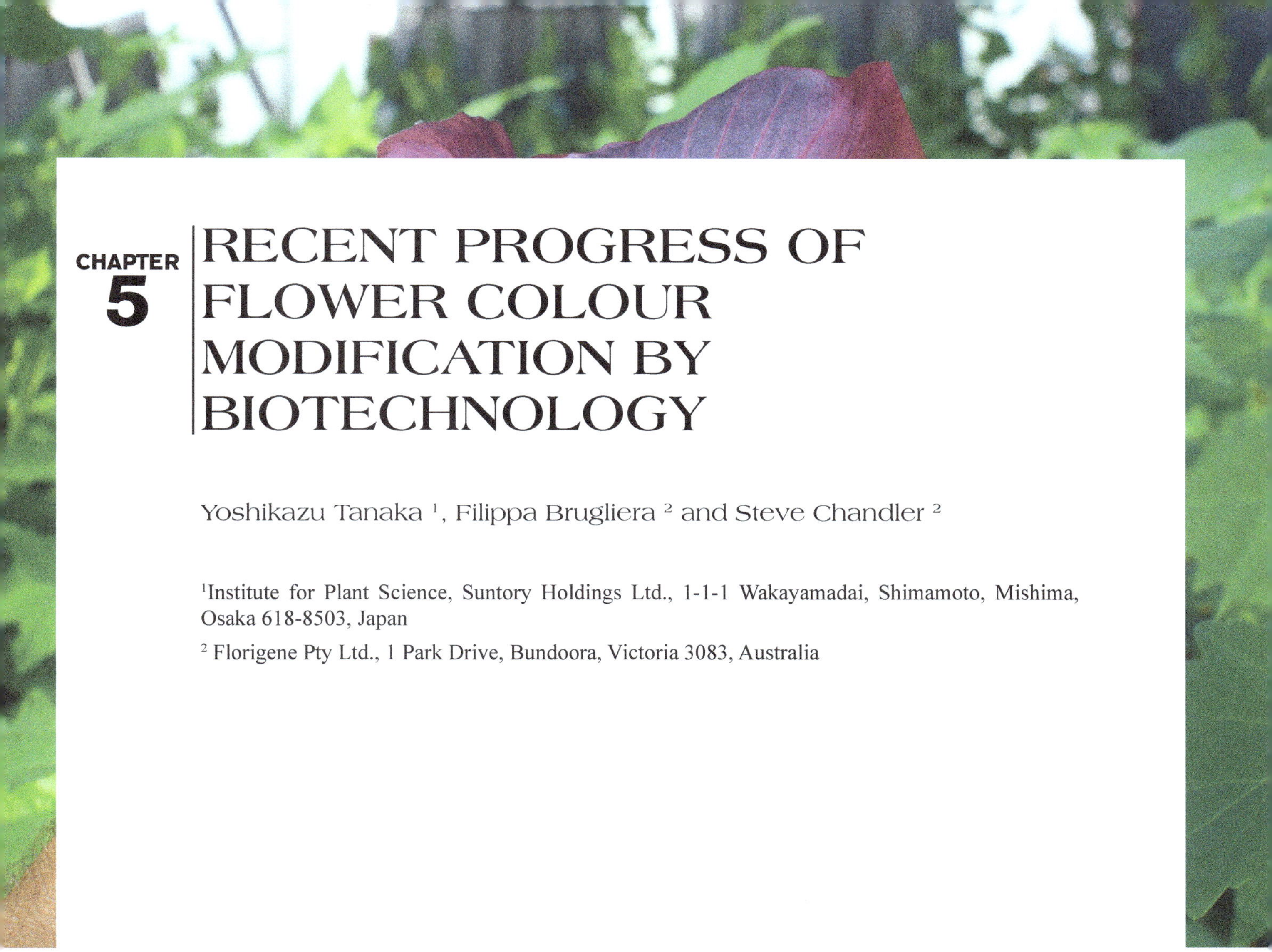

CHAPTER 5

RECENT PROGRESS OF FLOWER COLOUR MODIFICATION BY BIOTECHNOLOGY

Yoshikazu Tanaka [1], Filippa Brugliera [2] and Steve Chandler [2]

[1]Institute for Plant Science, Suntory Holdings Ltd., 1-1-1 Wakayamadai, Shimamoto, Mishima, Osaka 618-8503, Japan

[2] Florigene Pty Ltd., 1 Park Drive, Bundoora, Victoria 3083, Australia

ABSTRACT

Genetically-modified, colour-altered varieties of the important cut-flower crop carnation have now been commercially available for nearly ten years. In this review we describe the manipulation of the anthocyanin biosynthesis pathway that has lead to the development of these varieties and how similar manipulations have been successfully applied to both pot plants and another cut-flower species, the rose. From this experience it is clear that down- and up-regulation of the flavonoid and anthocyanin pathway is both possible and predictable. The major commercial benefit of the application of this technology has so far been the development of novel flower colours through the development of transgenic varieties that produce, uniquely for the target species, anthocyanins derived from delphinidin. These anthocyanins are ubiquitous in nature, and occur in both ornamental plants and common food plants. Through the extensive regulatory approval processes that must occur for the commercialization of genetically modified organisms, we have accumulated considerable experimental and trial data to show the accumulation of delphinidin based anthocyanins in the transgenic plants poses no environmental or health risk.

INTRODUCTION

Flavonoids, carotenoids and betalains are major floral pigments [1]. Between these three groups, the flavonoids contribute most to the range and type of coloured pigments in plants. As genetic engineering techniques have primarily targeted the flavonoid pathway as a means to alter flower colour, we will concentrate only on flavonoid pathway modification in this review. Flavonoids consist of more than 10 classes of compounds. Anthocyanins confer orange, red, magenta, violet and blue colours. Aurones and chalcones are yellow pigments while flavones and flavonols are colourless or very pale yellow, at least to the human eye.

The flavonoid biosynthetic pathway leading to floral pigment accumulation have been well characterized and the genes encoding relevant enzymes and transcriptional factors have been isolated [1,2]. Many studies of flower colour modification by genetic engineering have been performed as outlined in a recent review [3]. In this review, we will describe recent progress towards a further understanding of the contribution of flavonoid biosynthesis and cell physiology in the regulation of flower colour and discuss some examples of flower colour modification in important flower species.

Citation: Yoshikazu Tanaka, Filippa Brugliera, and Steve Chandler, Recent Progress of Flower Colour Modification by Biotechnology, doi: 10.3390/ijms10125350.

RECENT PROGRESS; ANTHOCYANIN BIOSYNTHESIS AND FLOWER COLOUR

Anthocyanin Structure and Colour

The final flower colour in those species in which colour is primarily derived from anthocyanins is determined by a combination of various factors. As well as anthocyanin structure, type and concentration, co-existing compounds (co-pigments), metal ion type and concentration, pH of vacuoles, anthocyanin localization and shapes of surface cells all contribute to final flower colour [4]. Through combinations of these factors, plant species have evolved flower colours to attract insect pollinators.

Though hundreds of anthocyanins have been reported [5], they are primarily based upon six common anthocyanidins (chromophores of anthocyanins); pelargonidin, cyanidin, peonidin, delphinidin, petunidin and malvidin. In terms of biosynthesis, since peonidin is derived from cyanidin and petunidin and malvidin are both derived from delphinidin, there are only three major anthocyanidins; pelargonidin, cyanidin and delphinidin (Scheme 1). Blue flowers tend to have delphinidin and its derivatives and intense red flowers tend to have pelargonidin as the anthocyanidin base. An increase in the number of hydroxyl groups on the B-ring (Scheme 1) imparts a bluer colour to the anthocyanins derived from the anthocyanidin, while methylation of the 3' or 5'-hydroxyl group (Scheme 1) results in a slight reddening.

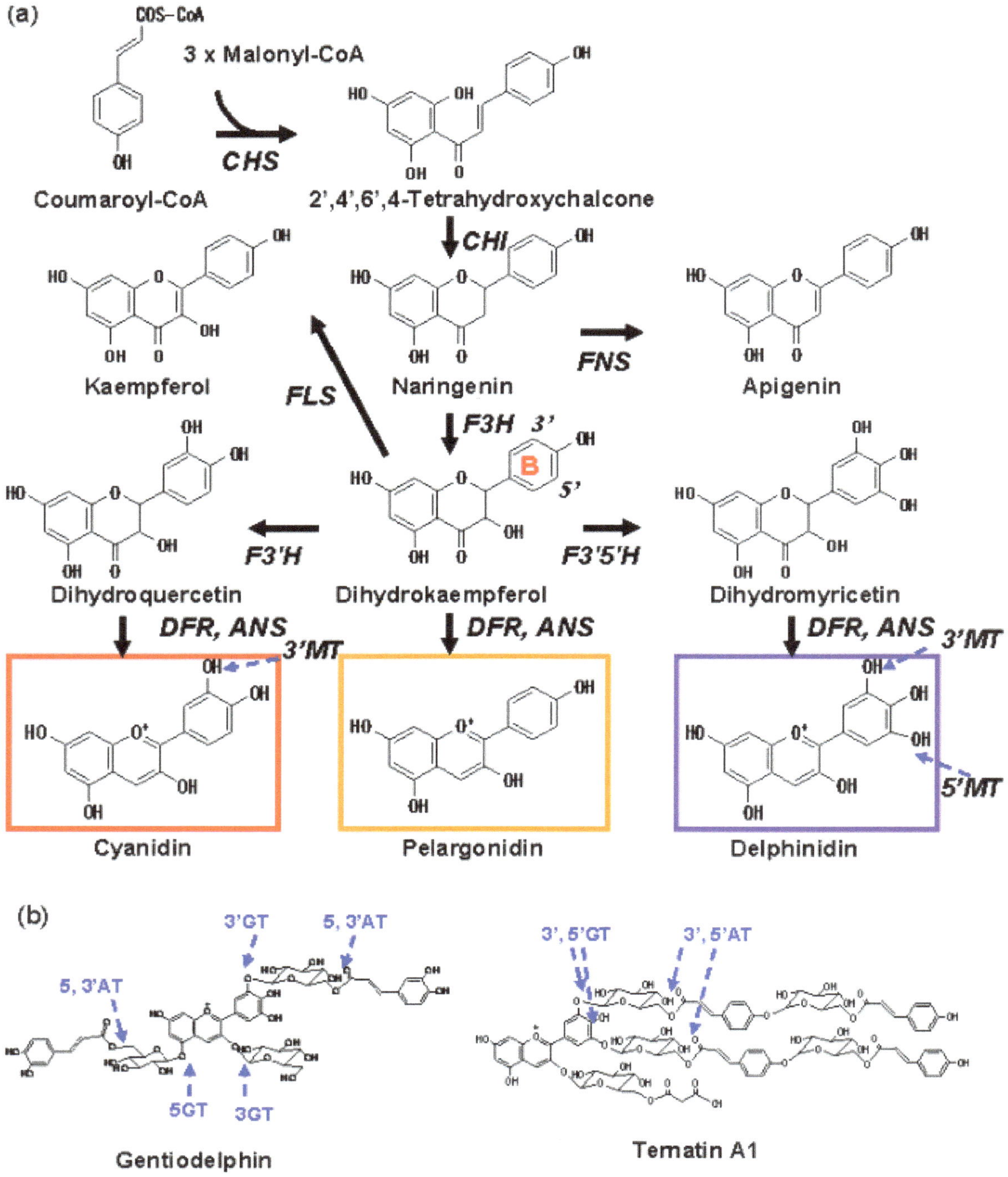

Scheme 1: (a) The biosynthetic pathway leading to the biosynthesis of the anthocyanidin. Anthocyanidin is further modified with glycosyl, acyl, or methyl groups catalysed by glycosyltransferase, acyltransferase and methyltransferase as described in the text. The .

Anthocyanin Modification

Anthocyanidins are modified with glycosyl or acyl moieties in a species specific manner by specific glycosyltransferases and acyltransferases. Aromatic acylation of anthocyanins shifts their colour toward blue and increases their stability.

Anthocyanins containing plural aromatic acyl groups are called polyacylated anthocyanins and exhibit blue colour as a result of intra-molecular stacking [4,6]. A well studied example is a gentian anthocyanin, gentiodelphin, in which the caffeoyl moiety at the 3' position of gentiodelphin has been shown to contribute to intra-molecular stacking, resulting in the stable blue colour of the flowers [7]. In terms of biosynthesis, acyl groups are transferred to glycosyl moieties of anthocyanins and thus glycosylation precedes acylation.

Glycosylation of Anthocyanins

The anthocyanin biosynthetic pathway leading to anthocyanidin 3-glucosides is well conserved and many plant species produce anthocyanidin 3,5-diglucosides and their respective derivatives. In general, 3-glucosylation occurs prior to 5-glucosylation and the two glucosylations are separately catalyzed, by UDP-glucosyl dependent anthocyanidin 3-glucosyltransferase (3GT) and anthocyanin 5-glucosyltransferase (5GT), respectively. Many 3GT and 5GT genes have been isolated and they form two distinct clusters within the family 1 glycosyltransferase [1]. Rose petals predominantly contain cyanidin or pelargonidin 3,5-diglucoside and a relatively small amount of cyanidin or pelargonidin 3-glucoside. Ogata *et al.* reported that *Rosa hybrida* petals have a unique UDP-glucose dependent anthocyanidin 5,3-glucosyltransferase and have isolated the corresponding cDNA clone [8]. Unusually, the enzyme catalyzes 5-glucosylation of anthocyanidins followed by 3-glucosylation of anthocyanidin 5-glucoside. In addition rose petals also contain an anthocyanidin 3-glucosyltransferase belonging to the 3GT cluster of genes [9].

Interestingly, butterfly pea (*Clitoria ternatea*) contains an anthocyanin 3′,5′-glucosyltransferase (3′,5′-GT [10], Scheme 1b) gene whose amino acid sequence is closely related to 3GT [11]. This indicates the 3′,5′-GT gene possibly evolved from the 3GT gene in butterfly pea. On the other hand the gentian anthocyanin 3′-glucosyltranferase (3′GT) belongs to a different GT clade (3′,7GT clade) from anthocyanidin 3GT [12] suggesting anthocyanin 3′-GT activity independently acquired in butterfly pea and gentian.

Acylation of Anthocyanins

The gene encoding an enzyme that catalyzes the transfer of aromatic acyl groups to anthocyanins was first isolated from gentian, *i.e.*, hydroxycinnamoyl CoA: anthocyanin 5-glucoside hydroxycinnamoyl CoA transferase (5AT) [13]. Extensive biochemical analysis *in vitro*revealed that the same enzyme catalyzes aromatic acylation of glucosyl moieties at both the 5 and 3' positions [14]. A hydroxycinnamoyl CoA: anthocyanin 3-glucoside hydroxycinnamoyl CoA transferase gene has now been isolated from perilla (*Perilla frutescens*) [15]. Several aliphatic acyltransferase genes have been also isolated as summarized [16]. All belong to the BAHD (benzylalcohol *O*-acetyltransferase (BEAT), anthocyanin *O*-hydroxycinnamoyltransferase (AHCT), anthranilate *N*-hydroxycinnamoyl/benzoyltransferase (HCBT) and deacetylvindoline 4-*O*-acetyltransferase (DAT)) type acyl CoA dependent acyltransferase group [17].

Recently, a number of acylglucose dependent anthocyanin acyltransferase genes have been isolated as shown below. All have serine carboxypeptidase-like sequences. *Arabidopsis* sinapoylglucose acyltransferase is required for the synthesis of the sinapoylated anthocyanins such as cyanidin 3-(6-(4-glucosyl-*p*-coumaroyl)-2-(2-sinapoyl-xylosyl)glucosyl)-5-(6-malonyl-glucoside) [18] while transfer of *p*-coumaroyl and malonyl moieties is catalyzed by BAHD type acyl CoA dependent [16]. Butterfly pea produces blue polyacylated anthocyanins such as ternatin A1 (Scheme 1b) and its derivatives and it has been shown that losing acyl groups resulted in mauve colour rather than blue [6,19]. Noda *et al.* purified a butterfly pea aromatic acyltransferase that catalyzed the transfer of an aromatic acyl group from acylglucose to both the 3' and 5'-glucosides of anthocyanins and isolated the corresponding cDNA clone [20]. Carnation petals generally accumulate macrocyclic anthocyanins 3,5-di-*O*-glucoside-(6″,6"'-malyl diester) of cyanidin or pelargonidin. A malylglucose dependent anthocyanin malyltransferase has been partially purified from carnation petals that have malylated anthocyanins [21] and the corresponding cDNA clone has been also isolated [22]. These studies indicate that plants have acquired enzymes exhibiting suitable substrate specificity through gene duplication and neo-functionalization of various pre-existing genes.

Methylation of Anthocyanins

Anthocyanin methylation adds structural and colour diversity to anthocyanins. The cDNA clones corresponding to *S*-adenosyl methionine dependent anthocyanin methyltransferase (MT) have been isolated from petunia and torenia flowers [23] and from grape berries [24]. Interestingly these genes belong to the cation dependent type II MT group, whereas flavone or flavonol MTs usually belong to type I MT.

Metal Ions

The role of metal ions in blue flower colours have been clarified in recent years and has been summarized in reviews [4,25]. Shiono *et al.*determined the X-ray chrystallographic structure of protocyanin (*Centaurea cyanus* (corn flower) anthocyanin flavone complex) to show that protocyanin consists of six molecule of cyanidin 3-(6-succinylglucoside)-5-glucoside, six molecules of apigenin 7-glucuronide-4′-(6-malonylglucoside), one Fe^{3+}, one Mg^{2+} and two Ca^{2+} ions. The X-ray crystallographic structure of commelinin from*Commelina communis* (Asiatic dayflower) has been also determined. *In vitro* reconstituted commelinin consists of six molecules of delphinidin 3-(6-*p*-coumaroylglucoside)-5-(6-malonylglucoside), six molecules of flavone glucoside (6-*C*-glucosylgenkwanin 4'-glucoside), and two Mg^{2+} ions [26], while native commelinin consists of six molecules of anthocyanins, six molecules of flavone glucoside, and four Mg^{2+} ions [27]. The two newly identified Mg atoms are proposed to stabilize the superstructure of commelinin.

Extensive chemical studies of hydrangea petal-like sepals revealed the presence of delphinidin-based anthocyanins, 5-*O*-acylquinic acid and Al^{3+} ions. A higher vacuolar pH (around 4.0 rather than 3.0) is necessary for the blue flower colours [4,28]. The lower sections of the petals of some tulip cultivars are blue while the remainder of the petal is purple. Shoji *et al.* revealed that the cells in the blue region contain about 9.5 mM of Fe^{3+} (25 times higher than purple coloured cells) while there are no significant differences in anthocyanin type (delphinidin 3-rutinoside) or concentration, flavonol type (quercetin glucosides) or concentration or vacuolar pH. The amount of Fe^{3+} ions therefore appears to play a critical role in development of the blue colour [29]. Further investigation has recently revealed that a homologue of *Arabidopsis VIT1* (vacuolar iron transporter) is highly expressed in the blue part of the tulip petals [30].

Regulation of Vacuolar pH

Anthocyanins are a red colour under low pH environments and a blue colour under neutral or alkaline pH environments. Since anthocyanins localize in vacuole, vacuolar pH is important for anthocyanin colour. Vacuolar pH is generally regulated by vacuolar ATPase and pyrophosphase in plant cells. Additionally morning glory and petunia have been recently shown to have unique pumps to carry H^+ to or from the vacuoles as shown below.

The petal colour of the Japanese morning glory (*Ipomoea nil*) changes from purple to blue during flower opening. Mutants, in which this colour change does not occur, are called *purple*. The vacuolar pH of purple and blue cells of petals *I. tricolor* were shown to be pH 6.6 and 7.7, respectively, while there was no changes in anthocyanin type or concentration in these cells [31]. *Purple* was shown to encode a protein that was homologous to a sodium proton antiporter and was highly expressed before flower opening [32]. The antiporter was later shown to exchange K^+ with H^+ rather than Na^+ with H^+ and this exchange contributes to flower colour change to blue and petal opening [33].

In petunia, mutation of any of the *PH1* to *PH7* loci results in flowers with a blue color rather than a purple colour and increased petal pH homogenates. This is the result of inhibition of vacuolar acidification.*PH4* encodes a *Myb* type transcription factor and the role of this factor in the regulation of vacuolar pH and anthocyanin biosynthesis has been discussed in a recent review [34]. *PH5* has been shown to encode a P_{3A}-H^+-ATPase localized on the vacuolar membrane while other members of the P_{3A}-H^+-ATPase family are believed to be plasma-membrane-localized proton pumps [35].

B-RING HYDROXYLATION

F3′H AND F3′5′H; KEY PLAYERS OF FLOWER COLOUR MODIFICATION

As described above, more and more potentially useful molecular tools to engineer blue or red flower colours are becoming available. However, only modification of B-ring hydroxylation has been practically demonstrated in changing flower colour. B-ring hydroxylation of flavonoids is catalyzed by the two cytochrome P450 type mono-oxygenases, flavonoid 3′-hdroxylase (F3′H) and flavonoid 3′,5′-hydroxylase (F3′5′H). (Figure 1). In cytochrome P450 nomenclature (http://drnelson.utmem.edu/CytochromeP450.html), they are designated as CYP75B and CYP75A, respectively. The genes encoding these two enzymes were first isolated from petunia, followed by many plant species. Phylogenetic analysis suggests that the genes encoding F3′H and F3′5′H diverted before speciation of flowering plants [36]. However, some chrysanthemum family species (aster, osteospermum, cineraria) which produce delphinidin based anthocyanins contain a CYP75B type F3′5′H and not the CYP75A type, indicating the chrysanthemum family once lost the F3′5′H gene during their evolution. F3′5′H function was reacquired by gene duplication and neo-functionalization of the CYP75B-type gene [37]. It has been also shown that only a few amino acid substitutions in the C-terminal region could functionally change F3′H to F3′5′H [38] illustrating the relatively high plasticity of the flavonoid biosynthetic pathway.

Figure 1: Commercialized transgenic carnation (a) and rose (b) cultivars producing delphinidin and having blue hue that hybridization breeding has not achieved. The transgenic carnation and rose also contain selectable marker genes; for herbicide resistance in .

Natural Mutations of F3`H and F3`5`H

The effect of B-ring hydroxylation on flower colour was clearly shown in mutants of these genes, as described below. The locus *magenta* (*mg*) in *I. nil* results in production of magenta flowers containing pelargonidin derivatives while the wild-type produces blue flowers containing peonidin derivatives, indicating that *Mg* regulates flavonoid 3′-hydroxylation. Molecular analysis of *mg* revealed that nucleotide substitution generated a stop codon, disrupting F3′H function. In *I. purplea*, a similar colour change was caused by a transposon insertion in the F3′H gene [39]. Species of *Ipomoea* from the *Mina* section, such as *I. quamoclit*, exhibit red flowers producing pelargonidin derived anthocyanins, which are pollinated by hummingbirds. However, blue flowered *Ipomoea* species are pollinated by bees. F3′H gene expression is dramatically reduced in *I. quamoclit* (however, the coding sequence has a F3′H catalytic activity *in vitro*) and its DFR has lost the ability to catalyze dihydroquercetin, a precursor of cyanidin [40].

Blue gentian accumulates delphinidin derivatives, whilst pink gentian (*Gentiana scabra*) cultivars which were bred from two independent mutants accumulate cyanidin based anthocyanins. In the pink varieties,*F3'5'H* transcripts were interrupted due to the insertion of transposable elements into the F3′5′H gene [41].

FLOWER COLOUR MODIFICATION USING GENETIC ENGINEERING

Lessons from Model Plant Experiments

Petunia, tobacco and torenia have been model species for the study of flower colour modification via genetic engineering, largely because they are easy to transform. Petunia does not accumulate pelargonidin based anthocyanins because its DFR does not catalyze dihydrokaempferol [42] (Scheme 1a). The first successful experiment aimed at manipulation of anthocyanin metabolism in plants was the production of pelargonidin based anthocyanins achieved by expression of a maize *DFR* in a petunia variety that was deficient in F3′5′H and F3′H activities [43].

Since this first experiment many kinds of transgenic plants with altered flower colours have been reported. Transgenic plants with white flowers have been obtained for many plant species by down regulating anthocyanin biosynthetic genes, as summarized in a recent review [3]. In torenia, transcription of double strand RNA (RNAi) has been shown to be a more effective method to down-regulate a target gene than antisense or sense suppression [44]. In this case however, the white colour was not always stable. This was especially notable when the plants were grown in an outdoor field trial in Australian summer conditions (unpublished results of Florigene Pty. Ltd.). In some cases therefore, alternative methods to knock out the function of a gene is more desirable. Direct genome modification utilizing zinc finger nucleases [45,46] may pave the way to achieve irreversible gene disruption, and so stably altered phenotypes.

In order to redirect flavonoid biosynthesis to an extent necessary to achieve a desirable flower colour change, it is not only the over-expression of the gene of a key enzyme on a new pathway that is necessary but also selection of a host that has the proper genetic background. This selection is to minimize competition of endogenous pathways with the introduced enzyme or to allow down-regulation a competing pathway. It is also important to choose a host cultivar with good commercial characteristics. In the next sections we will refer to actual examples of colour modification of flower colour, focusing on the commercially important cut flowers, carnation and rose.

Develpoment of the Moon® Series of Carnation

Over-expression of a petunia F3′5′H gene under the control of a constitutive promoter in a pelargonidin producing carnation variety produces petals in which delphinidin derivatives contribute to about 70% of total anthocyanins [47]. However, there was only a slight colour change toward blue. In order to increase the content of delphinidin derived anthocyanins it was necessary to avoid competition for substrates between two key endogenous enzymes of the anthocyanin pathway (DFR and F3′H) and the introduced enzyme (F3′5′H). To achieve this white carnation cultivars were selected that were specifically deficient in the DFR gene [48]. Expression of petunia F3′5′H (under the control of a promoter region from the snapdragon CHS gene) and petunia DFR (under the control of a constitutive promoter) genes in one such DFR mutant resulted in exclusive accumulation of delphinidin derivatives and significant colour change toward blue (FLORIGENE®Moondust™, Figure 1; the first genetically modified floricultural crop to be sold in the world). Expression of a pansy F3′5′H gene (under the control of a promoter region from the snapdragon CHS gene) and a petunia *DFR-A* gene (under the control of its own promoter and terminator regions) resulted in transgenic plants which also exclusively accumulated delphinidin but at a higher concentration. These flowers had a dark violet colour (FLORIGENE®Moonshadow™, Figure 1).

The same gene combinations were subsequently introduced into a white standard-type carnation (also a mutant in DFR). In these transgenic plants different amounts of delphinidin based pigments accumulated in the petals, depending on the transgenic events. Four events that exhibited stable colour were selected and have now been sold for more than eight years FLORIGENE®Moonvista™, FLORIGENE®Moonshade™, FLORIGENE®Moonlite™ and FLORIGENE®Moonaqua™ (Figure 1, [49]).

Petunia has a cytochrome b_5 that specifically transfers electrons to F3′5′H by which petunia can efficiently synthesize 3′,5′-hydroxylated flavonoids [50]. Expression of a petunia F3′5′H (*Hf1*) gene along with a petunia cytochrome b_5 gene in a carnation cultivar producing cyanidin derivatives resulted in efficient production of delphinidin based anthocyanins and subsequent colour change petals [51,52]. In this case, it was not necessary to use a carnation line mutant in DFR to achieve exclusive accumulation of delphinidin. The result indicated that enhancement of electron transfer system to F3′5′H sufficiently out competed the endogenous F3′H (and DFR) activities.

Regulatory Issues and Production of the Moon® Carnation

The transgenic carnations described above are grown in Ecuador, Colombia and Australia, and there is now a ten year history of safe use. Production is now consolidated in South America, from where there are good air-freight distribution systems to Europe, Japan and North America. In the course of obtaining the various regulatory approvals necessary for global trading in the transgenic carnation, it has become accepted by regulators that accumulation of delphinidin-based anthocyanins in transgenic carnation plants poses no environmental or health risk [53–55]. In this section we briefly describe the basis for this conclusion.

The Biology of the Parent Organism

Carnations are double-flowered cultivars of hybrids involving two or more *Dianthus* species, one of which is likely to be *Dianthus caryophyllus* [56–58]. The centre of biodiversity for *Dianthus* is southern Europe and the greatest range of *Dianthus* species are found in the south eastern European countries. *D. caryophyllus* is only found wild in coastal Mediterranean areas. Despite decades of cultivation, and plantings in parks and gardens, carnation has not become a weed, or escaped from cultivation, anywhere in the world. We have carried out detailed surveys in and around flower production areas in Colombia, including the location at which the transgenic carnation is currently grown and composted. No carnation plants have been found outside of cultivation [59]. The cultivated carnation has no capacity to escape from cultivation as the crop possesses no vegetative propagation mechanisms and there are no opportunities for seed-set.

Carnation is not a toxic plant and there are very limited reports of allergenic effects relating to the handling of carnation flowers [60,61]. As carnation is insect pollinated in the wild and does not produce air-borne pollen, respiratory allergenicity has not been reported.

The Modified Phenotype

Delphinidin based anthocyanins are ubiquitous in both ornamental plants and common food plants. Homologues of the anthocyanin biosynthesis genes in transgenic carnation are found in all plants producing delphinidin-related anthocyanins, including raw foods such as blueberries, black currants, grapes, eggplants, red currants, cherries and cranberries.

Dietary surveys estimate that the daily consumption of delphinidin is ~2.5–25 mg per person [62] and higher in some individual diets because of the high concentrations of certain foods. For example, in Finland the estimated intake is approximately 80 mg per day [63]. The levels of delphinidin in the transgenic carnation flowers are within the range of these and other common, widely cultivated plants [54,64,65].

Potential for Gene Flow from Transgenic Carnation to Native Plants

An assessment of the potential for gene flow is central to the risk assessment associated with the release of a genetically modified plant [66,67]. In the case of cultivated carnation, there are no realistic avenues for gene flow to occur.

The dispersal of genes from either cultivated plants or imported cut flowers of carnation could theoretically be via vegetative spread leading to the formation of wild populations, formation of seed by a recipient plant (fertilized by pollen dispersed from the transgenic plants) or the formation and dispersal of seed on the transgenic plants. The possibility of any of these events occurring is negligible.

Vegetative spread. Carnation does not spread vegetatively as it does not produce organs such as stolons, root-borne shoots, tubers, bulbs or runners. Cuttings must be struck in optimized conditions and roots do not form on discarded tissues. According to the numerous floras we have reviewed, including refuse sites [68] cultivated carnation has never been found growing wild, and efforts to locate wild carnation in production areas have been unsuccessful [59]. In experiments carried out in Colombia we have found that carnation deliberately planted in uncultivated habitats is unable to survive in competition with other plants.

Formation of seed by a recipient plant, fertilized by transgenic pollen. Introgression of genes from cultivated crops to wild relatives is well known [69] and in the case of transgenic carnation, which carries a herbicide resistance gene, it has been necessary to evaluate the potential of introgression to wild*Dianthus* populations. This evaluation has concluded that there is no realistic potential for introgression because;

In nature, *Dianthus* is insect-pollinated, and pollen is not spread by wind. Hybridization of *Dianthus* is only carried out by the *Lepidoptera* (butterflies, moths) which have proboscis long enough (up to 2.5 cm) to reach the nectaries, which are located right at the base of the flower in all *Dianthus* species [70]. Cultivated carnations are double-flowered and the many petals present on the flowers make it impossible for insects to access any pollen. No hybrid between carnation and any other *Dianthus*species has ever been recorded in the wild.

Some cultivars of the cultivated carnation produce little or no pollen [71,72].

Assuming, theoretically, an insect were to access a carnation flower in production, or in a vase in someone's home, the probability of subsequently fertilizing a recipient, flowering, *Dianthus* plant is limited. The*Dianthus* genus is not native to the production areas in South America, and though *Dianthus* species related to carnation are occasionally found they are rare—no*Dianthus* species were found in surveys of flower growing areas [59].

Formation and dispersal of seed. The process of seed development takes at least 5 weeks in carnation [73,74] and as cut flowers only survive 3 to 4 weeks at most in the hands of the consumer, seed development on a cut flower imported for consumption is not possible. In the production areas, the same argument can be raised, because the purpose of production is of course flower harvest, which in its own right eliminates the possibility of seed formation in the crop.

Comparative Trials

Comparative trials between the transgenic lines and the parent variety used for transformation are important, in order to ensure there have been no unexpected effects of the transformation process, induced by somaclonal variation. In the case of the transgenic carnation such trials have been carried out in Japan, Australia, Colombia and Europe. The purpose of the trials has been to select lines in which there have been no significant morphological changes in comparison to the controls, particularly in characters which would have a negative impact on the commercial quality of the flower. In carnation, the primary characteristics determining quality are stem length, vase life, number of petals per flower, flower size and, for spray carnations, number of flower buds per stem.

Transgenic Rose

Delphinidin Production and Flower Colour Modification

Roses are the most important cut flower commercially and have played a major role in human culture and artistry from ancient times. Wild roses usually have pink (these flowers contain anthocyanins derived from cyanidin) or white flowers. The cultivated roses we know today (*Rosa hybrida*) are the product of extensive inter-specific hybridization utilising wild species including yellow-flowered (producing carotenoids) and orange-flowered (producing pelargonidin related anthocyanins) species. However, despite the huge range of flower colours that have been bred, rose lack any varieties in the bluish range of flower colour because the genus *Rosa* does not have the biochemical pathway leading to delphinidin-based anthocyanins.

Expression of a F3′5′H gene from petunia, gentian or butterfly pea in rose resulted in no or little delphinidin accumulation in the petals of transgenic plants, even though these genes were shown to be functional in petunia, carnation or yeast. In contrast, expression of pansy (*Viola spp*) F3′5′H genes in rose resulted in a significant amount of delphinidin derived anthocyanins accumulating in petals of the transgenic plants [75]. We do not know the reason why only the pansy F3’5’H genes worked well in rose.

Rose cultivars that have higher vacuolar pH, large amount of flavonols (co-pigments) and weak or no F3’H activity were selected in order to enhance the blue hue of the transgenic petals and to achieve high content of delphinidin. Expression of pansy F3′5′H genes in such cultivars resulted in transgenic lines in which 95% of the anthocyanidins was delphinidin. The colour of the flowers in these lines were of a significantly bluer hue than any conventionally bred cultivar [76]. In recent experiments we have expressed a torenia anthocyanin 5AT gene in addition to the pansy F3′5′H. However, only a fraction of anthocyanins were modified and the acylation did not significantly affect the flower colour. Nevertheless, two selected lines from these experiments (WKS82-130-4-1 and WKS82-130-9-1) were selected for commercial release and processed through the regulatory procedures that we will refer to later (Section 4.4.2).

In order to out compete the endogenous rose DFR for substrates it was targeted for down regulation via RNAi mediated silencing .The resulting transgenic plants, in which pansy F3′5′H and Dutch iris DFR genes were also expressed, produced flowers that accumulated delphinidin-based anthocyanins exclusively with a concomitant colour change towards blue. In breeding experiments, pollen donated from a transgenic line was used in hybridization with a deep red, cyanidin producing rose cultivar. Progeny carrying the transgenes also exclusively contained delphinidin-derived anthocyanins however the amount of delphinidin varied amongst the progeny [76]. The flower colours obtained amongst the progeny was mostly magenta hues, which was primarily attributed to the lower vacuolar pH of the parent. Though, the results show that the complete conversion of the pathway (*i.e.*, exclusive accumulation of delphinidin-related anthocyanins) is heritable irrespective of genetic background, this specific modification of the pathway also conferred growth retardation and so was not suitable for commercialization.

As we mentioned above, blue flower colour is usually the additive result of various factors, such as anthocyanin modification, co-pigment concentration, vacuolar pH and metal ion type and concentration. Targeting these factors in addition to delphinidin production in transgenic carnation and rose is expected to yield a bluer flower colour.

Regulatory Procedures of Transgenic Rose

Regulatory procedures and practice vary to a large degree between countries and territories. In Japan, the transgenic rose is not considered a food or feed and so is largely examined in the context of possible effects on biodiversity in Japan, consistent with the application of the principles of under the Cartagena Biosafety protocol (http://www.bch.biodic.go.jp/english/law.html).

Japan has a number of wild *Rosa* species, such as *Rosa multiflora*, *R. lutea* and *R. rugosa.* These have also been utilized to breed *R. hybrida*. Japanese regulators were therefore concerned that the pollen from transgenic rose could hybridize with populations of wild roses. It was therefore considered desirable to show that the wild roses do not hybridize with cultivated roses under natural conditions. Though this is predictably unlikely because *R. hybrida* is generally tetraploid whilst wild roses are diploids, a genetic marker is available to allow for detection of hybrids in the field. This marker gene in cultivated rose, which regulates the ever-blooming character derived from *R. chinesis*[77] was not found in thousands of seeds of wild roses found growing close to cultivated *R. hybrida* in Japan, showing that wild roses and *R. hybrida* are unlikely to hybridize under natural condition.

WKS82-130-4-1 and WKS82-130-9-1 plants did not show significant difference from their host in terms of growth character, morphology, phytoalexin assays and so on except flower colour in glasshouse and field trials. Using *in situ* hybridization techniques and crossing experiments, it was possible to show that only L1 cells (the epidermal cells where anthocyanins are localized) of the transgenic rose plants contained the transgenes and L2 cells (producing pollen) did not. The lack of the transgene in the pollen cells is obvious barrier to potential gene flow to wild roses. On the basis of these results, general release permissions were issued for the two rose lines in Japan. In late 2009, one of these lines has been commercially produced and launched in Japan as the Suntory blue rose APPLAUSE™.

Genetic Modification for Red Flowers

Accumulation of pelargonidin based anthocyanins in flowers confers an intense red or orange colour, which is very desirable commercially. Attempts have therefore been made to obtain such colours by genetic engineering. In petunia (a variety accumulating cyanidin based anthocyanins), down regulation of the F3′H gene in conjunction with over-expression of rose *DFR* yielded transgenic petunia whose flowers accumulated pelargonidin based anthocyanins [78]. Down regulation of F3′5′H and F3′H genes and over-expression of pelargonium DFR in blue torenia (which normally produces delphinidin based anthocyanins) resulted in transgenics with pink flowers accumulating anthocyanins based on pelargonidin (manuscripts in preparation). Down regulation of F3′H and flavonol synthase and overexpression of gerbera DFR in tobacco yielded red coloured flowers accumulating pelargonidin based anthocyanins [79]. However down regulation of F3′H was not complete as the transgenic tobacco still produced significant amount of cyanidin-based pigments. In osteospermum a two step transformation method has been used in which down regulation of F3′5′H was followed by overexpression of gerbera DFR. This particular transformation resulted in pelargonidin accumulation in flowers and a flower colour change toward red. In contrast, overexpression of gerbera or strawberry DFR alone did not yield flowers accumulating pelargonidin [80]. As pelargonidin was accumulated even though F3'H activity was not down regulated, F3′H activity is weak in osteospermum.

Some cut flower species such as gentian and iris do not accumulate pelargonidin derived pigments and thus lack flowers with an intense red colour. Red gentian cultivars (Showtime Diva, Showtime Starlet, Showtime Spotlight) have been made by interspecies hybridization [81]. Genetic modification methods should be an alternative solution to achieving these new flower colours.

Though down-regulation of the F3′H and F3′5′H genes in combination with over expression of a correctly identified DFR gene should generate gentian flowers producing pelargonidin based pigments. Sophisticated optimization of gene expression or pathway engineering may be necessary to accumulate a large amount of pelargonidin, and so an intense red flower colour. Furthermore, *Agrobacterium*-mediated transformation of gentian occurs at a low efficiency and the promoter of the 35S cauliflower mosaic virus, a commonly used constitutive promoter, is prone silencing by methylation in gentian [82]. Such difficulties often occur in non model species but may be overcome. For example, a pale flower coloured gentian, which produced less acylated anthocyanins was made from a blue one by down regulation of F3′5′H and 5,3′AT. In this case, the *rol*C promoter was utilized [83].

CONCLUDING REMARKS

We have described some successful examples of colour modification. Recently genes regulating flower colour such as pH and metal transport genes have been identified and these may prove useful molecular tools for modification of flower colour in the future. Of course, efficient transformation systems have to be developed for target cultivars of interest and transgene expression also needs to be optimized. A considerable amount of try and error is inevitably required to achieve a satisfactory final result.

In our experience, genetic modification of an ornamental plant can be a successful venture, from both a scientific and a commercial perspective. The "Moon" series of colour modified carnations have been sold now for nearly a decade and millions of flowers have entered the traditional growing, distribution and retail chains for cut-flowers. There is no reason to think transgenic rose flowers will not be equally as readily accepted in the marketplace and to date there has been no negative response from consumers to the genetically modified rose. The transgenic varieties have proven to be genetically very stable during mass scale vegetative propagation and there have been no unexpected effects on either the environment or on the health of those handling the flowers. The beneficial health effects of anthocyanins are in fact well recognized; an understanding that is leading to the development of genetically modified, functional foods, with enhanced anthocyanin profiles [84].

What then is the major obstacle to dozens of other genetically modified ornamental products entering the marketplace, given that the technology for trait manipulation is in place, and that many important horticultural crops can be transformed? The answer lies in the barriers that the regulatory systems in many parts of the world place on the freedom to trial and develop GM varieties of ornamentals. As is the case for cut-flowers, ornamental products are an internationally traded commodity and until there is an internationally agreed, equitable process, for regulating genetically modified plant products it will continue to prove very difficult to release ornamental products, due to the costs and expertise required for commercial development. To ease this burden the regulatory requirements for non-food varieties, such as ornamentals, should be reduced.

ACKNOWLEDGMENTS

The authors thank past and present colleagues in Suntory and Florigene for their contribution to the effort to develop and commercialize transgenic plants. Due to the constraints of the length of this article we are unfortunately unable to cite all relevant research and references.

REFERENCES

1. Tanaka Y, Sasaki N, Ohmiya A. Plant pigments for coloration: Anthocyanins, betalains and carotenoids. Plant J. 2008;54:733–749.
2. Grotewold E. The genetics and biochemistry of floral pigments.Annu. Rev. Plant Biol. 2006;57:761–780.
3. Tanaka Y, Ohmiya A. Seeing is believing: Engineering anthocyanin and carotenoid biosynthetic pathways. Curr. Opin. Biotechnol.2008;19:190–197.
4. Yoshida K, Mori M, Kondo T. Blue flower color development by anthocyanins: From chemical structure to cell physiology. Nat. Prod. Rep. 2009;26:884–915.
5. Veitch NC, Grayer RJ. Flavonoids and their glycosides, including anthocyanins. Nat. Prod. Rep. 2008;25:555–611.
6. Honda T, Saito N. Recent progress in the chemistry of polyacylated anthocyanins as flower color pigment. Heterocycles. 2002;56:633–692.
7. Yoshida K, Toyama Y, Kameda K, Kondo T. Contribution of each caffeoyl residue of the pigment molecules of gentiodelphin to blue color development. Phytochemistry. 2000;54:85–92.
8. Ogata J, Kanno Y, Itoh Y, Tsugawa H, Suzuki M. Plant biochemistry: Anthocyanin biosynthesis in roses. Nature. 2005;435:757–758.
9. Tanaka Y, Katsumoto Y, Fukui Y, Mizutani M, Togami J. Flavonoid glycosyltransferase and utilization

thereof. Patent Publication Number WO/2007/094521, August232007

10. Kogawa K, Kato N, Kazuma K, Noda N, Suzuki M. Purification and characterization of UDP-glucose: Anthocyanin 3′,5′-*O*-glucosyltransferase from *Clitoria Ternatea.* Planta. 2007;227:1501–1509.
11. Noda N, Kato N, Kogawa K, Kazuma K, Suzuki M. Cloning and characterization of the gene encoding anthocyanin 3′,5′-*O*-glucosyltransferase involved in ternatin biosynthesis from blue petals of butterfly pea (*Clitoria Ternatea*) Plant Cell Physiol. 2004;45:s132.
12. Fukuchi-Mizutani M, Okuhara H, Fukui Y, Nakao M, Katsumoto Y, Yonekura-Sakakibra K, Kusumi T, Hase T, Tanaka Y. Biochemical and molecular characterization of a novel UDP-glucose: Anthocyanin 3′-*O*-glucosyltransferase, a key enzyme for blue anthocyanin biosynthesis, from gentian. Plant J. 2003;132:1652–1663.
13. Fujiwara H, Tanaka Y, Yonekura-Sakakibara K, Fukuchi-Mizutani M, Nakao M, Fukui Y, Yamaguchi M, Ashikari T, Kusumi T. cDNA cloning, gene expression and subcellular localization of anthocyanin 5-aromatic acyltransferase from *Gentiana Triflora.* Plant J. 1998;16:421–431.
14. Tanaka Y, Katsumoto Y, Mizutani M, Fukui Y, Togami J. Method of stabilizing, and bluing, of anthocyanin pigment using gene coding for transferase of aromatic acyl to 3′-position of anthocyanin. Patent Publication Number WO/2006/046780, May42006
15. Yonekura-Sakakibara K, Tanaka Y, Fukuchi-Mizutani M, Fujiwara H, Fukui Y, Ashikari T, Murakami Y, Yamaguchi M, Kusumi T. Molecular and biochemical characterization of hydroxycinnamoyl-CoA: Anthocyanin 3-*O*-glucoside-6″-*O*-hydroxycinnamoyltransferase from *Perilla Frutescens.* Plant Cell Physiol. 2000;132:1652–1663.
16. Luo J, Nishiyama Y, Fuell C, Taguchi G, Elliott K, Hill L, Tanaka Y, Kitayama M, Yamazaki M, Bailey P, Parr A, Michael AJ, Saito K, Martin C. Convergent evolution in the BAHD family of acyl transferases: Identification and characterization of anthocyanin acyl transferases from *Arabidopsis Thaliana.* Plant J. 2007;50:678–695.
17. D'Auria JC. Acyltransferases in plants: A good time to be BAHD.Curr. Opin. Plant. Biol. 2006;9:331–340.
18. Fraser CM, Thompson MG, Shirley AM, Ralph J, Schoenherr JA, Sinlapadech T, Hall MC, Chapple C. Related arabidopsis serine carboxypeptidase-like sinapoylglucose acyltransferases display distinct but overlapping substrate specificities. Plant Physiol. 2007;144:19860–19999.
19. Kazuma K, Noda N, Suzuki M. Flavonoid composition related to petal color in different lines of *Clitoria Ternatea.* Phytochemistry.2003;64:1133–1139.
20. Noda N, Kazuma K, Sasaki T, Furukawa K, Suzumi M. (Aomori prefecture) novel aromatic acyl group transferase gene. Patent Publication Number WO/2007/046148, April262007
21. Abe Y, Tera M, Sasaki N, Okamura M, Umemoto N, Momose M, Kawahara N, Kamakura H, Goda Y, Nagasawa K, Ozeki Y. Detection of 1-*O*-malylglucose: Pelargonidin 3-*O*-glucose-6-*O*-malyltransferase activity in carnation (*Dianthus Caryophyllus*) Biochem. Biophys. Res. Commun. 2008;373:473–477.
22. Umemoto N, Momose M. Novel protein having acyltransferase activity and gene encoding the same. Patent Publication Number WO/2008/082006, July102008
23. Brugliera F, Linda D, Koes R, Tanaka Y. Genetic sequences having methyltransferase activity and uses therefor. Patent Publication Number WO/03/062428, July312003
24. Hugueney P, Provenzano S, Verries C, Ferrandino A, Meudec E, Batelli G, Merdinoglu D, Cheynier V, Schubert A, Ageorges A. A novel cation-dependent O-methyltransferase involved in anthocyanin methylation in grapevine. Plant Physiol. 2009;150:2057–2070.
25. Takeda K. Blue metal complex pigments involved in blue flower color. Proc. Jpn. Acad.: Ser. B. 2006;82:142–154.
26. Kondo T, Yoshida K, Nakagawa A, Kawai T, Tamura H, Goto T. Structural basis of blue-colour development in flower petals: Structure determination of commelinin from Commelina communis. Nature.1992;358:515–518.
27. Shiono M, Matsugaki N, Takeda K. Structure of commelinin, a blue complex pigment from the blue flowers of *Commelina Communis.*Proc. Jpn. Acad.: Ser. B. 2008;84:452–456.
28. Yoshida K, Toyama-Kato Y, Kameda K, Kondo T. Sepal color variation of Hydrangea macrophylla and vacuolar pH measured with a proton-selective microelectrode. Plant Cell Physiol. 2003;44:262–268.
29. Shoji K, Miki N, Nakajima N, Monomoi K, Kato C, Yoshida K. Perianth bottom-specific blue color development in Tulip cv. Murasakizuisho requires ferrc ions. Plant Cell Physiol. 2007;48:243–251.
30. Momonoi K, Yoshida K, Mano S, Takahashi H, Nakamori C, Shoji K, Nitta A, Nishimura M. A vacuolar iron transporter in tulip, TgVit1, is responsible for blue coloration in petal cells through iron accumulation. Plant J. 2009;59:437–447.
31. Yoshida K, Kondo T, Okazaki Y, Katou K. Cause of blue petal colour. Nature. 1995;373:291.
32. Fukada-Tanaka S, Inagaki Y, Yamaguchi T, Saito N, Iida S. Colouring-enhancing protein in blue petals. Nature. 2000;407:581.
33. Yoshida K, Miki N, Nomonoi K, Kawachi M, Katou K, Okazaki Y, Uozumi N, Maeshima M, Kondo T. Synchrony between flower opening and petal-color change from red to blue in morning glory, Ipomoea tricolor cv. Heavenly blue. Proc. Jpn. Acad. Ser. B. 2009;85:187–197.
34. Quattrocchio F, Verweij W, Kroon A, Spelt C, Mol J, Koes R. *PH4*of petunia is an *R2R3 MYB* protein that activates vacuolar acidification through interactions with *basic-helix-loop-helix* transcription factors of the anthocyanin pathway. Plant Cell. 2006;18:1274–1291.
35. Verweij W, Spelt C, Di Sansebastiano G-P, Vermeer J, Reale L, Ferranti F, Koes R, Quattrocchio F. An H^+ P-ATPase on the tonoplast determines vacuolar pH and flower colour. Nature Cell Biol.2008;10:1456–1462.

36. Tanaka Y. Flower colour and cytochromes P450. Phyochem. Rev.2006;5:283–291.

37. Seitz C, Eder C, Deiml B, Kellner S, Martens S, Forkmann G. Cloning, functional identification and sequence analysis of flavonoid 3′-hydroxylase and flavonoid 3,5′-hydroxylase cDNA reveals independent evolution of flavonoid 3′,5′-hydroxylase in the Asteraceae family. Plant Mol. Biol. 2006;61:365–381.

38. Seitz C, Ameres S, Forkmann G. Identification of the molecular basis for the functional difference between flavonoid 3′-hydroxylase and flavonoid 3′,5′-hydroxylase. FEBS Lett. 2007;581:3429–3434.

39. Hoshino A, Morita Y, Choi JD, Saito N, Toki K, Tanaka Y, Iida S. Spontaneous mutations of the flavonoid 3′-hydroxylase gene conferring reddish flowers in the three morning glory species. Plant Cell Physiol.2003;44:990–1001.

40. Zufall RA, Rausher MD. Genetic changes associated with floral adaptation restrict future evolutionary potential. Nature. 2004;428:847–850.

41. Nakatsuka T, Nishihara M, Mishiba K, Hirano H, Yamamura S. Two different transposable elements inserted in flavonoid 3′,5′-hydroxylase gene contribute to pink flower coloration in *Gentiana Scabra*. Mol. Genet. Genomics. 2006;275:231–241.

42. Forkmann G, Ruhau B. Distinct substrate specificity of dihydroflavonol 4-reductase from flowers of *Petunia Hybrida*. Z. Naturforsch. 1987;42C:1146–1148.

43. Meyer P, Heidemann I, Forkmann G, Saedler H. A new petunia flower colour generated by transformation of a mutant with a maize gene. Nature. 1987;330:677–678.

44. Nakamura N, Fukuchi-Mizutani M, Suzuki K, Miyazaki K, Tanaka Y. RNAi suppression of the anthocyanidin synthase gene in *Torenia Hybrida* yields white flowers with higher frequency and better stability than antisense and sense suppression. Plant Biotechnol. 2006;23:13–17.

45. Shukla VK, Doyon Y, Miller JC, DeKelver RC, Moehle EA, Worden SE, Mitchell JC, Arnold NL, Gopalan S, Meng X, Choi VM, Rock JM, Wu YY, Katibah GE, Zhifang G, McCaskill D, Simpson MA, Blakeslee B, Greenwalt SA, Butler HJ, Hinkley SJ, Zhang L, Rebar EJ, Gregory PD, Urnov FD. Precise genome modification in the crop species *Zea Mays* using zinc-finger nucleases. Nature. 2009;459:437–441.

46. Townsend JA, Wright DA, Winfrey RJ, Fu F, Maeder ML, Joung JK, Voytas DF. High-frequency modification of plant genes using engineered zinc-finger nucleases. Nature. 2009;459:442–445.

47. Holton TA, Tanaka Y. Transgenic flowering plants. Patent Publication Number WO/94/28140, 1994

48. Holton TA. Transgenic plants exhibiting altered flower color & methods for producing same. Patent Publication Number WO/96/36716, 1996

49. Mol J, Cornish E, Mason J, Koes R. Novel coloured flowers. Curr. Opin. Biotechnol. 1999;10:198–201.

50. de Vetten N, Ter Horst J, Vab Schaik H-P, de Boer A, Mol J, Koes R. A cytochrome b5 is required for full activity of flavonoid 3′,5′-hydroxylase, a cytochrome P450 involved in the formation of blue flower colors. Proc. Natl. Acad. Sci. USA. 1999;96:778–783.

51. Brugliera F, Tull D, Holton TA, Karan M, Treloar N, Simpson K, Skurczynska J, Mason JG. Introduction of a cytochrome b5 enhances the activity of flavonoid 3′,5′-hydroxylase (a cytochrome P450) in transgenic carnation. Int Plant Mol Biol Rep (Suppl) 2000;18:S6–S8.

52. Tanaka Y, Katsumoto Y, Brugliera F, John M. Genetic engineering in floriculture. Plant Cell Tissue Organ Cult. 2005;80:1–24.

53. EFSA Opinion of the Scientific Panel on Genetically Modified Organisms on a request from the Commission related to the notification (Reference C/NL/04/02) for the placing on the market of the genetically modified carnation Moonlite 123.2.38 with a modified colour, for import of cut flowers for ornamental use, under Part C of Directive 2001/18/EC from Florigene. EFSA J. 2006;362:1–19.

54. EFSA Opinion of the Scientific Panel on Genetically Modified Organisms on a request from the Commission related to the notification (Reference C/NL/06/01) for the placing on the market of the genetically modified carnation Moonaqua 123.8.12 with a modified colour, for import of cut flowers for ornamental use, under Part C of Directive 2001/18/EC from Florigene. EFSA J. 2008;662:1–25.

55. Florigene Pty Ltd Inclusion of dealings with GM carnation lines, modified for flower colour, on the GMO register, 2006.http://www.health.gov.au/internet/ogtr/publishing.nsf/Content/register-3/$FILE/reg001risk.pdf (accessed December 10, 2009).

56. Hughes S. Carnations and Pinks, the Complete Guide. The Crowood Press; Marlborough, UK: 1991.

57. Allwood MC. Carnations, Pinks and All Dianthus. 4th ed. Allwood Bros Ltd; Sussex, UK: 1954.

58. Holley WD, Baker R. Carnation Production: Including the History, Breeding, Culture and Marketing of Carnations. WC Brown Co. Ltd; Dubuque, IA, USA: 1963.

59. Chandler SF, Artunduaga IRS, Jaramillo LEL, Osorio CAP. The production of transgenic carnation and rose in Colombia; A survey for wild populations and related species. 10th International Symposium on the Biosafety of Genetically Modified Organisms; Wellington, New Zealand. November 16–21, 2008.

60. Sanchez-Guerrero IM, Escudero AI, Bartolom B, Palacios R. Occupational allergy to carnation. J. Allergy Clin. Immuno.1999;104:181–185.

61. Vidal C, Polo F. Occupational allergy caused by *Dianthus Caryophyllus*, *Gypsophila panicularа* and *Lilium longiflorum*. Allergy.1998;53:995–998.

62. Wu X, Beecher GR, Holden JM, Haytowitz DB, Gebhardt SE, Prior RL. Concentrations of anthocyanins in common foods in the United States and estimation of normal consumption. J. Agri. Food Chem.2006;54:4069–4075.

63. Heinonen M. Review: Antioxidant activity and antimicrobial effect of berry phenolics—A finnish perspective. Mol. Nutr. Food Res.2007;51:684–691.

64. Catalano G, Fossen T, Andersen ØM. Petunidin 3-*O*-β-rhamnopyranoside-5-*O*-β-glucopyranoside and other anthocyanins from flowers of *Vicia Villosa*. J. Agri. Food Chem. 1998;49:4568–4570.

65. Ando T, Saito N, Tatsuzawau F, Kakefudau T, Yamakageu EK, Ohtani T, Koshi-ishiu M, MatsusakeuH Y, Kokubunu I, Watanabeu H, Tsukamotou T, Uedau Y, Hashimoto G, Marchesid E, Asakura K, Harar R, Sekir F. Floral anthocyanins in wild taxa of *Petunia*. Biochem. Syst. Ecol. 1999;27:623–650.

66. Chandler SF, Dunfell J. Gene flow, risk-assessment and the environmental release of transgenic plants. CRC Crit. Rev. Plant Sci.2008;27:25–49.

67. Wilkinson MJ, Ford CS. Estimating the potential for ecological harm from gene flow to crop wild relatives. Collect. Biosafety Rev.2007;3:42–63.

68. Kim KD. Plant invasion and management in turf-dominated waste landfills in South Korea. Land Degrad. Develop. 2002;13:257–267.

69. Ellstrand NC. Current knowledge of gene flow in plants: Implications for transgene flow. Phil. Trans. R. Soc. Lond. B.2003;358:1163–1170.

70. Hickey M, King C. 100 Families of Flowering Plants. Cambridge University Press; Cambridge, UK: 1981.

71. Mehlquist GAL, Ober D, Sagawa Y. Somatic mutations in the carnation, *Dianthus Caryophyllus*. Proc. Natl. Acad. Sci. USA.1954;40:432–436.

72. Kho YO, Baer J. The effect of temperature on pollen production in carnations. Euphytica. 1973;22:467–470.

73. Sparnaaij LD, Beeger GW. The improvement of seed production for breeding purposes in the glasshouse carnation (*Dianthus Caryophyllus*) Euphytica. 1973;22:274–278.

74. Gatt MK, Hammett KRW, Markham KR, Murray BG. Yellow pinks: Interspecific hybridization between Dianthus plumarius and related species with yellow flowers. Scientia Hort. 1998;77:207–218.

75. Brugliera F, Tanaka Y, Mason J. Flavonoid 3′,5′-hydroxylase gene sequences and uses therefor. Patent Publication Number US7612257, March112004

76. Katsumoto Y, Mizutani M, Fukui Y, Brugliera F, Holton T, Karan M, Nakamura N, Yonekura-Sakakibara K, Togami J, Pigeaire A, Tao G-Q, Nehra N, Lu C-Y, Dyson B, Tsuda S, Ashikari T, Kusumi T, Mason J, Tanaka Y. Engineering of the rose flavonoid biosynthetic pathway successfully generated blue-hued flowers accumulating delphinidin.Plant Cell Physiol. 2007;48:1589–1600.

77. Iwata H. Protein and gene participating in perpetual blooming of angiosperm. Patent Publication Number WO/2004/070036, 2004

78. Tsuda S, Fukui Y, Nakamura N, Katsumoto Y, Yonekura-Sakakibara K, Fukuchi-Mizutani M, Ohira K, Ueyama Y, Ohkawa H, Holton TA, Kusumi T, Tanaka Y. Flower color modification of *Petunia Hybrida*commercial varieties by metabolic engineering. Plant Biotechnol.2004;21:377–386.

79. Nakatsuka T, Abe Y, Kakizaki Y, Yamamura S, Nishihara M. Production of red-flowered plants by genetic engineering of multiple flavonoid biosynthetic genes. Plant Cell Rep. 2007;26:1951–1959.

80. Seitz C, Vitten M, Steinbach P, Hartl S, Hirsche J, Rhthje W, Treutter D, Forkmann G. Redirection of anthocyanin synthesis in*Osteospermum Hybrida* by a two-enzyme manipulation strategy.Phytochemistry. 2007;68:824–833.

81. Eason JR, Debenham M, McLachlan A, Morgan E. Novel red-flowered *Gentiana*: An emerging export cut flower crop from New Zealand. Acta Hort. 2007;755:259–266.

82. Mishiba K, Nishihara M, Nakatsuka T, Abe Y, Hirano H, Yokoi T, Kikuchi A, Yamamura S. Consistent transcriptional silencing of 35S-driven transgenes in gentian. Plant J. 2005;44:541–556.

83. Nakatsuka T, Mishiba KI, Kubota A, Abe Y, Yamamura S, Nakamura N, Tanaka Y, Nishihara M. Genetic engineering of novel flower colour by suppression of anthocyanin modification genes in gentian. J Plant Physiol. 2009 in press.

84. Butelli E, Titta L, Giorgio M, Mock H-P, Matros A, Peterek S, Schijlen EGWM, Hall R, Bovy AG, Luo J, Martin C. Enrichment of tomato fruit with health-promoting anthocyanins by expression of select transcription factors. Nature Biotechnol. 2008;26:1301–1308.

CHAPTER 6

GENETIC ENGINEERING: A PROMISING TOOL TO ENGENDER PHYSIOLOGICAL, BIOCHEMICAL, AND MOLECULAR STRESS RESILIENCE IN GREEN MICROALGAE

Freddy Guihéneuf[1], Asif Khan[2] and Lam-Son P. Tran[3,4]

[1]Botany and Plant Science, School of Natural Sciences, Ryan Institute, National University of Ireland Galway, Galway, Ireland

[2]Research Group Germline Biology, Centre for Organismal Studies (COS), Heidelberg University, Heidelberg, Germany

[3]Plant Abiotic Stress Research Group & Faculty of Applied Sciences, Ton Duc Thang University, Ho Chi Minh City, Vietnam

[4]Signaling Pathway Research Unit, RIKEN Center for Sustainable Resource Science, Tsurumi, Japan

ABSTRACT

As we march into the 21st century, the prevailing scenario of depleting energy resources, global warming and ever increasing issues of human health and food security will quadruple. In this context, genetic and metabolic engineering of green microalgae complete the quest toward a continuum of environmentally clean fuel and food production. Evolutionarily related, but unlike land plants, microalgae need nominal land or water, and are best described as unicellular autotrophs using light energy to fix atmospheric carbon dioxide (CO2) into algal biomass, mitigating fossil CO2 pollution in the process. Remarkably, a feature innate to most microalgae is synthesis and accumulation of lipids (60–65% of dry weight), carbohydrates and secondary metabolites like pigments and vitamins, especially when grown under abiotic stress conditions. Particularly fruitful, such an application of abiotic stress factors such as nitrogen starvation, salinity, heat shock, etc., can be used in a biorefinery concept for production of multiple valuable products. The focus of this mini-review underlies metabolic reorientation practices and tolerance mechanisms as applied to green microalgae under specific stress stimuli for a sustainable pollution-free future. Moreover, we entail current progress on genetic engineering as a promising tool to grasp adaptive processes for improving strains with potential biotechnological interests.

INTRODUCTION

Due to their taxonomic and biochemical diversity, microalgae symbolize an unconventional source of molecules (Stengel et al., 2011). For example, the inherent nature of microalgae to accumulate lipids, carbohydrates, and secondary metabolites under abiotic stress can be applied in a biorefinery concept for generation of value-added compounds (Markou and Nerantzis, 2013). Nonetheless, transgenic microalgae impact diverse businesses such as energy, human and animal nutraceuticals, pharmaceuticals, health, beauty, and exquisite chemicals (Rasala et al., 2014). As naturalized for bacteria, yeasts, and plants, genetic engineering, therefore, constitutes promising strategy for studying abiotic stress responses in microalgae through novel phenotypes and specific traits. Conversely, microalgae as model photosynthetic organisms represent an ideal experimental system to study major plant processes, as mimicking algal strategies to sense, respond and cope with abiotic stress can radically improve growth and metabolite productivity in plants. Over the years, significant advances in genetic manipulation of green microalgae have been

Citation: Freddy Guihéneuf, Asif Khan, and Lam-Son P. Tran, Genetic Engineering: A Promising Tool to Engender Physiological, Biochemical, and Molecular Stress Resilience in Green Microalgae, doi: 10.3389/fpls.2016.00400.

achieved (Mayfield and Golden, 2015), as exemplified through production of omega-3 fatty acids, carotenoids, biofuels, as well as improved photosynthetic growth (Gimpel et al., 2015). Despite tremendous advances in sequenced genome, in particular for chlorophytes and heterokonts (i.e., diatoms), major obstacles feature need of additional well-established transformation vectors, potential inability to engineer and specifically localize transgenic proteins to sub-cellular locations and robust expression of multiple nuclear-encoded transgenes (Gong et al., 2011; Scranton et al., 2015). In this minireview, we discuss recent examples highlighting futuristic use of genetic engineering to design strains with potential biotechnological interests. Moreover, we surmise major abiotic stress resilience strategies envisaged in unicellular eukaryotic green microalgae.

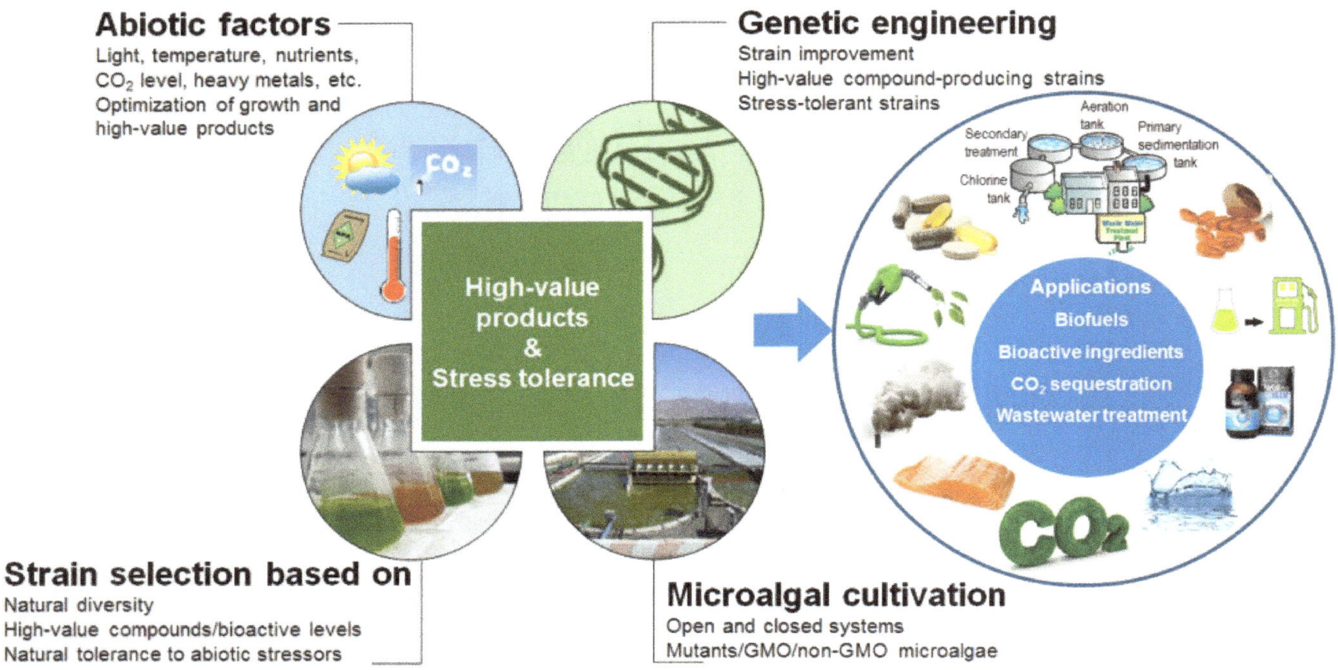

Figure 1: Illustrated scheme for improving microalgal strains for potential biotechnological interests via genetic engineering. GMO, genetically modified organism.

CURRENT PROGRESS ON GENETIC AND METABOLIC ENGINEERING OF GREEN MICROALGAE

What We Know and What Is Available?

The pursuit of stress-tolerant strains has seen an unprecedented surge of algal transgenesis for recombinant proteins, enhanced photosynthesis, and key metabolite pathways, as well as production of precious nutraceuticals, pharmaceuticals, and biofuels (Gangl et al., 2015). Coupled to this, advanced cultivation of novel species with genome-wide omics (genomics, transcriptomics, lipidomics, and metabolomics) has facilitated successful transformation of microalgae (e.g., *Chlamydomonas reinhardtii*, *Phaeodactylum tricornutum*, *Thalassiosira pseudonana*, *Porphyridium sp.*, *Nannochloropsis gaditana*, *Ostreococcus tauri*, *Haematococcus pluvialis*, etc.). Previously, lack of genome and transcriptome data, as well as efficient tools for algal transformation prevented their genetic manipulation. Consequently, algal transformation systems had to be revisited, developed anew, including methods to introduce transgenic DNA, identification of suitable promoters, distinct selectable markers and expression vectors, as well as needs for codon usage optimization (Qin et al., 2012; Hlavova et al., 2015).

Despite gigantic successes being achieved, only a few algal species still show firm and stable expression of foreign proteins. Increasing evidence suggests intrinsic algal gene silencing mechanisms; albeit repressive histone H3 lysine methylation, DNA cytosine methylation, RNA interference and microRNA-mediated gene regulation for instable foreign transcript or protein degradation (Kim et al., 2015). Likewise, the knowledge of such transgene evading mechanisms will be useful in regulating efficient protein production in metabolically engineered microalgae. Additionally, the emergence of fast and reliable next-generation sequencing techniques have sped up genome sequencing programs and transcriptomic research (Morozova and Marra, 2008). As a result, it is now possible to directly correlate the ensuing transcriptomic, proteomic, and metabolomic profile with respect to pre-determined environmental stimuli, bridging the gap between microalgal genotype and observed phenotype (Dong and Chen, 2013). In future, a combination of transcriptomic, proteomic, and metabolomic studies will be crucial in demarcating gateways to metabolic flux for boosting algal strain-engineering strategies (Jones and Mayfield, 2012). In addition, a battery of bioinformatic softwares like Algal Functional Annotation Tool, GreenCut2, AlgaGEM, Bag2D, PredAlgo (de Oliveira Dal'Molin et al., 2011; Karpowicz et al., 2011; Lopez et al., 2011;Tardif et al., 2012; Yao et al., 2015), and *in silico* kinetic and metabolic reconstruction models[1] are now available for gene/protein characterization, gene metabolic alterations, and predicting proteins subcellular localizations at a whole genome level in various algal species. For example, an *in silico* metabolic reconstruction map allowed characterization of light-induced metabolic response for >1000 genes in*Chlamydomonas* (Chang et al., 2011). Likewise, the *Chlamydomonas*resource center[2] harbors a collection of ~300 plasmids and >2700 strains for sustaining research within diverse aspects of algal cell biology. Additionally, a genome-wide *Chlamydomonas* insertional mutant library[3] containing >20,000 mutants covering 10,301 genes (58% of the genome) and cataloged at *Chlamydomonas* resource center has been recently reported (Li et al., 2016).

Genetic Engineering: Progress and Future Prospects

With an eye toward the future, methods to engineer entire biochemical pathways or multigene traits via homologous/heterologous recombination and development of optimized cloning vectors are underway, as demonstrated by stable integration and subcellular localization of reporter proteins in the chloroplast and nuclear genomes of *C. reinhardtii* (Noor-Mohammadi et al., 2012; Rasala et al., 2014;Scranton et al., 2016). Such a strategy can be advantageous for foreign gene integration, omitting positional side effects and transcriptional/post-transcriptional gene silencing processes associated with nuclear expression (Kim et al., 2015). In addition, successful yet simple nuclear gene targeting systems have been described in*Chlamydomonas* (Zorin et al., 2009). Novel permutations of promoter and 5′UTR, like 16S rRNA promoter and *atpA* 5′UTR have also been illustrated for sufficient heterologous transgene expression (Rasala et al., 2011) and robust expression of transgenic proteins within the *C. reinhardtii* chloroplast (Tissot-Lecuelle et al., 2014). Examples of metabolic engineering in microalgae comprise codon optimized expression of genes *hemH* and *lba* in the chloroplast of *C. reinhardtii*for optimum bio-hydrogen generation (Wu et al., 2011). Recently, synthetic promotors were generated to drive robust nuclear gene expression in *Chlamydomonas* (Scranton et al., 2016).

In search of optimal methods for exogenous DNA delivery within rigid algal cell walls, a zinc oxide nanowire array microdevice system was recently elaborated for *C. reinhardtii* (Bae et al., 2015) with recorded efficiency 6.52 × 10^4- and 9.66 × 10^4-fold higher than traditional glass bead beating and electroporation methods. Besides, a biocontainment method for transplastomic microalgae was reported via codon reassignment in the chloroplast of *C. reinhardtii* (Young and Purton, 2015). Such a codon reassignment strategy can be potentially useful for solving imminent problems in chloroplast engineering of genes whose products are toxic to *Escherichia coli* cells. Similarly, a self-cloning-based (cloning DNA from a donor into a recipient, between which the natural exchange of DNA is possible) positive selection system for breeding and genetic transformation of *Pseudochoricystis ellipsoidea*was developed (Kasai et al., 2015). Such recombinant microalgae are considered natural (non-GMO) in Japan and are not governed under the Cartagena domestic law with regards to GMO biosafety (Lusser et al., 2012; USDA Japan Reports, 2014). Such algal molecular breeding techniques may form basis for future large-scale cultivation in outdoor open pond cultivation (Ho et al., 2014).

Finally, reports for high level targeted genome editing in *C. reinhardtii*via zinc finger nucleases (Sizova et al., 2013), transcription activator-like effectors (TALE; Gao et al., 2014), and lately clustered regularly interspersed short palindromic repeat (CRISPR) system are emerging (Jiang et al., 2014). CRISPR-associated protein 9 (CAS9) is an RNA-guided DNA nuclease successfully practiced for targeted mutations into eukaryotic genomes (Dominguez et al., 2016). Although stable transformants of *C. reinhardtii* expressing the CAS9 protein could not be recovered, in future, such studies would serve as gold standard for dedicated nuclear, chloroplast, and mitochondria genome-editing strategies.

MICROALGAL STRESS RESPONSE STRATEGIES

Abiotic Stress Tolerance in Green Microalgae

Owing to incredible metabolic plasticity of microalgae (Stengel et al., 2011), numerous studies have outlined ramifications of abiotic stress influencing algal biology and metabolism. High light, temperature, salinity, metals, CO_2levels, and nutrient depletion affect growth, photosynthesis, and biochemical composition in a species-specific manner (Perales-Vela et al., 2006; Solovchenko and Khozin-Goldberg, 2013).

One of the most detrimental abiotic stressor damaging algae is high light intensity, when absorption of light energy surpasses the capacity for light utilization in photosynthesis. Several mechanisms termed as non-photochemical quenching (NPQ) exist, to avoid this photodamage (Allorent et al., 2013; Mayank et al., 2015). NPQ consists of three components: qE (energy-dependent quenching), qI (photoinhibitory quenching), and qT (light-state transition; Eberhard et al., 2008;Depauw et al., 2012). The qE increases the thermal dissipation of excessive light energy and represents the major photo-protective process characterizing vascular plants, green microalgae and diatoms (Müller et al., 2001; Roháček et al., 2014). In model green alga *C. reinhardtii*, NPQ is stimulated by variations in pH in the thylakoid lumen, in response to high light intensity, triggering the xanthophyll cycle (XC) and activating the PSII LHC stress-related (LhcSR) proteins (Finazzi et al., 2006; Maruyama et al., 2014; Wobbe et al., 2016). The latter initiates an onset of NPQ by perceiving lumenal pH environment and in a concentration-dependent manner aggravates fluorescence quenching; the XC plays key role in boosting light harvesting and photoprotection (Goss and Jakob, 2010). Nevertheless, studies of their NPQ and XC activities have suggested species-specific differences in process of excessive energy dissipation between six green microalgae (Quaas et al., 2015). Of interest, microalgal species like *Dunalliela* spp. and *H. pluvialis* display the capacity to synthetize and accumulate tremendous amounts of secondary carotenoids (β-carotene and astaxanthin, respectively) under stress (Lamers et al., 2012; Scibilia et al., 2015). In photosynthetic organisms, response to high light involves enormous changes in expression of light-regulated genes and their regulatory components (Lemoine and Schoefs, 2010). Over the years, dedicated gene expression studies with microalgae exposed to different light regimes were undertaken, unveiling peculiar molecular actors associated with light stress responses (e.g., *Lhcx* gene group in diatoms, known to be closely related to *LhcSR* genes of *C. reinhardtii*;Maruyama et al., 2014).

Microalgae further picture a suite of adaptive mechanisms in response to changes in salinity, i.e., passive "osmometer" behavior. Such a profile largely depends on physical and chemical characteristics of the cell wall membrane, regulation of water fluxes, mechanisms to equilibrate intracellular ionic levels using ion-selective channels and carriers, and synthesis or degradation of low-molecular-weight organic solutes (Kirst, 1990). The halotolerant green alga, *Dunaliella salina*, has been described as a model alga for deducing response of plant cells to salt stress (Cowan et al., 1992). Employing a proteomic approach for salinity stress induced proteins in *Dunaliella*, Liska et al. (2004)decoded up-regulation of major Calvin cycle enzymes, starch mobilization enzymes and enzymes involved in redox energy generation; factors involved in protein production and degradation; and an analog of bacterial Na^{++}-redox transporters. The results are suggestive of *Dunaliella's* response to high salinity by enhancing photosynthetic CO_2 consumption and allocating carbon and energy budget for synthesis of glycerol, its osmo-protective component. On the same lines, salt-stress up-regulated the carotenoid ketolase (*BKT*) in*Chlorella zogingiensis* and enhanced the accumulation of canthaxanthin and astaxanthin. High salinity also stimulated the generation of reactive oxygen species, which in turn triggered the up-regulation of distinct carotenogenic genes and increased build-up of antioxidant carotenoids (Li et al., 2009).

Changes in temperature are known to strongly affect growth, photosynthetic activity and biochemical composition in microalgae (Geider, 1987; Renaud et al., 2002; Claquin et al., 2008). Indeed, higher temperatures often result in decline in protein production and rise in lipid and carbohydrates levels. An increase in polyunsaturated fatty acids (PUFAs) might be one of the ways microalgae adjust to low-temperature and preserve membrane fluidity (Jiang and Gao, 2004). Interestingly, few microalgae exhibit high temperature tolerance up to 40–42°C (Hanagata et al., 1992; Sakai et al., 1995; de-Bashan et al., 2008). Heat shock proteins (HSPs) are molecular chaperones, crucial during heat shock response and ensuing adaptive homeostases against environmental stresses. 17 small HSPs has been identified from complete genome sequences of five diverse algae: *C. reinhardtii*,*Cyanidioschyzon merolae*, *Ostreococcus lucimarinus*, *O. tauri*, and *T. pseudonana* (Waters and Rioflorido, 2007). However, sequence analysis reveals that the diversity and abundance of algal small HSPs does not fully correlate with acclimation to extreme temperatures. Recently, two small *HSP*s have been shown to be up-regulated and activated at a threshold temperature of ~36°C in *C. reinhardtii*(Kobayashi et al., 2014). These results stress the advantage of an exclusive temperature-sensing system in the evolution and resilience to elevated temperatures among microalgae.

Mechanisms of CO_2 Concentration and Heavy Metal Tolerance

Rise in atmospheric O_2 and decrease in oceanic, dissolved inorganic carbon during eco-genesis enabled photoautotrophic organisms to formulate sophisticated CO_2-concentrating mechanisms (CCMs). Such mechanisms enhance CO_2 uptake by increasing CO_2 concentration around ribulose-1,5-bisphosphate carboxylase/oxygenase (Rubisco), an enzyme of the chloroplast stroma that catalyzes the entry of CO_2 into the Calvin–Benson–Bassham cycle to maintain sufficient photosynthesis (Raven, 2010). To date, numerous published reports have described low-CO_2-acclimation/adaptation within some cyanobacteria and unicellular eukaryotes; but, information regarding high-CO_2 tolerance processes remains elusive. One CO_2 mitigation strategy consists of selecting and developing strains that stand rigorously high CO_2 levels to couple microalgal-mediated CO_2 fixation and production of biofuels and other secondary metabolites (Bhola et al., 2014). Since CO_2 tolerance varies greatly between species, microalgae have been arbitrarily divided into CO_2-sensitive (inhibited by <2–5% CO_2) and CO_2-tolerant (surviving up to 20% CO_2) groups (Miyachi et al., 2003). An example of such CO_2-tolerant species includes chlorophytes *Chlorella* sp. KR-1 and *Chlorococcum littorale*that demonstrate rapid growth at 40 and 60% CO_2, respectively (Iwasaki et al., 1996; Sung et al., 1999). Principal factors responsible for high CO_2 tolerance in such extremophile symbiotic species constitute state transitions in photosynthetic apparatus that augment ATP generation, enhanced activity of H^+-ATPase for pumping protons out of the cell, hasty shutdown of CCMs, adjustment of fatty acid composition and diversion of excess photosynthates to generation of energy-rich compounds, such as triacylglycerols (Solovchenko and Khozin-Goldberg, 2013).

Microalgae also possess molecular machinery that allows them to distinguish between non-essential and essential heavy metals (Perales-Vela et al., 2006). This ability makes microalgae better suited for treatment of heavy metal polluted water, employing algal-based biotechnologies for waste-water remediation (Suresh Kumar et al., 2015). Indeed, microalgae take up and store toxic metals from the environment (Perales-Vela et al., 2006). For instance, microalgae can preferentially synthesize peptides capable to bind heavy metals. These peptides, as part of organometallic complexes, are additionally partitioned inside vacuoles to aid pertinent control of the cytoplasmic concentration of metal ions, thus nullifying their potential toxic effect (Cobbett and Goldsbrough, 2002). *C. reinhardtii* has been described as a model photosynthetic eukaryotic for investigating heavy metal tolerance or homeostasis due to molecular and genetic tools available for this species (Hanikenne, 2003). For example, phytochelatins (PC), low-molecular weight peptides, containing sulfur rich cysteine amino acid have been elicited as major intracellular cadmium (Cd) chelators in*C. reinhardtii* (Hu et al., 2001). In *Dunaliella tertiolecta*, PC, inducibly synthesized by zinc treatment, can function both in detoxifying heavy metals and alleviation of oxidative stress (Tsuji et al., 2002). Thioredoxins (TRXs) are also known to accord heavy metal detoxification in *Chlamydomonas* as exemplified by two *TRX* genes being differentially stimulated by Cd and mercury (Lemaire et al., 1999). As glutathione (GSH)-heavy metal adducts serve as substrates for PC synthase, an earlier study centered on two strains of*Scenedesmus acutus* (wild type and Cr^{6+} tolerant) reported the constitutive cysteine concentration higher in the Cr^{6+}-tolerant strain (Torricelli et al., 2004). When the cells were subjected to Cd^{2+}, the tolerant strain had increased levels of reduced GSH and class III metallothioneins (MtIII) as compared to the wild-type strain. Finally, proline (Pro) has also been reported to play a dominant role in alleviating environmental stress including heavy metal stress in plants, microorganisms and microalgae (Siripornadulsil et al., 2002).

Clearly, abiotic stress has the ability to reorient a holistic and sometimes specific microalgal metabolism as an efficient means for increasing the production of selected compounds. Keeping this in mind we highlight in the next section recent advances in microalgal engineering which, when combined with sophisticated physiological and biophysical approaches, may enhance microalgal stress resilience under disparate conditions.

Examples of Stress-Resilience in Microalgae via Genetic Engineering

Despite most engineering strategies favoring increased production of high value compounds (e.g., antioxidant pigments and PUFAs) and biofuel-molecules (e.g., hydrogen and triacylgycerol), recent works have been driven to develop resilient strains with applications like CO_2sequestration and heavy metal biomitigation. For example, a moth bean δ1-pyrroline-5-carboxylate synthetase (*P5CS*) gene has been expressed in *C. reinhardtii*, with transformants displaying 80% higher free-Pro levels, rapid growth at toxic Cd concentrations, and tremendous binding at four-fold higher Cd concentration compared to wild-type cells (Siripornadulsil et al., 2002). The results propose role of free-Pro as an antioxidant in Cd-stressed cells with resulting higher GSH levels facilitating enhanced phytochelatin synthesis and sequestration of Cd. Expeditious transgenesis has been reported using such microalgae to enhance heavy metal sensitivity and binding specificity for contaminated wastewaters and sediments (Rajamani et al., 2007). Much interest has also been diverted toward engineering small and large subunits of Rubisco (rbcS and rbcL) as targets for increasing net CO_2 fixation (photosynthesis) and promoting CO_2sequestration through improvement of growth (Genkov et al., 2010;Whitney et al., 2011). In this regard, *Chlamydomonas* is a notable host for genetic manipulation, as changes can be made to both *rbcS* and *rbcL*genes. For instance, hybrid Rubiscos have been reported, by combining plant (*Arabidopsis*, spinach and sunflower) small (rbcS) subunits with algal large (rbcL) subunits via transformation of a *C. reinhardtii* mutant deficient in *rbcS* gene (Genkov et al., 2010). Although the hybrid enzymes show 3–11% increase in CO_2/O_2 affinity, they retain optimal V_{max} values and catalytically proficient Rubisco. The hybrid strains, however, lack chloroplast pyrenoids and display reduced photosynthesis. As asserted by Whitney et al. (2011), future research to engineer and test superior Rubiscos will considerably rely on algal model systems, particularly *Chlamydomonas*.

Table 1: Examples of green algal species aptly transformed to date together with methods applied and trait(s) enhanced for proficient metabolic or genetic engineering.

Algal species	Transfection method	Modification	Genes	Trait(s) engineered	Reference
Chlamydomonas reinhardtii	Glass beads	RNAi three target genes	*DGAT*	24 and 37% reduction in TAGs with two genes 34% increase in TAGs with one gene	Deng et al., 2012
	Glass beads	Nuclear overexpresion	*PSY*	2.6-fold increase in lutein	Couso et al., 2011
	-	Insertional mutagenesis	*CHT7*	Repressor of cellular quiescence under nitrogen-replete conditions	Tsai et al., 2014
	Biolistics	Chloroplast expression	*hemH* & *lba*	Increase in hydrogen production	Wu et al., 2011
	Electroporation	Nuclear expression	*P5CS* (moth bean)	80% higher free-Pro levels, rapid growth under Cd toxicity, four-fold more binding to Cd	Siripornadulsil et al., 2002
	Glass beads	RNAi *LHC* gene family	20 genes encoding LHCI, LHCII, CP26 and CP29 down-regulated	Antenna-size reduction Increased competence for cell cultivation under high light intensity	Mussgnug et al., 2007
Chlorella ellipsoidea	Electroporation	Nuclear overexpression	Lipogenesis Transcription factors	52% increase in total lipids	Zhang et al., 2014
Chorella minutissima	Electroporation	Nuclear overexpression	Five genes encoding TAG enzymes	Twofold increase in TAGs with all five genes No change with individual gene	Hsieh et al., 2012
Dunaliella salina	Electroporation	RNAi	*PDS*	72% reduction of mRNA	Sun et al., 2008
Haematococcus pluvialis	Biolistics	Nuclear overexpression	*PDS*	26% increase in astaxanthin	Steinbrenner and Sandmann, 2006
Nannochloropsis oceanica	Electroporation	Nuclear overexpression	*Δ12-des*	Nitrogen-starvation induced increase deposition of polyunsaturated fatty acids (PUFAs) in TAG	Kaye et al., 2015

CHT7, compromised hydrolysis of triacylglycerols 7; Δ12-des, Δ12-desaturase; DGAT, Acyl-CoA:diacylglycerol acyltransferase; hemH, ferrochelatase gene; lba, leghemoglobin gene; LHC, light-harvesting antenna complexes; P5CS, δ1-pyrroline-5-carboxylate synthetase; PDS, phytoene desaturase; PSY, phytoene synthase; TAG, triacylglycerol.

CONCLUSION

Genetic manipulation of microalgae can provide feasible respite by meeting the immediate and long-term demands for food and fuel production on a sustainable basis. A clear step toward the realization of this goal is the development of microalgae-microbial fuel cells (mMFCs) that channelize the solar energy to electrical energy via algal metabolic pathways (Lee et al., 2015). Besides, genetic engineering of microalgae would supplement land plants and industrial biotechnology for value-added products, thereby saving land and air pollution. Already much is achieved through reinforced tools, including rapid sequencing of algal genomes, control of environmental stress conditions, as well as mining of new genes for enhancing lipid and bioactive productions, reducing photo-inhibition, and manipulation of Rubisco- and Calvin-cycle enzymes for metabolic engineering. The field of algal genetic engineering will require renewed attention both at basic and applied levels to envision a prosperous future.

CONFLICT OF INTEREST STATEMENT

The authors declare that the research was conducted in the absence of any commercial or financial relationships that could be construed as a potential conflict of interest.

REFERENCES

1. Allorent G., Tokutsu R., Roach T., Peers G., Cardol P., Girard-Bascou J., et al. (2013). A dual strategy to cope with high light in*Chlamydomonas reinhardtii*. *Plant Cell* 25 545–557. 10.1105/tpc.112.108274
2. Bae S., Park S., Kim J., Choi J. S., Kim K. H., Kwon D., et al. (2015). Exogenous gene integration for microalgal cell transformation using a nanowire-incorporated microdevice. *ACS Appl. Materials Interf.* 7 27554–27561. 10.1021/acsami.5b09964
3. Bhola V., Swalaha F., Ranjith Kumar R., Singh M., Bux F. (2014). Overview of the potential of microalgae for CO_2sequestration. *Int. J. Environ. Sci. Technol.* 11 2103–2118. 10.1007/s13762-013-0487-6
4. Chang R. L., Ghamsari L., Manichaikul A., Hom E. F. Y., Balaji S., Fu W., et al. (2011). Metabolic network reconstruction of Chlamydomonas offers insight into light-driven algal metabolism. *Mol. Syst. Biol.* 7:518 10.1038/msb.2011.52
5. Claquin P., Probert I., Lefebvre S., Veron B. (2008). Effects of temperature on photosynthetic parameters and TEP production in eight species of marine microalgae. *Aquatic Microbial Ecology*51 1–11. 10.3354/ame01187
6. Cobbett C., Goldsbrough P. (2002). Phytochelatins and metallothioneins: roles in heavy metal detoxification and homeostasis. *Annu. Rev. Plant Biol.* 53 159–182. 10.1146/annurev.arplant.53.100301.135154
7. Couso I., Vila M., Rodriguez H., Vargas M. A., León R. (2011).Overexpression of an exogenous phytoene synthase gene in the unicellular alga *Chlamydomonas reinhardtii* leads to an increase in the content of carotenoids. *Biotechnol. Prog.* 27 54–60. 10.1002/btpr.527
8. Cowan A. K., Rose P. D., Horne L. G. (1992). Dunaliella salina: a model system for studying the response of plant cells to stress.*J. Exp. Bot.* 43 1535–1547. 10.1093/jxb/43.12.1535
9. de-Bashan L. E., Trejo A., Huss V. A. R., Hernandez J.-P., Bashan Y. (2008). Chlorella sorokiniana UTEX 2805, a heat and intense, sunlight-tolerant microalga with potential for removing ammonium from wastewater. *Bioresour. Technol.* 99 4980–4989. 10.1016/j.biortech.2007.09.065
10. de Oliveira Dal'Molin C. G., Quek L.-E., Palfreyman R. W., Nielsen L. K. (2011). AlgaGEM – a genome-scale metabolic reconstruction of algae based on the Chlamydomonas *reinhardtii*genome. *BMC Genomics* 12:S5 10.1186/1471-2164-12-s4-s5
11. Deng X.-D., Gu B., Li Y.-J., Hu X.-W., Guo J.-C., Fei X.-W. (2012). The roles of acyl-CoA: *Diacylglycerol acyltransferase* 2 genes in the biosynthesis of triacylglycerols by the green algae*Chlamydomonas reinhardtii*. *Mol. Plant* 5 945–947. 10.1093/mp/sss040
12. Depauw F. A., Rogato A., d'Alcalá M., Falciatore A. (2012).Exploring the molecular basis of responses to light in marine diatoms. *J. Exp. Bot.* 63 1575–1591. 10.1093/jxb/ers005
13. Dominguez A. A., Lim W. A., Qi L. S. (2016). Beyond editing: repurposing CRISPR-Cas9 for precision genome regulation and interrogation. *Nat. Rev. Mol. Cell Biol.* 17 5–15. 10.1038/nrm.2015.2
14. Dong Z., Chen Y. (2013). Transcriptomics: advances and approaches. *Sci. China Life Sci.* 56 960–967. 10.1007/s11427-013-4557-2
15. Eberhard S., Finazzi G., Wollman F.-A. (2008). The dynamics of photosynthesis. *Annu. Rev. Genet.* 42 463–515. 10.1146/annurev.genet.42.110807.091452
16. Finazzi G., Johnson G. N., Dall'Osto L., Zito F., Bonente G., Bassi R., et al. (2006). Nonphotochemical quenching of chlorophyll fluorescence in *Chlamydomonas reinhardtii.Biochemistry* 45 1490–1498. 10.1021/bi0521588
17. Gangl D., Zedler J. A. Z., Rajakumar P. D., Martinez E. M. R., Riseley A., Włodarczyk A., et al. (2015). Biotechnological exploitation of microalgae. *J. Exp. Bot.* 66 6975–6990. 10.1093/jxb/erv426
18. Gao H., Wright D. A., Li T., Wang Y., Horken K., Weeks D. P., et al. (2014). TALE activation of endogenous genes in*Chlamydomonas reinhardtii*. *Algal Res.* 5 52–60. 10.1016/j.algal.2014.05.003
19. Geider R. J. (1987). Light and temperature dependence of the carbon to chlorophyll a ratio in microalgae and cyanobacteria: implications for physiology and growth of phytoplankton. *New Phytol.* 106 1–34. 10.1111/j.1469-8137.1987.tb04788.x
20. Genkov T., Meyer M., Griffiths H., Spreitzer R. J. (2010).Functional hybrid rubisco enzymes with plant small subunits and algal large subunits: engineered rbcS cDNA for expression in Chlamydomonas. *J. Biol. Chem.* 285 19833–19841. 10.1074/jbc.M110.124230

21. Gimpel J. A., Henríquez V., Mayfield S. P. (2015). Metabolic engineering of eukaryotic microalgae: potential and challenges come with great diversity. *Front. Microbiol.* 6:137610.3389/fmicb.2015.01376

22. Gong Y., Hu H., Gao Y., Xu X., Gao H. (2011). Microalgae as platforms for production of recombinant proteins and valuable compounds: progress and prospects. *J. Industrial Microbiol. Biotechnol.* 38 1879–1890. 10.1007/s10295-011-1032-6

23. Goss R., Jakob T. (2010). Regulation and function of xanthophyll cycle-dependent photoprotection in algae. *Photosynth. Res.* 106103–122. 10.1007/s11120-010-9536-x

24. Hanagata N., Takeuchi T., Fukuju Y., Barnes D. J., Karube I. (1992). Tolerance of microalgae to high CO_2 and high temperature. *Phytochemistry* 31 3345–3348. 10.1016/0031-9422(92)83682-O

25. Hanikenne M. (2003). *Chlamydomonas reinhardtii* as a eukaryotic photosynthetic model for studies of heavy metal homeostasis and tolerance. *New Phytol.* 159 331–340. 10.1046/j.1469-8137.2003.00788.x

26. Hlavova M., Turoczy Z., Bisova K. (2015). Improving microalgae for biotechnology - From genetics to synthetic biology. *Biotechnol. Adv.* 33 1194–1203. 10.1016/j.biotechadv.2015.01.009

27. Ho S.-H., Ye X., Hasunuma T., Chang J.-S., Kondo A. (2014).Perspectives on engineering strategies for improving biofuel production from microalgae - a critical review. *Biotechnol. Adv.*32 1448–1459. 10.1016/j.biotechadv.2014.09.002

28. Hsieh H.-J., Su C.-H., Chien L.-J. (2012). Accumulation of lipid production in Chlorella minutissima by triacylglycerol biosynthesis-related genes cloned from *Saccharomyces cerevisiae*and *Yarrowia lipolytica*. *J. Microbiol.* 50 526–534. 10.1007/s12275-012-2041-5

29. Hu S., Lau K. W. K., Wu M. (2001). Cadmium sequestration in*Chlamydomonas reinhardtii*. *Plant Sci.* 161 987–996. 10.1016/S0168-9452(01)00501-5

30. Iwasaki I., Kurano N., Miyachi S. (1996). Effects of high-CO_2stress on photosystem II in a green alga, *Chlorococcum littorale*, which has a tolerance to high CO_2. *J. Photochem. Photobiol. B Biol.* 36 327–332. 10.1016/S1011-1344(96)07385-X

31. Jiang H., Gao K. (2004). Effects of lowering temperature during culture on the production of polyunsaturated fatty acids in the marine diatom *Phaeodactylum tricornutum* (Bacillariophyceae).*J. Phycol.* 40 651–654. 10.1111/j.1529-8817.2004.03112.x

32. Jiang W., Brueggeman A. J., Horken K. M., Plucinak T. M., Weeks D. P. (2014). Successful transient expression of Cas9 and single guide RNA genes in *Chlamydomonas reinhardtii*.*Eukaryot. Cell* 13 1465–1469. 10.1128/EC.00213-14

33. Jones C. S., Mayfield S. P. (2012). Algae biofuels: versatility for the future of bioenergy. *Curr. Opin. Biotechnol.* 23 346–351. 10.1016/j.copbio.2011.10.013

34. Karpowicz S. J., Prochnik S. E., Grossman A. R., Merchant S. S. (2011). The GreenCut2 resource, a phylogenomically derived inventory of proteins specific to the plant lineage. *J. Biol. Chem.*286 21427–21439. 10.1074/jbc.M111.233734

35. Kasai Y., Oshima K., Ikeda F., Abe J., Yoshimitsu Y., Harayama S. (2015). Construction of a self-cloning system in the unicellular green alga *Pseudochoricystis ellipsoidea*. *Biotechnol. Biofuels* 81–12. 10.1186/s13068-015-0277-0

36. Kaye Y., Grundman O., Leu S., Zarka A., Zorin B., Didi-Cohen S., et al. (2015). Metabolic engineering toward enhanced LC-PUFA biosynthesis in Nannochloropsis oceanica: overexpression of endogenous Δ12 desaturase driven by stress-inducible promoter leads to enhanced deposition of polyunsaturated fatty acids in TAG. *Algal Res.* 11 387–398. 10.1016/j.algal.2015.05.003

37. Kim E.-J., Ma X., Cerutti H. (2015). Gene silencing in microalgae: mechanisms and biological roles. *Bioresour. Technol.*184 23–32. 10.1016/j.biortech.2014.10.119

38. Kirst G. O. (1990). Salinity tolerance of eukaryotic marine algae.*Annu. Rev. Plant Physiol. Plant Mol. Biol.* 41 21–53. 10.1146/annurev.pp.41.060190.000321

39. Kobayashi Y., Harada N., Nishimura Y., Saito T., Nakamura M., Fujiwara T., et al. (2014). Algae sense exact temperatures: small heat shock proteins are expressed at the survival threshold temperature in *Cyanidioschyzon merolae* and *Chlamydomonas reinhardtii*. *Genome Biol. Evol.* 6 2731–2740. 10.1093/gbe/evu216

40. Lamers P. P., Janssen M., De Vos R. C. H., Bino R. J., Wijffels R. H. (2012). Carotenoid and fatty acid metabolism in nitrogen-starved *Dunaliella salina*, a unicellular green microalga. *J. Biotechnol.* 162 21–27. 10.1016/j.jbiotec.2012.04.018

41. Lee D.-J., Chang J.-S., Lai J.-Y. (2015). Microalgae–microbial fuel cell: a mini review. *Bioresour. Technol.* 198 891–895. 10.1016/j.biortech.2015.09.061

42. Lemaire S., Keryer E., Stein M., Schepens I., Issakidis-Bourguet E., Gérard-Hirne C., et al. (1999). Heavy-metal regulation of thioredoxin gene expression in *Chlamydomonas reinhardtii*.*Plant Physiol.* 120 773–778. 10.1104/pp.120.3.773

43. Lemoine Y., Schoefs B. (2010). Secondary ketocarotenoid astaxanthin biosynthesis in algae: a multifunctional response to stress. *Photosynth. Res.* 106 155–177. 10.1007/s11120-010-9583-3

44. Li X., Zhang R., Patena W., Gang S. S., Blum S. R., Ivanova N., et al. (2016). An indexed, mapped mutant library enables reverse genetics studies of biological processes in *Chlamydomonas reinhardtii*. *Plant Cell* 28 367–387. 10.1105/tpc.15.00465

45. Li Y., Huang J., Sandmann G., Chen F. (2009). High-light and sodium chloride stress differentially regulate the biosynthesis of astaxanthin in *Chlorella zofingiensis* (Chlorophyceae). *J. Phycol.*45 635–641. 10.1111/j.1529-8817.2009.00689.x

46. Liska A. J., Shevchenko A., Pick U., Katz A. (2004). Enhanced photosynthesis and redox energy production contribute to salinity tolerance in Dunaliella as revealed by homology-based proteomics. *Plant Physiol.* 136 2806–2817. 10.1104/pp.104.039438

47. Lopez D., Casero D., Cokus S. J., Merchant S. S., Pellegrini M. (2011). Algal functional annotation tool: a web-based analysis suite to functionally interpret large gene lists using integrated annotation and expression data. *BMC Bioinform.* 12:28210.1186/1471-2105-12-282

48. Lusser M., Parisi C., Plan D., Rodríguez-Cerezo E. (2012).Deployment of new biotechnologies in plant breeding. *Nat. Biotechnol.* 30 231–239. 10.1038/nbt.2142

49. Markou G., Nerantzis E. (2013). Microalgae for high-value compounds and biofuels production: a review with focus on cultivation under stress conditions. *Biotechnol. Adv.* 31 1532–1542. 10.1016/j.biotechadv.2013.07.011

50. Maruyama S., Tokutsu R., Minagawa J. (2014). Transcriptional regulation of the stress-responsive light harvesting complex genes in *Chlamydomonas reinhardtii*. *Plant Cell Physiol.* 551304–1310. 10.1093/pcp/pcu068

51. Mayank A., Venkatesh J., Tran L. S. P. (2015). Regulation of photosynthesis during abiotic stress-induced photoinhibition.*Mol. Plant* 8 1304–1320. 10.1016/j.molp.2015.05.005

52. Mayfield S., Golden S. (2015). Photosynthetic bio-manufacturing: food, fuel, and medicine for the 21st century.*Photosynth. Res.* 123 225–226. 10.1007/s11120-014-0063-z

53. Miyachi S., Iwasaki I., Shiraiwa Y. (2003). Historical perspective on microalgal and cyanobacterial acclimation to low- and extremely high-CO_2 conditions. *Photosynth. Res.* 77 139–153. 10.1023/a:1025817616865

54. Morozova O., Marra M. A. (2008). Applications of next-generation sequencing technologies in functional genomics.*Genomics* 92 255–264. 10.1016/j.ygeno.2008.07.001

55. Müller P., Li X.-P., Niyogi K. K. (2001). Non-photochemical quenching. A response to excess light energy. *Plant Physiol.* 1251558–1566. 10.1104/pp.125.4.1558

56. Mussgnug J. H., Thomas-Hall S., Rupprecht J., Foo A., Klassen V., McDowall A., et al. (2007). Engineering photosynthetic light capture: impacts on improved solar energy to biomass conversion. *Plant Biotechnol. J.* 5 802–814. 10.1111/j.1467-7652.2007.00285.x

57. Noor-Mohammadi S., Pourmir A., Johannes T. W. (2012). Method to assemble and integrate biochemical pathways into the chloroplast genome of *Chlamydomonas reinhardtii*. *Biotechnol. Bioeng.* 109 2896–2903. 10.1002/bit.24569

58. Perales-Vela H. V., Peña-Castro J. M., Cañizares-Villanueva R. O. (2006). Heavy metal detoxification in eukaryotic microalgae.*Chemosphere* 64 1–10. 10.1016/j.chemosphere.2005.11.024

59. Qin S., Lin H., Jiang P. (2012). Advances in genetic engineering of marine algae. *Biotechnol. Adv.* 30 1602–1613. 10.1016/j.biotechadv.2012.05.004

60. Quaas T., Berteotti S., Ballottari M., Flieger K., Bassi R., Wilhelm C., et al. (2015). Non-photochemical quenching and xanthophyll cycle activities in six green algal species suggest mechanistic differences in the process of excess energy dissipation. *J. Plant Physiol.* 172 92–103. 10.1016/j.jplph.2014.07.023

61. Rajamani S., Siripornadulsil S., Falcao V., Torres M., Colepicolo P., Sayre R. (2007). "Phycoremediation of heavy metals using transgenic microalgae," in *Transgenic Microalgae as Green Cell Factories*, eds León R., Galván A., Fernández E., editors. (New York, NY: Springer;), 99–109.

62. Rasala B. A., Chao S.-S., Pier M., Barrera D. J., Mayfield S. P. (2014). Enhanced genetic tools for engineering multigene traits into green algae. *PLoS ONE* 9:e9402810.1371/journal.pone.0094028

63. Rasala B. A., Muto M., Sullivan J., Mayfield S. P. (2011).Improved heterologous protein expression in the chloroplast of*Chlamydomonas reinhardtii* through promoter and 5' untranslated region optimization. *Plant Biotechnol. J.* 9 674–683. 10.1111/j.1467-7652.2011.00620.x

64. Raven J. A. (2010). Inorganic carbon acquisition by eukaryotic algae: four current questions. *Photosynth. Res.* 106 123–134. 10.1007/s11120-010-9563-7

65. Renaud S. M., Thinh L.-V., Lambrinidis G., Parry D. L. (2002).Effect of temperature on growth, chemical composition and fatty acid composition of tropical Australian microalgae grown in batch cultures. *Aquaculture* 211 195–214. 10.1016/S0044-8486(01)00875-4

66. Roháček K., Bertrand M., Moreau B., Jacquette B., Caplat C., Morant-Manceau A., et al. (2014). Relaxation of the non-photochemical chlorophyll fluorescence quenching in diatoms: kinetics, components and mechanisms. *Philos. Trans. R. Soc. Lond. B Biol. Sci.* 369:20130241 10.1098/rstb.2013.0241

67. Sakai N., Sakamoto Y., Kishimoto N., Chihara M., Karube I. (1995). Chlorella strains from hot springs tolerant to high temperature and high CO_2. *Energy Convers. Manag.* 36 693–696. 10.1016/0196-8904(95)00100-R

68. Scibilia L., Girolomoni L., Berteotti S., Alboresi A., Ballottari M. (2015). Photosynthetic response to nitrogen starvation and high light in *Haematococcus pluvialis*. *Algal Res.* 12 170–181. 10.1016/j.algal.2015.08.024

69. Scranton M. A., Ostrand J. T., Fields F. J., Mayfield S. P. (2015).Chlamydomonas as a model for biofuels and bio-products production. *Plant J.* 82 523–531. 10.1111/tpj.12780

70. Scranton M. A., Ostrand J. T., Georgianna D. R., Lofgren S. M., Li D., Ellis R. C., et al. (2016). Synthetic promoters capable of driving robust nuclear gene expression in the green alga*Chlamydomonas reinhardtii*. *Algal Res.* 15 135–142. 10.1016/j.algal.2016.02.011

71. Siripornadulsil S., Traina S., Verma D. P. S., Sayre R. T. (2002).Molecular mechanisms of Proline-mediated tolerance to toxic heavy metals in transgenic microalgae. *Plant Cell* 14 2837–2847. 10.1105/tpc.004853

72. Sizova I., Greiner A., Awasthi M., Kateriya S., Hegemann P. (2013). Nuclear gene targeting in *Chlamydomonas* using engineered zinc-finger nucleases. *Plant J.* 73 873–882. 10.1111/tpj.12066

73. Solovchenko A., Khozin-Goldberg I. (2013). High-CO_2 tolerance in microalgae: possible mechanisms and implications for biotechnology and bioremediation. *Biotechnol. Lett.* 35 1745–1752. 10.1007/s10529-013-1274-7

74. Steinbrenner J., Sandmann G. (2006). Transformation of the green alga *Haematococcus pluvialis* with a phytoene desaturase for accelerated astaxanthin biosynthesis. *Appl. Environ. Microbiol.* 72 7477–7484. 10.1128/AEM.01461-06

75. Stengel D. B., Connan S., Popper Z. A. (2011). Algal chemodiversity and bioactivity: sources of natural variability and implications for commercial application. *Biotechnol. Adv.* 29483–501. 10.1016/j.biotechadv.2011.05.016

76. Sun G., Zhang X., Sui Z., Mao Y. (2008). Inhibition of pds gene expression via the RNA interference approach in *Dunaliella salina* (Chlorophyta). *Mar. Biotechnol.* 10 219–226. 10.1007/s10126-007-9056-7

77. Sung K.-D., Lee J.-S., Shin C.-S., Park S.-C., Choi M.-J. (1999).CO2 fixation by *Chlorella* sp. *KR*-1 and its cultural characteristics. *Bioresour. Technol.* 68 269–273. 10.1016/S0960-8524(98)00152-7

78. Suresh Kumar K., Dahms H.-U., Won E.-J., Lee J.-S., Shin K.-H. (2015). Microalgae – A promising tool for heavy metal remediation. *Ecotoxicol. Environ. Saf.* 113 329–352. 10.1016/j.ecoenv.2014.12.019

79. Tardif M., Atteia A., Specht M., Cogne G., Rolland N., Brugière S., et al. (2012). PredAlgo: a new subcellular localization prediction tool dedicated to green algae. *Mol. Biol. Evol.* 293625–3639. 10.1093/molbev/mss178

80. Tissot-Lecuelle G., Purton S., Dubald M., Goldschmidt-Clermont M. (2014). "Synthesis of recombinant products in the chloroplast," in *Plastid Biology*, eds Theg S. M., Wollman F.-A., editors. (New York, NY: Springer;), 517–557.

81. Torricelli E., Gorbi G., Pawlik-Skowronska B., di Toppi L. S., Corradi M. G. (2004). Cadmium tolerance, cysteine and thiol peptide levels in wild type and chromium-tolerant strains of*Scenedesmus acutus* (Chlorophyceae). *Aquat. Toxicol.* 68 315–323. 10.1016/j.aquatox.2004.03.020

82. Tsai C.-H., Warakanont J., Takeuchi T., Sears B. B., Moellering E. R., Benning C. (2014). The protein compromised hydrolysis of triacylglycerols 7 (CHT7) acts as a repressor of cellular quiescence in *Chlamydomonas*. *Proc. Natl. Acad. Sci. U.S.A.* 11115833–15838. 10.1073/pnas.1414567111

83. Tsuji N., Hirayanagi N., Okada M., Miyasaka H., Hirata K., Zenk M. H., et al. (2002). Enhancement of tolerance to heavy metals and oxidative stress in *Dunaliella tertiolecta* by Zn-induced phytochelatin synthesis. *Biochem. Biophys. Res. Commun.* 293653–659. 10.1016/S0006-291X(02)00265-6

84. USDA Japan Reports (2014). *United States Department of Agriculture. Japan takes step Forward to Improve its GE Product Review Process*. Available at:usdajapan.org/en/reports/reports2014.html

85. Waters E., Riofiorido I. (2007). Evolutionary analysis of the small heat shock proteins in five complete algal genomes. *J. Mol. Evol.* 65 162–174. 10.1007/s00239-006-0223-7

86. Whitney S. M., Houtz R. L., Alonso H. (2011). Advancing our understanding and capacity to engineer nature's CO_2-sequestering enzyme. *Rubisco. Plant Physiol.* 155 27–35. 10.1104/pp.110.164814

87. Wobbe L., Bassi R., Kruse O. (2016). Multi-level light capture control in plants and green algae. *Trends Plant Sci.* 21 55–68. 10.1016/j.tplants.2015.10.004

88. Wu S., Xu L., Huang R., Wang Q. (2011). Improved biohydrogen production with an expression of codon-optimized hemH and lba genes in the chloroplast of *Chlamydomonas reinhardtii.Bioresour. Technol.* 1022610–2616. 10.1016/j.biortech.2010.09.123

89. Yao L., Tan T. W., Ng Y.-K., Ban K. H. K., Shen H., Lin H., et al. (2015). RNA-Seq transcriptomic analysis with Bag2D software identifies key pathways enhancing lipid yield in a high lipid-producing mutant of the non-model green alga *Dunaliella tertiolecta*. *Biotechnol. Biofuels* 8 1–16. 10.1186/s13068-015-0382-0

90. Young R. E. B., Purton S. (2015). Codon reassignment to facilitate genetic engineering and biocontainment in the chloroplast of *Chlamydomonas reinhardtii*. *Plant Biotechnol. J.*[Epub ahead of print].10.1111/pbi.12490

91. Zhang J., Hao Q., Bai L., Xu J., Yin W., Song L., et al. (2014).Overexpression of the soybean transcription factor GmDof4 significantly enhances the lipid content of *Chlorella ellipsoidea.Biotechnol. Biofuels* 7 1–16. 10.1186/s13068-014-0128-4

92. Zorin B., Lu Y., Sizova I., Hegemann P. (2009). Nuclear gene targeting in *Chlamydomonas* as exemplified by disruption of the PHOT gene. *Gene* 432 91–96. 10.1016/j.gene.2008.11.028

CHAPTER 7

A COMPARATIVE ANALYSIS OF INSERTIONAL EFFECTS IN GENETICALLY ENGINEERED PLANTS: CONSIDERATIONS FOR PRE-MARKET ASSESSMENTS

J. Schnell C. Girard P. Macdonald[1], C. Girard[2], M. Steele A. Savoie[3], A. Savoie[4], J. Bean L. Bourbonnie`re[5], L. Bourbonnie`re[6], M. Neuspiel[7], N. Dormann C. Pearson[8], C. Pearson[9]

[1]Plant and Biotechnology Risk Assessment Unit, Canadian Food Inspection Agency, 1400 Merivale Road, Ottawa, ON K1A 0Y9, Canada

[2]e-mail: cecile.girard@inspection.gc.ca

[3]Animal Feed Division, Canadian Food Inspection Agency, 59 Camelot Drive, Ottawa, ON K1A 0Y9, Canada

[4]e-mail: annie.savoie@inspection.gc.ca

[5]Food Directorate, Health Canada, 251 Sir Frederick Banting Driveway, Ottawa, ON K1A 0K9, Canada

[6]e-mail: luc.bourbonniere@hc-sc.gc.ca

[7]Office of the Chief Science Operating Officer, Canadian Food Inspection Agency, 1400 Merivale Road, Ottawa, ON K1A 0Y9, Canada

[8]Plant Biosafety Office, Canadian Food Inspection Agency, 59 Camelot Drive, Ottawa, ON K1A 0Y9, Canada

[9]e-mail: cindy.pearson@inspection.gc.ca

ABSTRACT

During genetic engineering, DNA is inserted into a plant's genome, and such insertions are often accompanied by the insertion of additional DNA, deletions and/or rearrangements. These genetic changes are collectively known as insertional effects, and they have the potential to give rise to unintended traits in plants. In addition, there are many other genetic changes that occur in plants both spontaneously and as a result of conventional breeding practices. Genetic changes similar to insertional effects occur in plants, namely as a result of the movement of transposable elements, the repair of double-strand breaks by non-homologous end-joining, and the intracellular transfer of organelle DNA. Based on this similarity, insertional effects should present a similar level of risk as these other genetic changes in plants, and it is within the context of these genetic changes that insertional effects must be considered. Increased familiarity with genetic engineering techniques and advances in molecular analysis techniques have provided us with a greater understanding of the nature and impact of genetic changes in plants, and this can be used to refine pre-market assessments of genetically engineered plants and food and feeds derived from genetically engineered plants.

Citation: Jaimie Schnell, Marina Steele, Jordan Bean, Margaret Neuspiel, Cécile Girard, Nataliya Dormann, Cindy Pearson, Annie Savoie, Luc Bourbonnière, and Philip Macdonald, A comparative analysis of insertional effects in genetically engineered plants: considerations for pre-market assessments, doi: 10.1007/s11248-014-9843-7.

INTRODUCTION

Genetic engineering is frequently used to introduce new DNA sequences that contain one or more genes into plants, leading to the development of plants that exhibit novel traits. Given the complexity of plant cells and the current limitations of genetic engineering, unintended traits may result from genetic engineering in addition to the novel trait. Unintended traits are phenotypic changes in the plant that can materialize as a consequence of genetic changes such as DNA insertions, deletions and rearrangements, all of which can take place during genetic engineering, and the associated tissue culture process. These changes may include the introduction of a new plant characteristic, loss of a previously expressed plant characteristic or expression of a characteristic that is outside the range of what is typically observed in the plant species. The safety of novel traits intentionally introduced into plants during development is assessed during the pre-market assessment of genetically engineered plants, however, unintended traits are by their nature unexpected, making it difficult to test for them directly.

In most countries, a comparative approach used in the pre-market assessment of genetically engineered (GE) plants and foods and feeds derived from GE plants (hereafter, GE plants, foods and feeds) provides a structure that enables the identification and evaluation of unintended traits. The comparative approach was initially developed for the pre-market assessment of genetically engineered organisms in 1986, and has been adopted and used by national regulatory bodies for over 15 years to conduct pre-market assessments on more than a hundred GE plants, foods and feeds, many of which have become integral parts of the agricultural production systems and food and feed chains of more than 20 countries (Codex Alimentarius Commission 2003; OECD 1986,1992, 2003). The comparative approach is important because it recognizes that conventional plant varieties, foods and feeds have a history of safe use, and allows the pre-market assessment of the GE plant, food or feed to focus on differences between the GE plant, food or feed and a counterpart in order to determine potential safety concerns. The counterpart is typically the isogenic line closest to the GE plant with an established history of safe use, which is usually a conventional plant (i.e., not genetically engineered). This approach acknowledges the pre-existing nutritional profile, toxins, allergens and anti-nutrients of conventional plants, foods and feeds as well as the impact of the conventional plant and its associated agricultural practices on the environment.

Pre-market assessments, based on the comparative approach, function by defining the novel trait expressed in the GE product, identifying any unintended traits, and assessing the potential of these traits to adversely affect the environment, or impact the safety and/or nutritional quality of the food or the safety and/or efficacy of the feed derived from the GE plant. To accomplish this, pre-market assessments include a molecular characterization of the GE plant that considers the inserted DNA, expressed gene products and their impact on the GE plant, and a comparison of the composition and phenotype of the GE plant to that of a counterpart. The choice of which components and phenotypic characteristics to examine is, in part, determined by the nature of the novel trait, but it is generally broader to allow for the identification of unintended traits as well.

One source of unintended traits in GE plants is insertional effects, which are the changes to the transformed plant's genome that result from the process of inserting DNA by genetic engineering (Kessler et al. 1992; König et al. 2004). In addition to the intended DNA, in many cases additional DNA is also inserted. The additional DNA can originate from either the plant genome, the DNA being inserted or it can simply be filler DNA of unknown origin (Forsbach et al. 2003; Kim et al. 2003; Kohli et al. 1999; Krysan et al. 2002; Makarevitch et al.2003; Sallaud et al. 2003; Windels et al. 2003). Also common are small deletions in the plant genomic DNA flanking the site of insertion. These deletions are typically less than 100 bp, although larger deletions are also occasionally observed (Forsbach et al. 2003; Kim et al. 2003; Krysan et al. 2002; Makarevitch et al. 2003). Large scale rearrangements may also occur during genetic engineering.*Agrobacterium tumefaciens*-mediated transformation and biolistics-mediated transformation are two commonly used approaches for creating genetically engineered plants. Both of these approaches can result in complex genetic inserts containing multiple copies and/or re-arrangements of both the DNA intended to be inserted and the host plant DNA (Dan and Ow 2011; Stewart et al. 2011). For instance, in arabidopsis (*Arabidopsis thaliana*) transformed using *A. tumefaciens*, chromosomal translocations have been documented, where the flanking genomic DNA on either side of the inserted DNA mapped to two different chromosomes (Forsbach et al. 2003; Nacry et al. 1998). These insertional effects can alter patterns of gene expression in the plant or change the nature of RNA and/or proteins expressed by endogenous genes, either of which could result in an unintended trait in the plant.

The comparative approach to the pre-market assessment of GE plants, foods and feeds continues to be valid. However, there is now increased familiarity with genetic engineering techniques, as well as a growing body of scientific research both on the process of genetic engineering and on GE plants, foods and feeds themselves. In addition, advancements in molecular analysis techniques, particularly high-throughput sequencing and global profiling technologies, have given us an unprecedented understanding of the nature and the impact of genetic changes in plants. This raises the question of whether the comparative approach may be further refined.

It is, therefore, an opportune time to review and contextualize our understanding of insertional effects and their potential to result in unintended traits and to investigate how these insertional effects may be best evaluated within the current comparative approach to pre-market assessments. A simultaneous review of genetic changes that occur spontaneously as well as those that occur during conventional breeding will provide a useful comparison for insertional effects in keeping with the comparative approach to pre-market assessments. If insertional effects are found to be similar to other genetic changes that occur in plants, they should present a similar level of risk. It would then be reasonable to re-evaluate the information required for the comparative pre-market assessment of GE plants, foods and feeds.

SCOPE

This review will be limited to a discussion of potential unintended traits resulting from insertional effects. Other potential sources of unintended traits exist, including pleiotropic effects, which occur when a gene within the inserted DNA sequence confers multiple traits, and position effects, which are variations in the expression of genes within the inserted DNA that are dependent on the site of insertion.

Insertional effects and Unintended Traits

Although it is sometimes assumed that insertional effects are in and of themselves an indication that a GE plant will display an unintended trait, this is not an accurate conclusion. In a T-DNA insertional mutagenesis study in rice (*Oryza sativa*), examination of 22,665 field-grown T1 lines identified only 4,065 lines with a visible mutant trait (Chern et al. 2007). A study by El Ouakfaoui and Miki (2005) investigated the impact of the insertion of two marker genes on global gene expression profiles in four arabidopsis lines. In three independent experiments, reproducible changes in gene expression were not observed for any genes (with the exception of the introduced genes themselves) demonstrating that genes can be inserted without altering the global gene expression profile (El Ouakfaoui and Miki 2005).

Since not all insertional effects will result in an unintended trait, the question then becomes: what factors will determine whether an insertional effect will cause an unintended trait in a GE plant? In most cases, the DNA will need to be inserted within or near to an endogenous gene. The likelihood of this occurring will vary from species to species based on genome composition. For instance, gene rich regions are expected to account for 17 % of the maize (*Zea mays*) genome with an average gene density of 1 gene per 40 kb (Barakat et al. 1997; Messing et al. 2004). In contrast, in rice, gene rich regions account for 24 % of the genome and the average gene density is one gene every 4 to 5 kb (Barakat et al. 1997; Yu et al. 2002). Although some studies have suggested that transgene integration in *A. tumefaciens*-mediated transformation is biased towards actively transcribed genes, results of other studies indicate that at least some of this perceived insertion site bias is due to selection in insert identification (Francis and Spiker 2005; Kim and Gelvin 2007).

If the insertion occurs within or near to an endogenous gene or regulatory element, either the expression of associated endogenous gene(s) or the nature of the RNA(s) and/or protein(s) produced must then be affected in order for an insertional effect to result in an unintended trait. The expression of an endogenous gene can change if insertions, deletions and rearrangements within the gene render it non-functional, completely abolishing expression, or alter its pattern or level of expression. Regulatory elements within the inserted DNA can also influence the expression of nearby endogenous genes. This has been exploited in activation tagging, where DNA is inserted that contains a moderate to strong enhancer or promoter, such as the CaMV 35S promoter, with the goal of transcriptionally activating nearby genes (Chern et al. 2007; Jeong et al. 2002). In one such study, it was found that the average distance between the inserted enhancer and the activated gene was 4.4 kb, although this distance could be as great as 8 kb (Ichikawa et al. 2003). The second case, wherein the expressed sequence of the RNA(s) and/or protein(s) of an endogenous gene is changed, could occur by a number of mechanisms, for example due to insertions, deletions and/or rearrangements within the gene resulting in it being truncated, alternatively spliced, or if a novel chimeric gene is created.

Even if there are changes in the expression of an endogenous gene or the nature of the gene product that is expressed, these will not necessarily result in an unintended trait. The levels of expression of a given gene will vary based on genetic background and environment, so any changes within this range of expression may fall within normal variation and therefore not be biologically significant. For example, the expression of thousands of plant genes are frequently altered in response to stress (El Ouakfaoui and Miki 2005; Kawaura et al. 2006; Zheng et al. 2010). Even in the absence of stress, many genes are differentially expressed in plants grown at different locations and under different environmental conditions (Barros et al. 2010).

Furthermore, in cases where the function of the endogenous gene is compromised by insertional effects, it may still not result in an unintended trait because of the high level of gene redundancy in plant genomes. Genomic analysis of arabidopsis and rice has revealed gene redundancy throughout approximately one-third to one-half of the genome (Goff et al. 2002; The Arabidopsis Genome Initiative 2000; Yu et al. 2002). The buffering effect of duplicate genes is a legacy of polyploidization, which can be found in the ancestry of most plant species (Adams and Wendel 2005; Comai 2005). Duplicate copies of a damaged or aberrantly expressed gene can mask potentially detrimental phenotypes. For example, in 4,000 greenhouse-grown arabidopsis lines with insertions of a transposon in gene-coding regions, only 139 showed a visible mutant phenotype (Kuromori et al. 2006). In another example, only about 2 % of a mutant soybean (*Glycine max*) population generated by fast neutron (FN) radiation showed an abnormal visual phenotype (Bolon et al. 2011).

Finally, changes to the nature of the RNA and/or protein that is expressed from an endogenous gene may also fail to result in an unintended trait due to the RNA and protein quality control systems active in plant cells. These quality control systems actively target aberrant RNA and protein for degradation (Buchberger et al. 2010; Doma and Parker 2007), minimizing their impact on the plant phenotype.

In conclusion, the relationship between genotype and phenotype in plants is complex, and the role of the environment cannot be ignored. In many ways, plants are buffered against the consequences of genomic changes by the high level of gene redundancy in plant genomes and by quality control systems active in plants. All of these factors influence whether or not an insertional effect will produce an unintended trait. Therefore, while insertional effects, i.e. the changes to the transformed plant's genome that result from the process of inserting DNA by genetic engineering, are unavoidable during genetic engineering, unintended traits are not always a consequence of these genetic changes. This conclusion is supported by the fact that detailed review of the nutritional composition of over 100 GE plants in the United States and Canada failed to identify adverse effects of genetic engineering on the composition of these plants (Price and Underhill 2013).

Genetic changes Similar to Insertional effects Occur in Plants

Plant genomes change over time in response to natural selection and genetic drift. Those spontaneous genetic changes that confer traits that better allow a plant to survive and adapt are maintained by the species. The genomes of crops have additionally been shaped by humans through conventional breeding, through the selection of desired traits that have arisen spontaneously or been created through mutagenesis. Despite the genetic changes associated with conventional breeding, there is a history of safe use of conventional crops. It is therefore useful to determine whether the genetic changes that occur in plants, either spontaneously or through conventional breeding, are similar to insertional effects as a consequence of genetic engineering.

The Movement of Transposable Elements

The impact of the insertion of DNA through genetic engineering is probably most closely paralleled by the movement of transposable elements. Transposable elements can be found in the genomes of all plants. In arabidopsis, they account for at least 10 % of the genome, in rice at least 24.9 %, and in maize at least 57 % (Messing et al. 2004; The Arabidopsis Genome Initiative 2000; Yu et al. 2002). Typically ranging in size from hundreds to thousands of base pairs and, for autonomous elements, containing one or more open reading frames, the impact of their insertion into novel genomic loci is similar to that of the insertion of DNA via genetic engineering (Bennetzen 2000). In fact, many transposable elements were first identified by the mutant traits they generated and this phenomenon has been exploited in large scale insertional mutagenesis studies (Greco et al. 2001; Kuromori et al.2006; Walbot 2000). Like DNA insertions via genetic engineering, if a transposable element inserts within or near a plant gene, it can inactivate the gene, alter its level or pattern of expression, or alter the nature of the RNA and/or protein that is expressed by that gene (Casacuberta and Santiago 2003). This can lead to the expression of novel traits. For instance, transposition of a *doppia* transposable element resulted in the formation of the *R-r:std* allele in maize, within which the *R-s* subcomplex is expressed in the aleurone layer of seeds, leading to its pigmentation (May and Dellaporta 1998; Walker et al.1995). New cultivars of fruit have arisen from sports where the underlying genetic change is the insertion of a transposable element in close proximity to a plant gene. Research on grape (*Vitis vinifera*) berry colour has identified a connection between a lack of anthocyanin pigment (i.e. white berry) phenotype and the insertion of *Gret1*, a 10 422 bp long retrotransposon, in the promoter region of *VvmybA1*, a gene located in the berry colour locus that encodes a transcription factor that induces the anthocyanin pathway (Kobayashi et al. 2004). Insertion of this retrotransposon separates the *VvmybA1* coding sequence from its promoter and impairs its transcription, thus preventing anthocyanin production. The opposite effect has been observed in blood oranges (*Citrus sinensis*), where the insertion of a Copia-like retrotransposon adjacent to the *Ruby Myb* gene gave rise to the induction of the anthocyanin biosynthesis pathway (Butelli et al. 2012).

As is the case with the insertion of DNA through genetic engineering, the movement of transposable elements is also associated with insertions, deletions and rearrangements. Although movement of transposable elements between plant species has been reported (El Baidouri et al. 2014), the movement of transposable elements within a genome is much more common and would make a greater contribution to genetic change. DNA transposons often leave behind a footprint at the site of their excision, which may include small deletions, inversions, and insertions, including sequence from the transposable element itself (Bennetzen 2000; Rinehart et al. 1997). Rearrangements can result from aberrant transposition, which is known to occur with DNA transposons such as the *Ac/Ds* family (Gray 2000). As with the insertion of DNA through genetic engineering, these insertions, deletions and rearrangements that are associated with the movement of transposable elements can alter the nature and/or regulation of plant genes. In many cases, transposable element activities have played a significant part in plant evolution (Lisch 2013). For example, an in-depth analysis of transposable elements in sequenced crop genomes identified 51 transposable element-induced phenotypic changes associated with domestication or diversification of cultivated plants (Vitte et al. 2014).

The extent of movement of transposable elements in cultivated plants is not yet fully known. Most transposable elements are likely inactive, but some are known to be active in certain tissues or during certain developmental stages (Bennetzen 2000; Grandbastien 1998). Transposable elements may also become activated in response to stress, major genomic rearrangements, such as chromosome breaks, polyploidization, and hybridization, as well as during protoplast isolation, and cell and tissue culture (Bennetzen 2000; Grandbastien1998; Parisod et al. 2010; Shan et al. 2005; Wang et al. 2010; Zou et al.2011). In contrast to sporadic activation, some transposable elements, such as those identified in rice, appear to be continuously active under normal conditions of growth (Moon et al. 2006; Naito et al. 2006). Estimates of the frequency of movement of transposable elements vary. For example, approximately one new insertion of the *nDart*transposable element was found in every four plants in a study of backcrosses with the rice cultivar Hoshinoyume (Fujino et al. 2005). In contrast, a single rice line was reported to accumulate dozens of de novo *mPing* transposable elements per generation (Naito et al. 2006).

Repair of Double-Strand Breaks by Non-Homologous end-Joining

Double-strand breaks may be caused by physical stress on chromosomes, DNA replication across a nick, DNA crossing over during homologous recombination, or by reactive oxygen species, which are produced by the plant cell as a by-product of metabolism, in response to pathogen attack, or by exposure to environmental pollutants, ionizing radiation and ultraviolet light.

The main pathway for repair of double-strand breaks in plants is non-homologous end-joining (NHEJ). NHEJ is an error-prone recombination pathway, and deletions, insertions and rearrangements are often observed at the repair site where the two DNA ends have been rejoined (Gorbunova and Levy 1999). These are similar to the deletions, insertions and rearrangements that are often observed at the site of DNA insertion in plants having undergone genetic engineering. As with genetic engineering, if these deletions, insertions and rearrangements involve endogenous genes, the nature and/or regulation of these genes can be affected.

A hallmark of NHEJ repair sites is the presence of short repeats, sometimes referred to as microhomologies, that are typically 1–8 bp in length (Gorbunova and Levy 1999; Morita et al. 2009). These will be present when the repair is mediated by annealing of the exposed ends at these repeats. Interestingly, microhomologies are also often observed at sites where DNA has been inserted by genetic engineering, between the inserted DNA and the flanking plant genome sequences (Gorbunova and Levy 1999; Somers and Makarevitch 2004). This is because both*Agrobacterium*-mediated and particle bombardment transformation methods rely on the plant's DNA repair pathways, in particular NHEJ, to introduce the DNA, although the exact mechanisms have not yet been fully elucidated (Makarevitch et al. 2003; Mayerhofer et al. 1991; Somers and Makarevitch 2004; Takano et al. 1997). Furthermore, the footprints created by excision of DNA transposons also contain these microhomologies because, once again, NHEJ is involved in repairing the double-strand break caused by excision (Huang and Dooner 2012; Rubin and Levy 1997). The common mechanism that underlies the generation of the insertions, deletions and rearrangements found at sites of DNA insertion via genetic engineering, double-strand break repair and DNA transposon excision is evidence that these insertional effects are not unique to genetic engineering.

The generation of double-strand breaks and their repair is also a key feature of conventional breeding. As the phylogenetic distance between a cultivated crop and its wild relative becomes wider, it becomes more difficult to create hybrids and introgress desired traits. In such cases, traits can be introgressed through chromosomal translocations. Infrequently, such translocations may occur spontaneously, but in many cases they must be induced, for instance by using ionizing radiation to cause double-strand breaks (Fedak 1999; Friebe et al. 1996). Ionizing radiation is also used to induce double-strand breaks in order to create mutations. The resulting mutants typically have deletions ranging in size from tens to millions of base pairs, as well as rearrangements, including inversions and chromosomal translocations (Cecchini et al.1998; Li et al. 2001; Morita et al. 2009; Nambara et al. 1994; Sato et al.2009; Schuermann et al. 2005; Shirley et al. 1992; Sun et al. 1992; Yan et al. 2007). Again, microhomologies are often observed at repair sites. A comparison of genetic changes found at sites of double-strand break repair resulting from genetic engineering, cleavage with a restriction endonuclease and mutagenesis with gamma irradiation can be found in Table 1, further highlighting their similarity.

Table 1: A comparison of genetic changes found at sites of double-strand break repair resulting from genetic engineering (Forsbach et al. 2003), cleavage with a restriction endonuclease (Salomon and Puchta 1998) and mutagenesis with gamma irradiation (Morita et 2009)

	Microhomology		Insertions		Deletions		Rearrangements	
	Number of sites/total sites analyzed	Size range (bp)	Number of sites/total sites analyzed	Size range (bp)	Number of sites/total sites analyzed	Size range (bp)	Number of sites/total sites analyzed	Size range (bp)
Forsbach et al. (2003)	66/186[a]	1–8	109/186[a]	1–1,700	64/73	1–300	3/73	20, translocations
Salomon and Puchta (1998)	13/17	1–15	18/28	4–2,355	27/28	1–1,322	0/28	N/A
Morita et al. (2009)	16/21	1–8	1/21	1	21/21	1–129,700	2/21	1,284,800; 3,208,500

The number of sites at which microhomologies, insertions, deletions and rearrangements were identified is shown versus the total number of sites analyzed, as well as the size range of each

[a] Sequences at the left and right borders of the T-DNA insertions were analyzed separately

Intracellular Transfer of Organelle DNA

The presence of non-coding sequences in plant nuclear genomes that are homologous to plastid and mitochondrial sequences (nuclear organelle DNAs) suggests that organelle DNAs have repeatedly been inserted into the nuclear genome. In some cases it is clear that organelle DNA was inserted intact in the nuclear genome while in other cases rearrangements are apparent. In rice and arabidopsis, approximately 25 % of nuclear organelle DNAs occur within genes (Richly and Leister 2004), and in some cases sequences have become functional parts of coding regions (Noutsos et al. 2007). Once again, microhomologies found at the sites of insertion of nuclear organelle DNAs indicate that these sequences are inserted by NHEJ, and consistent with this mechanism, deletions and insertions were observed at sites of insertion (Huang et al. 2004). The insertion of organelle DNA into the nuclear genome is therefore similar to the insertion of DNA during genetic engineering. Like the insertion of DNA through genetic engineering, the nuclear organelle DNAs have the potential to alter the level or pattern of expression of endogenous genes or alter the nature of the RNA and/or protein expressed by endogenous genes.

The frequency with which organelle DNA is transferred to the nucleus is not well characterized, but there is evidence in tobacco (*Nicotiana tabacum*) that novel nuclear organelle DNAs are formed under typical conditions of growth. Huang et al. (2003) transformed the tobacco chloroplast genome with a nucleus-specific marker gene and found that the marker transferred to the nucleus at a rate of 1 in every 16,000 progeny. This number is presumed to be an underestimation of the true transfer rate, since it required transfer of the marker gene in a functional state. Wang et al. (2012) further demonstrated that the rate of transfer of DNA from the chloroplast to the nucleus increased up to tenfold in response to mild temperature stresses.

Other Sources of Genetic change in Plants Due to Inserted DNA

There are additional possible sources of genetic change in plants that are similar to insertional effects. These changes have been previously reviewed (Bock 2010; Liu et al. 2012) and while it is not the intent of this publication to review all sources of genetic change in plants in detail, a few examples will be provided. Pararetroviruses are double stranded

DNA viruses that infect plants. Although these viruses lack integrase enzymes, endogenous pararetrovirus DNA had been found in various plant genomes such as rice, tobacco, tomato (*Solanum lycopersicum*), potato (*Solanum tuberosum*), banana and plantain (*Musa*spp.), and petunia (*Petunia* spp.) (Liu et al. 2012; Staginnus et al. 2007; Staginnus and Richert-Pöggeler 2006). The mechanism by which these sequences are inserted into plant genomes has not been elucidated but is considered to likely involve NHEJ (Liu et al. 2012). Another example is processed pseudogenes. Pseudogenes are nonfunctional genomic sequences with significant sequence similarity to functional RNA or protein-coding genes (Vanin 1985). Processed pseudogenes, which are derived from retrotransposition events where double-stranded cDNAs derived from reverse transcription events are integrated into the genome (Zou et al. 2009), represent a source of inserted DNA in plants. Finally,*A. tumefaciens* and closely related bacteria are capable of transferring bacterial genes into plant genes, and this ability long predates genetic engineering (Gelvin 2000).

As we have illustrated here, the insertion of DNA via genetic engineering is similar to the movement of transposable elements and the transfer of organelle DNA and other DNA sequences to the nucleus; and the insertions, deletions and rearrangements that are often observed at the site of DNA insertion by genetic engineering are no different than those observed at sites where double-strand breaks have been repaired, which is reflective of their common underlying mechanism. This information demonstrates, therefore, that insertional effects are similar to other genetic changes that occur spontaneously in plants and during conventional breeding. By virtue of their association with the novel trait, insertional effects will be present throughout a variety as long as the novel trait is maintained. In contrast, spontaneous genetic changes will arise in a single plant and are less likely to become established in a variety. Similar to insertional effects, with conventional breeding many of the genetic changes will be closely linked to the desired trait that is being selected, and as a result these will be present throughout a variety as long as that desired trait is maintained. Therefore the strongest parallel can be drawn between the insertional effects of genetic engineering and the genetic changes associated with conventional breeding.

OTHER GENETIC CHANGES ALSO OCCUR IN PLANTS

In the previous section, we considered those genetic changes that are most similar to insertional effects. However, it is also important to consider other types of genetic changes that occur both spontaneously and as a result of conventional breeding in plants, because it is within this background of genetic change that insertional effects occur. Although these other genetic changes may not be similar in nature to insertional effects, they have the potential to be similar in impact.

Spontaneous Genetic Change

Mutations occur spontaneously in all living organisms. For instance, in arabidopsis, spontaneous mutations occur at a rate of 7×10^{-9}, which translates into 1.75 mutations per generation per diploid plant (Ossowski et al. 2010; Weber et al. 2012). If this rate of genetic change were consistent in plants with larger genomes, the number of mutations per generation would be correspondingly larger. For example, based on the genome size of soybean (Schmutz et al. 2010), more than 10 mutations per generation would be expected to spontaneously occur. This level of mutation is consistent with reports that single-nucleotide-polymorphisms (SNPs) were detected every 48–2,000 base pairs in wheat (*Triticum aestivum*), soybean, and maize, with coding regions containing fewer SNPs than noncoding regions (Weber et al. 2012). These mutations may occur during DNA replication, either due to mispairing, which results in single base substitutions, or strand slippage, which results in small insertions and deletions. Spontaneous lesions can also form, where bases are either lost or modified, causing them to mispair and again leading to single base substitutions. We discussed already the ability of environmental pollutants, ionizing radiation and ultraviolet light to cause double-strand breaks. These can also alter the structure of DNA directly, leading to single base substitutions.

Spontaneous genetic changes can also occur during homologous recombination, which, like NHEJ, can repair double-strand breaks, but also introduces genetic variation during meiosis. During homologous recombination, copy number variants can result from misalignment between repeated sequences on homologous chromosomes, leading to unequal crossing over. In arabidopsis, copy number variants were found to occur at a rate of $1–3 \times 10^{-6}$ for a given gene cluster, suggesting that a copy number variant could be formed as often as 1 in every 700 seeds (Gaut et al. 2007; Jelesko et al. 2004). Intrachromosomal recombination can also occur, and the outcome can include sequence deletions, inversions, and gene conversion. Recombination between homologous sequences located at different genomic locations can lead to large chromosomal rearrangements, including translocations (Gaut et al.2007).

In addition to the impacts of transposable elements on plant genomes already discussed, transposable elements can further alter plant genomes in ways for which there is no parallel observed during genetic engineering. For example, transposable elements will occasionally amplify genomic DNA sequences (Bennetzen 2000). This can lead to increases in gene copy number if the genomic DNA sequence contains a gene. Such amplifications can give rise to novel traits, as was observed for the tomato variety Sun1642, for which an elongated fruit trait has been attributed to the duplication of the gene *SUN* during the movement of a retrotransposon (Xiao et al. 2008). In addition, the capture of genes and gene fragments by transposable elements can lead to the creation of new genes through exon shuffling (Bennetzen 2005). In rice, many of the *Mutator*-like DNA elements (MULEs) contain gene fragments, often from different chromosomal loci, and there is evidence that at least some of these gene fragments are being expressed (Jiang et al. 2004). Likewise in soybean, the *Tgm-Express1* transposon (of the CACTA super family) carries multiple captured gene fragments. The presence of this transposon in Intron 2 of the flavanone 3-hydroxylase gene (*F3H*) (i.e. the *wp* mutant flower colour gene) results in the generation of several chimeric transcripts from alternative splicing of the pre-mRNA (Zabala and Vodkin 2007). The expression of these transcripts (including a transcript identical to that of the wild type gene) most likely contributes to the resultant mutant (i.e. pink) flower phenotype.

Finally, on a larger scale, plants may sometimes loose or gain whole chromosomes. This occurs during cell division, when chromosomes are not distributed properly between cells. Similarly, mitotic or meiotic failures or the fusion of unreduced gametes can lead to whole genome duplications. There is evidence that most plants have undergone at least two, and as many as three to four, whole genome duplications at some point in their evolution, suggesting that this process has had a major impact on plant evolution (Adams and Wendel 2005; Jiao et al. 2011; Leitch and Leitch 2008). Whole genome duplications are typically followed by a series of rearrangements, including extensive sequence loss, homologous recombination, translocations, as well as adjustments to gene expression (Adams and Wendel 2005). Sorghum (*Sorghum bicolour*) and rice have largely collinear genomic maps with maize, permitting comparison of their chromosomal arrangements. A comparison of select genomic regions from these three species suggests that at least 50 % of duplicate genes in maize were lost following polyploidization (Lai et al. 2004).

Clearly there are a number of mechanisms at work in plants that can bring about genetic changes. The cumulative impact of these genetic changes can be illustrated by comparative genomic analyses, which have demonstrated that plant genomes are remarkably diverse even within a given species. The maize genome, in particular, is known to be highly polymorphic. In a survey of the 3′ ends of 18 genes within 36 elite maize inbred lines, SNPs were found to occur on average every 60.8 bp and small insertions and deletions every 126 bp, ranging in size from 1 to 400 bp (Ching et al. 2002). Larger scale rearrangements are also common between individuals within a species, with genes found in different locations, different copy numbers or even absent in some individuals and present in others (i.e. presence/absence variation). A comparative genomic hybridization analysis of 19 maize genotypes and 14 teosinte genotypes using a custom microarray with 32,487 maize genes identified 1,065 genes with copy number variation and 2,824 genes that were present in some genotypes and absent in others (Swanson-Wagner et al. 2010). Similarly, a comparison of 2,094 contigs from the rice varieties Nipponbare and 93-11, which accounted for approximately 10 % of the whole genome, revealed 156 genes that were asymmetrically located, 91 with copy number variation, and 82 genes that were present in one variety and absent in the other (Ding et al. 2007). Recently, the use of whole genome sequencing has revealed that the potato, maize, and soybean genomes possess dozens to thousands of presence/absence variations between individual varieties of their respective species (Lai et al. 2010; Lam et al. 2010; McHale et al. 2012; Potato Genome Sequencing Consortium 2011).

Furthermore, gene movement is extensive in large plant genomes such as wheat, barley (*Hordeum vulgare*), and their relatives. Comparisons between syntenic (i.e. collinear) regions of the wheat, *Brachypodium*, rice, and sorghum genomes show that the movement (or duplication) of syntenic genes to non-syntenic positions are most likely resultant of transposable element activity and double-strand break repair (Wicker et al. 2010). These mechanisms frequently result in the creation of a pseudogene that may or may not be functional in its new position in the genome. However, if the non-syntenic gene maintains the functional activity of its homolog (i.e. the gene from which it was duplicated), and the homolog is somehow lost, the non-syntenic gene may be selected for and maintained at its new position, further eroding collinearity between related plant species (Wicker et al. 2010, 2011).

Some types of genetic change, such as DNA replication errors, spontaneous lesions, and the formation of copy number variants during recombination, constantly occur in plants. In contrast, the amount of DNA damage from exposure to external mutagens is highly dependent on environmental conditions, and the activity of transposable elements will similarly be dependent on a number of internal and external factors, as already discussed. While large changes such as polyploidization and other ploidy changes occur infrequently, they do illustrate the remarkable ability of plants to undergo large-scale genomic changes. The frequency of polyploidization events in the evolutionary history of flowering plants suggests that large-scale genomic changes in some cases offer a selective advantage.

Conventional Breeding

The most commonly applied conventional breeding technique is the introgression of a desired trait either from another individual within the same species, via intraspecific crossing, or from another species, typically a wild relative, via interspecific crossing. Genes for pest and disease resistance are more commonly targeted from these wild relatives. A study of 19 important commercial crops revealed evidence of introgression from 60 wild relatives into 13 of these crops (Hajjar and Hodgkin 2007). Another study on the use of introgression of disease resistance genes into wheat reported that six different genera have been used successfully as sources of these genes in wheat (Jones et al. 1995). Using these approaches, it is unlikely that only the gene responsible for the desired trait will be introduced into the final plant. Often, large amounts of DNA sequence surrounding the gene of interest will be transferred simultaneously, resulting in undesired traits being introduced along with the desired trait; this is known as linkage drag. Since some genes may be present in some individuals but not in others, intraspecific crosses have the potential to not only introduce novel alleles, but also novel genes. Linkage drag can be minimized by repeated backcrossing to eliminate more and more of the introgressed DNA sequence, but this process can be time-consuming, and it may not be possible to completely eliminate all of the undesired traits. For instance, the *Mi* locus in tomato introgressed from *S. peruvianum* can only be reduced to 650 kb and contains a number of genes, including a transposase, several with similarity to the arabidopsis transport inhibitor response-1 (*TIR-1*), and one with similarity to an arabidopsis transcription factor (jumonji) family protein (Seah et al. 2007). In another study, a wide variation was observed in the sizes of introgressed regions surrounding the *Tm-2* gene of *S. peruvianum*, which was introgressed independently intro various tomato cultivars, with one cultivar containing an entire chromosome arm (Young and Tanksley 1989).

As the phylogenetic distance between a cultivated crop and its wild relative becomes wider, it becomes more difficult to create hybrids and introgress the desired traits. Techniques such as embryo rescue, somatic hybridization and chromosome doubling may be employed in order to first establish viable hybrids. As already discussed, ionizing radiation may be used to create double-strand breaks in order to induce chromosomal translocations. Hybridization between distantly related species and chromosome doubling can have genome-wide consequences, including complete or partial elimination of the chromosomes of one of the two parents, and genome rearrangements, including translocations and recombination (Du et al. 2008; Liu and Li2007). The resulting hybrids are often asymmetric, with the majority of the genome originating from one parent with single chromosomes or chromosome fragments belonging to the other parent (Du et al. 2008; Faure et al. 2002; Liu and Li 2007; Tu et al. 2009; Xu et al. 2007). Three recombinant inbred lines (RILs) derived from the asymmetric hybridization between rice and its wild relative *Zizania latifolia*contained less than 0.1 % of *Z. latifolia* DNA, and yet genomic variation (in the form of SNPs and small insertions and deletions) was observed at approximately 30 % of the loci examined (Wang et al.2005). In addition, hybridization can sometimes result in the activation of transposable elements, the activity of which can further alter the genome, as already discussed (Shan et al. 2005; Wang et al. 2010; Zou et al. 2011).

Mutagenesis is another conventional breeding technique that creates novel genetic variation from which desired traits can be selected. The FAO/IAEA Mutant Variety Database (mvgs.iaea.org/AboutMutantVarities.aspx) currently lists 3,218 mutant plant varieties that are released worldwide. In these varieties, mutations were most frequently induced by ionizing radiation (predominantly using gamma rays and X-rays), although chemical mutagens were occasionally used (Ahloowalia et al. 2004). As already discussed, ionizing radiation can induce double-strand breaks. These breaks are repaired by NHEJ and often lead to deletions and rearrangements, but it can also alter bases leading to single base substitutions. For instance, gamma ray-induced single base substitutions in the rice *GA20ox-2* gene are responsible for the *sd1* mutation found in Japanese and American high-yielding semi-dwarf rice varieties that are widely commercialized (Ashikari et al. 2002). The popularity of chemical mutagens, particularly ethyl methanesulfonate (EMS), has increased in recent years. Chemical mutagens predominantly induce single base substitutions (Cooper et al. 2008; Greene et al. 2003; Olsen et al. 1993; Talamè et al. 2008; Uauy et al. 2009; Xin et al. 2008). EMS alters guanine residues such that they pair with thymine as opposed to cytosine. Following DNA repair, the G/C pair may be converted to an A/T (Greene et al. 2003). Other alkylating agents such as dimethyl sulphate and diethyl sulphate are expected to have similar effects (Hoffmann 1980).

Finally, in both conventional breeding and genetic engineering, plants can be regenerated from cell or tissue culture, and this laboratory technique can give rise to genomic level changes known as somaclonal variation. Somaclonal variation is both an unintended effect of cell and tissue culture and a potential source of novel traits for selection. Somaclonal variation can take the form of single base substitutions, insertions, deletions, rearrangements or changes in chromosome number (Jiang et al. 2011; Karp and Maddock 1984; Lee and Phillips1987; van den Bulk et al. 1990). In addition, transposable elements may be activated by cell and tissue culture. For instance, in regenerated rice plants, between 5 and 30 transposed copies of the retrotransposon*Tos17* could be detected, and the number of copies was found to increase with increased time spent in culture (Hirochika et al. 1996).

We have illustrated here that plant genomes are constantly changing in small ways, through errors in DNA replication or spontaneous lesions. They have also changed dramatically on a limited number of occasions in their evolutionary history through polyploidization. These genetic changes are necessary for their survival and adaptation and plants have a remarkable capacity to undergo major genetic change. In addition to these spontaneous genetic changes, plant genomes are further shaped by conventional breeding, which includes the introgression of foreign genes through breeding with wild relatives (Hajjar and Hodgkin 2007; Jones et al. 1995). Genomic comparisons have elegantly illustrated the cumulative impact of conventional breeding on plant genomes. This aspect has been exploited by plant breeders, creating novel varieties in conventional breeding, and recognized by seed production systems that include tolerance for off-types (http://seedgrowers.ca/).

For instance, a comparison of the genomes of 20 rice varieties and landraces determined that the modern indica and japonica varieties each had on average 8.4 Mb of DNA introgressed from other varietal groups and the more traditional aus varieties had an average of 17.5 Mb of introgressed DNA (McNally et al. 2009). In tomato, a study of sequence polymorphisms between fresh-market, processing, vintage and landrace varieties identified lower levels of variation in vintage and landrace varieties that was reflective of the reduced genetic diversity that developed during domestication as a result of genetic bottlenecks (Sim et al. 2009). Broader levels of variation were observed for the fresh-market and processing varieties, reflective of more recent tomato breeding practices in which new genetic variation is introgressed from wild species.

In conclusion, compared to genetic engineering, conventional breeding techniques can have a much larger impact on the plant genome. Hybridization, mutagenesis, and somaclonal variation all give rise to genome-wide genetic change. Similarly, introgression of a desired trait is frequently accompanied by additional unintended traits due to linkage drag, as well as introducing as many, if not more, genetic elements that could influence genes near to the site of introgression, compared to the insertion of DNA by genetic engineering. This premise is supported by several global gene expression profiling studies that illustrate that conventional breeding is more likely to impact the function of the genome than genetic engineering. For instance, mutagenesis of rice induced more changes in gene expression than the insertion of DNA by genetic engineering (Batista et al. 2008). Differences in gene expression are also typically greater between different conventionally bred cultivars than they are between a genetically engineered plant and an untransformed comparator, as demonstrated in wheat and soybean (Baudo et al. 2006; Cheng et al.2008).

GENETIC CHANGES AND RISK

As has already been discussed, the relationship between genotype and phenotype is complex and it is also tempered by the environment. Genetic changes may be introduced into plants spontaneously or through conventional breeding or genetic engineering. The buffering capabilities of plant genomes and the quality control systems in plant cells will prevent many of these genetic changes from giving rise to discernible changes in a plant's phenotype. Even in those cases where genetic changes introduced during conventional breeding or genetic engineering do give rise to unintended traits, the unintended traits may not impact the safety of foods or feeds derived from the plant or the risk to the environment from cultivation of the plant. Any risks will depend on whether the unintended trait has an adverse impact on food or feed safety or the environment, such as production of a novel toxin or a change in weediness, and on how the event containing the unintended trait is developed and processed.

While it is theoretically possible that a new toxin, anti-nutrient or allergen might be introduced into a plant, the potential for this to occur is low. In fact, there are no documented cases where conventional breeding has resulted in the production of a novel toxin or allergen in a crop (Steiner et al. 2013). It is recognized that genetic changes from any type of crop development could alter levels of endogenous toxins or anti-nutrients in plants. However, in most commercially grown crops, these are well characterized and are often monitored to avoid development of crops with negative attributes.

Similarly, while traits that contribute to weediness can be introduced into crop varieties during the conventional breeding process, feral crop populations are rare (Warwick and Stewart 2005), suggesting that the potential for conventional breeding to make a crop weedy is low. This is particularly true for highly domesticated crops that are largely dependent upon cultivation for reproduction.

Another important consideration is how the use of plants—meaning their development, production, storage, processing and preparation—may play a role in managing the risks associated with genetic changes from either genetic engineering or conventional breeding. Cultivar development typically requires upwards of 10 years and involves the evaluation of thousands of plants, resulting in the selection of one or very few final cultivars. Throughout breeding and seed production, selection is applied to eliminate off-types, which are those plants that show an unintended trait. As mentioned above, specific tests may also be conducted on those plants that are known to produce toxins or anti-nutrients in order to eliminate lines that produce unacceptably high levels of such compounds. For instance, glycoalkaloid levels in potato tubers and glucosinolates in canola may be measured. Another important end goal of the selection process is uniformity, which helps to ensure that a cultivar performs consistently once commercially released. The end result of this selection process is that unintended traits which may have occurred are frequently eliminated.

The processing and preparation of foods and feeds may also play a role in managing the risks associated with genetic changes. For example, processing conditions that involve heat or pressure may significantly reduce the levels of toxins and/or anti-nutrients in the food or feed before consumption, so that any genetic changes that alter the levels of such compounds do not present a safety concern. As long as the plants continue to be used in the same way as their conventional counterparts, the risks will be managed to the same extent.

Within the realm of conventional breeding, unintended traits are evaluated primarily for their potential to improve a cultivar. The focus of development for most crops not developed by genetic engineering is on the introduction of desirable characteristics for producers, processors and end-users, without compromising pre-existing characteristics, and while maintaining uniformity. The safety of the genetic changes which occur as a result of conventional breeding is rarely considered, and restricted to known toxins and anti-nutrients. Despite this, examples of adverse effects on human or animal health or the environment are extremely rare, and when they have occurred, they have all involved known compounds, not novel ones (Steiner et al. 2013). There is a history of safe use in agriculture as well as food and feed of cultivars developed using conventional breeding. As a result of this history of safe use, in many countries pre-market assessments are not conducted on new cultivars developed using conventional breeding.

CONCLUSIONS

A number of important conclusions may be drawn from this review of insertional effects and their potential to result in unintended traits. First, insertional effects are an unavoidable consequence of genetic engineering, but the introduction of unintended traits is not. Secondly, a comparison of insertional effects with other genetic changes that occur in plant genomes reveals that genetic changes similar to insertional effects occur spontaneously and during conventional breeding, for example by the movement of transposable elements, the transfer of organelle DNA to the nucleus and as a consequence of the repair of double-strand breaks by NHEJ. A broader consideration of genetic changes that either occur spontaneously or are generated during conventional breeding illustrates that plant genomes are constantly changing (McClintock 1984; Weber et al. 2012), and therefore that insertional effects introduced through genetic engineering make a relatively small contribution to the final genetic make-up of given plant varieties. Finally, the impact of genetic changes in a plant, including but not limited to insertional effects, is influenced by the low potential that these genetic changes will result in a phenotypic change in the plant, and by the low potential for this phenotypic change to have an adverse effect on the plant, or the resulting food and feed safety.

It was hypothesized that if insertional effects are found to be similar to other genetic changes that occur in plants, they should present a similar level of risk. We have illustrated here that the insertional effects associated with genetic engineering are similar to the genetic changes that occur in conventionally bred plants. Based on this similarity, insertional effects should present a similar level of risk as genetic changes associated with conventional breeding. In light of this conclusion, it is reasonable to re-evaluate the information required for the pre-market assessment of genetically engineered plants, foods and feeds. In doing so, it will be important to maintain consistency with international guidelines, such as Codex Alimentarius (2003).

Although these conclusions are relevant to the consideration of all transformation events, a re-evaluation of this information will be particularly useful for retransformants, which are plants that have been transformed with the identical construct(s) as a previously authorized plant of the same species that confers the same trait, and for which the safety of the trait has been previously assessed (CFIA 2012a, b). Retransformants are often generated for vegetatively propagated plants where it can be difficult to transfer a novel trait through sexual crosses, and therefore different cultivars are often independently transformed. Performing individual pre-market assessments on each transformant can be resource-intensive for both crop developers and regulatory authorities. If the genetic elements in the DNA that have been inserted in a retransformant are identical in sequence and order to those inserted in the original transformant, and genes are expressed at a similar level, the only potential difference between the original transformant that was assessed and a retransformant would be insertional effects. In such situations, the conclusions of this paper could be used to inform the pre-market assessment and determine appropriate information requirements.

REFERENCES

1. Adams KL, Wendel JF. Polyploidy and genome evolution in plants. Curr Opin Plant Biol. 2005;8:135–141.
2. Ahloowalia BS, Maluszynski M, Nichterlein K. Global impact of mutation-derived varieties Euphytica. 2004;135:187–204.
3. Ashikari M et al (2002) Loss-of-function of a rice gibberellin biosynthetic gene, *GA20 oxidase* (*GA20ox-2*) led to the rice 'Green Revolution'. Breed Sci 52
4. Barakat A, Carels N, Bernardi G. The distribution of genes in the genomes of Gramineae. Proc Natl Acad Sci USA. 1997;94:6857–6861.
5. Barros E, Lezar S, Anttonen MJ, van Dijk JP, Röhlig RM, Kok EJ, Engel K-H. Comparison of two GM maize varieties with a near-isogenic non-GM variety using transcriptomics, proteomics and metabolomics. Plant Biotechnol J. 2010;8:436–451.
6. Batista R, Saibo N, Lourenço T, Oliveira MM. Microarray analyses reveal that plant mutagenesis may induce more transcriptomic changes than transgene insertion. Proc Natl Acad Sci USA. 2008;105:3640–3645.
7. Baudo MM, Lyons R, Powers S, Pastori GM, Edwards KJ, Holdsworth MJ, Shewry PR. Transgenesis has less impact on the transcriptome of wheat grain than conventional breeding. Plant Biotechnol J. 2006;4:369–380.
8. Bennetzen JL. Transposable element contributions to plant gene and genome evolution. Plant Mol Biol. 2000;42:251–269.
9. Bennetzen JL. Transposable elements, gene creation and genome rearrangement in flowering plants. Curr Opin Genet Dev.2005;15:621–627.
10. Bock R. The give-and-take of DNA: horizontal gene transfer in plants. Trends Plant Sci. 2010;15:11–22.
11. Bolon Y-T, et al. Phenotypic and genomic analyses of a fast neutron mutant population resource in soybean. Plant Physiol.2011;156:240–253.
12. Buchberger A, Bukau B, Sommer T. Protein quality control in the cytosol and the endoplasmic reticulum: brothers in arms. Mol Cell. 2010;40:238–252.
13. Butelli E, et al. Retrotransposons control fruit-specific, cold-dependent accumulation of anthocyanins in blood oranges. Plant Cell Online. 2012;24:1242–1255.
14. Casacuberta JM, Santiago N. Plant LTR-retrotransposons and MITEs: control of transposition and impact on the evolution of plant genes and genomes. Gene. 2003;311:1–11.
15. Cecchini E, Mulligan BJ, Covey SN, Milner JJ. Characterization of gamma irradiation-induced deletion mutations at a selectable locus in Arabidopsis. Mutat Res. 1998;401:199–206.
16. CFIA (2012a) Directive 94-08 (Dir 94-08). Assessment criteria for determining environmental safety of plants with novel traits
17. CFIA (2012b) RG1 regulatory guidance: feed registration procedures and labelling standards
18. Cheng KC, Beaulieu J, Iquira E, Belzile FJ, Fortin MG, Strömvik MV. Effect of transgenes on global gene expression in soybean is within the natural range of variation of conventional cultivars. J Agric Food Chem. 2008;56:3057–3067.
19. Chern C-G, et al. A rice phenomics study—phenotype scoring and seed propagation of a T-DNA insertion-induced rice mutant population. Plant Mol Biol. 2007;65:427–438.
20. Ching A, et al. SNP frequency, haplotype structure and linkage disequilibrium in elite maize inbred lines. BMC Genet.2002;3:19.
21. Codex Alimentarius Commission (2003) Guideline for the conduct of food safety assessment of foods derived from recombinant-DNA plants. CAC/GL 45-2003
22. Comai L. The advantages and disadvantages of being polyploid.Nat Rev Genet. 2005;6:836–846.
23. Cooper JL, et al. TILLING to detect induced mutations in soybean. BMC Plant Biol. 2008;8:9.
24. Dan Y, Ow DW (2011) Plant transformation technology revolution in last three decades: historical technology developments in plant transformation, vol 1. Bentham Science
25. Ding J, Araki H, Wang Q, Zhang P, Yang S, Chen J-Q, Tian D.Highly asymmetric rice genomes BMC Genom. 2007;8:154.
26. Doma MK, Parker R. RNA quality control in eukaryotes. Cell.2007;131:660–668.
27. Du X-Z, Ge X-H, Zhao Z-G, Li Z-Y. Chromosome elimination and fragment introgression and recombination producing intertribal partial hybrids from *Brassica napus* × *Lesquerella fendleri* crosses. Plant Cell Rep. 2008;27:261–271.
28. El Baidouri M, et al. Widespread and frequent horizontal transfers of transposable elements in plants. Genome Res.2014;24:831–838.
29. El Ouakfaoui S, Miki B. The stability of the Arabidopsis transcriptome in transgenic plants expressing the marker genes*nptII* and *uidA*. Plant J. 2005;41:791–800.
30. Faure N, Serieys H, Bervillé A, Cazaux E, Kaan F. Occurrence of partial hybrids in wide crosses between sunflower (*Helianthus annuus*) and perennial species *H. mollis* and *H. orgyalis*. Theor Appl Genet. 2002;104:652–660.
31. Fedak G. Molecular aids for integration of alien chromatin through wide crosses. Genome. 1999;42:584–591.

32. Forsbach A, Schubert D, Lechtenberg B, Gils M, Schmidt R. A comprehensive characterization of single-copy T-DNA insertions in the *Arabidopsis thaliana* genome. Plant Mol Biol.2003;52:161–176.

33. Francis KE, Spiker S. Identification of *Arabidopsis thaliana*transformants without selection reveals a high occurrence of silenced T-DNA integrations. Plant J. 2005;41:464–477.

34. Friebe B, Jiang J, Raupp WJ, McIntosh RA, Gill BS. Characterization of wheat-alien translocations conferring resistance to diseases and pests: current status. Euphytica.1996;91:59–87.

35. Fujino K, Sekiguchi H, Kiguchi T. Identification of an active transposon in intact rice plants. Mol Genet Genomics.2005;273:150–157.

36. Gaut BS, Wright SI, Rizzon C, Dvorak J, Anderson LK. Recombination: an underappreciated factor in the evolution of plant genomes. Nat Rev Genet. 2007;8:77–84.

37. Gelvin SB. Agrobacterium and plant genes involved in T-DNA transfer and integration. Annu Rev Plant Biol. 2000;51:223–256.

38. Goff SA, et al. A draft sequence of the rice genome (*Oryza sativa*L ssp *japonica*) Science. 2002;296:92–100.

39. Gorbunova V, Levy AA. How plants make ends meet: DNA double-strand break repair. Trends Plant Sci. 1999;4:263–269.

40. Grandbastien M-A. Activation of plant retrotransposons under stress conditions. Trends Plant Sci. 1998;3:181–187.

41. Gray YHM. It takes two transposons to tango: transposable-element-mediated chromosomal rearrangements. Trends Genet.2000;16:461–468.

42. Greco R, et al. Transposon insertional mutagenesis in rice. Plant Physiol. 2001;125:1175–1177.

43. Greene EA, et al. Spectrum of chemically induced mutations from a large-scale reverse-genetic screen in Arabidopsis.Genetics. 2003;164:731–740.

44. Hajjar R, Hodgkin T. The use of wild relatives in crop improvement: a survey of developments over the last 20 years.Euphytica. 2007;156:1–13.

45. Hirochika H, Sugimoto K, Otsuki Y, Tsugawa H, Kanda M. Retrotransposons of rice involved in mutations induced by tissue culture. Proc Natl Acad Sci USA. 1996;93:7783–7788.

46. Hoffmann GR. Genetic effects of dimethyl sulfate, diethyl sulfate, and related compounds. Mutat Res. 1980;75:63–129.

47. Huang JT, Dooner HK. The spectrum and frequency of self-inflicted and host gene mutations produced by the transposon *Ac*in maize. Plant Cell. 2012;24:4149–4162.

48. Huang CY, Ayliffe MA, Timmis JN. Direct measurement of the transfer rate of chloroplast DNA into the nucleus. Nature.2003;422:72–76.

49. Huang CY, Ayliffe MA, Timmis JN. Simple and complex nuclear loci created by newly transferred chloroplast DNA in tobacco.Proc Natl Acad Sci USA. 2004;101:9710–9715.

50. Ichikawa T, et al. Sequence database of 1172 T-DNA insertion sites in Arabidopsis activation-tagging lines that showed phenotypes in T1 generation. Plant J. 2003;36:421–429.

51. Jelesko JG, Carter K, Thompson W, Kinoshita Y, Gruissem W. Meiotic recombination between paralogous *RBCSB* genes on sister chromatids of *Arabidopsis thaliana*. Genetics.2004;166:947–957.

52. Jeong D-H, et al. T-DNA insertional mutagenesis for activation tagging in rice. Plant Physiol. 2002;130:1636–1644.

53. Jiang N, Bao Z, Zhang X, Eddy SR, Wessler SR (2004) Pack-MULE transposable elements mediate gene evolution in plants. Nature 431:569–573.http://www.nature.com/nature/journal/v431/n7008/suppinfo/nature02953_S1.html

54. Jiang C, et al. Regenerant *Arabidopsis* lineages display a distinct genome-wide spectrum of mutations conferring variant phenotypes. Curr Biol. 2011;21:1385–1390.

55. Jiao Y, et al. Ancestral polyploidy in seed plants and angiosperms. Nature. 2011;473:97–100.

56. Jones SS, Murray TD, Allan RE. The development of disease resistance in wheat. Annu Rev Phytopathol. 1995;33:429–443.

57. Karp A, Maddock SE. Chromosome variation in wheat plants regenerated from cultured immature embryos. Theor Appl Genet.1984;67:249–255.

58. Kawaura K, Mochida K, Yamazaki Y, Ogihara Y. Transcriptome analysis of salinity stress responses in common wheat using a 22 k oligo-DNA microarray. Funct Integr Genomics.2006;6:132–142.

59. Kessler D, Taylor M, Maryanski J, Flamm E, Kahl L. The safety of foods developed by biotechnology. Science (New York, NY)1992;256:1747–1749.

60. Kim SI, Gelvin SB. Genome-wide analysis of Agrobacterium T-DNA integration sites in the Arabidopsis genome generated under non-selective conditions. Plant J. 2007;51:779–791.

61. Kim S-R, Lee J, Jun S-H, Park S, Kang H-G, Kwon S, An G. Transgene structures in T-DNA-inserted rice plants. Plant Mol Biol. 2003;52:761–773.

62. Kobayashi S, Goto-Yamamoto N, Hirochika H. Retrotransposon-induced mutations in grape skin color Science. 2004;304:982.

63. Kohli A, Griffiths S, Palacios N, Twyman RM, Vain P, Laurie DA, Christou P. Molecular characterization of transforming plasmid rearrangements in transgenic rice reveals a recombination hotspot in the CaMV 35S promoter and confirms the predominance of microhomology mediated recombination.Plant J. 1999;17:591–601.

64. König A, et al. Assessment of the safety of foods derived from genetically modified (GM) crops. Food Chem Toxicol.2004;42:1047–1088.

65. Krysan PJ, Young JC, Jester PJ, Monson S, Copenhaver G, Preuss D, Sussman MR. Characterization of T-DNA insertion sites in *Arabidopsis thaliana* and the

implications for saturation mutagenesis OMICS. J Integr Biol. 2002;6:163–174.

66. Kuromori T, et al. A trial of phenome analysis using 4000 *Ds*-insertional mutants in gene-coding regions of Arabidopsis. Plant J. 2006;47:640–651.

67. Lai J, et al. Gene loss and movement in the maize genome.Genome Res. 2004;14:1924–1931.

68. Lai J, et al. Genome-wide patterns of genetic variation among elite maize inbred lines. Nat Genet. 2010;42:1027–1030.

69. Lam H-M, et al. Resequencing of 31 wild and cultivated soybean genomes identifies patterns of genetic diversity and selection. Nat Genet. 2010;42:1053–1059.

70. Lee M, Phillips RL. Genomic rearrangements in maize induced by tissue culture. Genome. 1987;29:122–128.

71. Leitch AR, Leitch IJ. Genomic plasticity and the diversity of polyploid plants. Science. 2008;320:481–483.

72. Li X, et al. A fast neutron deletion mutagenesis-based reverse genetics system for plants Plant J. 2001;27:235–242.

73. Lisch D. How important are transposons for plant evolution? Nat Rev Genet. 2013;14:49–61.

74. Liu M, Li Z-Y. Genome doubling and chromosome elimination with fragment recombination leading to the formation of*Brassica rapa*-type plants with genomic alterations in crosses with *Orychophragmus violaceus*. Genome. 2007;50:985–993.

75. Liu R, Koyanagi KO, Chen S, Kishima Y. Evolutionary force of AT-rich repeats to trap genomic and episomal DNAs into the rice genome: lessons from endogenous pararetrovirus. Plant J.2012;72:817–828.

76. Makarevitch I, Svitashev SK, Somers DA. Complete sequence analysis of transgene loci from plants transformed via microprojectile bombardment. Plant Mol Biol. 2003;52:421–432.

77. May BP, Dellaporta SL. Transposon sequences drive tissue-specific expression of the maize regulatory gene *R-s*. Plant J.1998;13:241–247.

78. Mayerhofer R, et al. T-DNA integration: a mode of illegitimate recombination in plants. EMBO J. 1991;10:697–704.

79. McClintock B. The significance of responses of the genome to challenge. Science. 1984;226:792–801.

80. McHale LK, et al. Structural variants in the soybean genome localize to clusters of biotic stress-response genes. Plant Physiol.2012;159:1295–1308.

81. McNally KL, et al. Genomewide SNP variation reveals relationships among landraces and modern varieties of rice. Proc Natl Acad Sci USA. 2009;106:12273–12278.

82. Messing J, et al. Sequence composition and genome organization of maize. Proc Natl Acad Sci USA. 2004;101:14349–14354.

83. Moon S, et al. Identification of active transposon *dTok*, a member of the *hAT* family, in rice. Plant Cell Physiol. 2006;47:1473–1483.

84. Morita R, Kusaba M, Iida S, Yamaguchi H, Nishio T, Nishimura M. Molecular characterization of mutations induced by gamma irradiation in rice. Genes Genet Syst. 2009;84:361–370.

85. Nacry P, Camilleri C, Courtial B, Caboche M, Bouchez D. Major chromosomal rearrangements induced by T-DNA transformation in Arabidopsis. Genetics. 1998;149:641–650.

86. Naito K, et al. Dramatic amplification of a rice transposable element during recent domestication. Proc Natl Acad Sci USA.2006;103:17620–17625.

87. Nambara E, Keith K, McCourt P, Naito S. Isolation of an internal deletion mutant of the *Arabidopsis thalianaABI3* Gene. Plant Cell Physiol. 1994;35:509–513.

88. Noutsos C, Kleine T, Armbruster U, DalCorso G, Leister D. Nuclear insertions of organellar DNA can create novel patches of functional exon sequences. Trends Genet. 2007;23:597–601.

89. OECD . Recombinant-DNA safety considerations. Paris: Organisation for Economic Co-Operation and Development; 1986.

90. OECD . Safety considerations for biotechnology. Paris: Organisation for Economic Co-operation and Development; 1992.

91. OECD . Principles for the risk analysis of foods derived from modern biotechnology. Paris: Organisation for Economic Co-operation and Development; 2003.

92. Olsen O, Wang X, von Wettstein D. Sodium azide mutagenesis: preferential generation of A.T → G.C transitions in the barley*Ant18* gene. Proc Natl Acad Sci USA. 1993;90:8043–8047.

93. Ossowski S, et al. The rate and molecular spectrum of spontaneous mutations in *Arabidopsis thaliana*. Science.2010;327:92–94.

94. Parisod C, et al. Impact of transposable elements on the organization and function of allopolyploid genomes. New Phytol.2010;186:37–45.

95. Potato Genome Sequencing Consortium Genome sequence and analysis of the tuber crop potato. Nature. 2011;475:189–195.

96. Price WD, Underhill L. Application of laws, policies, and guidance from the united states and canada to the regulation of food and feed derived from genetically modified crops: interpretation of composition data. J Agric Food Chem.2013;61:8349–8355.

97. Richly E, Leister D. NUPTs in sequenced eukaryotes and their genomic organization in relation to NUMTs. Mol Biol Evol.2004;21:1972–1980.

98. Rinehart TA, Dean C, Weil CF. Comparative analysis of non-random DNA repair following *Ac* transposon excision in maize and *Arabidopsis*. Plant J. 1997;12:1419–1427.

99. Rubin E, Levy AA. Abortive gap repair: underlying mechanism for *Ds* element formation. Mol Cell Biol. 1997;17:6294–6302.

100. Sallaud C, et al. Highly efficient production and characterization of T-DNA plants for rice (*Oryza sativa* L.) functional genomics.Theor Appl Genet. 2003;106:1396–1408.

101. Salomon S, Puchta H. Capture of genomic and T-DNA sequences during double-strand break repair in somatic plant cells. EMBO J. 1998;17:6086–6095.

102. Sato Y, Morita R, Katsuma S, Nishimura M, Tanaka A, Kusaba M. Two short-chain dehydrogenase/reductases, NON-YELLOW COLORING 1 and NYC1-LIKE, are required for chlorophyll *b*and light-harvesting complex II degradation during senescence in rice. Plant J. 2009;57:120–131.

103. Schmutz J et al (2010) Genome sequence of the palaeopolyploid soybean. Nature 463:178–183.http://www.nature.com/nature/journal/v463/n7278/suppinfo/nature08670_S1.html

104. Schuermann D, Molinier J, Fritsch O, Hohn B. The dual nature of homologous recombination in plants. Trends Genet.2005;21:172–181.

105. Seah S, Telleen AC, Williamson VM. Introgressed and endogenous *Mi-1* gene clusters in tomato differ by complex rearrangements in flanking sequences and show sequence exchange and diversifying selection among homologues. Theor Appl Genet. 2007;114:1289–1302.

106. Shan X, et al. Mobilization of the active MITE transposons*mPing* and *Pong* in rice by introgression from wild rice (*Zizania latifolia* Griseb.) Mol Biol Evol. 2005;22:976–990.

107. Shirley BW, Hanley S, Goodman HM. Effects of ionizing radiation on a plant genome: analysis of two Arabidopsis*transparent testa* mutations. Plant Cell. 1992;4:333–347.

108. Sim S-C, Robbins MD, Chilcott C, Zhu T, Francis DM. Oligonucleotide array discovery of polymorphisms in cultivated tomato (*Solanum lycopersicum* L.) reveals patterns of SNP variation associated with breeding. BMC Genom. 2009;10:466.

109. Somers DA, Makarevitch I. Transgene integration in plants: poking or patching holes in promiscuous genomes? Curr Opin Biotechnol. 2004;15:126–131.

110. Staginnus C, Richert-Pöggeler KR. Endogenous pararetroviruses: two-faced travelers in the plant genome. Trends Plant Sci.2006;11:485–491.

111. Staginnus C, et al. Endogenous pararetroviral sequences in tomato (*Solanum lycopersicum*) and related species. BMC Plant Biol. 2007;7:24.

112. Steiner H-Y, et al. Evaluating the potential for adverse interactions within genetically engineered breeding stacks. Plant Physiol. 2013;161:1587–1594.

113. Stewart CN, Touraev A, Citovsky V, Tzfira T. Plant transformation technologies. New York: Wiley; 2011.

114. Swanson-Wagner RA, Eichten SR, Kumari S, Tiffin P, Stein JC, Ware D, Springer NM. Pervasive gene content variation and copy number variation in maize and its undomesticated progenitor.Genome Res. 2010;20:1689–1699.

115. Takano M, Egawa H, Ikeda J-E, Wakasa K. The structures of integration sites in transgenic rice. Plant J. 1997;11:353–361.

116. Talamè V, Bovina R, Sanguineti MC, Tuberosa R, Lundqvist U, Salvi S. TILLMore, a resource for the discovery of chemically induced mutants in barley. Plant Biotechnol J. 2008;6:477–485.

117. The Arabidopsis Genome Initiative (2000) Analysis of the genome sequence of the flowering plant *Arabidopsis thaliana*. Nature 408:796–815.http://www.nature.com/nature/journal/v408/n6814/suppinfo/408796a0_S1.html

118. T-p Sun, Goodman HM, Ausubel FM. Cloning the Arabidopsis*GA1* locus by genomic subtraction. Plant Cell. 1992;4:119–128.

119. Tu Y, Sun J, Ge X, Li Z. Chromosome elimination, addition and introgression in intertribal partial hybrids between *Brassica rapa*and *Isatis indigotica*. Ann Bot. 2009;103:1039–1048.

120. Uauy C, et al. A modified TILLING approach to detect induced mutations in tetraploid and hexaploid wheat. BMC Plant Biol.2009;9:115.

121. van den Bulk RW, Löffler HJM, Lindhout WH, Koornneef M. Somaclonal variation in tomato: effect of explant source and a comparison with chemical mutagenesis. Theor Appl Genet.1990;80:817–825.

122. Vanin EF. Processed pseudogenes: characteristics and evolution.Annu Rev Genet. 1985;19:253–272.

123. Vitte C, Fustier M-A, Alix K, Tenaillon MI (2014) The bright side of transposons in crop evolution. Brief Funct Genomics:elu002

124. Walbot V. Saturation mutagenesis using maize transposons. Curr Opin Plant Biol. 2000;3:103–107.

125. Walker EL, Robbins TP, Bureau TE, Kermicle J, Dellaporta SL. Transposon-mediated chromosomal rearrangements and gene duplications in the formation of the maize *R-r* complex. EMBO J.1995;14:2350–2363.

126. Wang Y-M, Dong Z-Y, Zhang Z-J, Lin X-Y, Shen Y, Zhou D, Liu B. Extensive *de novo* genomic variation in rice induced by introgression from wild rice (*Zizania latifolia* Griseb.) Genetics.2005;170:1945–1956.

127. Wang N, et al. Transpositional reactivation of the *Dart*transposon family in rice lines derived from introgressive hybridization with *Zizania latifolia*. BMC Plant Biol.2010;10:190.

128. Wang D, Lloyd AH, Timmis JN. Environmental stress increases the entry of cytoplasmic organellar DNA into the nucleus in plants. Proc Natl Acad Sci USA. 2012;109:2444–2448.

129. Warwick SI, Stewart CN (2005) Crops come from wild plants—how domestication, transgenes, and linkage together shape ferality. In: Gressel J (ed) Crop ferality and volunteerism. CRC Press, Boca Raton, pp 9–30. doi:10.1201/9781420037999.ch2

130. Weber N, Halpin C, Hannah LC, Jez JM, Kough J, Parrott W. Crop genome plasticity and its relevance to food and feed safety of genetically engineered breeding stacks. Plant Physiol.2012;160:1842–1853.

131. Wicker T, Buchmann JP, Keller B. Patching gaps in plant genomes results in gene movement and erosion of colinearity.Genome Res. 2010;20:1229–1237.

132. Wicker T, et al. Frequent gene movement and pseudogene evolution is common to the large and complex genomes

of wheat, barley, and their relatives. Plant Cell Online. 2011;23:1706–1718.

133. Windels P, De Buck S, Van Bockstaele E, De Loose M, Depicker A. T-DNA integration in Arabidopsis chromosomes. Presence and origin of filler DNA sequences. Plant Physiol.2003;133:2061–2068.

134. Xiao H, Jiang N, Schaffner E, Stockinger EJ, van der Knaap E. A retrotransposon-mediated gene duplication underlies morphological variation of tomato fruit. Science.2008;319:1527–1530.

135. Xin Z, Wang ML, Barkley NA, Burow G, Franks C, Pederson G, Burke J. Applying genotyping (TILLING) and phenotyping analyses to elucidate gene function in a chemically induced sorghum mutant population. BMC Plant Biol. 2008;8:103.

136. Xu C-Y, Wan-Yan R-H, Li Z-Y. Origin of new *Brassica* types from a single intergeneric hybrid between *B. rapa* and*Orychophragmus violaceus* by rapid chromosome evolution and introgression. J Genet. 2007;86:249–257.

137. Yan C, Yan S, Zeng X, Zhang Z, Gu M. Fine mapping and isolation of *Bc7(t)*, allelic to *OsCesA4*. J Genet Genomics.2007;34:1019–1027.

138. Young N, Tanksley S. RFLP analysis of the size of chromosomal segments retained around the Tm-2 locus of tomato during backcross breeding. Theor Appl Genet. 1989;77:353–359.

139. Yu J, et al. A draft sequence of the rice genome (*Oryza sativa* L. ssp. *indica*) Science. 2002;296:79–92.

140. Zabala G, Vodkin L. Novel exon combinations generated by alternative splicing of gene fragments mobilized by a CACTA transposon in glycine max. BMC Plant Biol. 2007;7:38.

141. Zheng J, et al. Genome-wide transcriptome analysis of two maize inbred lines under drought stress. Plant Mol Biol. 2010;72:407–421.

142. Zou C, Lehti-Shiu MD, Thibaud-Nissen F, Prakash T, Buell CR, Shiu S-H. Evolutionary and expression signatures of pseudogenes in Arabidopsis and rice. Plant Physiol. 2009;151:3–15.

143. Zou J, et al. *De novo* genetic variation associated with retrotransposon activation, genomic rearrangements and trait variation in a recombinant inbred line population of *Brassica napus* derived from interspecific hybridization with *Brassica rapa*. Plant J. 2011;68:212–224.

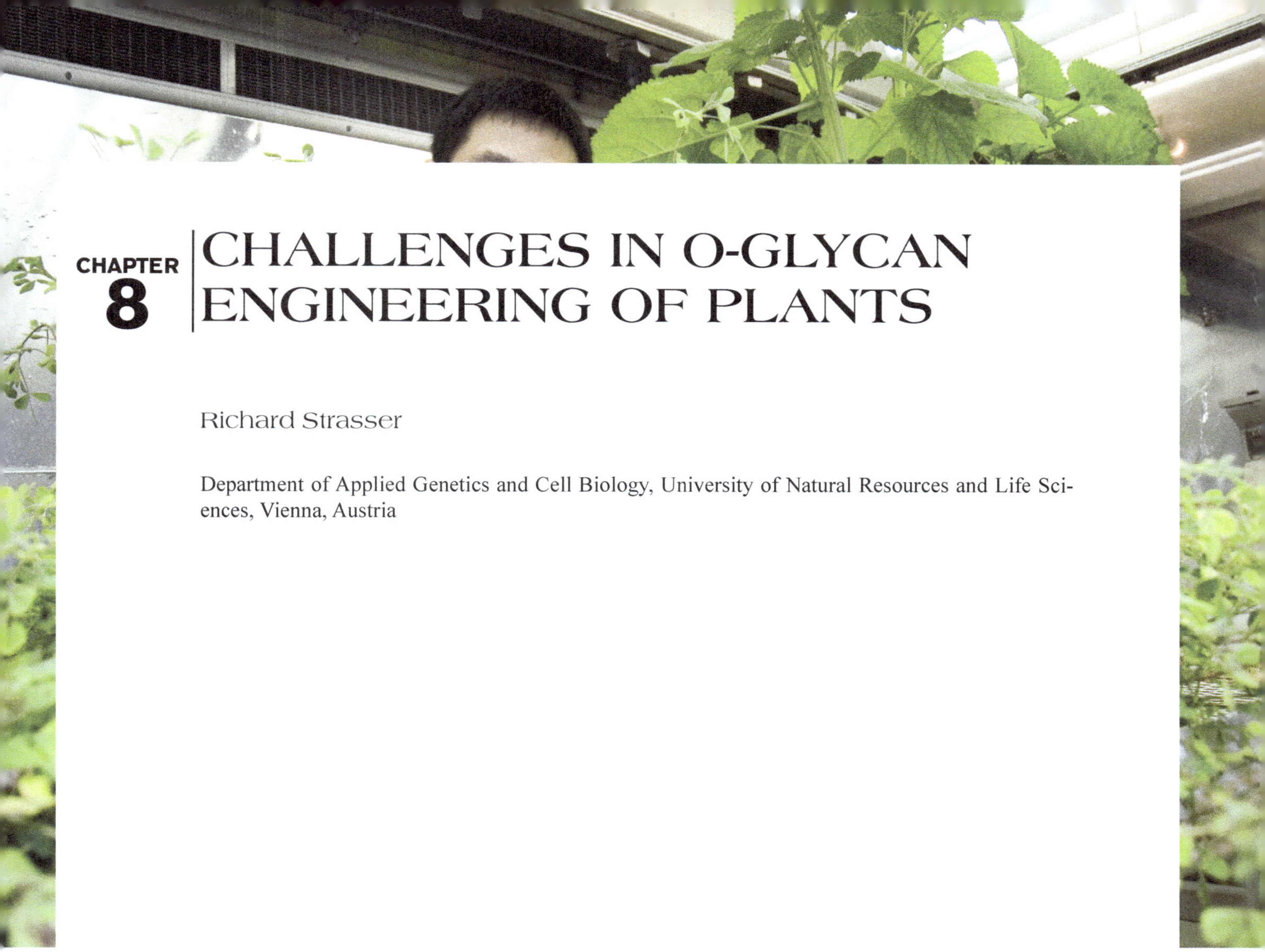

CHAPTER 8 | CHALLENGES IN O-GLYCAN ENGINEERING OF PLANTS

Richard Strasser

Department of Applied Genetics and Cell Biology, University of Natural Resources and Life Sciences, Vienna, Austria

ABSTRACT

Plants are attractive alternative expression hosts for the production of recombinant proteins. Many therapeutic proteins are glycosylated with N- and O-glycosylation being the most prevalent forms of protein glycosylation. While N-glycans have already been modified in plants toward the formation of homogenous mammalian-type glycoforms with equal or improved biological function compared to mammalian-cell culture produced glycoproteins little attention has been paid to the modification of O-linked glycans. Recently, the first step of mammalian O-glycan biosynthesis has been accomplished in plants. However, as outlined in this short review there are important issues that have to be addressed in the future. These include: (i) elimination of potentially immunogenic or allergenic carbohydrate epitopes containing arabinosides or arabinogalactans, (ii) a detailed investigation of the interplay between engineered N- and O-glycosylation pathways to avoid competition for common metabolites like UDP-GlcNAc, and (iii) a deeper understanding of signals and mechanisms for distribution of glycan processing enzymes, which is a prerequisite for complete and homogenous glycosylation of recombinant proteins.

INTRODUCTION

Recombinant pharmaceuticals are the fastest growing class of novel medicine with monoclonal antibodies, hormones like erythropoietin (EPO) and growth factors as the major drivers (Aggarwal, 2010). The majority of the protein-based therapeutics are glycosylated (Durocher and Butler, 2009) and the impact of different glycoforms on the function of recombinant proteins is well documented (Jefferis, 2009). Despite these facts, current mammalian cell culture-based expression systems cannot produce customized glycoforms on recombinant proteins. One reason for this inability lies in the complex mammalian glycome that consists of hundreds of glycosylation enzymes (Cummings, 2009) and produces an unwanted background glycosylation in expression hosts. Consequently, there is a growing demand for production systems that allow the control of glycosylation and generate defined homogenous glycans (Rich and Withers, 2009). Plants are an attractive alternative because they are cost-effective, highly scalable, free from human pathogens, and, importantly, can carry out post-translational modifications like glycosylation similar to mammals (Ma et al., 2003; Gomord et al., 2010; Nagels et al., 2012). *N*- and *O*-glycans are the two major carbohydrates on recombinant protein therapeutics. Both glycosylation types differ in their linkage to the

Citation: Richard Strasser, Challenges in O-glycan engineering of plants, doi: 10.3389/fpls.2012.00218.

protein backbone, their biosynthesis and sugar composition. Engineering of the *N*-glycosylation pathway of various plant species led to remarkable success resulting not only in the removal of immunogenic sugar moieties like β1,2-linked xylose and core α1,3-linked fucose (Koprivova et al., 2004; Strasser et al., 2004, 2008; Cox et al., 2006; Shin et al., 2011) but also in the complete reconstruction of mammalian glycosylation pathways in plants (Palacpac et al., 1999;Bakker et al., 2001; Rouwendal et al., 2009; Strasser et al., 2009;Castilho et al., 2010, 2011; Nagels et al., 2011). In contrast to that, comparatively little attempts have so far been made to modify the *O*-glycosylation machinery of plants. Here, I give a short overview of recent developments toward the production of humanized *O*-glycan structures and summarize challenges for the future.

MUCIN-TYPE *O*-GLYCAN BIOSYNTHESIS

O-glycosylation is a common post-translational modification of serine (Ser)/threonine (Thr) residues of secreted and membrane-bound mammalian proteins. *O*-glycosylation is fundamentally different from*N*-glycosylation as a typical consensus amino acid sequence has not been clearly identified yet (Bennett et al., 2012) and *O*-glycan biosynthesis occurs in a stepwise fashion involving the sequential transfer of single sugar residues by distinct glycosyltransferases. The initiation of mucin-type *O*-glycan formation, which is the most common *O*-linked glycan in humans, encompasses the transfer of an *N*-acetylgalactosamine (GalNAc) residue from the nucleotide sugar UDP-GalNAc to hydroxyl side chains of Ser/Thr. This specific reaction is catalyzed by a family of more than 20 different polypeptide GalNAc-transferases. Initiation of mucin-type *O*-glycosylation in the Golgi suggests that the *O*-linked structures are generated entirely post-translationally, after folding and subunit assembly has been completed in the endoplasmic reticulum. Due to the possibility of various chain elongation and branching steps there is considerably heterogeneity of mucin-type *O*-glycans. Typically these carbohydrates can be further elongated by incorporation of galactose, fucose, *N*-acetylglucosamine (GlcNAc) and sialic acid residues in different linkages (Tarp and Clausen, 2008). Single *O*-linked glycans are for example found on EPO or on interleukin-2, while other human proteins like mucin-1 (MUC1) or immunoglobulin A (IgA) contain clusters of *O*-glycans. Since specific *O*-glycans are more abundant on tumors it has been proposed that peptides containing defined *O*-glycan structures can be used as anti-cancer vaccines to elicit an immune reaction against epitopes present on cancer cells (Springer, 1984; Ju et al., 2011).

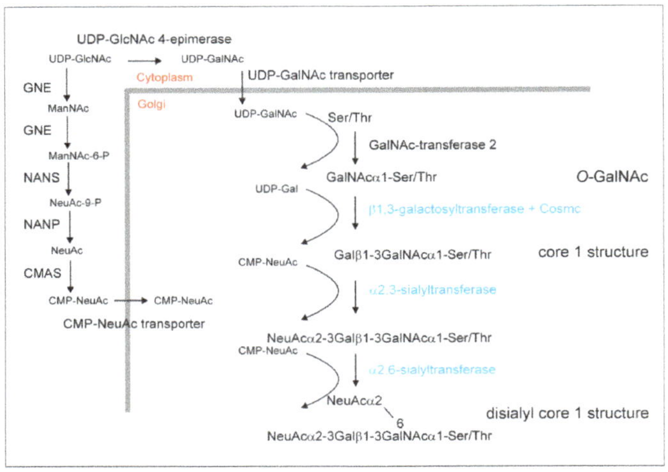

Figure 1: *O*-glycosylation engineering steps required to produce the disialylated core 1 structure which is, for example, the major *O*-glycan on Ser-126 of human EPO (Tsuda et al., 1990) In addition to providing CMP-NeuAc in the Golgi and initiation of *O*-glycosylation...

Plants lack the machinery to produce mammalian-type *O*-glycosylation (Daskalova et al., 2010; Yang et al., 2012) and recombinant EPO produced in *Nicotiana benthamiana* does not contain any *O*-linked GalNAc residues (Castilho et al., 2011). To initiate mucin-type *O*-glycan formation the corresponding mammalian GalNAc-transferase that transfers a single GalNAc residue to Ser/Thr residues has to be expressed in plants. Apart from the GalNAc-transferase and the acceptor substrate *O*-GalNAc formation depends on the presence of the nucleotide sugar in the same subcellular compartment. The glycoengineering of the *O*-glycosylation initiation step has been reported recently by two groups. Daskalova et al. (2010) could show by lectin blotting that the transfer of single GalNAc residues to a transiently expressed *O*-glycosylation substrate containing human MUC1 tandem repeats requires the expression of three non-plant proteins in *N. benthamiana*. To generate sufficient amounts of the UDP-GalNAc donor substrate and transport it into the Golgi apparatus, a microbial UDP-GlcNAc 4-epimerase and a UDP-GlcNAc/UDP-GalNAc transporter from nematodes were stably expressed together with human GalNAc-transferase 2. A similar approach was used by Yang et al. (2012) to engineer *O*-GalNAc residues into three different reporter proteins transiently expressed in *N. benthamiana*. Structural analysis by mass spectrometry confirmed the presence of GalNAc residues in a MUC1 tandem repeat containing peptide (Yang et al., 2012). Transient expression of two different human GalNAc-transferases with partly overlapping substrate specificities resulted in high-density *O*-GalNAc formation on multiple sites of the recombinant MUC1 peptide. Interestingly, *O*-glycans were also detected without ectopic expression of a UDP-GalNAc transporter indicating that the expression of two non-plant proteins is sufficient to initiate mucin-type*O*-glycosylation in plants. A possible explanation for the discrepancy between the two studies comes from differences in the described expression levels. Both groups used transient expression of their *O*-glycosylation substrates in *N. benthamiana*, but in the first study the expression of the MUC1 reporter construct was driven by the viral-based magnICON vector system and resulted in higher expression levels (Daskalova et al., 2010; Yang et al., 2012). It is therefore possible that at lower expression an endogenous plant transporter could provide sufficient amounts of UDP-GalNAc while under high productivity conditions the nucleotide sugar pool in the Golgi is rate limiting.

Apart from *O*-GalNAc formation no further elongation or branching reactions have been reported so far in plants. One important elongation is the transfer of a galactose residue in β1,3-linkage to *O*-GalNAc to generate a core 1 *O*-glycan structure. This reaction is catalyzed in the Golgi by core 1 β1,3-galactosyltransferase. For efficient folding human core 1 β1,3-galactosyltransferase is dependent on the assistance of a specific molecular chaperone termed Cosmc (Ju and Cummings, 2002). An ortholog of Cosmc is not present in plants and consequently it is expected that the efficient formation of core 1 *O*-glycans on plant-produced recombinant proteins requires the expression of both mammalian proteins. Alternatively, the expression of an invertebrate core 1 β1,3-galactosyltransferase that is functional in the absence of a specific chaperone and has been successfully used in an *O*-glycan engineering approach in yeast (Amano et al., 2008) could be exploited for this specific elongation step. The core 1 structure is the major *O*-glycan present on IgA1 molecules and hypogalactosylated IgA1 variants result in the elicitation of autoantibodies against galactose-deficient IgA1 proteins (Wada et al., 2010). This finding highlights the need for expression systems with controlled *O*-glycan formation on recombinant proteins to avoid adverse immunologic reactions.

Another widespread modification of mammalian *O*-glycans is the attachment of sialic acid in different linkages. Typical disialylated structures are found on glycoprotein hormones like EPO and small amounts are present in the hinge region of the IgA1 heavy chain (Mattu et al., 1998). The efficient generation of sialylated *O*-glycans involves the expression of the machinery for CMP-*N*-acetylneuraminic acid (CMP-NeuAc) biosynthesis and transport of this nucleotide sugar from the cytoplasm to the Golgi lumen and distinct sialyltransferases catalyzing the transfer of sialic acid residues from CMP-NeuAc to *O*-linked glycans. While the mammalian sialic acid biosynthesis pathway has been successfully constructed in *Arabidopsis thaliana* and *N. benthamiana* (Castilho et al., 2008, 2010), the *O*-glycan specific mammalian sialyltransferases required for the generation of sialylated *O*-glycans have so far not been expressed in plants.

ENDOGENOUS PLANT *O*-GLYCOSYLATION

Mammalian *O*-GalNAc attachment sites are exposed on the protein surface and frequently contain nearby proline residues. Expression of recombinant glycoproteins with putative *O*-glycosylation sites in plants revealed the presence of plant-specific glycosylation. In particular, plants convert the proline residues to hydroxyproline and attach arabinose residues to recombinant proteins (Karnoup et al., 2005;Pinkhasov et al., 2011). These non-human glycans can seriously hamper the broad use of plant-made therapeutics since single arabinosyl residues linked to hydroxyprolines can constitute an IgE binding epitope and thus could play a role in allergic reactions (Leonard et al., 2005). It is currently unclear whether the plant-specific glycosylation of hydroxyprolines and mammalian glycosyltransferases compete for adjacent acceptor sites (Daskalova et al., 2010; Pinkhasov et al., 2011;Yang et al., 2012). Hydroxyproline residues have been found on MUC1 peptides expressed in *N. benthamiana* in the presence and absence of the *O*-glycosylation machinery necessary for the transfer of GalNAc residues (Yang et al., 2012). The effect of mammalian-type *O*-glycosylation on arabinosylation is unclear since no arabinose residues were detected on the expressed recombinant protein. Moreover, the prolyl 4-hydroxylases (P4Hs) that are responsible for the conversion of proline to hydroxyproline in tobacco and related species are not well described (Yuasa et al., 2005). However, several members of the *A. thaliana* P4H family have been characterized recently (Velasquez et al., 2011), which makes it now practicable to screen for P4H candidates that hydroxylate specific proline residues on recombinant glycoproteins and eliminate them from expression hosts similar to the successful removal of immunogenic sugar residues from *N*-glycans (Strasser et al., 2004, 2008). Such a strategy very likely requires the elimination of several P4Hs with overlapping substrate specificities and due to possible effects on cell wall assembly (Velasquez et al., 2011) it might be necessary to perform tissue-specific knockouts or knockdowns to avoid problems with biomass formation and overall growth of the plants. Apart from preventing P4Hs expression there have also been attempts to block the formation of hydroxyproline residues using inhibitors. The ferrous chelator 2,2′-dipyridyl, which is a potent inhibitor of P4Hs was recently used in tobacco seedlings to reduce arabinogalactosylation of endogenous proteins (Moriguchi et al., 2011). Such a chemical inhibition strategy could be quite useful for transient expression but is less suitable for the stable expression of *O*-glycosylated recombinant proteins.

N- AND *O*-GLYCAN ENGINEERING

Many valuable recombinant glycoproteins contain both *N*- and *O*-glycans. For example, IgA1 molecules carry in the hinge region a number of potential *O*-glycosylation sites, while other domains of the heavy chain are subjected to *N*-glycosylation (Ju et al., 2011). Human EPO contains three *N*-linked glycans along with a single *O*-linked glycan at Ser-126 and also other human plasma proteins that might be future candidates for plant-based production systems display several *N*- and *O*-glycans that contribute to their function. Consequently, we are facing the challenge to simultaneously modify both pathways in plants to achieve complete human-like glycosylation. Engineering of *N*-glycosylation on recombinant proteins involves the elimination of β1,2-xylose and core α1,3-fucose, the incorporation of β1,4-galactose as well as terminal sialic acid residues and branching of *N*-glycans.

For both *N*- and *O*-glycosylation engineering, UDP-GlcNAc is the common starting metabolite for the generation of nucleotide sugars (UDP-GalNAc, CMP-NeuAc) required as donor substrates for the different glycosyltransferases like GalNAc-transferases and sialyltransferases. Moreover, UDP-GlcNAc is also used as nucleotide sugar by endogenous *N*-acetylglucosami- nyltransferases and mammalian *N*-acetylglucosaminyltransferases that initiate branching of *N*-glycans when expressed in plants (Strasser et al., 1999;Castilho et al., 2011; Nagels et al., 2011). In mammals it is well documented that nucleotide sugar biosynthesis is tightly regulated and subjected to feedback regulation. So far there is no evidence that the engineered changes in the flux of nucleotide sugars are deleterious for plants. Castilho et al. (2008) have shown that stable transformed *A. thaliana* plants tolerate the generation of high amounts of CMP-sialic acid and its precursor without any drastic effect on plant growth and development. Importantly, neither in *A. thaliana* nor in *N. benthamiana*does the expression of the CMP-sialic acid pathway proteins result in alterations of GlcNAc levels on complex *N*-glycans from recombinant proteins. Consistent with that, no phenotypic changes have been reported in any plant that has been subjected to *N*- or *O*-glycan engineering (Castilho et al., 2008, 2010; Daskalova et al., 2010; Nagels et al., 2011; Yang et al., 2012) indicating that plants can efficiently maintain intracellular pools of different nucleotide sugars for the synthesis of various glycoconjugates. Although GalNAc has not been detected in any plant glycan so far, the presence of the corresponding nucleotide sugar UDP-GalNAc has been shown in *A. thaliana*suspension cells and in fenugreek endosperm tissue (Alonso et al., 2010). In both species the UDP-GlcNAc levels are three to four times higher than the UDP-GalNAc levels. However, the corresponding concentrations in tissues that are used for expression of recombinant proteins like *N. benthamiana* leaves and the steady-state concentration of UDP-GalNAc or other nucleotide sugars in the Golgi are completely unknown. In view of the fact that the recent studies on *O*-glycan engineering did not analyze *N*-glycans of recombinant or endogenous proteins its effect on the *N*-glycosylation capacity of plants remains to be shown. Clearly, further research is needed to better understand the nucleotide sugar interconversion reactions and mechanisms that control their biosynthesis and subcellular distribution, which will be necessary to achieve complete and highly homogenous glycosylation in plant expression systems that are subjected to intense *N*- and *O*-glycoengineering.

GOLGI ORGANIZATION OF GLYCOSYLATION ENZYMES

All secreted recombinant glycoproteins pass through the Golgi where they acquire their final glycosylation by Golgi-resident enzymes that are distributed in sequential Golgi cisternae. The targeting signals and mechanisms that regulate the organization of glycosylation enzymes in the early secretory pathway of plants are still poorly understood (Schoberer and Strasser, 2011). A common feature of the Golgi-located glycosylation enzymes is their type II membrane protein topology. For several plant glycosyltransferases and glycosidases it has been demonstrated that the Golgi-targeting information is present in their *N*-terminal cytoplasmic, transmembrane, and stem (CTS) region without any contribution from the large luminal catalytic domain (Essl et al., 1999; Saint-Jore-Dupas et al., 2006; Schoberer and Strasser, 2011). Studies on the generation of β1,4-galactosylated *N*-glycans revealed that sub-Golgi targeting of the human β1,4-galactosyltransferase is slightly different between plant and mammalian cells leading to aberrant glycosylation (Palacpac et al., 1999; Bakker et al., 2001;Strasser et al., 2009). Similar results were observed for mammalian *N*-acetylglucosaminyltransferases (Castilho et al., 2011; Nagels et al., 2011). Replacement of their *N*-terminal domains with targeting regions from Golgi-resident plant glycosylation enzymes resulted in a more homogenous glycosylation pattern on recombinant proteins. It is apparent from these results that further approaches to elongate and branch *O*-glycans need to address the precise localization of the glycosylation enzymes within the Golgi. Given the sequential nature of*O*-glycan biosynthesis it appears essential to organize glycosyltransferases into some kind of assembly line along the Golgi apparatus. Such an ordered distribution of glycosylation enzymes across Golgi stacks has been experimentally shown for endogenous plant enzymes involved in *N*-glycan processing (Schoberer et al., 2010). However, known targeting regions might not be sufficient to fulfill this goal as there is, for example, currently only one plant-derived CTS region for *trans*-Golgi localization of glycosyltransferases available (Strasser et al., 2007; Schoberer and Strasser, 2011). It is therefore fundamental to increase our understanding of the glycosylation enzyme organization and cargo transport within the Golgi. Especially, more efforts are required to decipher sub-Golgi targeting motifs present in the CTS region of glycosylation enzymes and identify key molecular players that control their subcellular localization.

CONCLUSION

Developments in the last 10 years have demonstrated that plants are well suited for the production of recombinant proteins with homogenous human-like *N*-glycosylation. The recent progress in initiation of mucin-type *O*-glycan formation indicates that plants are also amenable to engineering of defined mammalian-type *O*-glycosylation. However, as summarized here there are still limitations associated mainly with endogenous plant-specific glycosylation, cross-talk between engineered pathways and our restricted tools for compartmentalization of Golgi-located glycan processing enzymes that have to be overcome in the future. With the rapid progress in this field, the combination of *N*- and *O*-glycan modification steps is only a matter of time. Furthermore, current efforts in characterization of hydroxyproline formation and Golgi-targeting sequences will help to optimize the current engineering approaches and ultimately will result in the generation of efficient plant-based production platforms for glycoprotein therapeutics.

ACKNOWLEDGMENTS

This work was supported by a grant from the Federal Ministry of Transport, Innovation and Technology (bmvit) and Austrian Science Fund (FWF): TRP 242-B20.

REFERENCES

1. Aggarwal S. (2010). What's fueling the biotech engine – 2009–2010. *Nat. Biotechnol.* 28 1165–1171
2. Alonso A. P., Piasecki R. J., Wang Y., LaClair R. W., Shachar-Hill Y. (2010). Quantifying the labeling and the levels of plant cell wall precursors using ion chromatography tandem mass spectrometry. *Plant Physiol.* 153 915–924
3. Amano K., Chiba Y., Kasahara Y., Kato Y., Kaneko M. K., Kuno A., Ito H., Kobayashi K., Hirabayashi J., Jigami Y., Narimatsu H. (2008). Engineering of mucin-type human glycoproteins in yeast cells. *Proc. Natl. Acad. Sci. U.S.A.* 105 3232–3237
4. Bakker H., Bardor M., Molthoff J., Gomord V., Elbers I., Stevens L., Jordi W., Lommen A., Faye L., Lerouge P., Bosch D. (2001).Galactose-extended glycans of antibodies produced by transgenic plants. *Proc. Natl. Acad. Sci. U.S.A.* 98 2899–2904

5. Bennett E. P., Mandel U., Clausen H., Gerken T. A., Fritz T. A., Tabak L. A. (2012). Control of mucin-type *O*-glycosylation: a classification of the polypeptide GalNAc-transferase gene family.*Glycobiology* 22 736–756
6. Castilho A., Gattinger P., Grass J., Jez J., Pabst M., Altmann F., Gorfer M., Strasser R., Steinkellner H. (2011). *N*-glycosylation engineering of plants for the biosynthesis of glycoproteins with bisected and branched complex *N*-glycans. *Glycobiology* 21 813–823
7. Castilho A., Pabst M., Leonard R., Veit C., Altmann F., Mach L., Glössl J., Strasser R., Steinkellner H. (2008). Construction of a functional CMP-sialic acid biosynthesis pathway in *Arabidopsis*.*Plant Physiol.* 147 331–339
8. Castilho A., Strasser R., Stadlmann J., Grass J., Jez J., Gattinger P., Kunert R., Quendler H., Pabst M., Leonard R., Altmann F., Steinkellner H. (2010). In planta protein sialylation through overexpression of the respective mammalian pathway. *J. Biol. Chem.* 285 15923–15930
9. Cox K., Sterling J., Regan J., Gasdaska J., Frantz K., Peele C., Black A., Passmore D., Moldovan-Loomis C., Srinivasan M., Cuison S., Cardarelli P., Dickey L. (2006). Glycan optimization of a human monoclonal antibody in the aquatic plant *Lemna minor*. *Nat. Biotechnol.* 24 1591–1597
10. Cummings R. D. (2009). The repertoire of glycan determinants in the human glycome. *Mol. Biosyst.* 5 1087–1104
11. Daskalova S. M., Radder J. E., Cichacz Z. A., Olsen S. H., Tsaprailis G., Mason H., Lopez L. C. (2010). Engineering of*N.benthamiana* L. plants for production of *N*-acetylgalactosamine-glycosylated proteins – towards development of a plant-based platform for production of protein therapeutics with mucin type *O*-glycosylation. *BMC Biotechnol.*10 62 10.1186/1472-6750-10-62
12. Durocher Y., Butler M. (2009). Expression systems for therapeutic glycoprotein production. *Curr. Opin. Biotechnol.* 20700–707
13. Essl D., Dirnberger D., Gomord V., Strasser R., Faye L., Glössl J., Steinkellner H. (1999). The N-terminal 77 amino acids from tobacco *N*-acetylglucosaminyltransferase I are sufficient to retain a reporter protein in the Golgi apparatus of *Nicotiana benthamiana* cells. *FEBS Lett.* 453 169–173
14. Gill D. J., Clausen H., Bard F. (2011). Location, location, location: new insights into *O*-GalNAc protein glycosylation.*Trends Cell Biol.* 21 149–158
15. Gomord V., Fitchette A. C., Menu-Bouaouiche L., Saint-Jore-Dupas C., Plasson C., Michaud D., Faye L. (2010). Plant-specific glycosylation patterns in the context of therapeutic protein production. *Plant Biotechnol. J.* 8 564–587
16. Jefferis R. (2009). Glycosylation as a strategy to improve antibody-based therapeutics. *Nat. Rev. Drug Discov.* 8 226–234
17. Ju T., Cummings R. D. (2002). A unique molecular chaperone Cosmc required for activity of the mammalian core 1 beta 3-galactosyltransferase. *Proc. Natl. Acad. Sci. U.S.A.* 99 16613–16618
18. Ju T., Otto V. I., Cummings R. D. (2011). The Tn antigen-structural simplicity and biological complexity. *Angew. Chem. Int. Ed. Engl.* 50 1770–1791
19. Karnoup A. S., Turkelson V., Anderson W. H. (2005). O-linked glycosylation in maize-expressed human IgA1. *Glycobiology* 15965–981
20. Koprivova A., Stemmer C., Altmann F., Hoffmann A., Kopriva S., Gorr G., Reski R., Decker E. L. (2004). Targeted knockouts of*Physcomitrella* lacking plant-specific immunogenic *N*-glycans.*Plant Biotechnol. J.* 2 517–523
21. Leonard R., Petersen B. O., Himly M., Kaar W., Wopfner N., Kolarich D., van Ree R., Ebner C., Duus J., Ferreira F., Altmann F. (2005). Two novel types of *O*-glycans on the mugwort pollen allergen Art v 1 and their role in antibody binding. *J. Biol. Chem.*280 7932–7940
22. Ma J. K., Drake P. M., Christou P. (2003). The production of recombinant pharmaceutical proteins in plants. *Nat. Rev. Genet.* 4794–805
23. Mattu T. S., Pleass R. J., Willis A. C., Kilian M., Wormald M. R., Lellouch A. C., Rudd P. M., Woof J. M., Dwek R. A. (1998). The glycosylation and structure of human serum IgA1, Fab, and Fc regions and the role of *N*-glycosylation on Fc alpha receptor interactions. *J. Biol. Chem.* 273 2260–2272
24. Moriguchi R., Matsuoka C., Suyama A., Matsuoka K. (2011).Reduction of plant-specific arabinogalactan-type *O*-glycosylation by treating tobacco plants with ferrous chelator 2,2′-dipyridyl.*Biosci. Biotechnol. Biochem.* 75 994–996
25. Nagels B., Van Damme E. J., Pabst M., Callewaert N., Weterings K. (2011). Production of complex multiantennary *N*-glycans in*Nicotiana benthamiana* plants. *Plant Physiol.* 155 1103–1112
26. Nagels B., Weterings K., Callewaert N, Van Damme E. J. (2012).Production of plant made pharmaceuticals: from plant host to functional protein. *Crit. Rev. Plant Sci.* 31 148–180
27. Palacpac N., Yoshida S., Sakai H., Kimura Y., Fujiyama K., Yoshida T., Seki T. (1999). Stable expression of human beta1,4-galactosyltransferase in plant cells modifies N-linked glycosylation patterns. *Proc. Natl. Acad. Sci. U.S.A.* 96 4692–4697
28. Pinkhasov J., Alvarez M. L., Rigano M. M., Piensook K., Larios D., Pabst M., Grass J., Mukherjee P., Gendler S. J., Walmsley A. M., Mason H. S. (2011). Recombinant plant-expressed tumour-associated MUC1 peptide is immunogenic and capable of breaking tolerance in MUC1.Tg mice. *Plant Biotechnol. J.* 9991–1001
29. Rich J. R., Withers S. G. (2009). Emerging methods for the production of homogeneous human glycoproteins. *Nat. Chem. Biol.* 5 206–215
30. Rouwendal G. J., Florack D. E., Hesselink T., Cordewener J. H., Helsper J. P., Bosch D. (2009). Synthesis of Lewis X epitopes on plant *N*-glycans. *Carbohydr. Res.* 344 1487–1493

31. Saint-Jore-Dupas C., Nebenführ A., Boulaflous A., Follet-Gueye M., Plasson C., Hawes C., Driouich A., Faye L., Gomord V. (2006). Plant *N*-glycan processing enzymes employ different targeting mechanisms for their spatial arrangement along the secretory pathway. *Plant Cell* 18 3182–3200

32. Schoberer J., Runions J., Steinkellner H., Strasser R., Hawes C., Osterrieder A. (2010). Sequential depletion and acquisition of proteins during Golgi stack disassembly and reformation. *Traffic*11 1429–1444

33. Schoberer J., Strasser R. (2011). Sub-compartmental organization of Golgi-resident *N*-glycan processing enzymes in plants. *Mol. Plant* 4 220–228

34. Shin Y. J., Chong Y. J., Yang M. S., Kwon T. H. (2011).Production of recombinant human granulocyte macrophage-colony stimulating factor in rice cell suspension culture with a human-like *N*-glycan structure. *Plant Biotechnol. J.* 9 1109–1119

35. Springer G. F. (1984). T and Tn, general carcinoma autoantigens.*Science* 224 1198–1206

36. Strasser R., Altmann F., Mach L., Glössl J., Steinkellner H. (2004). Generation of *Arabidopsis thaliana* plants with complex*N*-glycans lacking beta1,2-linked xylose and core alpha1,3-linked fucose. *FEBS Lett.* 561 132–136

37. Strasser R., Bondili J., Vavra U., Schoberer J., Svoboda B., Glössl J., Léonard R., Stadlmann J., Altmann F., Steinkellner H., Mach L. (2007). A unique beta1, 3-galactosyltransferase is indispensable for the biosynthesis of *N*-glycans containing Lewis a structures in *Arabidopsis thaliana*. *Plant Cell* 19 2278–2292

38. Strasser R., Castilho A., Stadlmann J., Kunert R., Quendler H., Gattinger P., Jez J., Rademacher T., Altmann F., Mach L., Steinkellner H. (2009). Improved virus neutralization by plant-produced anti-HIV antibodies with a homogeneous beta1,4-galactosylated *N*-glycan profile. *J. Biol. Chem.* 284 20479–20485

39. Strasser R., Mucha J., Schwihla H., Altmann F., Glössl J., Steinkellner H. (1999). Molecular cloning and characterization of cDNA coding for beta1,2N-acetylglucosaminyltransferase I (GlcNAc- TI) from *Nicotiana tabacum*. *Glycobiology* 9 779–785

40. Strasser R., Stadlmann J., Schähs M., Stiegler G., Quendler H., Mach L., Glössl J., Weterings K., Pabst M., Steinkellner H. (2008). Generation of glyco-engineered *Nicotiana benthamiana*for the production of monoclonal antibodies with a homogeneous human-like *N*-glycan structure. *Plant Biotechnol. J.* 6 392–402

41. Tarp M. A., Clausen H. (2008). Mucin-type *O*-glycosylation and its potential use in drug and vaccine development. *Biochim. Biophys. Acta* 1780 546–563

42. Tsuda E., Kawanishi G., Ueda M., Masuda S., Sasaki R. (1990).The role of carbohydrate in recombinant human erythropoietin.*Eur. J. Biochem.* 188 405–411

43. Velasquez S. M., Ricardi M. M., Dorosz J. G., Fernandez P. V., Nadra A. D., Pol-Fachin L., Egelund J., Gille S., Harholt J., Ciancia M., Verli H., Pauly M., Bacic A., Olsen C. E., Ulvskov P., Petersen B. L., Somerville C., Iusem N. D., Estevez J. M. (2011). *O*-glycosylated cell wall proteins are essential in root hair growth. *Science* 332 1401–1403

44. Wada Y., Dell A., Haslam S. M., Tissot B., Canis K., Azadi P., Bäckström M., Costello C. E., Hansson G. C., Hiki Y., Ishihara M., Ito H., Kakehi K., Karlsson N., Hayes C. E., Kato K., Kawasaki N., Khoo K. H., Kobayashi K., Kolarich D., Kondo A., Lebrilla C., Nakano M., Narimatsu H., Novak J., Novotny M. V., Ohno E., Packer N. H., Palaima E., Renfrow M. B., Tajiri M., Thomsson K. A., Yagi H., Yu S. Y., Taniguchi N. (2010). Comparison of methods for profiling *O*-glycosylation: Human Proteome Organisation Human Disease Glycomics/Proteome Initiative multi-institutional study of IgA1. *Mol. Cell. Proteomics*9 719–727

45. Yang Z., Drew D. P., Jørgensen B., Mandel U., Bach S. S., Ulvskov P., Levery S. B., Bennett E. P., Clausen H., Petersen B. L. (2012). Engineering mammalian mucin-type *O*-glycosylation in plants. *J. Biol. Chem.* 287 11911–11923

46. Yuasa K., Toyooka K., Fukuda H., Matsuoka K. (2005). Membrane-anchored prolyl hydroxylase with an export signal from the endoplasmic reticulum. *Plant J.* 41 81–94

CHAPTER 9

DEVELOPMENT AND PROPERTIES OF GENETICALLY ENCODED PH SENSORS IN PLANTS

Alexandre Martinière[1], Guilhem Desbrosses[2], Hervé Sentenac[1] and Nadine Paris[1]

[1]Biochimie et Physiologie Moléculaire des Plantes, Institut de Biologie Intégrative des Plantes, UMR 5004 CNRS/UMR 0386 INRA/Montpellier SupAgro/Université Montpellier 2, Montpellier, France

[2]Laboratory of Tropical and Mediterranean Symbioses (UMR113, Université Montpellier 2, Institut de Recherche pour le Développement,Cirad Montpellier SupAgro, Institut National de la Recherche Agronomique), Université Montpellier 2, Montpellier, France

ABSTRACT

Fluorescent proteins (FPs) have given access to a large choice of live imaging techniques and have thereby profoundly modified our view of plant cells. Together with technological improvements in imaging, they have opened the possibility to monitor physico-chemical changes within cells. For this purpose, a new generation of FPs has been engineered. For instance, pHluorin, a point mutated version of green fluorescent protein, allows to get local pH estimates. In this paper, we will describe how genetically encoded sensors can be used to measure pH in the microenvironment of living tissues and subsequently discuss the role of pH in (i) exocytosis, (ii) ion uptake by plant roots, (iii) cell growth, and (iv) protein trafficking.

INTRODUCTION

Genetically encoded fluorescent sensors are becoming extremely powerful tools for functional imaging. For instance, the classic calcium sensor introduced by Roger Tsien's lab, cameleon, is now extensively used in the plant field (Miyawaki et al., 1997; Romoser et al., 1997). Such calcium sensors have allowed major breakthroughs regarding, for instance, the molecular dialog between plant roots and symbiotic microorganisms (Sieberer et al., 2012; Singh and Parniske, 2012). More recently, dedicated strategies have led to the development of very promising biosensors, which allow to probe the local concentration of disaccharides, phosphate, amino acid, or ammonium (reviewed inGjetting et al., 2013; Jones et al., 2013). In this perspective issue, we will focus on a specific type of sensors optimized for pH measurement.

Citation: Alexandre Martinière,Guilhem Desbrosses, Hervé Sentenac, and Nadine Paris, Development and properties of genetically encoded pH sensors in plants, doi: 10.3389/fpls.2013.00523.

PROPERTIES OF GENETICALLY ENCODED PH SENSORS

Classical fluorescent dyes have been used to estimate pH in plants. For example, fluorescein coupled to dextran allows pH recording in the close vicinity of roots (Monshausen et al., 2009, 2011) or shoots (Geilfus and Mühling, 2012). The 2′,7′-Bis(2-carboxyethyl)-5(6)-carboxyfluorescein (BCECF), which was first introduced by Rink et al. (1982), is also classically used to measure plant vacuolar pH, especially in its membrane permeable form as acetomethyl ester (BCECF-AM;Krebs et al., 2010). Nevertheless, such approaches are invasive and do not allow measuring pH in other intracellular cell structures. Genetically encoded pH sensors have been designed to overcome these limitations.

The chromophore of green fluorescent protein (GFP) is formed by two autocatalytic reactions (cyclization and oxidation) between Ser65, Tyr66, and Gly67 (review in Wachter, 2007). It absorbs light at two maxima: 395 nm when protonated and 475 nm when depronated (Bizzarri et al., 2009). This shift in absorption has been extensively used to record pH in living organisms (Kneen et al., 1998; Llopis et al., 1998), and optimized versions of GFP for pH measurement have been generated. Ecliptic and superecliptic pHluorins display almost no fluorescence when protonated, allowing sensitive detection of biological processes associated with pH increase. However, the equilibrium between the protonated and deprotonated states of these proteins is also affected by temperature and ionic strength. Furthermore, the light emitted by such sensor protein is dependent on its own concentration. Thus a variation in fluorescence signal can result either from a change in pH or in concentration of the sensor. To circumvent this issue, ratiometric pH sensors have been developed providing a pH readout independent of the probe concentration. Among such sensors, one of the most popular is pHluorin. It was generated by introducing point mutations in *Aequorea victoria* GFP (AvGFP) sequence, especially S202H, which conferred ratiometric behavior to the sensor (Miesenböck et al., 1998; Bizzarri et al., 2009). For example, pHluorin displays a strong increase in absorption at 475 nm and, simultaneously, a strong decrease in absorption at 395 nm when the pH is shifted from 7.5 to 5.5. The pHluorin sensor has a pKa of 6.9 and the dynamic range of fluorescent signal is from pH 5.4 to 8.4. This pH range is suitable for most plant subcellular compartments, with the exception of the apoplast, which is too acidic, and the vacuole, where GFP is degraded (Tamura et al., 2003). Later, derivatives of this protein were engineered to improve the level of expression. This is the case for ratiometric Venus Citrine (RaVC), whose codon usage has been optimized for expression in *Aspergillus* (Bagar et al., 2009), and the plant-solubility-modified ratiometric pHluorin (*PRpHluorin*), optimized for *Arabidopsis* (Shen et al., 2013). To avoid erroneous pH measurement due to chromatic aberrations, an alternative type of ratiometric sensor was generated from S65T GFP, where the emission is yellow-shifted upon alkalization. This dual emission GFP (deGFP) contains the point mutations T203C and H148G or H148C (Hanson et al., 2002). In addition, some sensors such as E^2GFP and E^1GFP combine dual emission and dual excitation features (Bizzarri et al., 2006; Arosio et al., 2007).

Table 1: Selected genetically encoded pH sensors available for *in vivo* pH measurements

Constructs	Type	Excitation(nm)	Emission(nm)	pH range	pKa	Delta ratio	Reference
pHluorin, pHluorin2*, RaVC**, PRpHluorin***, pHGFP****	Ratiometric	395/475	510	5.4–8.4	6.9	2.5	Miesenböck et al. (1998)
Ecliptic pHluorin, PEpHluorin***	Ecliptic	395/477	510	6.5–8	7.2	13.5	Miesenböck et al. (1998)
Superecliptic pHluorin	Ecliptic	477	510	5.5–8.5	7.2	13	Sankaranarayanan and Ryan (2001)
deGFP3	Ratiometric	396	460/515	6–9	7.3	20	Hanson et al. (2002)
E2GFP	Ratiometric	458/488	500/560	5–8.5	6.9/7.5	10	Bizzarri et al. (2006)
ptGFP	Ratiometric	390/475	540	3.8–8.2	7.3	36	Schulte et al. (2006)
pHlameleon 6	FRET-based ratiometric	458	481/533	5–7.5	5.6	9	Esposito et al. (2008)
pHusion	Ratiometric	488/558	510/600	4.5–8	5.8	6	Gjetting et al. (2012)
pHlash	BRET	n.a.	475/525	5.4–9	n.a.	3.3	Zhang et al. (2012)

Eightfold brighter version of pHluorin (Mahon, 2011)
***pHluorin with optimize codon usage for Aspergillus (Bagar et al., 2009)*
****pHluorin with optimize codon usage for Arabidopsis (Shen et al., 2013)*
*****pHGFP, chimera between smGFP (Davis and Vierstra, 1998) and pHluorin to prevent splicing (Moseyko and Feldman, 2001)*

The major limitation of using pHluorin and its derivates, especially in plant tissues, is their acidic quenching at an apparent pH near 4.5. Two strategies were used to improve pH sensing in acidic environments. The first strategy is to use a fluorescent protein (FP), *Pt*-GFP, from orange sea pen, *Ptilosarcus gurneyi* (Schulte et al., 2006). Although *Pt*-GFP has a pKa similar to that of pHluorin (pK_{PtGFP}: 7.3), this protein has a broader pH response range and is more stable in acidic conditions (Schulte et al., 2006). The second strategy is to make a tandem of FPs with highly different ranges of pH responsiveness. The measure of pH is then either given by Förster resonance energy transfer (FRET) between these two proteins, as for pHlameleon (Esposito et al., 2008) or by the ratio of fluorescence of the proteins. In particular, the tandem of enhanced GFP (eGFP) and monomeric red FP (mRFP1) is suitable for measuring pH in the apoplasm (Gjetting et al., 2012). Finally, a new pH biosensor was developed recently based on bioluminescence resonance energy transfer (BRET). This pHlash sensor is composed of a luciferase as a donor, fused to pH-sensitive Venus as acceptor (Zhang et al., 2012). Use of bioluminescence significantly improves the signal-to-noise ratio making BRET particularly useful for experimentation at the whole tissue or organism scale.

Regardless of the exact type of genetically encoded pH sensors, estimation of the local pH value from the fluorescence data is not straightforward. This requires calibration to characterize the relationship between the fluorescence properties of the sensor and pH. The calibration can be carried out either *in vitro* or *in vivo* (Gao et al., 2004; Schulte et al., 2006; Shen et al., 2013). *In vitro* calibration uses heterologously expressed sensors in buffers of different pH. Such a calibration does not take into account the ionic strength and buffering capacity of the cell and thus can lead to substantial artifacts (Schulte et al., 2006). *In vivo* calibration uses sensor-expressing plant cells that are treated with an ionophore, such as nigericin, in presence of a solution with sufficient buffering capacity to clamp the pH of the cellular environment containing the sensor. This approach should be favored whenever possible, but is still limited for several reasons. First, there is no objective way to ensure that the equilibrium has reached between the bathing media and the cell compartment where the sensor is located. This is especially true when the sensor is in the lumen of intracellular structures, meaning that the ionophore has to be effective on two successive membranes. Secondly, the ionophore modifies the ion content of the cell and consequently the native buffering capacity and ionic strength. Such calibration issues may explain the ongoing debate on the feasibility of measuring absolute pH values with sensors *in vivo*(Schulte et al., 2006; Bizzarri et al., 2009).

ANALYSIS OF EXOCYTOSIS EVENTS WITH PH SENSORS

Genetically encoded pH sensors can be used to investigate processes that give rise to, or rely on, changes in pH. During exocytosis, vesicle lumen undergoes a rapid alkalinization upon fusion to the plasma membrane. Since the pioneering study by Miesenböck et al. (1998), ecliptic pHluorin has been extensively used as an exocytosis indicator (Sankaranarayanan et al., 2000; Gandhi and Stevens, 2003; Taylor et al., 2011). For instance, in the synaptic cleft, Gandhi and Stevens (2003) were able to distinguish three types of vesicle behaviors at the plasma membrane. In another study, ecliptic pHluorin used to describe the dependency of vesicle fusion on actin filaments and cdc42, a protein controlling cell division (Alberts et al., 2006). Similar approaches would appear useful in the plant field. According to the classical view of ligand–receptor trafficking, a difference in pH is expected between the lumen of the *trans*-Golgi network (TGN) and the apoplastic face of the plasma membrane (Robinson et al., 2012). Hence, the luminal pH of exocytic vesicles is predicted to be different before and after fusion with the plasma membrane. In recent reports, TGN compartments have been shown to have the most acidic lumen of the secretory pathway, with a pH around 6. On the other side, apoplastic pH is likely to be highly variable and dependent on tissue type and development stage (Gao et al., 2004; Staal et al., 2011; Geilfus and Mühling, 2012;Gjetting et al., 2012). Unfortunately, studies that concomitantly measure the pH in the TGN and in the apoplasm are still lacking.

Table 2: ummary of pH measurements in plant cells using pH sensors.

	Tobacco epidermis (Martinière et al., 2013)	*Arabidopsis* root tip (Martinière et al., 2013)	*Arabidopsis* protoplasts (Shen et al., 2013)
Cytosol	7.8 (pHluorin)	8 (pHluorin)	7.3 (PpHluorin)
ER	7.5 (pHluorin-hdel)	7.7 (pHluorin-hdel)	7.1 (PpHluorin-hdel)
Cis-Golgi			6.8 (ManI-PpHluorin)
Trans-Golgi	6.9 (ST-pHluorin)	6.8 (ST-pHluorin)	
TGN	6.1 (pHluorin-AtVSR2)	6 (pHluorin-AtVSR2)	6.2 (PpHluorin-AtVSR2 trunc)
PVC	6.8 (pHluorin-AtVSR2)		
Late PVC	7.1 (pHluorin-AtVSR2-IMAA)		
TGN plus recycling endosomes	6.5 (pHluorin-AtVSR2-YA)		6.3 (PpHluorin-truncAtVSR2-YA)
Vacuole	6 (BCECF-AM)	5.7 (BCECF-AM)	5.2 (Aleurain-PpHluorin)
Plastid stroma			7.2 (RecaA-PpHluorin)
Peroxisome			8.4 (PpHluorin-SLR)
Mitochondria			8.1 (mito-PpHluorin)
Nucleus			7.2 (NLS-PpHluorin)

pH WITHIN THE APOPLAST

In plants, the plasma membrane is energized by P-type proton pump ATPases. Direct evidence has been obtained that proton diffusion within the apoplast away from the membrane can be rate-limiting when compared to proton excretion by the proton pumps. Using weak acid influx for assaying pH, it has been shown that the local pH in the unstirred aqueous layers next to the plasmalemma can be much lower than that of the external bathing solution, by up to 2 pH units under conditions that maximize the rate of net proton excretion (Sentenac and Grignon, 1987; Grignon and Sentenac, 1991). A few studies on the dependency of membrane transport in roots on external pH have quantitatively assessed this steady state pH shift at the membrane surface due to proton excretion by using weak acid influx assays under the same conditions as those used for the transport measurements (Thibaud et al., 1998). Clearly, the results indicated that this shift in pH strongly contributed to energization of the cell membrane and had to be taken into account to explain, for instance, kinetic and thermodynamic features of nitrate, phosphate, or potassium transport (Thibaud et al., 1988), or the effect of $HCO-3$, a buffer that can be naturally present in the external solution (in calcareous soils), on the transport rate of these nutriments (Toulon et al., 1989) or on iron reduction at the cell surface (Toulon et al., 1992).

Different methods have been developed to estimate local pH values, using proton-selective micro-electrodes (Felle, 1998) or impermeable pH-sensitive dyes (Taylor et al., 1996; Monshausen et al., 2009, 2011), or collecting apoplasmic fluids (Mühling and Sattelmacher, 1995). However, such approaches provide estimates of the pH conditions at the organ surface or, at best, of the mean pH within the apoplast, and not of the local pH in close vicinity of the membrane, which is the actual pH sensed by the ion transporters. Another source of difficulties is the possibility of heterogeneities within the apoplast. For example, in the root cortex and for the inner cortical cells, the diffusion barrier that slows proton migration toward the external solution comprises the entire apoplastic continuum, from cortical cells surrounding the endodermis to the root epidermis surface. Furthermore, longitudinal heterogeneities in surface pH have been observed between the root apex, the elongation region and the fully differentiated zone, due to differences in the rate and direction of proton transport. The physiological consequences and roles of such radial and longitudinal differences, in terms of membrane energization and transport activity, have been poorly investigated. Clearly, genetically encoded pH sensors open novel perspectives in this domain of the root structure–function relationship. In this respect, Moseyko and Feldman (2001) and Gao et al. (2004) have developed pioneering approaches. For example, large pH variations were observed in response to salt stress. When challenged with 100 mM NaCl, root cells displayed a more alkaline apoplasm and a more acidic cytosol. These changes in pH might result from specific activation of H^+:Na^+ antiport activity mediated by the plasma membrane Salt-Overly-Sensitive1 (SOS1) antiporter in response to salinity constraint (Quintero et al., 2011). Such analyses clearly indicate that expression of pH sensor proteins in mutant plants could allow rapid progress in investigating the root structure–function relationship and its role in membrane energization, ion transport and plant adaptation to environmental stresses.

CELL WALL PH AND GROWTH

Cell capacity to control the apoplastic pH plays a major role also in plant growth and development. Indeed, according to the acid growth theory, cell wall "loosening" favored by low pH allows cell elongation (Rayle and Cleland, 1992). In a typical sequence of events, auxin stimulates proton secretion by plasmalemma proton pump ATPases, resulting in cell wall acidification, causing activation of expansins, which loosen cellulose microfibrils. As a result, cell walls become more extensible favoring turgor-driven cell elongation. In support to the acid growth theory, the elongation zone of the root generally displays a more acidic apoplasm than the fully differentiated zone (Staal et al., 2011). Intriguingly, a close-up analysis using a pH sensor reveals highly dynamic fluctuations of root surface pH with alkalinization and acidification periods and alkalinization bursts along the root longitudinal axis or alternating between opposing flanks (Monshausen et al., 2011). Such results indicate that a net influx of protons can occur in epidermal cells, as previously reported in other studies (O'Neill and Scott, 1983). Intriguingly, a recent report shows that treatment of root tips with indole-3-acetic acid (IAA) induces an increase in pH by about 0.5–0.8 pH units (Gjetting et al., 2012), which might be related to the periodical alkalinization previously reported in root tips (Monshausen et al., 2011).

pH MAPPING OF ORGANELLES AND PROTEIN TARGETING

Taking advantage of growing knowledge in protein targeting mechanisms, appropriate signal sequences can be fused to the coding sequence of the pH sensor so that this is targeted later to various cell membranes or compartments. Using such an approach, pH in various intracellular compartments was measured *in vivo* in mammalian and, more recently, in *Arabidopsis* protoplasts (Shen et al., 2013), tobacco leaf epidermal cells or *Arabidopsis* root tip cells (Martinière et al., 2013). These studies directly demonstrate a gradual acidification of 1.5–2 pH units in the lumen of the endomembrane system with ER displaying the most alkaline pH and vacuole the most acidic one. These variations could be the result of biological variation between the systems tested (intact cells versus protoplast) or could be due to differences in growing condition. For instance, pH homeostasis is known to be sensitive to light intensity (Martinière et al., 2013). Most compartment markers cycle at least between two subcellular localizations and there is still no way to target a sensor solely to either the lumen of the TGN, the prevacuolar compartment (PVC) or the late PVC. Fusion of pH sensors to the vacuolar sorting receptor (VSR) mutants Y165A and IMAA allowed to partially solve such localization issue, the former mutant being preferentially retained in the TGN and the latter localized in the LPVC. This allowed differentiation between TGN and the PVC and getting pH estimates for these two compartments (Martinière et al., 2013; Shen et al., 2013). However, it should be noted that, in addition to TGN, the Y165A VSR mutant is very likely to also label recycling endosomes and plasma membranes (daSilva et al., 2006; Saint-Jean et al., 2010). Furthermore, accumulation of the IMAA mutant in the LPVC is only transient (Saint-Jean et al., 2010).

In average, the population of VSR labeled compartments has a pH intermediate between the *trans*-Golgi and the vacuole in all the systems tested. But surprisingly, the TGN is actually more acidic, with a pH 6, than the PVC (pH 6.8) or even the late PVC This result fits well with the localization of V-type ATPase involved in pumping protons. Indeed, VHA-a1 protein is preferentially accumulated in the TGN (Dettmer et al., 2006) and is not detected in the multivesicular bodies (MVB; Robinson et al., 2012) or PVC. Accordingly to the proton leak model, while V-type ATPase acidifies a compartment, antiporters could be involved in proton extrusion (Demaurex, 2002). For instance, NHX5 and NHX6 Na^+/H^+antiports have been proposed to be involved in such a mechanism in plant endosomes (Bassil et al., 2011). Therefore, the relative amount of VHA-a1 and proton antiporters such as NHX5/6, in compartment on the way to the vacuole could have a preponderant role in proton homeostasis and substantial consequence in terms of protein sorting. In support to this hypothesis, is the mistargeting of a soluble vacuolar marker in *nhx5 nhx6* double knockout (Bassil et al., 2011) and the fact that the pH measured in the TGN and the PVC fits well with the distribution of VHA-a1 and NHX5/6 in tobacco leaf epidermal cells (Martinière et al., 2013).

The fact that the PVC is more alkaline than TGN is surprising and in contradiction with the VSR-mediated vacuolar sorting model where acidification in the PVC is thought to be responsible for a release of the ligand by the receptor (Kirsch et al., 1994; Ahmed et al., 2000;Shimada et al., 2003). In fact, the *in vitro* binding curve between VSR and the ligand is bell shaped with an optimum pH at 6 (Kirsch et al., 1994). This value corresponds to the pH measured *in vivo* in the TGN (Martinière et al., 2013) suggesting that binding of VSR to its ligand would be optimal in the TGN but significantly lower in more acidic and alkaline compartments. The use of ratiometric sensors to measure the endomembrane luminal pH therefore suggests that alkalinization rather than acidification would trigger ligand release from the receptor

VSR in the PVC (Martinière et al., 2013). Of course this model does not take into account other environmental parameters in TGN and PVC such as the calcium concentration that is also known to regulate ligand binding for a least some members of the VSR family (Watanabe et al., 2002,2004). Finally, it is also important to point out that the pH reported for the PVC or the TGN is an average from a large population of independent organelles with some variation of pH values. This difference is either due to technical limitations in imaging or may reflect pH variability between compartments of same identity, e.g., TGN or PVC. A model for VSR-mediated transport is still a matter of debate (Robinson and Pimpl, 2013).

It is likely that pH ratiometric sensors will open new perspectives in our understanding of the mobility, functional heterogeneity and remodeling capacities of organelles. For instance, the TGN has been assumed to undergo maturation as its morphology changes over time (Scheuring et al., 2011). Using pH sensors along with dynamic studies could provide evidence for these maturation events.

CONCLUDING REMARKS

Genetically encoded pH sensors have opened avenues for cell biology. They will provide valuable functional data with respect to cell compartmentation, protein trafficking, acidic growth, or ion diffusion within plant tissue. One of the future challenges will be to improve the definition of pH information provided by the sensor as cell nano-environments are anticipated. For instance, protein and lipid composition of membranes have a substantial effect on proton diffusion. As well, it is reasonable to anticipate a higher proton concentration in the vicinity of ATPase pumps and therefore heterogeneity on membrane surfaces with respect to pH.

Conflict of Interest Statement

The authors declare that the research was conducted in the absence of any commercial or financial relationships that could be construed as a potential conflict of interest.

REFERENCES

1. Ahmed S. U., Rojo E., Kovaleva V., Venkataraman S., Dombrowski J. E., Matsuoka K., et al. (2000). The plant vacuolar sorting receptor AtELP is involved in transport of NH(2)-terminal propeptide-containing vacuolar proteins in *Arabidopsis thaliana*. *J. Cell Biol.* 149 1335–134410.1083/jcb.149.7.1335
2. Alberts P., Rudge R., Irinopoulou T., Danglot L., Gauthier-Rouvière C., Galli T. (2006). Cdc42 and actin control polarized expression of TI-VAMP vesicles to neuronal growth cones and their fusion with the plasma membrane. *Mol. Biol. Cell* 17 1194–120310.1091/mbc.E05-07-0643
3. Arosio D., Garau G., Ricci F., Marchetti L., Bizzarri R., Nifosї R., et al. (2007). Spectroscopic and structural study of proton and halide ion cooperative binding to gfp. *Biophys. J.* 93 232–24410.1529/biophysj.106.102319
4. Bagar T., Altenbach K., Read N. D, Benčina M. (2009). Live-Cell imaging and measurement of intracellular pH in filamentous fungi using a genetically encoded ratiometric probe. *Eukaryot. Cell* 8 703–71210.1128/EC.00333-08
5. Bassil E., Ohto M., Esumi T., Tajima H., Zhu Z., Cagnac O., et al. (2011). The *Arabidopsis* intracellular Na+/H+ antiporters NHX5 and NHX6 are endosome associated and necessary for plant growth and development. *Plant Cell* 23 224–23910.1105/tpc.110.079426
6. Bizzarri R., Arcangeli C., Arosio D., Ricci F., Faraci P., Cardarelli F., et al. (2006). Development of a novel GFP-based ratiometric excitation and emission pH indicator for intracellular studies. *Biophys. J.* 90 3300–331410.1529/biophysj.105.074708
7. Bizzarri R., Serresi M., Luin S., Beltram F. (2009). Green fluorescent protein based pH indicators for in vivo use: a review.*Anal. Bioanal. Chem.* 393 1107–112210.1007/s00216-008-2515-9
8. daSilva L. L. P., Foresti O., Denecke J. (2006). Targeting of the plant vacuolar sorting receptor BP80 is dependent on multiple sorting signals in the cytosolic tail. *Plant Cell* 18 1477–149710.1105/tpc.105.040394
9. Davis S. J., Vierstra R. D. (1998). Soluble, highly fluorescent variants of green fluorescent protein (GFP) for use in higher plants. *Plant Mol. Biol.* 36 521–52810.1023/A:1005991617182
10. Demaurex N. (2002). pH Homeostasis of cellular organelles.*News Physiol. Sci.* 17 1–5
11. Dettmer J., Hong-Hermesdorf A., Stierhof Y.-D., Schumacher K. (2006). Vacuolar H^+-ATPase activity is required for endocytic and secretory trafficking in *Arabidopsis*. *Plant Cell* 18 715–73010.1105/tpc.105.037978
12. Esposito A., Gralle M., Dani M. A. C., Lange D., Wouters F. S. (2008). pHlameleons: a family of FRET-based protein sensors for quantitative pH imaging. *Biochemistry* 47 13115–1312610.1021/bi8009482
13. Felle H. H. (1998). The apoplastic pH of the *Zea mays* root cortex as measured with pH-sensitive microelectrodes: aspects of regulation. *J. Exp. Bot.* 49 987–99510.1093/jxb/49.323.987
14. Gandhi S. P., Stevens C. F. (2003). Three modes of synaptic vesicular recycling revealed by single-vesicle imaging. *Nature*423 607–61310.1038/nature01677
15. Gao D., Knight M. R., Trewavas A. J., Sattelmacher B., Plieth C. (2004). Self-reporting *Arabidopsis* expressing pH and $[Ca^{2+}]$ indicators unveil ion dynamics in the cytoplasm and in the apoplast under abiotic stress. *Plant Physiol.* 134 898–90810.1104/pp.103.032508
16. Geilfus C.-M, Mühling K. H. (2012). Transient alkalinization in the leaf apoplast of *Vicia faba* L. depends on NaCl stress intensity: an in situ ratio imaging study. *Plant Cell Environ.* 35578–587 10.1111/j.1365-3040.2011.02437.x
17. Gjetting K. S. K., Ytting C. K., Schulz A., Fuglsang A. T. (2012).Live imaging of intra- and extracellular pH in plants using pHusion, a novel genetically encoded biosensor. *J. Exp. Bot.* 633207–321810.1093/jxb/ers040

18. Gjetting S. K., Schulz A., Fuglsang A. T. (2013). Perspectives for using genetically encoded fluorescent biosensors in plants. *Front. Plant Sci.* 4:234 10.3389/fpls.2013.00234

19. Grignon C., Sentenac H. (1991). pH and ionic conditions in the apoplast. *Annu. Rev. Plant Physiol. Plant Mol. Biol.* 42 103–12810.1146/annurev.pp.42.060191.000535

20. Hanson G. T., McAnaney T. B., Park E. S., Rendell M. E. P., Yarbrough D. K., Chu S., et al. (2002). Green fluorescent protein variants as ratiometric dual emission pH sensors. 1. Structural characterization and preliminary application. *Biochemistry* 4115477–1548810.1021/bi026609p

21. Jones A. M., Grossmann G., Danielson J. Å, Sosso D., Chen L.-Q., Ho C.-H., et al. (2013). In vivo biochemistry: applications for small molecule biosensors in plant biology. *Curr. Opin. Plant Biol.* 16 389–39510.1016/j.pbi.2013.02.010

22. Kirsch T., Paris N., Butler J. M., Beevers L., Rogers J. C. (1994).Purification and initial characterization of a potential plant vacuolar targeting receptor. *Proc. Natl. Acad. Sci. U.S.A.* 913403–340710.1073/pnas.91.8.3403

23. Kneen M., Farinas J., Li Y., Verkman A. S. (1998). Green fluorescent protein as a noninvasive intracellular pH indicator.*Biophys. J.* 74 1591–159910.1016/S0006-3495(98)77870-1

24. Krebs M., Beyhl D., Görlich E., Al-Rasheid K. A. S., Marten I., Stierhof Y.-D., et al. (2010). *Arabidopsis* V-ATPase activity at the tonoplast is required for efficient nutrient storage but not for sodium accumulation. *Proc. Natl. Acad. Sci. U.S.A.* 107 3251–325610.1073/pnas.0913035107

25. Llopis J., McCaffery J. M., Miyawaki A., Farquhar M. G., Tsien R. Y. (1998). Measurement of cytosolic, mitochondrial, and Golgi pH in single living cells with green fluorescent proteins.*Proc. Natl. Acad. Sci. U.S.A.* 95 6803–680810.1073/pnas.95.12.6803

26. Mahon M. J. (2011). pHluorin2: an enhanced, ratiometric, pH-sensitive green florescent protein. *Adv. Biosci. Biotechnol.* 2 132–13710.4236/abb.2011.23021

27. Martinière A., Bassil E., Jublanc E., Alcon C., Reguera M., Sentenac H., et al. (2013). In vivo intracellular pH measurements in Tobacco and *Arabidopsis* reveal an unexpected pH gradient in the endomembrane system.*PlantCell*25 4028–4043 10.1105/tpc.113.116897

28. Miesenböck G., De Angelis D. A., Rothman J. E. (1998). Visualizing secretion and synaptic transmission with pH-sensitive green fluorescent proteins. *Nature* 394 192–19510.1038/28190

29. Miyawaki A., Llopis J., Heim R., McCaffery J. M., Adams J. A., Ikura M., et al. (1997). Fluorescent indicators for Ca^{2+} based on green fluorescent proteins and calmodulin. *Nature* 388 882–88710.1038/42264

30. Monshausen G. B., Bibikova T. N., Weisenseel M. H., Gilroy S. (2009). Ca^{2+} regulates reactive oxygen species production and pH during mechanosensing in *Arabidopsis* roots. *Plant Cell* 212341–235610.1105/tpc.109.068395

31. Monshausen G. B., Miller N. D., Murphy A. S., Gilroy S. (2011).Dynamics of auxin-dependent Ca^{2+} and pH signaling in root growth revealed by integrating high-resolution imaging with automated computer vision-based analysis. *Plant J.* 65 309–31810.1111/j.1365-313X.2010.04423.x

32. Moseyko N., Feldman L. J. (2001). Expression of pH-sensitive green fluorescent protein in *Arabidopsis thaliana*. *Plant Cell Environ.* 24 557–56310.1046/j.1365-3040.2001.00703.x

33. Mühling K. H., Sattelmacher B. (1995). Apoplastic ion concentration of intact leaves of field bean (*Vicia faba*) as influenced by ammonium and nitrate nutrition. *J. Plant Physiol.*147 81–8610.1016/S0176-1617(11)81417-3

34. O'Neill R. A., Scott T. K. (1983). Proton flux and elongation in primary roots of barley (*Hordeum vulgare* L.). *Plant Physiol.* 73199–20110.1104/pp.73.1.199

35. Quintero F. J., Martinez-Atienza J., Villalta I., Jiang X., Kim W.-Y., Ali Z., et al. (2011). Activation of the plasma membrane Na/H antiporter Salt-Overly-Sensitive 1 (SOS1) by phosphorylation of an auto-inhibitory C-terminal domain. *Proc. Natl. Acad. Sci. U.S.A.* 108 2611–261610.1073/pnas.1018921108

36. Rayle D. L., Cleland R. E. (1992). The Acid Growth Theory of auxin-induced cell elongation is alive and well. *Plant Physiol.* 991271–127410.1104/pp.99.4.1271

37. Rink T. J., Tsien R. Y., Pozzan T. (1982). Cytoplasmic pH and free Mg2+ in lymphocytes. *J. Cell Biol.* 95 189–19610.1083/jcb.95.1.189

38. Robinson D. G., Pimpl P. (2013). Receptor-mediated transport of vacuolar proteins: a critical analysis and a new model.*Protoplasma* [Epub ahead of print]10.1007/s00709-013-0542-7

39. Robinson D. G., Pimpl P., Scheuring D., Stierhof Y.-D., Sturm S., Viotti C. (2012). Trying to make sense of retromer. *Trends Plant Sci.* 17 431–43910.1016/j.tplants.2012.03.005

40. Romoser V. A., Hinkle P. M., Persechini A. (1997). Detection in living cells of Ca^{2+}-dependent changes in the fluorescence emission of an indicator composed of two green fluorescent protein variants linked by a calmodulin-binding sequence. A new class of fluorescent indicators. *J. Biol. Chem.* 272 13270–1327410.1074/jbc.272.20.13270

41. Saint-Jean B., Seveno-Carpentier E., Alcon C., Neuhaus J.-M., Paris N. (2010). The cytosolic tail dipeptide Ile-Met of the pea receptor BP80 is required for recycling from the prevacuole and for endocytosis. *Plant Cell* 22 2825–283710.1105/tpc.109.072215

42. Sankaranarayanan S., De Angelis D., Rothman J. E., Ryan T. A. (2000). The use of pHluorins for optical measurements of presynaptic activity. *Biophys. J.* 79 2199–220810.1016/S0006-3495(00)76468-X

43. Sankaranarayanan S., Ryan T. A. (2001). Calcium accelerates endocytosis of vSNAREs at hippocampal synapses. *Nat. Neurosci.* 4 129–13610.1038/83949

44. Scheuring D., Viotti C., Krüger F., Künzl F., Sturm S., Bubeck J., et al. (2011). Multivesicular bodies

mature from the trans-Golgi network/early endosome in *Arabidopsis*. *Plant Cell* 23 3463–348110.1105/tpc.111.086918

45. Schulte A., Lorenzen I., Böttcher M., Plieth C. (2006). A novel fluorescent pH probe for expression in plants. *Plant Methods* 2710.1186/1746-4811-2-7

46. Sentenac H., Grignon C. (1987). Effect of H excretion on the surface pH of corn root cells evaluated by using weak zacid influx as a pH probe. *Plant Physiol.* 84 1367–137210.1104/pp.84.4.1367

47. Shen J., Zeng Y., Zhuang X., Sun L., Yao X., Pimpl P., et al. (2013). Organelle pH in the *Arabidopsis* endomembrane system.*Mol. Plant* 6 1419–1437 10.1093/mp/sst079

48. Shimada T., Fuji K., Tamura K., Kondo M., Nishimura M., Hara-Nishimura I. (2003). Vacuolar sorting receptor for seed storage proteins in *Arabidopsis thaliana*. *Proc. Natl. Acad. Sci. U.S.A.*100 16095–1610010.1073/pnas.2530568100

49. Sieberer B. J., Chabaud M., Fournier J., Timmers A. C. J., Barker D. G. (2012). A switch in Ca^{2+} spiking signature is concomitant with endosymbiotic microbe entry into cortical root cells of*Medicago truncatula*. *Plant J.* 69 822–83010.1111/j.1365-313X.2011.04834.x

50. Singh S., Parniske M. (2012). Activation of calcium- and calmodulin-dependent protein kinase (CCaMK), the central regulator of plant root endosymbiosis. *Curr. Opin. Plant Biol.* 15444–45310.1016/j.pbi.2012.04.002

51. Staal M., De Cnodder T., Simon D., Vandenbussche F., Van der Straeten D., Verbelen J.-P., et al. (2011). Apoplastic alkalinization is instrumental for the inhibition of cell elongation in the*Arabidopsis* root by the ethylene precursor 1-aminocyclopropane-1-carboxylic acid. *Plant Physiol.* 155 2049–205510.1104/pp.110.168476

52. Tamura K., Shimada T., Ono E., Tanaka Y., Nagatani A., Higashi S.-I., et al. (2003). Why green fluorescent fusion proteins have not been observed in the vacuoles of higher plants. *Plant J.* 35545–55510.1046/j.1365-313X.2003.01822.x

53. Taylor D. P., Slattery J., Leopold A. C. (1996). Apoplastic pH in corn root gravitropism: a laser scanning confocal microscopy measurement. *Physiol. Plant.* 97 35–3810.1111/j.1399-3054.1996.tb00475.x

54. Taylor M. J., Perrais D., Merrifield C. J. (2011). A high precision survey of the molecular dynamics of mammalian clathrin-mediated endocytosis. *PLoS Biol.* 9:e100060410.1371/journal.pbio.1000604

55. Thibaud J. B., Davidian J. C., Sentenac H., Soler A., Grignon C. (1988). H^+ cotransports in corn roots as related to the surface pH shift induced by active H^+ excretion. *Plant Physiol.* 88 1469–147310.1104/pp.88.4.1469

56. Toulon V., Sentenac H., Thibaud J. B., Davidian J. C., Moulineau C., Grignon C. (1992). Role of apoplast acidification by the H(+) pump: effect on the sensitivity to pH and CO2 of iron reduction by roots of *Brassica napus* L. *Planta* 186 212–21810.1007/BF00196250

57. Toulon V., Sentenac H., Thibaud J.-B., Soler A., Clarkson D., Grignon C. (1989). Effect of HCO−3 concentration in the absorption solution on the energetic coupling of H^+-cotransports in roots of *Zea mays* L. *Planta* 179 235–24110.1007/BF00393694

58. Wachter R. M. (2007). Chromogenic cross-link formation in green fluorescent protein. *Acc. Chem. Res.* 40 120–12710.1021/ar040086r

59. Watanabe E., Shimada T., Kuroyanagi M., Nishimura M., Hara-Nishimura I. (2002). Calcium-mediated association of a putative vacuolar sorting receptor PV72 with a propeptide of 2S albumin.*J. Biol. Chem.* 277 8708–871510.1074/jbc.M109346200

60. Watanabe E., Shimada T., Tamura K., Matsushima R., Koumoto Y., Nishimura M., et al. (2004). An ER-localized form of PV72, a seed-specific vacuolar sorting receptor, interferes the transport of an NPIR-containing proteinase in *Arabidopsis* leaves. *Plant Cell Physiol.* 45 9–1710.1093/pcp/pch012

61. Zhang Y., Xie Q., Robertson J. B., Johnson C. H. (2012). pHlash: a new genetically encoded and ratiometric luminescence sensor of intracellular pH. *PLoS ONE* 7:e4307210.1371/journal.pone.0043072

CHAPTER 10

SALINITY TOLERANCE IN PLANTS QUANTITATIVE APPROACH TO ION TRANSPORT STARTING FROM HALOPHYTES AND STEPPING TO GENETIC AND PROTEIN ENGINEERING FOR MANIPULATING ION FLUXES

Vadim Volkov

Faculty of Life Sciences and Computing, London Metropolitan University, London, UK

ABSTRACT

Ion transport is the fundamental factor determining salinity tolerance in plants. The Review starts from differences in ion transport between salt tolerant halophytes and salt-sensitive plants with an emphasis on transport of potassium and sodium via plasma membranes. The comparison provides introductory information for increasing salinity tolerance. Effects of salt stress on ion transport properties of membranes show huge opportunities for manipulating ion fluxes. Further steps require knowledge about mechanisms of ion transport and individual genes of ion transport proteins. Initially, the Review describes methods to measure ion fluxes, the independent set of techniques ensures robust and reliable basement for quantitative approach. The Review briefly summarizes current data concerning Na+ and K+ concentrations in cells, refers to primary thermodynamics of ion transport and gives special attention to individual ion channels and transporters. Simplified scheme of a plant cell with known transport systems at the plasma membrane and tonoplast helps to imagine the complexity of ion transport and allows choosing specific transporters for modulating ion transport. The complexity is enhanced by the influence of cell size and cell wall on ion transport. Special attention is given to ion transporters and to potassium and sodium transport by HKT, HAK, NHX, and SOS1 proteins. Comparison between non-selective cation channels and ion transporters reveals potential importance of ion transporters and the balance between the two pathways of ion transport. Further on the Review describes in detail several successful attempts to overexpress or knockout ion transporters for changing salinity tolerance. Future perspectives are questioned with more attention given to promising candidate ion channels and transporters for altered expression. Potential direction of increasing salinity tolerance by modifying ion channels and transporters using single point mutations is discussed and questioned. An alternative approach from synthetic biology is to create new regulation networks using novel transport proteins with desired properties for transforming agricultural crops. The approach had not been widely

Salinity is among the most serious problems for modern agriculture with the estimated annual losses nowadays being over USD 12 billion (Pitman and Läuchli, 2002; Shabala, 2013). More than 20% of irrigated lands and up to 50% (Flowers, 1998) are affected by salinity; it essentially reduces the yield of agricultural crops since most of them are salt-sensitive glycophytes (Munns and Tester, 2008). The collapse of Sumer civilization about 4000 years ago was caused by improper agricultural techniques, which led to soil salinization and drop in the agricultural productivity in the area (Serrano, 1996; Pitman and Läuchli, 2002). However, progress in modern

Citation: Vadim Volkov, Salinity tolerance in plants. Quantitative approach to ion transport starting from halophytes and stepping to genetic and protein engineering for manipulating ion fluxes, doi: 10.3389/fpls.2015.00873.

agriculture and science rather allowed to set problems and pose questions than to get clear answers. We still do not know how to deal with salinity and to grow plants under salinization.

Effects of salinity on living plants include osmotic stress, toxic effects of high salt, mostly sodium ions, and corresponding oxidative stress (Flowers, 2004; Munns and Tester, 2008; Shabala, 2013). Osmotic stress is proportional to concentration of external salt solution, usually with osmotic pressures over 1 MPa. Toxic effect of sodium ions results from their rise in cytoplasm of plant cells. Developing in plant cells oxidative stress adds to the negative effects of salinity on the whole plants.

Ion transport via cell membranes is the basic factor determining salinity tolerance. Ion fluxes control ion concentrations; finally the values and regulation of the fluxes are essential for salinity tolerance. The review starts from brief description of ion transport in halophytes. Then it provides basic details and features of ion transport aiming to understand what could be altered and the potential results of the changes. Further discussion is directed to specific transporters and to genetically modified and artificially designed transport proteins for modulating ion transport under conditions of salinization.

ION TRANSPORT IN HALOPHYTES AND EFFECTS OF SALT STRESS ON MEMBRANE TRANSPORT IN GLYCOPHYTES

Brilliant examples of salinity tolerance are represented by halophytes, which are able to grow at high concentrations of salt, under irrigation by seawater and even under several times higher salt concentrations than in seawater (over 2 M of NaCl) (Flowers et al., 1977; Flowers and Colmer, 2008). Halophytes are usually considered like sodium tolerant plants. Indeed, NaCl is the main source of salinity in most areas, though chloride, sulfate, calcium, magnesium, and the other ions are involved constituting sometimes the main salt for soil and water salinization (Flowers et al., 1977). "Domestication" of halophytes is one more way to use salt-affected and salinized territories.

Figure 1: Halophytes Salicornia sp. (larger and bright green plants at the picture) andSuaeda maritima (smaller and grayish-green plants at the picture) are growing at the salt-affected soil near river Medway in the UK, where the area is flooded with mixed sea ...

Halophytes is an undisputable example and proof that salinity tolerance in plants is achievable. The main questions are:

- what are the salinity tolerance traits in halophytes;
- how to bring the relevant trait or multiple traits to agricultural plants;
- to which extent the transfer could be realized without essentially influencing the growth rate, agricultural productivity and quality of grain, fruits, edible parts, flowers or the other economically important elements/ features/traits of agricultural plants.

The simplest idea under salinity is to decrease sodium net influx and increase potassium net uptake via plasma membrane of epidermal root cells to alter mineral nutrition. Surprisingly, in most cases the pure straightforward and strict approach is not realized neither in halophytes, nor in glycophytes (Flowers, 2004; Maathuis et al., 2014). Halophytes imply several strategies to cope with high concentrations of salts including (1) sodium exclusion from roots, (2) accumulation of high sodium in shoots, (3) shedding specialized leaves, (4) localizing salts in vacuoles, (5) excreting salts via salt glands etc. The role and contribution of each strategy depend on the habitat and type of a halophyte plant (Flowers et al., 1977; Breckle, 2002; Lüttge, 2002;Flowers and Colmer, 2008; Munns and Tester, 2008; Shabala, 2013). So far the known transport systems in halophytes are basically the same like in glycophytes due to common ancestry and evolution (Flowers et al., 2010): the trait of salt tolerance emerged independently over 70 times in different groups of grasses (Bennett et al., 2013). The knowledge facilitates potential transfer of salinity tolerance traits to agriculturally important plants.

It is important to consider living cell as a complex system. There are networks of signaling events and metabolic reactions. Under salinity the increase in cytoplasmic Na^+ and reduction of K^+ result in changes of membrane potential, osmotic pressure, turgor pressure, calcium signaling, reactive oxygen species signaling, transcriptional regulation, alteration of gene expression, modification of protein expression pattern and spectra of siRNAs, signaling molecules, phytohormones and metabolites. The set of parameters including ion fluxes characterizes cell in a definite physiological state. Transition from one physiological state to the other one could be described by a trajectory in a multidimensional space, while the stable physiological states are considered as dynamic attractors. The volume and shape of the attractor in multidimensional space could be registered using means of "omics" that is RNA expression microarrays, proteomics, ionomics, metabolomics etc. Properties of the attractors are slowly studied and understood for biological systems (Spiller et al., 2010;Breeze et al., 2011), though the idea is initially well developed in physics, especially in plasma physics for non-equilibrium thermodynamic systems (Akhromeeva et al., 1989, 1992). Obviously and intuitively the biological systems with tens of thousands of participating expressed genes are more complex than physical systems. Salt treatment up- and down-regulates tens and hundreds of genes including transcription factors, genes of transporters and regulatory proteins (e.g., Taji et al., 2004; Volkov et al., 2004; Liu et al., 2013;Takahashi et al., 2015) when the cells are moving to a new distinct physiological state of gene expression, metabolic control and ion transport.

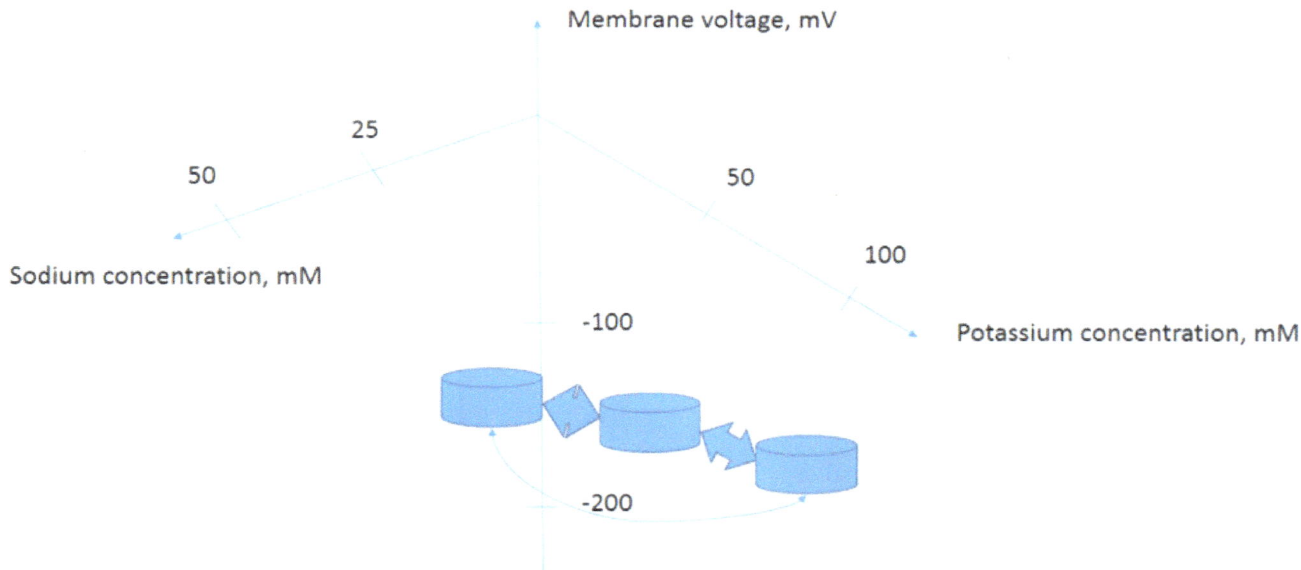

Figure 2: Proposed simplified model of several attractors for a plant cell with specific metabolic and regulatory networks determined by cytoplasmic ion concentrations and membrane potential. Changes in extracellular ion concentrations result in transition of cells ...

Comparison of ion fluxes via membranes between halophytes and glycophytes often demonstrates lower sodium uptake for halophytes (reviewed in: Flowers and Colmer, 2008). However, an evident problem in comparison is the high variability in ion transport between plant species because of growth rate, tissue-specific variability and the other physiological factors. It is reasonable to consider similar plants and achieve comparable values. Recent results on ion fluxes in glycophyte*Arabidopsis thaliana* and similar from the point of genome and morphology halophyte *Thellungiella halophila* demonstrated lower Na^+fluxes and higher K^+/Na^+ selectivity of ion currents in the roots and root protoplasts of the halophyte under salt treatment (**Figures** and , and in: Volkov et al., 2004; Volkov and Amtmann, 2006; Wang, 2006;Wang et al., 2006; Amtmann, 2009; resembling results for roots of the two plants: Alemán et al., 2009). Lower Na^+ accumulation and higher K^+/Na^+ ratio under salt treatment were also found in roots of leguminous halophyte *Melilotus indicus* compared to similar herbaceous glycophyte *Medicago intretexta* (Zahran et al., 2007).

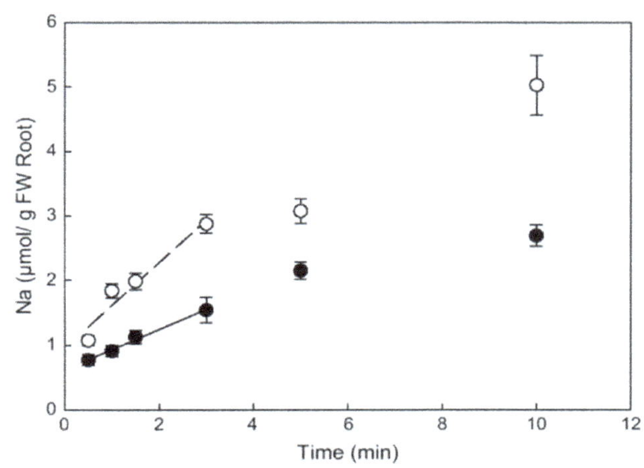

Figure 3: Kinetics of initial unidirectional Na^+influx into roots of *Arabidopsis thaliana* **(open circles) and***Thellungiella halophila* **(closed circles) as determined from** $^{22}Na^+$accumulation of individual plants from $^{22}Na^+$ **labeled nutrient solution with 100 mM** NaCl .

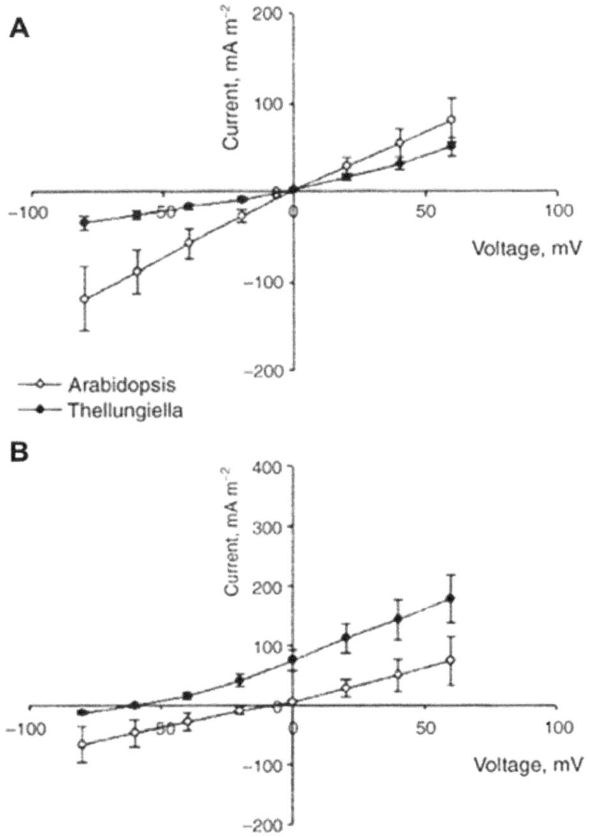

Figure 4: Comparison of current–voltage curves for whole-cell instantaneous currents in root protoplasts of *A. thaliana* **(open symbols)** and *T. halophila* **(closed symbols).** Currents are normalized to protoplast surface. The pipette solution was always 100 .

Other strategies may be involved depending on the level of salinity tolerance, plant morphology, habitat and the other environmental factors and evolutionary history. For example, salt tolerant *Plantago maritima* had similar sodium uptake rates by roots compared to salt-sensitive *P. media* (Erdei and Kuiper, 1979; de Boer, 1985); the salt-tolerance in the pair is rather associated with xylem transport and sodium accumulation in vacuoles of leaf cells.

Vacuolar membranes of several halophytes were also a subject of special investigation. Electrophysiological patch clamp study of vacuoles from leaves of *Suaeda maritima* did not find any unusual features to support the high salt tolerance of the halophyte (Maathuis et al., 1992). Patch clamp experiments to compare vacuoles from roots of*P. maritima* and *P. media* also did not reveal striking differences apart from an extra smaller ion channel conductance in the tonoplast of the halophyte; salt stress essentially reduced the open probability of larger non-selective between K^+ and Na^+ ion channel conductance in both species but did not change the properties of the conductance (Maathuis and Prins, 1990). However, comparison of tonoplast from suspension culture cells of halophytic sugar beet with glycophytic tomato revealed rectification properties of ion channels in vacuolar membrane of the halophyte (Pantoja et al., 1989). At positive voltages in outside-out configuration (corresponding to ion currents out of vacuole) the ion-channel-like conductance dropped 6.5 times presumably preventing the transport of ions from vacuole to cytoplasm; the conductance was not selective for K^+ over Na^+ in both species (Pantoja et al., 1989).

Complexity of ion transport within the whole plant under salinity is confirmed also in experiments with glycophytes. Ion transport could be essentially influenced by salt stress in a cell-specific manner. The information is required to consider the whole plant responses to salinity and to alter them in a desired direction. An example of changes in membrane electric conductance after salt stress in several types of cells from barley leaves is given in **Figure** .

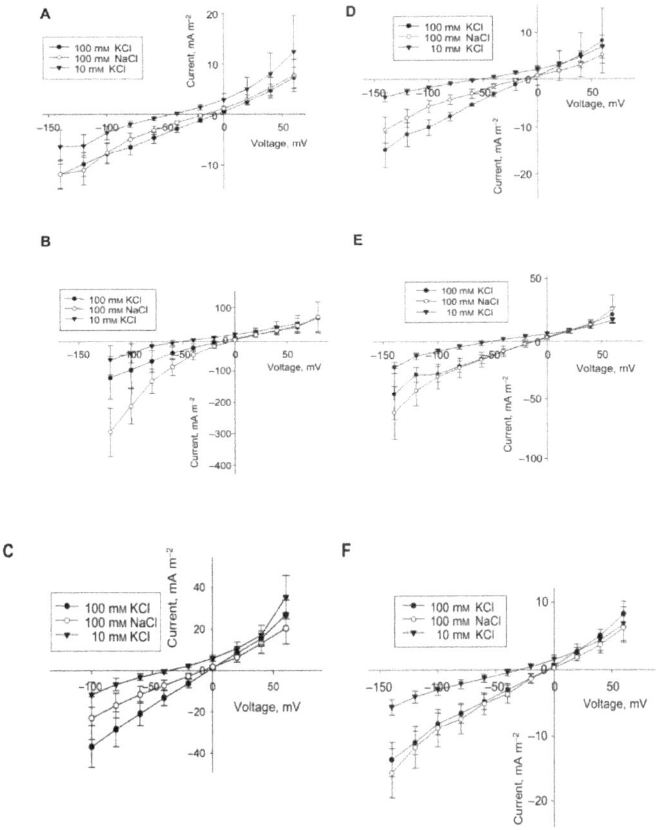

Figure 5: Effect of salt stress on instantaneous current in protoplasts from the elongation zone and emerged blade portion of the developing leaf 3 of barley *(Hordeum vulgare* **L.) (A–F).**Averaged current–voltage (I–V) relationship in protoplasts .

Basic Assumptions for Regulating Ion Transport in Plant Cells

Before pondering ways to modulate ion transport it is helpful to describe techniques to measure ion fluxes and then introduce basic important parameters, which could be influenced and manipulated to alter ion fluxes and intracellular ion concentrations. The knowledge is also required to explain, understand and predict the results with overexpression/knockout of individual transporters and ion pumps (see below).

Comparison between Different Methods for Measuring Ion Fluxes

Total activity of proton pumps, ion channels, transporters and potential "leak" pathways of ion transport (see below) is reflected in changing and regulated ion fluxes. The coinciding values of ion fluxes measured by several independent techniques are needed to provide reliable and robust basis for the further conclusions. Direct and indirect methods include (1) estimates based on kinetic measurements of ion concentrations (both in plants and in nutrient solution), (2) electrophysiological methods, (3) technique of measuring ion fluxes using vibrating ion-selective electrodes [microelectrode ion flux estimate/measurements (MIFE)], (4) measurements of unidirectional ion fluxes (e.g., radioactive isotope $^{22}Na^+$ for sodium, while Rb^+ is often used to imitate K^+ fluxes) and (5) several other methods including ion-selective fluorescent and non-fluorescent indicators, potentially NMR spectroscopy, ion-conductance scanning microscopy etc.

Table 1: Comparative summary for methods to measure ion fluxes in plants and their tissues and cells

Method or group of methods	Principle of the method	Spatial and temporal resolution, advantages	Disadvantages
Kinetic measurements of ion concentrations	Ion concentrations in plant tissues or in nutrient solution are measured in time, the concentration differences along the time points are plotted against time.	Resolution is at the level of whole plant or plant tissues, usually tens of minutes and hours are needed to register changes. Simple methods without a need of special equipment, convenient for most ions.	Measures net fluxes, not separate influx and flux outside of plant tissues. Low level of resolution.
MIFE: microelectrode ion flux estimate/ measurements	Tiny ion-selective microelectrode with tip around a μmeter vibrates within seconds in the vicinity of a cell or plant tissue and measures ion concentrations. Changing in time difference in concentrations is recalculated to ion fluxes.	Resolution at the level of individual cells within seconds, measurements may last for hours.	Requires special equipment. Ion-selective electrodes often interfere with organic compounds. Reliable ion-selective resins available for a few major ions only.
Measurements of unidirectional ion fluxes	Plant tissues or organs are loaded with radioactive ions or rare ions to imitate ions of interest. Unidirectional usually outward ion fluxes of the isotope or rare ion are measured then as changes in concentrations against time.	Spatial resolution at the level of whole plant or plant tissues, recordings from minutes lasting to hours are needed to register changes.	Requires radioactive isotopes and often complicated calculations with several proposed pools of ions.
Electrophysiological methods	Isolated cell membrane, piece of membrane or single cell within a plant preparation are subjected to set of physiological voltages and ion current is registered in the form of electric current.	Spatial resolution of single molecules or single cells. Temporal resolution from μseconds to minutes. High accuracy and possibility to find out specific molecules for transport of definite ions.	Indirect measurements, measure total electric current carried out by several ion species.
Fluorescent indicators	Fluorescent ion selective indicators	Resolution at the level of individual cells within tens of seconds to minutes, recordings over tens of minutes to hours.	Require specific protocols for loading, intrinsic autofluorescence and non-specific adsorption of fluorescent indicators could be drawbacks.

The distinction between the methods is partially arbitrary.

Estimates from ion concentrations are the sum of influx and efflux of ions from plant cells. The measurements finally provide the important physiological information about averaged net changes and fluxes. However, the existing concentrations are often already high, therefore it takes at least hours and sometimes days to determine kinetic changes of ion concentrations and obtain kinetic curves, usually without detailed peculiarities and high temporal resolution of minutes. For example, after 25 h of 100 mM NaCl treatment sodium contents in roots of glycophyte *A. thaliana* rose from 0.2 to 2% of dry weight (DW) of the roots (0.087–0.87 mmole Na^+ per g of DW), while in halophyte *T. halophila* sodium in roots increased from 0.15 to 1.2% (0.065–0.52 mmole Na^+ per g of DW) (Volkov et al., 2004). Fast initial rates of net Na^+ uptake by roots during the first 6 h were 0.064 μmole/(g root DW*min) in *Thellungiella* and 0.048 μmole/(g root DW*min) in*Arabidopsis*. After 24 h of salt treatment, the rates dropped to 0.0004 nmole/(g root DW*min) in *Thellungiella* and 0.003 nmole/(g root DW*min) in *Arabidopsis* (Wang, 2006).

To elucidate detailed mechanisms of ion transport it is essential to add the other methods. One more requirement is to compare ion fluxes measured by different techniques: changes of ion concentrations are expressed in moles/g of fresh or dry weight (FW or DW), while electrophysiological and MIFE measurements are normalized per a unit of surface area, A/m^2 and moles/(m^2*s), correspondingly. Rough estimate for conversion is that DW is about 10–15% of FW; for recalculation per surface area the size of roots or protoplasts is required. Estimates relating potassium flux in epidermal cells to ion electric currents were done for rye roots (White and Lemtiri-Chlieh, 1995). The net K^+ flux was estimated 1.0–1.9 μmole/(g root WF*h) and unidirectional K^+ flux was about 7 μmole/(g root WF*h) from mineral solution with 0.6 mM K^+ (White et al., 1991). Epidermal cells with diameter 26 μm were considered 8.3% of root volume; then the net fluxes in μmole/(g root WF*h) corresponded to 0.11–0.21 pmole/(cell*hour) or 3.1–5.9 pA/cell. Unidirectional flux of 7 μmole/(g root WF*h) corresponded to 0.77 pmole/(cell*hour) or 21.7 pA/cell (White and Lemtiri-Chlieh, 1995). Converting per surface area of the whole protoplasts, the fluxes of 0.11–0.21 pmole/(cell*hour) correspond to 14–27 nmole/(m^2*s) and 0.77 pmole/(cell*hour) ≈ 100 nmole/(m^2*s), the values reasonably confirmed by the other methods (see below). Conversion of the flux in pmoles to number of ions per a cell gives 0.77 pmole/(cell*hour) ≈ 1.3 * 10^8 ions/(cell*second) (see below).

Electrophysiological techniques of two-electrode voltage clamp or patch clamp measure ion currents across membrane of plant cells (protoplasts for patch clamp) under determined applied voltage via the membrane and provide current–voltage curves with resolution up to pA and μseconds (**Figures** and 55). For the example of patch clamp study with root protoplasts of *Arabidopsis* and *Thellungiella*, the inward K^+fluxes in external 100 mM KCl at -80 mV were 120 mA/m^2 = 120 fA/$μm^2$ ≈ 1.2 10^{-18} mole/($μm^2$*s) = 1.2 μmole/(m^2*s) for *Arabi-*

dopsis and 30 mA/m^2 ≈ 300 nmole/(m^2*s) for *Thellungiella*, correspondingly. Inward Na^+ fluxes under the same conditions, but in 100 mM NaCl in the external medium are 70 mA/m^2 ≈ 700 nmole/(m^2*s) for*Arabidopsis* and 15 mA/m^2 ≈ 150 nmole/(m^2*s) for *Thellungiella*(Volkov and Amtmann, 2006).

Comparisons of electric currents and ion fluxes were performed in a series of simultaneous measurements using patch clamp and MIFE for wheat root protoplasts, when MIFE measured H^+ fluxes (Tyerman et al., 2001). Proton fluxes in the experiments basically correlated to electric currents, though with large variation between protoplasts (Tyerman et al., 2001). Further on MIFE proved the lack of"dark" electroneutral fluxes for K^+ transport in the wheat root protoplasts (the situation was different for Ca^{2+}, large Ca^{2+} fluxes were not electrogenic): K^+ ion currents via outward and inward potassium channels nearly exactly corresponded to the K^+ ion fluxes measured by ion selective electrodes of MIFE (Tyerman et al., 2001; Gilliham et al., 2006). Some deviations from 1:1 ratio and variation between protoplasts were explained by uneven distribution of ion channels (Gilliham et al., 2006). Potentially atomic force microscopy or ion-conductance scanning microscope with further ion channel recordings (e.g., Hansma et al., 1989; Korchev et al., 1997; Lab et al., 2013 for ion-conductance scanning microscope) could be useful to explore the distribution of lipid rafts and clusters of ion channels at nanometre resolution.

Values of ion fluxes measured by patch clamp in protoplasts are usually higher than fluxes measured by MIFE for intact roots (e.g., Shabala and Lew, 2002; Chen et al., 2013), but coincide within orders of magnitude and essentially depend on composition of ambient medium, concentration of ions, membrane potential of cells and on multiple physiological factors (e.g., Ivashikina et al., 2001; Shabala et al., 2009). Experiments with MIFE are non-invasive and simpler; therefore provide huge opportunities with temporal resolution of seconds and spatial resolution within less than tens of microns for exploring physiological factors and conditions, which influence ion fluxes to and out of cells (Newman et al., 1987; Newman, 2001; Shabala, 2006; Sun et al., 2009; Shabala and Bose, 2012). Ion fluxes of K^+, Na^+, and other ions in the vicinity of roots were measured by MIFE and compared for different salt-tolerant and salt-sensitive cultivars and agricultural species (Chen et al., 2005; Cuin et al., 2008, 2012), along root zones of several plants (Garnett et al., 2001; Chen et al., 2005; Pang et al., 2006), for mutants in specific ion channels or transporters (e.g., Shabala et al., 2005; Demidchik et al., 2010), under treatment by physiologically active compounds (Cuin and Shabala, 2007a; Shabala et al., 2009; Pandolfi et al., 2010; Demidchik et al., 2011; Ordoñez et al., 2014), after generation of reactive oxygen species and salt stress (Cuin and Shabala, 2007b; Demidchik et al., 2010) and for numerous other conditions. MIFE is also a good method for studying ion transport in cell biology (Lew et al., 2006; Valencia-Cruz et al., 2009; Demidchik et al., 2010). The results provided huge volume of information about characteristics, kinetics and physiological features of ion fluxes under salt stress, helped to develop fast tests for salinity tolerance (Chen et al., 2005; Cuin et al., 2008). The results are described in hundreds of publications and several reviews (e.g., Newman, 2001; Shabala, 2006;Sun et al., 2009; Shabala and Bose, 2012), so are not covered here in more detail. Among the limitations of MIFE is the selectivity of ion-selective electrodes, which is influenced by interfering ions (e.g.,Knowles and Shabala, 2004) and sometimes affected by physiologically active compounds and proteins in the medium for measurements after interaction with the material of ion-selective electrodes (e.g., Chen et al., 2005), hence demanding more control checks.

Unidirectional fluxes of $^{22}Na^+$ for sodium and $^{42}K^+$ or Rb^+ (a tracer for K^+) provide fast kinetics of ion transport within tens of seconds as determined by the speed of sampling and changes in concentrations at the background of initial level without $^{22}Na^+$ and $^{42}K^+/Rb^+$ (e.g.,**Figure** for $^{22}Na^+$ fluxes). The method is used widely and had already provided essential advances in plant ion transport (e.g., MacRobbie and Dainty, 1958; Rains and Epstein, 1965; Epstein, 1966). The outward fluxes could be measured after loading plants with $^{22}Na^+$ or $^{42}K^+/Rb^+$and transferring then to different chemical solutions without the ions (e.g., Wang, 2006; Wang et al., 2006 for $^{22}Na^+$ for roots of *Arabidopsis*and *Thellungiella*). Details and the methodical procedures are well described with different modifications to get more information about compartmentation of the absorbed ions and to exclude potential sources of errors (Cheeseman, 1986 for analysis of compartmentation based on efflux kinetics; Britto et al., 2006; Wang, 2006 for $^{42}K^+$ in application for the tracer efflux by barley roots; Britto and Kronzucker, 2013 for comprehensive practical description of experimental procedures to measure potassium fluxes). The values of measured fluxes differ significantly, sometimes over 100 times depending on plant species, physiological conditions and ion concentrations (e.g., summarized for Na^+ fluxes in: Kronzucker and Britto, 2011).

of unidirectional fluxes is often complemented by the other methods to obtain better understanding of the processes. An interesting comparison for influx of K^+ ($^{86}Rb^+$) from 0.1 mM K_2SO_4 solution and net K^+ influx measured by external K^+ microelectrodes (prototype of MIFE) gave nearly the same values of fluxes: 2.6 µmole/(g FW*h) and 2.5 µmole/(g FW*h), correspondingly (Newman et al., 1987). Measurements of unidirectional $^{24}Na^+$, $^{42}K^+$ fluxes in barley together with membrane potential measurements and pharmacological profiling of the fluxes allowed to study high affinity transport of sodium from low µM – 50 mM solutions and provided predictions about possible mechanisms for the transport (Schulze et al., 2012).

Among methods to determine ion fluxes is the use of ion-selective fluorescent indicators for estimating cytoplasmic and vacuolar concentrations of Na^+ and K^+ and kinetics of their changes (e.g., mostly for protoplasts: Lindberg, 1995; Halperin and Lynch, 2003;D'Onofrio et al., 2005; Kader and Lindberg, 2005), measurements by intracellular ion-selective electrodes (e.g., Carden et al., 2003 and references there), ^{23}Na-NMR spectroscopy (e.g., Bental et al., 1988) and several others; the methods are not discussed in detail here.

Ion Concentrations in Cells

Ion concentrations are the reflection of net ion fluxes via plasma membrane. Certain range of ion concentrations, especially of K^+, Na^+, and Ca^{2+}, is vital for cell physiology and function of proteins. Typical potassium concentrations in cytoplasm of plant cells were measured independently by several methods (including ion-selective electrodes, fluorescent dyes and X-ray microanalysis) and range around 60–140 mM (Pitman et al., 1981; Hajibagheri et al., 1988; Hajibagheri and Flowers, 1989; Walker et al., 1995, 1998; Korolev et al., 2000; Cuin et al., 2003; Halperin and Lynch, 2003; Shabala et al., 2006; Hammou et al., 2014), though concentrations above 200 mM were estimated by efflux analysis (reviewed in Britto and Kronzucker, 2008) and as low as 12 mM and even lower K^+ concentrations were measured by ion-selective electrodes in root cells of potassium-deprived *Arabidopsis*plants (Armengaud et al., 2009). Higher potassium concentrations of 200–350 mM measured in cell sap by X-ray microanalysis or capillary electrophoresis are rather attributed to vacuolar compartment under sufficient potassium supply (Malone et al., 1991; Fricke et al., 1994;Bazzanella et al., 1998; Volkov et al., 2004). It is worth mentioning that in animal cells potassium seems to be among regulators of apoptotic enzymes activating them at K^+ concentrations below 50 mM (Hughes and Cidlowski, 1999).

Sodium cytoplasmic concentrations of plant cells are usually low reaching about 20-50 mM after several days of NaCl treatment (Carden et al., 2001, 2003; Halperin and Lynch, 2003). Higher sodium concentrations had also been measured depending on duration of salt stress, external sodium, concentration of the other ions and on plant species: cytoplasmic sodium concentrations over 200 mM were reported in salt-tolerant halophytes (Hajibagheri and Flowers, 1989;Halperin and Lynch, 2003; Flowers and Colmer, 2008; Kronzucker and Britto, 2011).

An interesting example is halotolerant alga *Dunaliella salina*, which is a good unicellular eukaryotic model for studying salinity tolerance within the range of 0.05–5.5 M NaCl (e.g., Katz and Avron, 1985). Cytoplasmic sodium concentrations about 90 mM (88 ± 28 mM) were reported in the alga using ^{23}Na-NMR spectroscopy (Bental et al., 1988). Na^+ concentrations were nearly the same (within the error of a few measurements) in the algal cells adapted to a wide range of external Na^+, from 0.1 to 4 M (Bental et al., 1988). Similar or even lower sodium concentrations below 100 mM were measured by the other methods for the alga under 0.5–4 M or 1–4 M sodium treatment (Katz and Avron, 1985; Pick et al., 1986). The small alga *D. salina* has length about 10–11 µm, width of 6 µm and volume around 200 fL (Masi and Melis, 1997) or even smaller dimensions with volume around 90–100 fL then (Katz and Avron, 1985). It probably possesses specific transport system to exclude Na^+, similar to Na^+-ATPase expected to function in the plasma membrane of the marine unicellular alga *Platymonas viridis* (Balnokin and Popova, 1994), alga *D. maritima*(Popova et al., 2005) and several other marine algae (reviewed in:Balnokin, 1993; Gimmler, 2000).

High sodium concentrations over 100 mM often have inhibiting effect on protein synthesis at least in salt-sensitive glycophytes (Hall and Flowers, 1973; Wyn Jones and Pollard, 1983; Flowers and Dalmond, 1992). Sodium is also (1) competing with potassium for allosteric sites of enzymes and (2) interacting with ion channels (for example, sodium ions change the gating of potassium outward rectifying currents in root protoplasts of halophyte plant *Thellungiella*, which are most likely carried by Shaker type potassium channels: Volkov and Amtmann, 2006). Moreover at the cellular level salt stress induces apoptosis (Katsuhara and Kawasaki, 1996; Huh et al., 2002; shortly reviewed inShabala, 2009; Demidchik et al., 2010).

Much higher sodium concentrations could be tolerated in vacuoles, one of the functions of the organelle is to sequester and isolate sodium. Concentrations of sodium in vacuoles may exceed 0.5–1 M being up to ten times over the cytoplasmic sodium concentrations (eg Flowers et al., 1977; Zhao et al., 2005; Flowers and Colmer, 2008) at the expenses of activity of specific ion-transport systems (reviewed in: Martinoia et al., 2012). Under salt treatment of 2 M NaCl for 85 days shoot tissue concentrations of sodium in halophytes *Tecticornia* were about 2 M, so presumably vacuolar Na^+ concentrations could be over 2 M in the halophytes (English and Colmer, 2013).

The reasons for sodium competing with potassium are that the ions have (1) the same electric charge, $1.6{*}10^{-19}$ coulombs, (2) similar cation radii in non-hydrated, about 0.1 nm = 1.0 Å for sodium cation and 0.14 nm = 1.4 Å for potassium (diameter being nearly 2–3% of cell membrane thickness) and (3) hydrated forms, about 3.6 Å for sodium and 3.2 Å for potassium ions (Nightingale, 1959; Collins, 1997; Mähler and Persson, 2012) and, hence, (4) similar surface electric charge densities, which differ about two times (twice higher for non-hydrated sodium according to charge and diameter of the ion). Therefore interactions of Na^+ and K^+ with amino acids of protein surfaces, active centers of enzymes, pockets of allosteric regulation or binding of proteins, selectivity filters of ion channels are similar and often differ only several times in selectivity of the interactions. The selectivity depends on the nature and number of interacting amino acids and their spatial location. Molecular dynamics simulations together with conductivity measurements for several proteins and oligopeptides demonstrated up to five times higher affinity of sodium over potassium to the protein surfaces (especially with numerous carboxyl groups;Vrbka et al., 2006). The phenomenon could explain higher destabilizing effect of sodium over potassium on proteins ("salting them out"), which was initially discovered with white proteinaceous part of hen's eggs by Hofmeister in 1888 (Hofmeister, 1888; Kunz et al., 2004). The effect could be the reason why potassium and not sodium is chosen and naturally selected for being the major intracellular monovalent cation, pumped into cells while pumping out sodium cations (Collins, 1997) though sea and ocean water contains more than 40 times higher concentration of sodium.

Under salt stress, for plants it is important to keep higher K^+/Na^+ ratio (Maathuis and Amtmann, 1999). It is essential, however, to mention that some proteins (due to specific amino acid composition or structural peculiarities) and processes from halophytes are able to withstand higher sodium concentrations without loss of activity (eg. Flowers and Dalmond, 1992; Premkumar et al., 2005); it seems to be the secondary evolutionary adaptation. It is also interesting that cell wall proteins of studied halophytes and also glycophytes did not change their activity within wide range of sodium concentrations, often from 0 to over 0.5–1 M (Thiyagarajah et al., 1996).

Driving Forces and Pathways for Ion Transport to Cells

Transport of ions is driven by physico-chemical forces including differences of ion concentrations (to be more precise, activities of ions) and differences in electric potential at the sides of membranes.

Membrane potential of plant cells is routinely measured by microelectrodes with tiny sharp tips around 0.1 µm in diameter after impalement of a plant cell of interest (e.g., described in: Blatt, 1991). Recently developed voltage-sensitive fluorescent proteins and dyes (reviewed in: Mutoh et al., 2012) could also be used for at least indications of membrane potential in cell tissues and populations of cells (Matzke and Matzke, 2013). Membrane potentials below -70 mV and above -200 to -220 mV are recorded by microelectrodes though values around -300 mV were also reported; more negative values are often measured in root cells compared to leaf ones (apart from leaf guard cells) (Higinbotham, 1973; L'Roy and Hendrix, 1980; Blatt, 1987; Walker et al., 1995, 1998; Carden et al., 2001, 2003; Shabala and Lew, 2002; Fricke et al., 2006; Murthy and Tester, 2006; Shabala et al., 2006; Volkov and Amtmann, 2006; Armengaud et al., 2009; Hammou et al., 2014). Vacuolar membrane potential is the same or 10–40 mV above the values for cytoplasm with pH about 2 or over units lower, about 5.0–6.1 or less in vacuoles compared to pH = 7.0–7.7 in cytoplasm (e.g., Walker et al., 1995; Carden et al., 2003; Cuin et al., 2003; Martinoia et al., 2012).

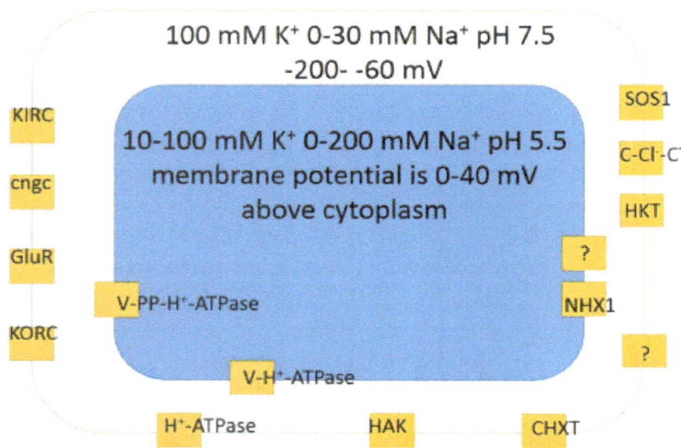

Figure 6: Basic scheme of membrane potentials, potassium K^+ **and sodium** Na^+concentrations and pH values in a generalized plant cell together with main ion transport systems ensuring potassium and sodium transport according to electrochemical forces.Different cell .

Thermodynamics of ion transport is described by several equations. Nernst equation applied to selectively permeable membrane links ion concentrations at the sides of membrane to the electric potential via the membrane under equilibrium conditions (when net flux of ions via the membrane is absent):

$$E_S = E_1 - E_2 = R*T/(Z_S*F)*\ln([S_2]/[S_1])$$

(Hille, 2001). Here E is electric potential, R is universal gas constant equal to 8.31 J/(K*mole), T is temperature in K, Z_S is the charge of ion S, F is Faraday constant equal to 96,500 s*A/mole, $[S_1]$ and $[S_2]$ are concentrations of ion S at the sides of the membrane. Basically, the diffusion of ion S due to different concentrations is equilibrated by distinction in electric potential, which is about ± 60 mV (slightly depending on temperature) per 10-fold difference in concentrations with sign determined by the ion charge. For potassium with typical 100 mM in cytoplasm of epidermal root cell and low 100 µM in soil solution the membrane potential to ensure uptake of K^+ should be below -180 mV to satisfy the electrochemically downhill transport of K^+ ions. Lower concentrations of potassium outside the cells may require co-transport of K^+ with the other ions (e.g., with H^+).

For several ion species with specific permeabilities via the membrane a more complicated Goldman–Hodgkin–Katz voltage equation is applicable; it takes into account permeabilities of ions. For sodium, potassium and chloride (obviously more ions to be considered and more components should be added) the equation will be:

$$E_{membrane} = R*T/F*\ln((P_{Na+}[Na^+]_{out} + P_{K+}[K^+]_{out} + P_{Cl-}[Cl^-]_{in})/(P_{Na+}[Na^+]_{in} + P_{K+}[K^+]_{in} + P_{Cl-}[Cl^-]_{out}))$$

where $E_{membrane}$ is membrane potential via the membrane (or $E_{reversal}$with zero net ion current via the membrane), P are permeabilities of the corresponding ions and [] stands for concentrations of the ions (Hille, 2001). Usually potassium permeability is dominating and membrane potential is close to $E_{reversal}$ of K^+, though may change and depend on cell type. Active plasma membrane proton pump H^+-ATPase (reviewed e.g., in: Palmgren, 2001) shifts membrane potential to more negative values compared to the calculated $E_{reversal}$ for K^+.

Transport of most ions including Na^+ and K^+ in plants occurs passively (following the electrochemical forces) via ion-selective proteinaceous pores of ion channels. Most ion channels can change their conformation from open to close states and vice versa (so called "gating") under applied voltages or after binding ligands and regulators. Another pathway is via proteinaceous transporters with slower transport rates. Ion transport via ion channels is electrogenic since ions carry electric charge, while transporters realize electrogenic or non-electrogenic transport, transporting one ion or co-transporting/antiporting several charged ions in one or opposite directions, correspondingly.

Co-transport of several ions or even small molecules may add extra energy for transport. For example, HAK transporters presumably co-transport K^+ together with H^+ (Banuelos et al., 1995;Rodriguez-Navarro, 2000; Grabov, 2007), which gives several orders of concentration differences extra due to transport of H^+ according to electric charge. Membrane potential of -180 mV potentially allows potassium uptake, when co-transported with $H^+$with, e.g., stoichiometry 1:1 under similar external pH to pH of cytoplasm, against 10^6 differences in K^+ concentrations (e.g., Rodriguez-Navarro, 2000). Higher concentrative capacity could be achieved using also pH differences or higher number of protons per K^+; cytoplasmic pH is about 7.5 and external low pH of 4 will add over three orders of concentration more. The surprising example is described for yeast*Schwanniomyces occidentalis*, which was reported to deplete external potassium to 0.03 µM, presumably taking up K^+ against 3,000,000 differences in concentrations (assuming over 100 mM of cytoplasmic K^+) due to HAK transporters (Banuelos et al., 1995). Much higher concentrations, around 80 µM, arising from K^+ contamination from agar and the other chemicals (Armengaud et al., 2009; Kellermeier et al., 2014) (though contamination from agar was estimated at 1–3 µM:Kellermeier et al., 2014) resulted in symptoms of severe potassium deficiency in *Arabidopsis* and essentially changed transcription profile in roots and shoots of the plants (Armengaud et al., 2004), so more detailed examination with special attention to transport systems of different species (e.g., Coskun and Kronzucker, 2013) is required. Potentially new transporters and ion channels from different organisms could be a source of diversity and comparison for the existing pool of membrane transport proteins and for creating novel artificial transporters.

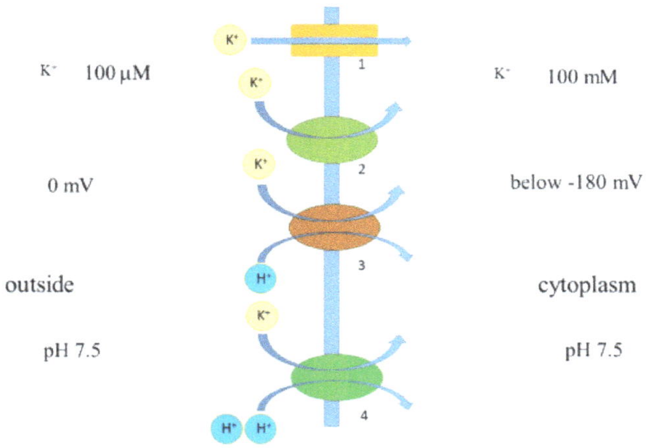

Figure 7: Simplified scheme demonstrating principles of ion transport via membrane. Voltage difference below -180 mV allows potassium transport against 1000-fold concentration difference via potassium-selective pore of ion channel 1. Similar thermodynamically favorable .

A set of transporters and ion channels is specific for cell types and organisms, includes tens and more distinct characterized so far proteins, which often form heteromers with variable properties and regulation (Figure ; e.g., reviewed in: Isayenkov, 2012). Detailed analysis of genome sequences of salt-sensitive model plant *Arabidopsis* revealed that about 5% of about 25,000 genes of the plant potentially encode membrane transport proteins; the genes of about 880 proteins are classified in 46 unique groups, while genes of cation channels/transporters predict for coding over 150 proteins (Mäser et al., 2001). Special databases include information about transport proteins, e.g., plant membrane transport database http://aramemnon.uni-koeln.de/; http://www.yeastgenome.org/ is a useful source of information for yeast proteins including yeast membrane transport proteins.

Cell Size-Volume-Surface/Volume Ratio and Effects of Cell Wall

Interestingly, cell surface/volume ratio has an effect on ion concentration under the same ion fluxes (Figure). The sizes of cells within a plant differ orders of magnitude. Cells of xylem parenchyma in roots of dicot *Arabidopsis* are less than 5 µm in diameter with length often below 20–30 µm (eg: Dolan et al., 1993; Ivanov, 1997; Kurup et al., 2005; Verbelen et al., 2006; Ivanov and Dubrovsky, 2013); the cells could be isolated and result in protoplasts of about 10 µm in diameter compared to larger 20 µm epidermal protoplasts from root elongation zone or 15–25 µm protoplasts from root cortex parenchyma cells and the other root tissues (Demidchik and Tester, 2002; Demidchik et al., 2002a; Volkov and Amtmann, 2006). The volumes would correspond to $4/3^*\pi^*R^3$ that is about 500 fL for 10 µm protoplasts and about 8 pL for 25 µm protoplasts.

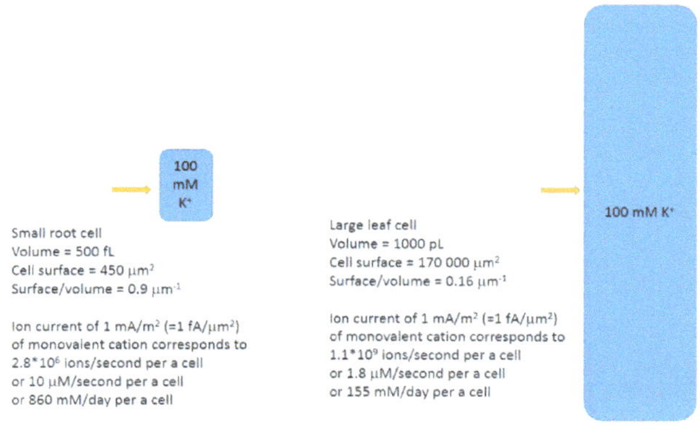

Figure 8: Simplified scheme of different plant cells and the ion fluxes for the cells.The same ion fluxes result in different changes of ion concentrations for cells of different sizes and surface/volume ratios. Ion fluxes are directly recalculated from number .

Cells in leaf epidermis of several monocots are quite large. Barley epidermal leaf cells could be up to 2 mm long and about 25–30 µm wide with nearly square cross-section, thus reaching volume over 1000 pL (Volkov et al., 2007). Several studies involved isolated protoplasts from barley leaf epidermal cells and reported 60 µm (100 pL) protoplasts (Dietz et al., 1992), 40 µm (from 20 to 80 µm) (30 pL, from to 4 to 250 pL) protoplasts (Karley et al., 2000), 25 pL protoplasts with large variations (over 10 times) in volume (Volkov et al., 2007). It is worth mentioning that about 99% of large epidermal leaf protoplasts could be occupied by vacuole (Winter et al., 1994).

Assuming volumes of usual plant cells within 500 fL to 1000 pL, the calculated surface to volume ratios will differ about 10 times for the cells: from about 0.9 μm^{-1} to 0.16 μm^{-1} for oblong cells within plant tissues. The larger cells need higher ion fluxes or larger time for the same concentration changes.

Effects of cell walls on ion transport are well known though not often remembered and studied in detail. They include changes of cell membrane potential depending on ion composition of ambient medium due to fixed electric charges in cell walls, assumed ion-rectifying properties of cell walls and expected effect of shrinking and swelling of cell walls on mechanosensitive ion channels. Effects of cell walls on ion buffering and ion concentrations have more experimental evidence. For example, protoplasts isolated from bean leaf mesophyll did not show NaCl-induced calcium efflux compared to mesophyll tissue (Shabala and Newman, 2000). Salt-induced H^+ efflux also differed between protoplasts compared to tissue, so, presumably, all the Ca^{2+} efflux over an hour of measurement was from calcium ion exchange stores of cell walls (Shabala and Newman, 2000).

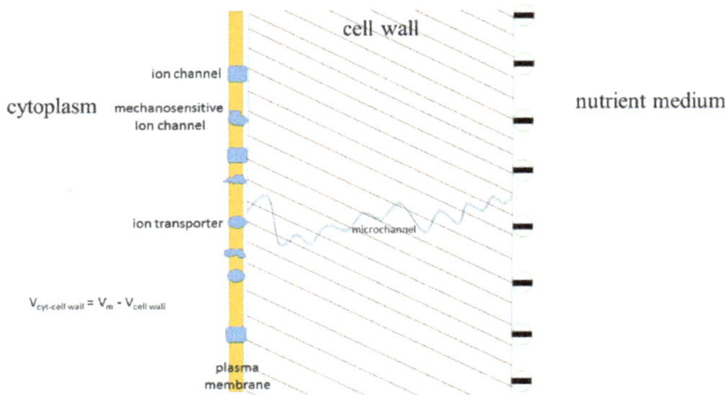

Figure 9: The scheme indicating expected and studied effects of cell walls on ion fluxes and cell plasma membrane potential. Plasma membrane of approximately 10 nm thickness is surrounded by the cell wall of about 250 nm thickness, the interior of the cell is under .

Ion Channels vs. Ion Transporters: More about Pathways of Ion Transport to Cells

One of disputable questions of ion transport is about the relative role of ion channels and transporters in transport. It is commonly accepted that ion channels in an open state/conformation allow passage of over 10^6–10^8 ions per second via a selective pore formed within a protein molecule. The diameter of the pore is determined by the molecular structure of ion channel, from 12 Å for potassium channel KcsA with narrow part of 4 Å in diameter (e.g., Doyle et al., 1998; Jiang et al., 2002; MacKinnon, 2004) to 15 Å and even 28 Å diameters of pores for the general bacterial porins with low selectivity and permitted passage for small hydrophilic molecules (about 6 Å pores for the highly selective porins) (e.g., Galdiero et al., 2012). The diameter of the pore and amino acids lining it essentially determine the ion selectivity of ion channel and potential number of passing ions per unit of time. The selectivity could be, for example, over 1,000 for K^+ over Na^+ in potassium selective ion channels or over 10 for Na^+ over K^+ in sodium selective channels due to special selectivity filters with conserved amino acids for specific channel types. Often amino acid sequence glycine-tyrosine-glycine (GYG) indicates selectivity for K^+, introducing mutations into the pore to change the amino acids converted potassium selective ion channels to non-selective ones (Heginbotham et al., 1992). The interactions of ions with the protein molecule of ion channel are not well understood yet and probably involve non-electrostatic ion–ion interactions, van der Waals forces, interaction with water molecules and numerous other interactions. Several methods of modeling and simulations of molecular dynamics are applied within at least the last 30 years; sharp increase in computing power allowed to include the lipid environment of membranes, pH and the known biochemical factors and regulators to the models (e.g., reviewed in: Maffeo et al., 2012).

Direct measurements are the basement for investigating ion fluxes via ion channels; they provide information about permeating ions, number of ions per second, selectivity and complex transitions of protein molecules of ion channels (gating) during the transport processes. Indeed, a small current of 1 pA corresponds to 10^{-12} A/($1.6{*}10^{-19}$coulombs) ≈ $6{*}10^6$ ions/second ($1.6{*}10^{-19}$ coulombs is elementary charge, a charge of monovalent cation), while most ion channels demonstrate much larger electric currents with complex voltage-dependent patterns of open–closed states.

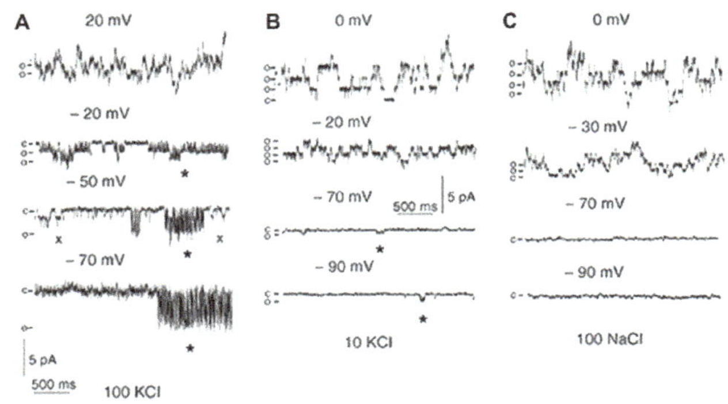

Figure 10: Single-channel recordings in outside-out patches of *T. halophila* **root protoplasts.** The pipette solution was 100 mM KCl. The bath solution contained 100 mM KCl (A), 10 mM KCl(B) or 100 mM NaCl (C). c, current level with no open channels; o, current levels .

Transporters could be considered as enzymes where conformational changes of a protein molecule are required for a complete transport cycle of ions (e.g., Gadsby, 2009). Turnover rate of the transporter is the number of complete transport cycles performed per second (e.g.,Longpré and Lapointe, 2011). The lower estimate for turnover rate of transporters is the activity of ion pumps. Plant plasma membrane H^+-ATPase pumps about 100 ions per a second (Sze et al., 1999). The value is comparable to 20–100 H^+ per second by yeast plasma membrane ATPase Pma1 (Serrano, 1988) and turnover rate of 160 $s^{-1}$of animal Na^+/K^+-ATPase (Skou, 1998). Similar or even lower turnover rates, from 3 to 60 s^{-1}, were shown for sodium/glucose cotransporter (Longpré and Lapointe, 2011 and references therein), while turnover rate about 500 s^{-1} was estimated for sucrose/H^+ co-transporter from maize ZmSUT1 (Carpaneto et al., 2010). The highest possible turnover rate for activity of ion transporters could be assessed from protein structure studies and frequency of conformational changes with estimated upper limit of 10^6 s^{-1} (Chakrapani and Auerbach, 2005), which seems to be an overestimated value. A more realistic value for the fast transporters is around 10,000 ions/second, when they are mutated and have accelerated turnover rate (Gadsby, 2009). A higher value of 100,000 ions per second was reported for Cl^-/H^+ antiporter ClC-5, which is rather a unique type of transporter similar to ClC channels (Zdebik et al., 2008).

Mechanisms of ion transport by transporters are less understood than transport via pores of ion channels, though the genes of transporters are sequenced and well-studied. Briefly, several mechanisms are expected for different transporters and described below.

HKT transporters with at least eight transmembrane domains could be similar to ion channels, they form a specific ion-selective pore with properties distinct from the pore of ion channels according to basic crystal structure analysis (Cao et al., 2011; reviewed in: Yamaguchi et al., 2013). HKT transporters can electrically resemble ion channels with similar IV curve, reversal potential of ion current mediated by HKT can shift following ion concentrations inside and outside the cell. The phenomenon is observed in *Xenopus* oocytes heterologously expressing different HKT transporters (Jabnoune et al.,

2009; Almeida et al., 2014;de Almeida, 2014), where rectification and Nernstian shift (according to Nernst equation for ion concentrations and voltages) in reversal potential were measured. Pore of HKT transporters has selectivity filter in the first transmembrane domain with conserved glycine for K^+-selective and serine for Na^+-selective HKT transporters (reviewed inYamaguchi et al., 2013; de Almeida, 2014). An extra amino acid constriction with arginine residue in the last transmembrane domain makes an additional energy barrier for ion transport (Kato et al., 2007;Cao et al., 2011; reviewed in: Yamaguchi et al., 2013; Benito et al., 2014; de Almeida, 2014). Selectivity of HKT transporters for Na^+ or K^+ could be altered by amino acid substitutions, while K^+/Na^+ symport or Na^+ uniport are exhibited by different HKT transporters and even varied from K^+/Na^+ symport to Na^+ uniport depending on the cation concentrations (Mian et al., 2011; Almeida et al., 2014; reviewed inAlmeida et al., 2013; Waters et al., 2013; de Almeida, 2014). Some HKT transporters symport K^+ with Na^+ at low K^+ concentrations (reviewed in Waters et al., 2013); presumably Na^+ ion adds energy for co-transport "pushing" K^+ via the amino acid constrictions (Benito et al., 2014), though the exact mechanisms of transport are still to be elucidated.

HAK transporters are not found in animals and Protista (Grabov, 2007). Presumably they co-transport K^+ together with H^+ (Banuelos et al., 1995; Rodriguez-Navarro, 2000; Grabov, 2007), but had not been crystallized so far while attempts to express them in Xenopus oocytes failed; therefore mechanisms of ion transport by HAK transporters are not well studied yet. Gene sequences and comparison of HAK transporters predict for 10–14 transmembrane domains (Greiner et al., 2011). Amino acid substitutions within the region between the second and third putative transmembrane domains of *Arabidopsis* HAK5 transporter essentially changed ion transport selectivity indicating for selectivity filter within the region (Alemán et al., 2014). Lack of putative specific pore similar to HKT transporters suggests that HAK transporters realize specific mechanism for K^+/H^+ symport (Alemán et al., 2014).

Nhx1 and SOS1 are cation/H^+ antiporters. SOS1 has 12 predicted transmembrane domains at the N-terminal part and long C-terminal tail composed of 700 amino acids (Shi et al., 2000), the protein forms homodimers (Núñez-Ramírez et al., 2012). Molecular mechanisms of Na^+/H^+ antiport by SOS1 are under investigation, though the known crystal structure of bacterial Na^+/H^+ antiporter NhaA (Hunte et al., 2005) could be a potential basement for understanding transport by both Nhx1 and SOS1. NhaA has 12 transmembrane domains, exists as a dimer, amino acid helices of the protein form negatively charged funnel-like structure, which leads to cytoplasm from the center of membrane and selectively attracts cations. Cation-bound Nha1 follows transformation and releases cation to the outer side being protonated at the same time at aspartate moieties. Deprotonation, release of H^+ to cytoplasmic side and return to the initial conformation completes the transport cycle (Hunte et al., 2005). The mechanism with conformation changes limited within a part of the protein makes NhaA one of the fastest transporters; NhaA has activity of catalytic center (turnover rate) about 89,000 s^{-1} (reviewed in Hunte et al., 2005).

Tonoplast Nhx1 could share the same transport mechanism. Discovery of Nhx1 initially provided a molecular basis for Na^+/H^+ antiport activity of vacuolar membrane. Na^+/H^+ antiport was measured in tonoplast of several plants: it included amiloride-sensitive transport in vesicles from *Beta vulgaris* (Blumwald and Poole, 1985) and from mesophyll cells of halophytic plant *Mesembryanthemum crystallinum*(Barkla et al., 1995), vesicles from roots of NaCl-grown salt-tolerant *P. maritima*, but not salt-sensitive *P. media* (Staal et al., 1991), in preparations from salt-grown barley roots (Garbarino and DuPont, 1988). Gene *AtNHX1* of the transporter was cloned from *Arabidopsis*and rescued some of the salt-sensitive yeast phenotypes under heterologous expression (Gaxiola et al., 1999). Moreover, overexpression of *AtNHX1* conferred salinity tolerance to *Arabidopsis*and significantly increased Na^+/H^+ antiport activity of vacuolar membrane (Apse et al., 1999). Expression of *AtNHX1* in yeast resulted in increased amiloride-sensitive electroneutral Na^+/H^+ exchange in yeast vacuolar vesicles (Darley et al., 2000). However, AtNHX1 with 9–11 putative transmembrane domains (reviewed in Rodríguez-Rosales et al., 2009) demonstrated also high K^+/H^+ exchange capacity depending on regulation by luminal C-terminal domain; the ratio of maximal rates of K^+ to Na^+ transport rose upon binding calmodulin in calcium and pH-dependent manner (Yamaguchi et al., 2003, 2005). Further evidence for role of *AtNHX1* in K^+ transport came from transgenic plants. Overexpression of *AtNHX1* in tomato plants conferred higher vacuolar K^+ under different growth conditions and increased salinity tolerance via retaining intracellular K^+ without influencing vacuolar Na^+ accumulation (Leidi et al., 2010). The simple model of AtNHX1 transporting and localizing excess Na^+ in vacuole was modified to more complex schemes. It was suggested that AtNHX1 is more important for K^+ transport to vacuole thus stimulating K^+uptake by roots, then K^+ ions recycle between cytoplasm and vacuole, while Na^+ is transported to vacuole and "locked" there (Jiang et al., 2010). Different effects of overexpressed *NHX* genes on vacuolar $K^+$and Na^+ concentrations under salt stress and increase in salt tolerance led to conclusions that NHX plays role in both Na^+ and K^+ vacuolar transport and K^+ homeostasis (reviewed in: Rodríguez-Rosales et al., 2009; Bassil et al., 2012; Yamaguchi et al., 2013; Bassil and Blumwald, 2014).

Unfortunately, ion fluxes via a single transporter (order of several fA or much lower) are below the resolution for electrophysiological measurements. However, potentially the ion currents via at least hundreds and rather thousands or millions of electrogenic ion transporters could be measured under specific conditions. A report about unitary conductance of Cl^-/H^+ antiporter ClC-5 is an exception, the conductance of 0.45 pS (ion current about 63 fA at 140 mV) for the transporter was determined from noise analysis of recordings with hundreds/thousands of transporters (Zdebik et al., 2008).

To study their properties, transporters are routinely heterologously expressed in large *Xenopus* oocytes. Detectable ion currents or fluxes of radioactive tracers are reported following the expression at the background of usually small endogenous electric currents of the oocytes. The successful expression in *Xenopus* oocytes was reported for several HKT transporters and cation-chloride cotransporters, attempts to record activity of HAK, SOS1, or Nhx transporters were less fruitful so far (Rodriguez-Navarro, 2000; Liu et al., 2001; Colmenero-

Flores et al., 2007; Jabnoune et al., 2009; Rodríguez-Rosales et al., 2009). Mature *Xenopus* oocytes used for heterologous expression are quite even and have diameter of 1 mm, so measurements could be easily normalized per surface area. A typical surface area of *Xenopus* oocyte is $4{*}\pi{*}500{*}500\ \mu m^2 \approx 4{*}3.14{*}250000 \approx 3{,}100{,}000\ \mu m^2$. Assuming that the very high recorded values of about -10 μA at -150 mV for heterologous expression of rice transporter HKT in *Xenopus* oocytes (Jabnoune et al., 2009) are reasonable and not due to incorrect folding/partial proteolysis/interaction with endogenous transport systems of *Xenopus* oocytes, we can recalculate the current per a unit of oocyte surface and compare with recordings from plant protoplasts: $-10\ \mu A/3{,}100{,}000\ \mu m^2 \approx 3\ pA/\mu m^2 = 3{,}000\ mA/m^2$ or about 1 nA per a small root protoplast with diameter around 10 μm. The values are very high and comparable or even much higher (see below) than the ones recorded from activity of ion channels using patch clamp.

Another theoretical estimate is useful for assessing activity of ion transporters expressed in *Xenopus* oocytes. Assuming the high turnover rate for the transporter around 10,000 and the very high expression of about 10,000 transporters/μm^2 (means a transporter per $10{*}10\ nm^2$, nearly maximal density due to the physical and steric limitations) we get the possible presumed current per oocyte of *Xenopus*: 10,000 ions per transporter per second $*$ 3,100,000 μm^2 $*$ 10,000 transporters/μm^2 = $3{,}1{*}10^{14}$ ions/second, then taking elementary charge of monovalent ion being $1.6{*}10^{-19}$ coulombs, the estimate gives $1.6{*}10^{-19}$ coulombs/ion $*3{,}1{*}10^{14}$ ions/second $\approx$ 50 μA per an oocyte. An excessive order of magnitude could be reasonably explained by a lower level of expression and lower transport rate of a transporter, so gives a reasonable agreement with experimental data and leads to several conclusions.

1. Transporters can provide sufficient ion currents for registered ion transport under conditions of salinity.
2. . Ion channels are not the only essential pathway for ion transport under salinity.
3. Balance between relative share of ion transport by ion channels or by ion transporters depends on abundances of the corresponding proteins (ion channels or transporters), their regulation and the other factors (composition of ion solutions, membrane potential etc.). Transporters can provide coupled transport of several ions and potentially may ensure fine-tuning of ion transport, while ion channels provide large ion fluxes when required.

The estimates indicate that total ion current via thousands/millions of electrogenic transporters could be measured and characterized in plant protoplasts using patch clamp in whole-cell configuration (recording sum of all ion currents via the whole membrane). Well-studied non-selective cation channels with low conductance carry small instantaneous currents; potential total current via numerous transporters could be of the same range. Ion current via ion-selective ion channels is described by Goldmann–Hodgkin–Katz equations based on assumption of independent passage of ions via channel pore (or constant electric field along the diffusion zone) (Hille, 2001), additional charges of lipid bilayer and the surface of ion channel can modify the ideal curves (Alcaraz et al., 2004). However, similar to Goldmann–Hodgkin–Katz curves were recorded for HKT transporters expressed in *Xenopus*oocytes (Jabnoune et al., 2009; de Almeida, 2014).

Recently sodium currents via AtHKT1;1 transporters were presumably measured in *Arabidopsis* root stelar protoplasts overexpressing AtHKT1;1 (Møller et al., 2009; Xue et al., 2011). Patch clamp experiments recorded about 30 mA/m^2 lower (more negative) currents in 10 mM external Na^+ and about 50 mA/m^2 lower currents in 25 mM external Na^+ at -120 mV, when compared to control protoplasts. The currents in AtHKT1;1 overexpressing protoplasts shifted the reversal potential according to external Na^+ concentrations, so confirmed Na^+selectivity. Further study (Xue et al., 2011) compared Na^+ and K^+currents in root stelar cell protoplasts from wild type (control) and*athtk1;1–4* mutant plants lacking AtHKT1;1. Potassium currents were similar (about -50 pA at -120 mV in 5 mM internal/50 mM external K^+), while sodium currents were about -50 pA at -120 mV in control protoplasts at 50 mM internal/50 mM external Na^+ compared to about -10 pA in *athtk1;1–4* mutant ones and demonstrated Nernstian shift according to reversal potential of sodium (Xue et al., 2011). The results pose questions for future study. Earlier it had been demonstrated that non-selective cation currents in similar root protoplasts of *Arabidopsis*(e.g., Demidchik and Tester, 2002 and see below) are slightly (1.5 times) more selective for potassium over sodium, so predictably*athtk1;1–4* mutant protoplasts should have 2–4 times higher sodium currents than measured in (Xue et al., 2011) due to expected non-selective cation currents.

The non-selective cation currents are studied well for root protoplasts, especially in *Arabidopsis* and carried by cyclic nucleotide gated channels (about 20 genes for *Arabidopsis*) and glutamate receptors (about 20 genes for *Arabidopsis*) (see below). One of the possible explanations for the paradox is to assume that nonselective cation currents in root stelar protoplasts of *Arabidopsis* are highly selective for potassium over sodium; the selectivity was shown for root protoplasts of *Thellungiella* (Volkov and Amtmann, 2006), salt-tolerant relative of*Arabidopsis* (Bressan et al., 2001; Inan et al., 2004; Amtmann, 2009). It is important to understand specific tissue and cell type expression of genes and proteins for non-selective cation channels and HKT transporters for characterizing their role in total ion currents. So far, more electrophysiological studies were performed with non-selective cation currents, while HKT transporters were mostly studied using molecular biology with the known genes in heterologous expression systems.

Special modeling for different pores of ion channels will help to understand better the peculiarities of ion currents via ion channels and HKT-like transporters. Pharmacological analysis and profiling of ion currents is also essential together with the further use of mutants (knock out or overexpression) and heterologous expression of the genes of interest in cell culture or in *Xenopus* oocytes. Non-selective ion currents are well characterized electrophysiologically and pharmacologically, especially for root protoplasts (White, 1993; White and Lemtiri-Chlieh, 1995; Roberts and Tester, 1997; Tyerman et al., 1997; Tyerman and Skerrett, 1998; Demidchik and Tester, 2002; reviewed in: Demidchik et al., 2002b; Volkov and Amtmann, 2006; reviewed in: Demidchik and Maathuis, 2007), less is known for transporters. Recently quinine (500 μM) was shown to have slight inhibiting effect on ion currents induced by HKT1;4 transporters from durum wheat after heterologous expression in *Xenopus* oocytes; Zn^{2+}, La^{3+}, Gd^{3+}, or amiloride had no effect (Ben Amar et al., 2014). For cation-chloride co-transporter from*A. thalina*, which was expressed in *Xenopus* oocytes and presumably transported Na^+:K^+:2Cl^- 100 μM bumetanide had an inhibiting effect on uptake of radioactive ions similar to analogous animal co-transporters (Colmenero-Flores et al., 2007). Amiloride is shown to inhibit Nhx1 vacuolar Na^+(cation)/H^+ antiporter (Barkla et al., 1990;Darley et al., 2000). Experiments with heterologous expression of rice HKT transporter OsHKT2;4 in *Xenopus* oocytes demonstrated channel-like behavior with single channel traces and inhibition by Ba^{2+}, La^{3+}, and Gd^{3+} (Lan et al., 2010), however, the properties were not typical for a transporter and further on the results were not confirmed and attributed to endogenous currents of the expression system (Sassi et al., 2012).

A further complication for understanding the pathways of membrane ion transport comes from electroporation experiments (e.g., Pakhomov et al., 2009; Wegner et al., 2011, 2013; Wegner, 2013, 2014). Nanopores with diameter about 1.0 -1.8 nm are formed in lipid membrane bilayer of several animal cell lines and plant protoplasts after (1) voltage pulses of 1 V and lower to -350 to +250 mV and (2) also during patch clamp experiments after applied holding voltages above +200–250 mV or below -300 to -250 mV (Pakhomov et al., 2009; Wegner et al., 2011, 2013; Wegner, 2013). The voltage-induced nanopores existed for minutes and demonstrated selectivity for cations (including even TEA^+) over anions (Cl^-) and slight selectivity for different cations (Ca^{2+} and Li^+ were the most permeable); certain similarity to behavior of non-selective cation channels was found (Wegner et al., 2011, 2013; Wegner, 2013). It is a question whether the nanopores appear under physiological conditions in plants (voltages below -300 mV were recorded for plant cells). Heterologous expression of specific ion channels together with use of mutants, elucidating detailed properties, pharmacological profiles and peculiarities of ion fluxes are required to exclude misinterpretation.

GENETIC ENGINEERING WITH NON-SPECIFIC OR TISSUE-SPECIFIC OVEREXPRESSION/ KNOCKOUT/DISRUPTION OF SPECIFIC TRANSPORTERS MODIFIES SALINITY TOLERANCE

Several obvious ways to achieve salinity tolerance include: (1) decreasing sodium conductance and increasing potassium/ sodium selectivity of plasma membrane of root epidermal cells; (2) increasing sodium efflux by root epidermal cells; (3) increasing sodium accumulation in vacuoles; (4) altering sodium and potassium loading and unloading to xylem and phloem depending on plant strategy to cope with salinity. The strategies had been realized by salt-tolerant plants or revealed in plants overexpressing genes of specific transporters.

Modification of gene activity started after the essential rise and success of molecular methods together with the identification and characterisation of individual ion channels and transporters. Lower sodium conductance and higher K^+/ Na^+ selectivity of root epidermis discovered in halophytes (see above) could be potentially reached in agriculturally important plants by RNA silencing of non-selective cation channels or modifying their expression pattern and regulation. However, still not much is known about the exact genes for the ion channels and importance of the individual genes in sodium uptake. Successful attempts to overexpress or knockout genes of vacuolar proton pump H^+-PPase, NHX, HKT, or SOS1-like transporters and to modulate the salinity tolerance of plants had already been reported.

Overexpression of the vacuolar H^+-pump would enhance the proton pumping activity at vacuolar membrane and thus permit to accumulate more Na^+ in vacuoles due to activity of Na^+(cation)/H^+ antiporters NHX. The choice of H^+-pyrophosphatase is explained by a single gene required for the protein, while the other vacuolar H^+-ATPase is composed of several subunits and needs correct overexpression of several genes (reviewed in e.g., Silva and Gerós, 2009). Overexpression of vacuolar H^+-PPase under control of strong non-specific viral *35S*promoter sharply increased salinity tolerance in *Arabidopsis*, to 250 mM of NaCl (Gaxiola et al., 2001). Overexpressing plants accumulated more sodium and potassium in their leaves and also demonstrated higher drought resistance. Further attempts to overexpress vacuolar H^+-PPases from different microbial (D'yakova et al., 2006) and plant species increased salinity tolerance in tobacco (D'yakova et al., 2006;Gao et al., 2006; Li et al., 2014), transgenic rice overexpressing also vacuolar transporter NHX1 (Zhao et al., 2006), in alfalfa (Bao et al., 2009), cotton (Pasapula et al., 2011), tomato (Bhaskaran and Savithramma, 2011), and sugarcane (Kumar et al., 2014). The experiments used *35S* promoter, NaCl concentration of 150–400 mM and reported higher Na^+ concentrations in leaves of overexpressing plants under salt treatment, while K^+ changes were not consistent between the species (Gao et al., 2006; Zhao et al., 2006; Bao et al., 2009; Bhaskaran and Savithramma, 2011; Li et al., 2014). Gene of vacuolar H^+-pyrophosphatase was among salinity tolerance determinants in barley (Shavrukov et al., 2013). Overexpression of vacuolar H^+-PPase had also an effect on the whole physiology of plants, for example, increasing root growth via probably auxin transport-associated genes, antioxidant enzymes activities and photosynthetic rate in tobacco (Li et al., 2014). Without salt stress the transgenic plants overexpressing H^+-pyrophosphatase under control of*35S* promoter demonstrated phenotypes either similar to non-transformed plants (Zhao et al., 2006; Bao et al., 2009; Bhaskaran and Savithramma, 2011), or exhibited improved morphological features (Kumar et al., 2014; Li et al., 2014) or lower osmotic potential in leaves (Gao et al., 2006). Salt treatment for non-transgenic plants resulted in both up- and down-regulation of vacuolar H^+-PPase in different species, therefore suggesting an important role of vacuolar H^+-ATPase in responses to the stress factor (reviewed in: Silva and Gerós, 2009).

Another candidates for overexpression are vacuolar *NHX* genes. Overexpression of *AtNHX1* increased salinity tolerance in *Arabidopsis*to 200 mM NaCl, the overexpressing plants accumulated more Na^+compared to wild type and demonstrated higher Na^+/H^+ exchange activity in isolated leaf vacuoles (Apse et al., 1999). The approach of overexpressing *AtNHX1* to improve salinity tolerance proved to be successful for tomato; the transgenic plants accumulated more sodium in leaves but not in fruits at 200 mM NaCl (Zhang and Blumwald, 2001). Cotton plants with *AtNHX1* from *Arabidopsis* (He et al., 2005), rice overexpressing *SsNHX1* from halophyte *Suaeda salsa* (Zhao et al., 2006), tomato with heterologous NHX from *Pennisetum glaucum*(Bhaskaran and Savithramma, 2011) also showed increased salinity tolerance. Overexpression of *NHX* did not influence the phenotype of plants under control conditions (Apse et al., 1999; Zhang and Blumwald, 2001; He et al., 2005; Zhao et al., 2006; Bhaskaran and Savithramma, 2011). The results with heterologous expression or overexpression of NHX transporters lead to conclusions that the gene is among determinants and potential candidates for engineering salinity tolerance (e.g., Rodríguez-Rosales et al., 2009; Peleg et al., 2011 with more references for successful overexpression of *NHX* to increase salinity tolerance in sugar beet, wheat, maize and the other plants). However, the overexpression of *NHX* was not tissue-specific and under the control of strong promoters, one report did not confirm increase in salinity tolerance in *Arabidopsis* overexpressing *AtNHX1* (Yang et al., 2009). Expression in a tissue-specific manner could be the next step for using NHX to increase salinity tolerance.

The amazing simplicity of the idea to play with the expression of known and functionally well characterized transporters and get salt tolerant or salt sensitive plants is applied to plasma membrane SOS1 Na^+/H^+ antiporters and Na^+ or Na^+/K^+ HKT transporters. SOS1 is expressed in (1) epidermal root cells where it participates in sodium efflux and in (2) xylem parenchyma cells where SOS1 may load Na^+ to xylem under moderate salinity and unloads Na^+ under high salinity or has more complex mode of xylem loading/unloading (Shi et al., 2000,2002; Pardo et al., 2006; Oh et al., 2007, 2009a; Olías et al., 2009).*Arabidopsis* mutants with defects in gene of SOS1 exhibited strong growth inhibition under salt treatment (Wu et al., 1996), which was rescued in *sos1* mutant by overexpression of *SOS1* gene under *35S*promoter (Shi et al., 2000). Overexpression of *SOS1* gene in wild type plants under *35S* promoter enhanced salinity tolerance of *Arabidopsis* at 100–200 mM NaCl (Shi et al., 2003; Yang et al., 2009), reduced sodium accumulation in shoots and sodium concentration in xylem sap (Shi et al., 2003). Further on overexpression of *SOS1* from *A. thaliana*increased salinity tolerance in transgenic tobacco (Yue et al., 2012) and in transgenic tall fescue (Ma et al., 2014). *SOS1* gene from durum wheat conferred salinity tolerance to *sos1* mutant of *Arabidopsis* (Feki et al., 2014). Interestingly, the effects of overexpression were observed under salt treatment, while in the absence of stress no differences were observed in growth or morphology between wild-type plants and the transgenic lines. Disruption of SOS1 activity by RNA interference in*Thellungiella* on the opposite resulted in the loss of tolerance of the halophyte indicating importance of Na^+ efflux and essential role of SOS1 in salinity tolerance (Oh et al., 2009a). RNA interference of*SOS1* significantly changed the whole transcriptome of *Thellungiella*(Oh et al., 2007) and vacuolar pH under salt treatment (Oh et al., 2009b) proving the complex nature of metabolic and regulatory networks in plants and yet the probabilistic chances of success in strict overexpression of specific transporters for salinity tolerance improvement. A more complicated situation emerges due to tissue-specific expression. SOS1 is important for long-distance ion transport and xylem loading/unloading in *Arabidopsis* (Shi et al., 2002; discussed in: de Boer and Volkov, 2003), sodium partioning between plant organs in tomato (Olías et al., 2009) and ion fluxes in root meristem zone (Guo et al., 2009), therefore attempts to express it in specific tissues could increase salinity tolerance to a higher extent.

Genetic modification of salinity tolerance using HKT transporters was also successful. Analysis of *Arabidopsis* plants with mutated *HKT* gene revealed higher salt sensitivity of the mutants under long term stress, higher sodium accumulation in their shoots under mild salinity treatment (Mäser et al., 2002) and suggested that HKT is involved in recirculation of sodium within plants (Berthomieu et al., 2003). Further study confirmed increased sodium in the shoots of *Arabidopsis hkt1;1*mutants and clarified that HKT is important for root accumulation of Na^+ and Na^+ uptake from xylem in *Arabidopsis* (Davenport et al., 2007). The next step was to create plants overexpressing *HKT* (Møller et al., 2009). *Arabidopsis* plants overexpressing *AtHKT* under the control of *35S* promoter were compared with plants specifically overexpressing *HKT* in cells of root stele. *Pro35S:HKT1;1* plants were salt sensitive probably due to higher Na^+ uptake by roots while tissue specific overexpression of *HKT* in stele increased salinity tolerance and reduced sodium accumulation in shoots (Møller et al., 2009). The approach was applied to rice where gene from *Arabidopsis AtHKT1;1*was heterologously expressed in root cortex. It resulted in lower shoot Na^+ concentrations, improved salinity tolerance and involved up- and down-regulation of several membrane transport genes including vacuolar H^+-pyrophosphatases (Plett et al., 2010). Overexpression of HKT had none (Møller et al., 2009; Plett et al., 2010; Mian et al., 2011) or slight inhibiting pleiotropic effect on growth without NaCl depending on type of promotor for expression and on plant line studied (Møller et al., 2009; Plett et al., 2010). HKT transporters proved to be important for Na^+ exclusion in wheat and were transferred from durum wheat to bread wheat by interspecific crossing; the genes gave beneficial effects including higher K^+/Na^+ ratio in leaves under saline conditions (James et al., 2011). Some plants including barley accumulate Na^+ in shoots; overexpression of barley *HvHKT2;1* under*35S* promoter in barley increased salinity tolerance at 100 mM NaCl, but opposite to *Arabidopsis* increased Na^+ concentration in xylem and Na^+ accumulation in barley leaves (Mian et al., 2011). Taken together the results set HKT transporters to potential candidates for engineering salinity tolerance and among the determinants of the trait (reviewed in:Horie et al., 2009; Almeida et al., 2013; Maathuis et al., 2014) together with the above mentioned NHX1, SOS1 and presumably new studied transporters, e.g., similar to CHX21 from *Arabidopsis* (Hall et al., 2006).

Genes which are important for salinity tolerance in the other groups of organisms and not present in higher plants could also be potential candidates for engineering the trait. Sodium pumping ATPase from moss *Physcomitrella patens* was cloned and expressed under *35S*promoter in rice; plasma membrane expression resulted in higher biomass of transgenic plants compared to control ones after 2 weeks of 50 mM salt treatment. Surprisingly, expression of Na^+-ATPase did not influence Na^+ and K^+ concentrations in transgenic compared to control plants under any tested conditions, so needs more investigation (Jacobs et al., 2011). Recently a new bacterial rhodopsin from *Krokinobacter eikastus* was discovered; it is the first light-driven Na^+ pump. This rhodopsin was crystallized and resolved using X-ray; the structural basis for Na^+ transport was revealed (Gushchin et al., 2015; Kato et al., 2015). The photo switchable sodium pump seems a good simple molecular tool. The initial results are already promising and involve the diverse pool of Na^+-ATPases and transporters from yeast, algae and microbes for improving salinity tolerance in plants.

Perspectives of Protein Engineering Structure-Function Studies and Potential Future for Expression of Novel Ion Channels, Pumps, and Transporters

Novel opportunities for increasing salinity tolerance in plants are arising with the development of new methods of molecular biology, understanding regulation networks from synthetic biology and growing knowledge about single point mutations changing specific amino acids within molecules of an ion channel or a transporter.

Single amino acid substitutions, e.g., within K^+ selectivity filter GYG of potassium channels, may change selectivity of the ion channels rendering them from K^+ selective to non-selective ones (e.g.,Heginbotham et al., 1992). Amino acid substitutions within a presumed pore region of HKT transporters are able to alter them from Na^+selective to Na^+ and K^+ permeable (e.g., Mäser et al., 2002; de Almeida, 2014). Moreover, the single point mutations could be determining for salinity tolerance, e.g., amino acid substitution V395L in rice transporter OsHKT1;5 presumably explained the salt sensitivity of the rice cultivar (Cotsaftis et al., 2012). Specific single point substitution in *Arabidopsis* HAK5 transporter (F130S) over 100 times increased affinity for K^+ under heterologous expression and reduced inhibition constants for Na^+ and Cs^+ (Alemán et al., 2014). Effects of single point mutations on the whole pattern of physiology and on phenotype are well known and better studied in human biomedical science when inherited diseases cystic fibrosis and sickle cell anemia are caused by amino acid substitutions in transport protein CFTR and in hemoglobin, correspondingly.

The knowledge about structure-function correlations of proteins allows to modify the selectivity and create the required properties of ion transport proteins. The way in the direction is to employ the existing and growing information about structure-function of different ion channels and transporters. Next step is to change their ion selectivity and gating properties according to the requirements using single amino acids mutations and to transform the plants of interest in a tissue-specific or cell-specific manner. Examples of tissue-specific transformation already exist (e.g., Møller et al., 2009; Plett et al., 2010) while the new methods and opportunities are progressing enormously (e.g., Brand et al., 2006; Oszvald et al., 2008 etc.). Recent opportunities to directly edit genome using CRISPR–Cas system could overcome some difficulties and directly modify the expressed genes of ion transport proteins (reviewed in e.g., Shan et al., 2013; Kumar and Jain, 2015).

An alternative approach from synthetic biology is not to modify the existing membrane transport proteins, but to create new ones with desired properties for the further cell-specific transformation. The idea is different from what could be assumed at a first glance. Existing biological organisms emerged over the process of long evolution, when previous "building blocks" and elements were used for the future development and often could not be essentially modified due to intrinsic links within organisms and biological systems. It leaves out the question of ideal design, which is mostly not present in biological organisms. Indeed, they are largely predetermined by the previous evolutional history with intrinsic evolutionary trajectories and evolved under multifactor environment (composition of atmosphere, illumination, temperature, salinity and mineral nutrients, water availability etc., while interactions with the other organisms and biotic interactions are often the most important). A simple example is related to temperature. Ion channels in homoeothermic animals like mammals or birds evolved over hundreds of millions of years under stable conditions and hence differ in many properties from ion channels in plants. Sodium and calcium selective ion channels had not been found in plants while in animals they ensure action potentials in neurons and cardiomyocytes. Specialized highly temperature-sensitive ion channel in animals provide temperature sensation (for example transient receptor potential channels, e.g., Ramsey et al., 2006; Myers et al., 2009). Plants are different, they rely on calcium signaling via non-selective cation channels, may have distinct groups of transporters with specific properties, have no known sodium-selective ion channels and use ion channels with low temperature sensitivity (e.g., plant potassium channels and their regulation are reviewed in: Dreyer and Uozumi, 2011; more general review: Hedrich, 2012). Obviously, cell signaling and regulation in plants have numerous specific peculiarities compared to homoeothermic animals. The differences could be reflected in protein structures, therefore simple comparisons may be misleading. Moreover, plant ion channels and transporters are often functioning at membrane voltages below -200 mV while animal membranes are usually not experiencing voltages below -100 mV. Structure and functioning of ion transport proteins from yeast, algae, Protista, and microbes could provide more insights for synthetic proteins for plants.

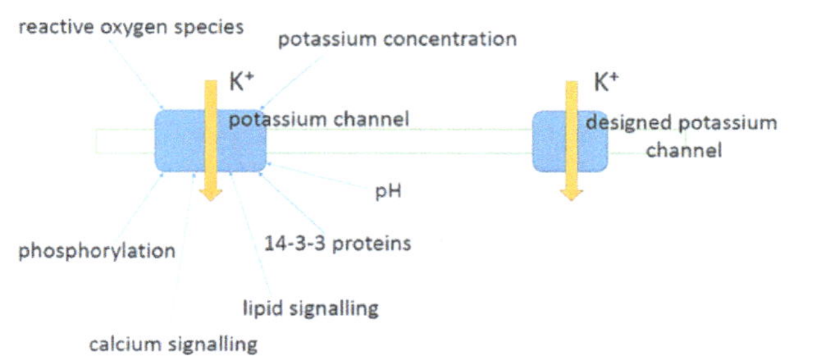

Figure 11: Novel artificially designed ion channels and ion transporters potentially provide an opportunity to avoid the present evolved protein regulation networks and may be useful for altering ion concentrations, membrane potential and signaling without essentially .

From the point of the above mentioned it seems that attempts to design novel artificial ion channels and transporters with known characterized ion selectivity filters and voltage sensors adding or excluding specific interacting regulatory elements for the proteins might be productive. The ion channels and transporters when expressed in cell-specific manner under controlled conditions and in defined numbers may avoid fine tuning of regulation and potentially could provide shortcuts in natural signaling networks. The appearing opportunities offer new chances to design salt tolerant plants with previously unknown features and wider ranges of regulation circuits and networks. The potential strategy to increase salinity tolerance includes

1. choice of plant,
2. understanding ion transport and features of salinity tolerance for the plant,
3. determining ion fluxes and ion conductances important for Na^+ and K^+ accumulation and compartmentation,
4. modeling ion fluxes and adding/removing in the models ion fluxes and conductances to ensure better nutrient supply and rise of salinity tolerance under salt stress,
5. correlating required changes in ion transport with potential membrane ion transport proteins to realize the ion fluxes,
6. designing specific membrane proteins,
7. expressing the membrane proteins in a tissue-specific manner,
8. checking the salinity tolerance of transgenic plants in laboratory and field experiments.

The proposed sequence of events is hypothetical so far and had not been realized yet. However, progress in molecular biology and new ways of thinking from synthetic biology may bring it to fruition and provide new discoveries on the path.

Design of transmembrane proteins is at the very beginning nowadays. Certain structural blocks of the proteins are known better and used for several applications. Voltage-sensors are intensively studied for voltage-gated ion channels and for transporters (Bezanilla, 2008). Moreover, optogenetics needs and already efficiently applies voltage-sensing structural blocks of proteins (Mutoh et al., 2012; Cao et al., 2013; Zou et al., 2014). Selectivity filters of ion channels and transporters are also known and under investigation. Addition of regulatory domains for binding depends on our current information and new endeavors to understand protein–protein interactions and regulation. The further tissue-specific expression and cell-specific localisation of the proteins to be realized via signal peptides and tissue-specific promoters. Problems, strategies and perspectives for engineering novel membrane proteins and re-engineering the existing ones are widely discussed and reviewed (e.g., Grosse et al., 2011;Subramanyam and Colecraft, 2015). *De novo* design, synthesis confirmed by X-ray structure of Zn^{2+}-transporting four-helix transmembrane protein bundle is a recent experimental advance (Joh et al., 2014). The Zn^{2+}-transport activity of the novel protein was confirmed in functional assays (Joh et al., 2014) proving feasibility of the approach and setting engineering of membrane proteins to a new higher level.

CONCLUSION AND PERSPECTIVES

Development of agriculture often coincides with salinization of the used land, which is later not available for agricultural plants and inhabited by natural halophyte plants. The problem exists since collapse of Sumer civilization about 4000 years ago due to improper agricultural techniques. Recent decades added modern plant physiology, biophysics, molecular and systems biology to pure agriculture and breeding for creating salt-tolerant crops and agriculturally important plants. New information about individual ion transport systems gave solid physical basis for improving salinity tolerance of plants. Emerging opportunities to overexpress individual genes allowed to sharply increase salinity tolerance in laboratory trials. Growing knowledge in protein engineering and synthetic biology sets novel aims and horizons for producing artificial proteins with predefined transport properties and for designing new regulation networks. Potentially the progress in the direction may lead to partially artificial plants with desired salinity tolerance. The next step would be to fill the huge gap between rapid success in laboratory experiments and field practice. Expectedly the future advances will help to release the problem of salinization of agricultural lands.

ACKNOWLEDGMENTS

The Author would like to thank colleagues who we worked together over several projects summarized in the Review, Dr Svetlana Bagirova for critical reading and comments on the Manuscript and to bring apologies for not citing all the relevant literature sources due to the limitations by the volume of the Review. The earlier preprint version of the Review entitled "Salinity tolerance in plants: attempts to manipulate ion transport" was uploaded to repository of electronic preprints arXiv at http://arxiv.org/abs/1411.1553 on the 6th of November 2014.

REFERENCES

1. Akhromeeva T. S., Kurdyumov S. P., Malinetskii G. G., Samarskii A. A. (1989). Nonstationary dissipative structures and diffusion-induced chaos in nonlinear media. *Physics. Rep.* 176 5–6. 10.1016/0370-1573(89)90001-X
2. Akhromeeva T. S., Kurdyumov S. P., Malinetskij G. G., Samarskij A. A. (1992). *Nonstationary Structures and Diffusion Chaos (Nestatsionarnye Struktury i Diffuznyj Khaos).* Moskva: Nauka Publisher; 544.
3. Alcaraz A., Nestorovich E. M., Aguilella-Arzo M., Aguilella V. M., Bezrukov S. M. (2004). Salting out the ionic selectivity of a wide channel: the asymmetry of OmpF. *Biophys. J.* 87 943–957. 10.1529/biophysj.104/043414
4. Alemán F., Caballero F., Ródenas R., Rivero R. M., Martínez V., Rubio F. (2014). The F130S point mutation in the *Arabidopsis*high-affinity K^+ transporter AtHAK5 increases K^+ over Na^+ and Cs^+ selectivity and confers Na^+ and Cs^+ tolerance to yeast under heterologous expression. *Front. Plant Sci.* 5:43010.3389/fpls.2014.00430
5. Alemán F., Nieves-Cordones M., Martínez V., Rubio F. (2009).Potassium/sodium steady-state homeostasis in *Thellungiella halophila* and *Arabidopsis thaliana* under long-term salinity conditions. *Plant Sci.* 176 768–774. 10.1016/j.plantsci.2009.02.020
6. Almeida P., Katschnig D., de Boer A. H. (2013). HKT transporters-state of the art. *Int. J. Mol. Sci.* 14 20359–20385. 10.3390/ijms141020359
7. Almeida P. M. F., de Boer G. J., de Boer A. H. (2014). Assessment of natural variation in the first pore domain of the tomato HKT1;2 transporter and characterization of mutated versions of SlHKT1;2 expressed in *Xenopus laevis* oocytes and via complementation of the salt sensitive athkt1;1 mutant. *Front. Plant Sci.* 5:600 10.3389/fpls.2014.00600

8. Amtmann A. (2009). Learning from evolution: thellungiella generates new knowledge on essential and critical components of abiotic stress tolerance in plants. *Mol. Plant* 2 3–12. 10.1093/mp/ssn094

9. Apse M. P., Aharon G. S., Snedden W. A., Blumwald E. (1999).Salt tolerance conferred by overexpression of a vacuolar Na^+/H^+antiport in *Arabidopsis. Science* 285 1256–1258. 10.1126/science.285.5431.1256

10. Armengaud P., Breitling R., Amtmann A. (2004). The potassium-dependent transcriptome of*Arabidopsis* reveals a prominent role of jasmonic acid in nutrient signaling. *Plant Physiol.* 136 2556–2576. 10.1104/pp.104.046482

11. Armengaud P., Sulpice R., Miller A. J., Stitt M., Amtmann A., Gibon Y. (2009). Multilevel analysis of primary metabolism provides new insights into the role of potassium nutrition for glycolysis and nitrogen assimilation in *Arabidopsis* roots. *Plant Physiol.* 150 772–785. 10.1104/pp.108.133629

12. Balnokin Y. V. (1993). Ionic homeostasis and osmoregulation in halotolerant microalgae. *Russian J. Plant Physiol.* 40 567–576.

13. Balnokin Y. V., Popova L. G. (1994). The ATP-driven Na^+-pump in the plasma membrane of the marine unicellular alga,*Platymonas viridis*. *FEBS Lett.* 343 61–64. 10.1016/0014-5793(94)80607-1

14. Banuelos M. A., Klein R. D., Alexander-Bowman S. J., Rodriguez-Navarro A. (1995). A potassium transporter of the yeast Schwanniomyces occidentalis homologous to the Kup system of *Escherichia coli* has a high concentrative capacity.*EMBO J.* 14 3021–3027.

15. Bao A. K., Wang S. M., Wu G. Q., Xi J. J., Zhang J. L., Wang C. M. (2009). Overexpression of the Arabidopsis H^+-PPase enhanced resistance to salt and drought stress in transgenic alfalfa (*Medicago sativa* L.). *Plant Sci.* 176 232–240. 10.1016/j.plantsci.2008.10.009

16. Barkla B. J., Charuk J. H., Cragoe E. J., Blumwald E. (1990).Photolabeling of tonoplast from sugar beet cell suspensions by [3H]5-(N-Methyl-N-Isobutyl)-Amiloride, an inhibitor of the vacuolar Na^+/H^+ antiport. *Plant Physiol.* 93 924–930. 10.1104/pp.93.3.924

17. Barkla B. J., Zingarelli L., Blumwald E., Smith J. (1995).Tonoplast Na^+/H^+ antiport activity and its energization by the vacuolar h+-atpase in the halophytic plant *Mesembryanthemum crystallinum* L. *Plant Physiol.* 109 549–556.

18. Bassil E., Blumwald E. (2014). The ins and outs of intracellular ion homeostasis: NHX-type cation/H^+ transporters. *Curr. Opin. Plant Biol.* 22 1–6. 10.1016/j.pbi.2014.08.002

19. Bassil E., Coku A., Blumwald E. (2012). Cellular ion homeostasis: emerging roles of intracellular NHX Na^+/H^+antiporters in plant growth and development. *J. Exp. Bot.* 635727–5740. 10.1093/jxb/ers250

20. Bazzanella A., Lochmann H., Tomos A. D., Bächmann K. (1998). Determination of inorganic cations and anions in single plant cells by capillary zone electrophoresis. *J. Chromatogr.* 809231–239. 10.1016/S0021-9673(98)00144-7

21. Ben Amar S., Brini F., Sentenac H., Masmoudi K., Véry A. A. (2014). Functional characterization in *Xenopus oocytes* of Na^+transport systems from durum wheat reveals diversity among two HKT1;4 transporters. *J. Exp. Bot.* 65 213–222. 10.1093/jxb/ert361

22. Benito B., Haro R., Amtmann A., Cuin T. A., Dreyer I. (2014).The twins K^+ and Na^+ in plants. *J. Plant Physiol.* 171 723–731. 10.1016/j.jplph.2013.10.014

23. Bennett T. H., Flowers T. J., Bromham L. (2013). Repeated evolution of salt-tolerance in grasses. *Biol. Lett.* 9 2013002910.1098/rsbl.2013.0029

24. Bental M., Degani H., Avron M. (1988). 23Na-NMR studies of the intracellular sodium ion concentration in the halotolerant alga*Dunaliella salina*. *Plant Physiol.* 87 813–817. 10.1104/pp.87.4.813

25. Berthomieu P., Conéjéro G., Nublat A., Brackenbury W. J., Lambert C., Savio C., et al. (2003). Functional analysis of AtHKT1 in *Arabidopsis* shows that Na^+ recirculation by the phloem is crucial for salt tolerance. *EMBO J.* 22 2004–2014. 10.1093/emboj/cdg207

26. Bezanilla F. (2008). How membrane proteins sense voltage. *Nat. Rev. Mol. Cell Biol.* 9 323–332. 10.1038/nrm2376

27. Bhaskaran S., Savithramma D. L. (2011). Co-expression of*Pennisetum glaucum* vacuolar Na^+/H^+ antiporter and *Arabidopsis*H^+-pyrophosphatase enhances salt tolerance in transgenic tomato. *J. Exp. Bot.* 62 5561–5570. 10.1093/jxb/err237

28. Blatt M. R. (1987). Electrical characteristics of stomatal guard cells: the ionic basis of the membrane potential and the consequence of potassium chlorides leakage from microelectrodes. *Planta* 170 272–287. 10.1007/BF00397898

29. Blatt M. R. (1991). "A primer in plant electrophysiological methods," in *Methods in Plant Biochemistry* ed. Hostettmann K., editor. (London: Academic Press;) 281–321.

30. Blumwald E., Poole R. J. (1985). Na^+/H^+ antiport in isolated tonoplast vesicles from storage tissue of *Beta vulgaris*. *Plant Physiol.* 78 163–167. 10.1104/pp.78.1.163

31. Brand L., Hörler M., Nüesch E., Vassalli S., Barrell P., Yang W., et al. (2006). A versatile and reliable two-component system for tissue-specific gene induction in *Arabidopsis*. *Plant Physiol.* 1411194–1204. 10.1104/pp.106.081299

32. Breckle S.-W. (2002). "Salinity, halophytes and salt affected natural ecosystems," in *Salinity: Environment-Plants-Molecules*eds Läuchli R., Lüttge U., editors. (The Netherlands: Kluwer Academic Publishers;) 53–77.

33. Breeze E., Harrison E., McHattie S., Hughes L., Hickman R., Hill C., et al. (2011). High-resolution temporal profiling of transcripts during *Arabidopsis* leaf senescence reveals a distinct chronology of processes and regulation. *Plant Cell* 23 873–894. 10.1105/tpc.111.083345

34. Bressan R. A., Zhang C., Zhang H., Hasegawa P. M., Bohnert H. J., Zhu J. K. (2001). Learning from the *Arabidopsis* experience. The next gene search paradigm. *Plant Physiol.* 127 1354–1360. 10.1104/pp.010752

35. Britto D. T., Kronzucker H. J. (2008). Cellular mechanisms of potassium transport in plants. *Physiol. Plant.* 133 637–650. 10.1111/j.1399-3054.2008.01067.x

36. Britto D. T., Kronzucker H. J. (2013). Flux measurements of cations using radioactive tracers. *Methods Mol. Biol.* 953 161–170. 10.1007/978-1-62703-152-3_10

37. Britto D. T., Szczerba M. W., Kronzucker H. J. (2006). A new, non-perturbing, sampling procedure in tracer exchange measurements. *J. Exp. Bot.* 57 1309–1314. 10.1093/jxb/erj105

38. Cao G., Platisa J., Pieribone V. A., Raccuglia D., Kunst M., Nitabach M. N. (2013). Genetically targeted optical electrophysiology in intact neural circuits. *Cell* 154 904–913. 10.1016/j.cell.2013.07.027

39. Cao Y., Jin X., Huang H., Derebe M. G., Levin E. J., Kabaleeswaran V., et al. (2011). Crystal structure of a potassium ion transporter, TrkH. *Nature* 471 336–340. 10.1038/nature09731

40. Carden D. E., Diamond D., Miller A. J. (2001). An improved Na^+-selective microelectrode for intracellular measurements in plant cells. *J. Exp. Bot.* 52 1353–1359. 10.1093/jexbot/52.359.1353

41. Carden D. E, Walker D. J., Flowers T. J., Miller A. J. (2003).Single-cell measurements of the contributions of cytosolic $Na^+$and K^+ to salt tolerance. *Plant Physiol.* 131 676–683. 10.1104/pp.011445

42. Carpaneto A., Koepsell H., Bamberg E., Hedrich R., Geiger D. (2010). Sucrose- and H^+-dependent charge movements associated with the gating of sucrose transporter ZmSUT1. *PLoS ONE* 5:e12605 10.1371/journal.pone.0012605

43. Chakrapani S., Auerbach A. (2005). A speed limit for conformational change of an allosteric membrane protein. *Proc. Natl. Acad. Sci. U.S.A.* 102 87–92. 10.1073/pnas.0406777102

44. Cheeseman J. M. (1986). Compartmental efflux analysis: an evaluation of the technique and its limitations. *Plant Physiol.* 801006–1011. 10.1104/pp.80.4.1006

45. Chen J., Xiong D. Y., Wang W. H., Hu W. J., Simon M., Xiao Q., et al. (2013). Nitric oxide mediates root K^+/Na^+ balance in a mangrove plant. *PLoS ONE* 8:e7154310.1371/journal.pone.0071543

46. Chen Z., Newman I., Zhou M., Mendham N., Zhang G., Shabala S. (2005). Screening plants for salt tolerance by measuring K^+flux: a case study for barley. *Plant Cell Environ.* 28 1230–1246. 10.1111/j.1365-3040.2005.01364.x

47. Collins K. D. (1997). Charge density-dependent strength of hydration and biological structure. *Biophys. J.* 72 65–76. 10.1016/S0006-3495(97)78647-8

48. Colmenero-Flores J. M., Martínez G., Gamba G., Vázquez N., Iglesias D. J., Brumós J., et al. (2007). Identification and functional characterization of cation-chloride cotransporters in plants. *Plant J.* 50 278–292. 10.1111/j.1365-313X.2007.03048.x

49. Coskun D., Kronzucker H. J. (2013). Complexity of potassium acquisition: how much flows through channels? *Plant Signal Behav.* 8:e24799 10.4161/psb.24799

50. Cotsaftis O., Plett D., Shirley N., Tester M., Hrmova M. (2012).A two-staged model of Na^+ exclusion in rice explained by 3D modeling of HKT transporters and alternative splicing. *PLoS ONE* 7:e39865 10.1371/journal.pone.0039865

51. Cuin T. A., Betts S. A., Chalmandrier R., Shabala S. (2008). A root's ability to retain K^+ correlates with salt tolerance in wheat.*J. Exp. Bot.* 59 2697–2706. 10.1093/jxb/ern128

52. Cuin T. A., Miller A. J., Laurie S. A., Leigh R. A. (2003). Potassium activities in cell compartments of salt-grown barley leaves. *J. Exp. Bot.* 54 657–661. 10.1093/jxb/erg072

53. Cuin T. A., Shabala S. (2007a). Amino acids regulate salinity-induced potassium efflux in barley root epidermis. *Planta* 225753–761. 10.1007/s00425-006-0386-x

54. Cuin T. A., Shabala S. (2007b). Compatible solutes reduce ROS-induced potassium efflux in *Arabidopsis* roots. *Plant Cell Environ.* 30 875–885. 10.1111/j.1365-3040.2007.01674.x

55. Cuin T. A., Zhou M., Parsons D., Shabala S. (2012). Genetic behaviour of physiological traits conferring cytosolic K^+/Na^+homeostasis in wheat. *Plant Biol.* 14 438–446. 10.1111/j.1438-8677.2011.00526.x

56. Darley C. P., van Wuytswinkel O. C., van der Woude K., Mager W. H., de Boer A. H. (2000). *Arabidopsis thaliana* and*Saccharomyces cerevisiae* NHX1 genes encode amiloride sensitive electroneutral Na^+/H^+ exchangers. *Biochem. J.* 351241–249. 10.1042/0264-6021:3510241

57. Davenport R. J., Muñoz-Mayor A., Jha D., Essah P. A., Rus A., Tester M. (2007). The Na^+ transporter AtHKT1;1 controls retrieval of Na^+ from the xylem in *Arabidopsis*. *Plant Cell Environ.* 30 497–507. 10.1111/j.1365-3040.2007.01637.x

58. de Almeida P. M. F. (2014). *The Role of HKT Transporters in Salinity Tolerance of Tomato.* Ph.D. thesis, Vrije Universiteit; Amsterdam: 219.

59. de Boer A. H. (1985). *Xylem/Symplast Ion Exchange: Mechanism and Function in Salt Tolerance and Growth.* Ph.D. thesis, Rijksuniversiteit Groningen; Groningen.

60. de Boer A. H., Volkov V. (2003). Logistics of water and salt transport through the plant: structure and functioning of the xylem. *Plant Cell Environ.* 26 87–101. 10.1046/j.1365-3040.2003.00930.x

61. Demidchik V., Bowen H. C., Maathuis F. J., Shabala S. N., Tester M. A., White P. J., et al. (2002a). *Arabidopsis thaliana* root non-selective cation channels mediate calcium uptake and are involved in growth. *Plant J.* 32 799–808. 10.1046/j.1365-313X.2002.01467.x

62. Demidchik V., Davenport R. J., Tester M. (2002b). Nonselective cation channels in plants. *Annu. Rev. Plant Biol.* 53 67–107. 10.1146/annurev.arplant.53.091901.161540
63. Demidchik V., Cuin T. A., Svistunenko D., Smith S. J., Miller A. J., Shabala S., et al. (2010). *Arabidopsis* root K^+-efflux conductance activated by hydroxyl radicals: single-channel properties, genetic basis and involvement in stress-induced cell death. *J. Cell Sci.* 123 1468–1479. 10.1242/jcs.064352
64. Demidchik V., Maathuis F. J. (2007). Physiological roles of nonselective cation channels in plants: from salt stress to signalling and development. *New Phytol.* 175 387–404. 10.1111/j.1469-8137.2007.02128.x
65. Demidchik V., Shang Z., Shin R., Colaço R., Laohavisit A., Shabala S., et al. (2011). Receptor-like activity evoked by extracellular ADP in *Arabidopsis* root epidermal plasma membrane. *Plant Physiol.* 156 1375–1385. 10.1104/pp.111.174722
66. Demidchik V., Tester M. (2002). Sodium fluxes through nonselective cation channels in the plasma membrane of protoplasts from *Arabidopsis* roots. *Plant Physiol.* 128 379–387. 10.1104/pp.010524
67. Dietz K. J., Schramm M., Betz M., Busch H., Dürr C., Martinoia E. (1992). Characterization of the epidermis from barley primary leaves: I. Isolation of epidermal protoplasts. *Planta* 187 425–430. 10.1007/BF00199959
68. Dolan L., Janmaat K., Willemsen V., Linstead P., Poethig S., Roberts K., et al. (1993). Cellular organisation of the *Arabidopsis thaliana* root. *Development* 119 71–84.
69. D'Onofrio C., Kader A., Lindberg S. (2005). Uptake of sodium in quince, sugar beet, and wheat protoplasts determined by the fluorescent sodium-binding dye benzofuran isophthalate. *J. Plant Physiol.* 162 421–428. 10.1016/j.jplph.2004.07.017
70. Doyle D. A., Cabral M. J., Pfuetzner R. A., Kuo A., Gulbis J. M., Cohen S. L., et al. (1998). The structure of the potassium channel: molecular basis of K^+ conduction and selectivity.*Science* 280 69–77. 10.1126/science.280.5360.69
71. Dreyer I., Uozumi N. (2011). Potassium channels in plant cells.*FEBS J.* 278 4293–4303. 10.1111/j.1742-4658.2011.08371.x
72. D'yakova E. V., Rakitin A. L., Kamionskaya A. M., Baikov A. A., Lahti R., Ravin N. V., et al. (2006). A study of the effect of expression of the gene encoding the membrane H^+-pyrophosphatase of *Rhodospirillum rubrum* on salt resistance of transgenic tobacco plants. *Doklady Biol. Sci.* 409 346–348. 10.1134/S0012496606040247
73. English J. P., Colmer T. D. (2013). Tolerance of extreme salinity in two stem-succulent halophytes (*Tecticornia* species). *Funct. Plant Biol.* 40 897–912.
74. Epstein E. (1966). Dual pattern of ion absorption by plant cells and by plants. *Nature* 212 1324–1327. 10.1038/2121324a0
75. Erdei L., Kuiper P. J. C. (1979). The effect of salinity on growth. Cation Content, Na^+-uptake and translocation in salt-sensitive and salt-tolerant *Plantago* Species. *Physiol. Plant.* 47 95–99. 10.1111/j.1399-3054.1979.tb03197.x
76. Evans A., Hall D., Pritchard J., Newbury H. J. (2012). The roles of the cation transporters CHX21 and CHX23 in the development of *Arabidopsis thaliana*. *J. Exp. Bot.* 63 59–67. 10.1093/jxb/err271
77. Feki K., Quintero F. J., Khoudi H., Leidi E. O., Masmoudi K., Pardo J. M., et al. (2014). A constitutively active form of a durum wheat Na^+/H^+ antiporter SOS1 confers high salt tolerance to transgenic *Arabidopsis*. *Plant Cell Rep.* 33 277–288. 10.1007/s00299-013-1528-9
78. Flowers T. J. (1998). Salinisation and horticultural production.*Sci. Hortic.* 78 1–4.
79. Flowers T. J. (2004). Improving crop salt tolerance. *J. Exp. Bot.*55 307–319. 10.1093/jxb/erh003
80. Flowers T. J., Colmer T. D. (2008). Salinity tolerance in halophytes. *New Phytol.* 179 945–963. 10.1111/j.1469-8137.2008.02531.x
81. Flowers T. J., Dalmond D. (1992). Protein synthesis in halophytes: the influence of potassium, sodium and magnesium in vitro. *Plant Soil* 146 153–161. 10.1007/BF00012008
82. Flowers T. J., Galal H. K., Bromham L. (2010). Evolution of halophytes: multiple origins of salt tolerance in land plants.*Funct. Plant Biol.* 37 604–612. 10.1071/FP09269
83. Flowers T. J., Troke P. F., Yeo A. R. (1977). The mechanism of salt tolerance in halophytes. *Annu. Rev. Plant Physiol.* 28 89–121. 10.1146/annurev.pp.28.060177.000513
84. Fricke W., Akhiyarova G., Wei W., Alexandersson E., Miller A., Kjellbom P. O., et al. (2006). The short-term growth response to salt of the developing barley leaf. *J. Exp. Bot.* 57 1079–1095. 10.1093/jxb/erj095
85. Fricke W., Leigh R. A., Tomos A. D. (1994). Epidermal solute concentrations and osmolality in barley leaves studied at the single-cell level. *Planta* 192 317–323. 10.1007/BF00198566
86. Gadsby D. C. (2009). Ion channels versus ion pumps: the principal difference, in principle. *Nat. Rev. Mol. Cell Biol.* 10344–352. 10.1038/nrm2668
87. Galdiero S., Falanga A., Cantisani M., Tarallo R., Della Pepa M. E., D'Oriano V., et al. (2012). Microbe-host interactions: structure and role of Gram-negative bacterial porins. *Curr. Protein Pept. Sci.* 13 843–854. 10.2174/138920371121308001 2
88. Gao F., Gao Q., Duan X., Yue G., Yang A., Zhang J. (2006).Cloning of an H^+-PPase gene from *Thellungiella halophila* and its heterologous expression to improve tobacco salt tolerance. *J. Exp. Bot.* 57 3259–3270. 10.1093/jxb/erl090
89. Garbarino J., DuPont F. M. (1988). NaCl Induces a Na^+/H^+Antiport in tonoplast vesicles from barley roots. *Plant Physiol.* 86231–236. 10.1104/pp.86.1.231
90. Garnett T. P., Shabala S. N., Smethurst P. J., Newman I. A. (2001). Simultaneous measurements of ammonium, nitrate and proton fluxes along the length of eucalypt roots. *Plant Soil* 23655–62. 10.1023/A:1011951413917

91. Gaxiola R. A., Li J., Undurraga S., Dang L. M., Allen G. J., Alper S. L., et al. (2001). Drought- and salt-tolerant plants result from overexpression of the AVP1 H^+-pump. *Proc. Natl. Acad. Sci. U.S.A.* 98 11444–11449. 10.1073/pnas.191389398

92. Gaxiola R. A., Rao R., Sherman A., Grisafi P., Alper S. L., Fink G. R. (1999). The *Arabidopsis thaliana* proton transporters. AtNhx1 and Avp1, can function in cation detoxification in yeast.*Proc. Natl. Acad. Sci. U.S.A.* 96 1480–1485. 10.1073/pnas.96.4.1480

93. Gilliham M., Sullivan W., Tester M., Tyerman S. D. (2006).Simultaneous flux and current measurement from single plant protoplasts reveals a strong link between K^+ fluxes and current, but no link between Ca^{2+} fluxes and current. *Plant J.* 46 134–144. 10.1111/j.1365-313X.2006.02676.x

94. Gimmler H. (2000). Primary sodium plasma membrane ATPases in salt-tolerant algae: facts and fictions. *J. Exp. Bot.* 51 1171–1178. 10.1093/jexbot/51.348.1171

95. Grabov A. (2007). Plant KT/KUP/HAK potassium transporters: single family - multiple functions. *Ann. Bot.* 99 1035–1041. 10.1093/aob/mcm066

96. Greiner T., Ramos J., Alvarez M. C., Gurnon J. R., Kang M., Van Etten J. L., et al. (2011). Functional HAK/KUP/KT-like potassium transporter encoded by chlorella viruses. *Plant J.* 68977–986. 10.1111/j.1365-313X.2011.04748.x

97. Grosse W., Essen L. O., Koert U. (2011). Strategies and perspectives in ion-channel engineering. *Chembiochem* 12 830–839. 10.1002/cbic.201000793

98. Guo K. M., Babourina O., Rengel Z. (2009). Na^+/H^+ antiporter activity of the SOS1 gene: lifetime imaging analysis and electrophysiological studies on *Arabidopsis* seedlings. *Physiol. Plant.* 137 155–165. 10.1111/j.1399-3054.2009.01274.x

99. Gushchin I., Shevchenko V., Polovinkin V., Kovalev K., Alekseev A., Round E., et al. (2015). Crystal structure of a light-driven sodium pump. *Nat. Struct. Mol. Biol.* 22 390–395. 10.1038/nsmb.3002

100. Hajibagheri M. A., Flowers T. J. (1989). X-ray microanalysis of ion distribution within root cortical cells of the halophyte *Suaeda maritima* (L.) Dum. *Planta* 177 131–134. 10.1007/BF00392163

101. Hajibagheri M. A., Flowers T. J., Collins J. C., Yeo A. R. (1988).A comparison of the methods of X-ray microanalysis. *J. Exp. Bot.* 39 279–290. 10.1093/jxb/39.3.279

102. Hall D., Evans A. R., Newbury H. J., Pritchard J. (2006). Functional analysis of CHX21: a putative sodium transporter in*Arabidopsis*. *J. Exp. Bot.* 57 1201–1210. 10.1093/jxb/erj092

103. Hall J. L., Flowers T. J. (1973). The effect of salt on protein synthesis in the halophyte *Suaeda maritima*. *Planta* 110 361–368. 10.1007/BF00387064

104. Halperin S. J., Lynch J. P. (2003). Effects of salinity on cytosolic Na^+ and K^+ in root hairs of *Arabidopsis thaliana*: in vivo measurements using the fluorescent dyes SBFI and PBFI. *J. Exp. Bot.* 54 2035–2043. 10.1093/jxb/erg219

105. Hammou K. A., Rubio L., Fernández J. A., García-Sánchez M. J. (2014). Potassium uptake in the halophyte *Halimione portulacoides* L. Aellen *Environ. Exp. Bot.* 107 15–24. 10.1016/j.envexpbot.2014.05.001

106. Hansma P. K., Drake B., Marti O., Gould S. A., Prater C. B. (1989). The scanning ion-conductance microscope. *Science* 243641–643. 10.1126/science.2464851

107. He C., Yan J., Shen G., Fu L., Holaday A. S., Auld D., et al. (2005). Expression of an Arabidopsis vacuolar sodium/proton antiporter gene in cotton improves photosynthetic performance under salt conditions and increases fiber yield in the field. *Plant Cell Physiol.* 46 1848–1854. 10.1093/pcp/pci201

108. Hedrich R. (2012). Ion Channels in Plants. *Physiol. Rev.* 921777–1811. 10.1152/physrev.00038.2011

109. Heginbotham L., Abramson T., MacKinnon R. (1992). A functional connection between the pores of distantly related ion channels as revealed by mutant K^+ channels. *Science* 258 1152–1155. 10.1126/science.1279807

110. Higinbotham N. (1973). Electropotentials of Plant Cells. *Annu. Rev. Plant Physiol.* 24 25–46. 10.1146/annurev.pp.24.060173.000325

111. Hille B. (2001). *Ion Channels and Excitable Membranes* 3rd Edn (Sunderland, MA: Sinauer Associates;) 814.

112. Hirsch R. E., Lewis B. D., Spalding E. P., Sussman M. R. (1998).A role for the AKT1 potassium channel in plant nutrition.*Science* 280 918–921. 10.1126/science.280.5365.918

113. Hofmeister F. (1888). Zur Lehre von der Wirkung der Salze.*Archiv Exp. Pathol. Pharmakol.* 24 247–260. 10.1007/BF01918191

114. Horie T., Hauser F., Schroeder J. I. (2009). HKT transporter-mediated salinity resistance mechanisms in *Arabidopsis* and monocot crop plants. *Trends Plant Sci.* 14 660–668. 10.1016/j.tplants.2009.08.009

115. Hughes F. M., Jr., Cidlowski J. A. (1999). Potassium is a critical regulator of apoptotic enzymes in vitro and in vivo. *Adv. Enzyme Regul.* 39 157–171. 10.1016/S0065-2571(98)00010-7

116. Huh G. H., Damsz B., Matsumoto T. K., Reddy M. P., Rus A. M., Ibeas J. I., et al. (2002). Salt causes ion disequilibrium-induced programmed cell death in yeast and plants. *Plant J.* 29 649–659. 10.1046/j.0960-7412.2001.01247.x

117. Hunte C., Screpanti E., Venturi M., Rimon A., Padan E., Michel H. (2005). Structure of a Na^+/H^+ antiporter and insights into mechanism of action and regulation by pH. *Nature* 435 1197–1202. 10.1038/nature03692

118. Inan G., Zhang Q., Li P., Wang Z., Cao Z., Zhang H., et al. (2004). Salt cress. A halophyte and cryophyte *Arabidopsis*relative model system and its applicability to molecular genetic analyses of growth and

development of extremophiles. *Plant Physiol.* 135 1718–1737. 10.1104/pp.104.041723

119. Isayenkov S. V. (2012). Physiological and molecular aspects of salt stress in plants. *Cytol. Genet.* 46 302–318. 10.3103/S0095452712050040
120. Ivanov V. B. (1997). Relationship between cell proliferation and transition to elongation in plant roots. *Int. J. Dev. Biol.* 41 907–915.
121. Ivanov V. B., Dubrovsky J. G. (2013). Longitudinal zonation pattern in plant roots: conflicts and solutions. *Trends Plant Sci.*18 237–243. 10.1016/j.tplants.2012.10.002
122. Ivashikina N., Becker D., Ache P., Meyerhoff O., Felle H. H., Hedrich R. (2001). K^+ channel profile and electrical properties of*Arabidopsis* root hairs. *FEBS Lett.* 508 463–469. 10.1016/S0014-5793(01)03114-3
123. Jabnoune M., Espeout S., Mieulet D., Fizames C., Verdeil J. L., Conéjéro G., et al. (2009). Diversity in expression patterns and functional properties in the rice HKT transporter family. *Plant Physiol.* 150 1955–1971. 10.1104/pp.109.138008
124. Jacobs A., Ford K., Kretschmer J., Tester M. (2011). Rice plants expressing the moss sodium pumping ATPase PpENA1 maintain greater biomass production under salt stress. *Plant Biotechnol. J.*9 838–847. 10.1111/j.1467-7652.2011.00594.x
125. James R. A., Blake C., Byrt C. S., Munns R. (2011). Major genes for Na^+ exclusion. Nax1 and Nax2 (wheat HKT1;4 and HKT1;5), decrease Na^+ accumulation in bread wheat leaves under saline and waterlogged conditions. *J. Exp. Bot.* 62 2939–2947. 10.1093/jxb/err003
126. Jiang X., Leidi E. O., Pardo J. M. (2010). How do vacuolar NHX exchangers function in plant salt tolerance? *Plant Signal Behav.* 5792–795. 10.4161/psb.5.7.11767
127. Jiang Y., Lee A., Chen J., Cadene M., Chait B. T., MacKinnon R. (2002). The open pore conformation of potassium channels.*Nature* 417 523–526. 10.1038/417523a
128. Joh N. H., Wang T., Bhate M. P., Acharya R., Wu Y., Grabe M., et al. (2014). De novo design of a transmembrane Zn^{2+}-transporting four-helix bundle. *Science* 6216 1520–1524. 10.1126/science.1261172
129. Kader M. A., Lindberg S. (2005). Uptake of sodium in protoplasts of salt-sensitive and salt-tolerant cultivars of rice. *J. Exp. Bot.* 56 3149–3158. 10.1093/jxb/eri312
130. Karley A. J., Leigh R. A., Sanders D. (2000). Differential ion accumulation and ion fluxes in the mesophyll and epidermis of barley. *Plant Physiol.* 122 835–844. 10.1104/pp.122.3.835
131. Kato H. E., Inoue K., Abe-Yoshizumi R., Kato Y., Ono H., Konno M., et al. (2015). Structural basis for Na^+ transport mechanism by a light-driven Na^+ pump. *Nature* 521 48–53. 10.1038/nature14322
132. Kato N., Akai M., Zulkifli L., Matsuda N., Kato Y., Goshima S., et al. (2007). Role of positively charged amino acids in the M2D transmembrane helix of Ktr/Trk/HKT type cation transporters.*Channels (Austin)* 1 161–171. 10.4161/chan.4374
133. Katsuhara M., Kawasaki T. (1996). Salt stress induced nuclear and DNA degradation in meristematic cells of barley roots. *Plant Cell Physiol.* 37 169–173. 10.1093/oxfordjournals.pcp.a028928
134. Katz A., Avron M. (1985). Determination of intracellular osmotic volume and sodium concentration in dunaliella. *Plant Physiol.* 78817–820. 10.1104/pp.78.4.817
135. Kellermeier F., Armengaud P., Seditas T. J., Danku J., Salt D. E., Amtmann A. (2014). Analysis of the root system architecture of arabidopsis provides a quantitative readout of crosstalk between nutritional signals. *Plant Cell* 26 1480–1496. 10.1105/tpc.113.122101
136. Knowles A., Shabala S. (2004). Overcoming the problem of non-ideal liquid ion exchanger selectivity in microelectrode ion flux measurements. *J. Membr. Biol.* 202 51–59. 10.1007/s00232-004-0719-2
137. Korchev Y. E., Bashford C. L., Milovanovic M., Vodyanoy I., Lab M. J. (1997). Scanning ion conductance microscopy of living cells. *Biophys. J.* 73 653–658. 10.1016/S0006-3495(97)78100-1
138. Korolev A. V., Tomos A. D., Bowtell R., Farrar J. F. (2000).Spatial and temporal distribution of solutes in the developing carrot taproot measured at single-cell resolution. *J. Exp. Bot.* 51567–577. 10.1093/jexbot/51.344.56
139. Kronzucker H. J., Britto D. T. (2011). Sodium transport in plants: a critical review. *New Phytol.* 189 54–81. 10.1111/j.1469-8137.2010.03540.x
140. Kumar T., Uzma A. S., Khan M. R., Abbas Z., Ali G. M. (2014).Genetic Improvement of sugarcane for drought and salinity stress tolerance using arabidopsis vacuolar pyrophosphatase (AVP1) gene. *Mol. Biotechnol.* 56 199–209. 10.1007/s12033-013-9695-z
141. Kumar V., Jain M. (2015). The CRISPR-Cas system for plant genome editing: advances and opportunities. *J. Exp. Bot.* 66 47–57. 10.1093/jxb/eru429
142. Kunz W., Henle J., Ninham B. W. (2004). Zur Lehre von der Wirkung der Salze' (about the science of the effect of salts): franz Hofmeister's historical papers. *Curr. Opin. Colloid Interface Sci.*9 19–37. 10.1016/j.cocis.2004.05.005
143. Kurup S., Runions J., Köhler U., Laplaze L., Hodge S., Haseloff J. (2005). Marking cell lineages in living tissues. *Plant J.* 42 444–453. 10.1111/j.1365-313X.2005.02386.x
144. Lab M. J., Bhargava A., Wright P. T., Gorelik J. (2013). The scanning ion conductance microscope for cellular physiology.*Am. J. Physiol. Heart Circ. Physiol.* 304 H1–H11. 10.1152/ajpheart.00499.2012
145. Lan W. Z., Wang W., Wang S. M., Li L. G., Buchanan B. B., Lin H. X., et al. (2010). A rice high-affinity potassium transporter (HKT) conceals a calcium-permeable cation channel. *Proc. Natl. Acad. Sci. U.S.A.* 107 7089–7094. 10.1073/pnas.1000698107
146. Leidi E. O., Barragán V., Rubio L., El-Hamdaoui A., Ruiz M. T., Cubero B., et al. (2010). The AtNHX1 exchanger mediates potassium compartmentation in vacuoles of

transgenic tomato.*Plant J.* 61 495–506. 10.1111/j.1365-313X.2009.04073.x

147. Lew R. R., Levina N. N., Shabala L., Anderca M. I., Shabala S. N. (2006). Role of a mitogen-activated protein kinase cascade in ion flux-mediated turgor regulation in fungi. *Eukaryot. Cell* 5480–487. 10.1128/EC.5.3.480-487.2006

148. Li X., Guo C., Gu J., Duan W., Zhao M., Ma C., et al. (2014). Overexpression of VP, a vacuolar H^+-pyrophosphatase gene in wheat (*Triticum aestivum* L.), improves tobacco plant growth under Pi and N deprivation, high salinity, and drought. *J. Exp. Bot.* 65 683–696. 10.1093/jxb/ert442

149. Lindberg S. (1995). In-situ determination of intracellular concentrations of K^+ in barley (*Hordeum vulgare* L. cv. Kara) using the K^+-binding fluorescent dye benzofuran isophthalate*Planta*. 195 525–529. 10.1007/BF00195710

150. Liu W., Fairbairn D. J., Reid R. J., Schachtman D. P. (2001).Characterization of two HKT1 homologues from *Eucalyptus camaldulensis* that display intrinsic osmosensing capability. *Plant Physiol.* 127 283–294.

151. Liu Y., Ji X., Zheng L., Nie X., Wang Y. (2013). Microarray analysis of transcriptional responses to abscisic acid and salt stress in *Arabidopsis thaliana*. *Int. J. Mol. Sci.* 14 9979–9998. 10.3390/ijms14059979

152. Longpré J. P., Lapointe J. Y. (2011). Determination of the Na^+/glucose cotransporter (SGLT1) turnover rate using the ion-trap technique. *Biophys. J.* 100 52–59. 10.1016/j.bpj.2010.11.012

153. L'Roy A., Hendrix D. L. (1980). Effect of salinity upon cell membrane potential in the marine halophyte, *Salicornia bigelovii*Torr. *Plant Physiol.* 65 544–549. 10.1104/pp.65.3.544

154. Lüttge U. (2002). "Mangroves," in *Salinity: Environment-Plants-Molecules* eds Läuchli R., Lüttge U., editors. (The Netherlands: Kluwer Academic Publishers;) 113–135.

155. Ma D. M., Xu W. R., Li H. W., Jin F. X., Guo L. N., Wang J., et al. (2014). Co-expression of the *Arabidopsis* SOS genes enhances salt tolerance in transgenic tall fescue (*Festuca arundinacea*Schreb.). *Protoplasma* 251 219–231. 10.1007/s00709-013-0540-9

156. Maathuis F. J., Prins H. B. (1990). Patch clamp studies on root cell vacuoles of a salt-tolerant and a salt-sensitive *Plantago*species. . *Plant Physiol.* 92 23–28. 10.1104/pp.92.1.23

157. Maathuis F. J. M., Ahmad I., Patishtan J. (2014). Regulation of Na^+ fluxes in plants. *Front. Plant Sci.* 5:46710.3389/fpls.2014.00467

158. Maathuis F. J. M., Amtmann A. (1999). K^+ nutrition and Na^+toxicity: the basis of cellular K^+/Na^+ ratios. *Ann. Bot.* 84 123–133. 10.1006/anbo.1999.0912

159. Maathuis F. J. M., Flowers T. J., Yeo A. R. (1992). Sodium-chloride compartmentation in leaf vacuoles of the halophyte*Suaeda maritima* (L) Dum. and its relation to tonoplast permeability. *J. Exp. Bot.* 43 1219–1223. 10.1093/jxb/43.9.1219

160. MacKinnon R. (2004). Potassium channels and the atomic basis of selective ion conduction (Nobel Lecture). *Angew. Chem. Int. Ed. Engl.* 43 4265–4277. 10.1002/anie.200400662

161. MacRobbie E. A., Dainty J. (1958). Ion transport in Nitellopsis obtusa. *J. Gen. Physiol.* 42 335–353. 10.1085/jgp.42.2.335

162. Maffeo C., Bhattacharya S., Yoo J., Wells D., Aksimentiev A. (2012). Modeling and simulation of ion channels. *Chem. Rev.* 1126250–6284. 10.1021/cr3002609

163. Mähler J., Persson I. (2012). A study of the hydration of the alkali metal ions in aqueous solution. *Inorg. Chem.* 51 425–438. 10.1021/ic2018693

164. Malone M., Leigh R. A., Tomos A. D. (1991). Concentrations of vacuolar inorganic ions in individual cells of intact wheat leaf epidermis. *J. Exp. Bot.* 42 305–309. 10.1093/jxb/42.3.305

165. Martinoia E., Meyer S., De Angeli A., Nagy R. (2012). Vacuolar transporters in their physiological context. *Annu. Rev. Plant Biol.*63 183–213. 10.1146/annurev-arplant-042811-105608

166. Mäser P., Eckelman B., Vaidyanathan R., Horie T., Fairbairn D. J., Kubo M., et al. (2002). Altered shoot/root Na^+ distribution and bifurcating salt sensitivity in *Arabidopsis* by genetic disruption of the Na^+ transporter AtHKT1. *FEBS Lett.* 531 157–161. 10.1016/S0014-5793(02)03488-9

167. Mäser P., Thomine S., Schroeder J. I., Ward J. M., Hirschi K., Sze H., et al. (2001). Phylogenetic relationships within cation transporter families of *Arabidopsis*. *Plant Physiol.* 126 1646–1667. 10.1104/pp.126.4.1646

168. Masi A., Melis A. (1997). Morphological and molecular changes in the unicellular green alga *Dunaliella salina* grown under supplemental UV-B radiation: cell characteristics and Photosystem II damage and repair properties. *BBA Bioenerget.*1321 183–193. 10.1016/S0005-2728(97)00054-6

169. Matzke A. J., Matzke M. (2013). Membrane "potential-omics": toward voltage imaging at the cell population level in roots of living plants. *Front. Plant Sci.* 4:311 10.3389/fpls.2013.00311

170. Mian A., Oomen R. J., Isayenkov S., Sentenac H., Maathuis F. J., Véry A. A. (2011). Over-expression of an Na^+-and K^+-permeable HKT transporter in barley improves salt tolerance. *Plant J.* 68468–479. 10.1111/j.1365-313X.2011.04701.x

171. Møller I. S., Gilliham M., Jha D., Mayo G. M., Roy S. J., Coates J. C., et al. (2009). Shoot Na^+ exclusion and increased salinity tolerance engineered by cell type-specific alteration of Na^+transport in *Arabidopsis*. *Plant Cell* 21 2163–2178. 10.1105/tpc.108.064568

172. Munns R., Tester M. (2008). Mechanisms of salinity tolerance.*Annu. Rev. Plant Biol.* 59 651–681. 10.1146/annurev.arplant.59.032607.092911

173. Murthy M., Tester M. (2006). Cation currents in protoplasts from the roots of a Na^+ hyperaccumulating mutant of *Capsicum annuum*. *J. Exp. Bot.* 57 1171–1180. 10.1093/jxb/erj115

174. Mutoh H., Akemann W., Knöpfel T. (2012). Genetically engineered fluorescent voltage reporters ACS. *Chem. Neurosci.* 3585–592. 10.1021/cn300041b

175. Myers B. R., Sigal Y. M., Julius D. (2009). Evolution of thermal response properties in a cold-activated TRP channel. *PLoS ONE*4:e5741 10.1371/journal.pone.0005741

176. Newman I. A. (2001). Ion transport in roots: measurement of fluxes using ion-selective microelectrodes to characterize transporter function. *Plant Cell Environ.* 24 1–14. 10.1046/j.1365-3040.2001.00661.x

177. Newman I. A., Kochian L. V., Grusak M. A., Lucas W. J. (1987).Fluxes of H^+ and K^+ in corn roots: characterization and stoichiometries using ion-selective microelectrodes. *Plant Physiol.* 84 1177–1184. 10.1104/pp.84.4.1177

178. Nightingale E. R., Jr. (1959). Phenomenological theory of ion solvation. Effective radii of hydrated ions. *J. Phys. Chem.* 631381–1387. 10.1021/j150579a011

179. Núñez-Ramírez R., Sánchez-Barrena M. J., Villalta I., Vega J. F., Pardo J. M., Quintero F. J., et al. (2012). Structural insights on the plant salt-overly-sensitive 1 (SOS1) Na^+/H^+ antiporter. *J. Mol. Biol.* 424 283–294. 10.1016/j.jmb.2012.09.015

180. Oh D. H., Gong Q., Ulanov A., Zhang Q., Li Y., Ma W., et al. (2007). Sodium stress in the halophyte *Thellungiella halophila*, and transcriptional changes in a thsos1-RNA iterference line. *J. Integr. Plant Biol.* 49 1484–1496. 10.1111/j.1672-9072.2007.00548.x

181. Oh D. H., Leidi E., Zhang Q., Hwang S. M., Li Y., Quintero F. J., et al. (2009a). Loss of halophytism by interference with SOS1 expression. *Plant Physiol.* 151 210–222. 10.1104/pp.109.137802

182. Oh D. H., Zahir A., Yun D. J., Bressan R. A., Bohnert H. J. (2009b). SOS1 and halophytism. *Plant Signal. Behav.* 4 1081–1083. 10.4161/psb.4.11.9786

183. Olías R., Eljakaoui Z., Li J., De Morales P. A., Marín-Manzano M. C., Pardo J. M., et al. (2009). The plasma membrane Na^+/H^+antiporter SOS1 is essential for salt tolerance in tomato and affects the partitioning of Na^+ between plant organs. *Plant Cell Environ.* 32 904–916. 10.1111/j.1365-3040.2009.01971.x

184. Ordoñez N. M., Marondedze C., Thomas L., Pasqualini S., Shabala L., Shabala S., et al. (2014). Cyclic mononucleotides modulate potassium and calcium flux responses to H2O2 in*Arabidopsis* roots. *FEBS Lett.* 588 1008–1015. 10.1016/j.febslet.2014.01.062

185. Oszvald M., Gardonyi M., Tamas C., Takacs I., Jenes B., Tamas L., et al. (2008). Development and characterization of a chimaeric tissue-specific promoter in wheat and rice endosperm.*In Vitro Cell. Dev. Biol. Plant* 44 1–7. 10.1007/s11627-007-9082-1

186. Pakhomov A. G., Bowman A. M., Ibey B. L., Andre F. M., Pakhomova O. N., Schoenbach K. H. (2009). Lipid nanopores can form a stable, ion channel-like conduction pathway in cell membrane. *Biochem. Biophys. Res. Commun.* 385 181–186. 10.1016/j.bbrc.2009.05.035

187. Palmgren M. G. (2001). PLANT PLASMA MEMBRANE H^+-ATPases: powerhouses for nutrient uptake. *Annu. Rev. Plant. Physiol. Plant Mol. Biol.* 52 817–845. 10.1146/annurev.arplant.52.1.817

188. Pandolfi C., Pottosin I., Cuin T., Mancuso S., Shabala S. (2010).Specificity of polyamine effects on Nacl-induced ion flux kinetics and salt stress amelioration in plants. *Plant Cell Physiol.*51 422–434. 10.1093/pcp/pcq007

189. Pang J. Y., Newman I., Mendham N., Zhou M., Shabala S. (2006). Microelectrode ion and O2 fluxes measurements reveal differential sensitivity of barley root tissues to hypoxia. *Plant Cell Environ.* 29 1107–1121. 10.1111/j.1365-3040.2005.01486.x

190. Pantoja O., Dainty J., Blumwald E. (1989). Ion channels in vacuoles from halophytes and glycophytes. *FEBS Lett.* 255 92–96. 10.1016/0014-5793(89)81067-1

191. Pardo J. M., Cubero B., Leidi E. O., Quintero F. J. (2006). Alkali cation exchangers: roles in cellular homeostasis and stress tolerance. *J. Exp. Bot.* 57 1181–1199. 10.1093/jxb/erj114

192. Pasapula V., Shen G., Kuppu S., Paez-Valencia J., Mendoza M., Hou P., et al. (2011). Expression of an *Arabidopsis* vacuolar H^+-pyrophosphatase gene (AVP1) in cotton improves drought- and salt tolerance and increases fibre yield in the field conditions. *Plant Biotechnol. J.* 9 88–99. 10.1111/j.1467-7652.2010.00535.x

193. Peleg Z., Apse M. P., Blumwald E. (2011). Engineering salinity and water-stress tolerance in crop plants: getting closer to the field. *Adv. Bot. Res.* 57 405–443. 10.1016/B978-0-12-387692-8.00012-6

194. Pick U., Karni L., Avron M. (1986). Determination of ion content and ion fluxes in the halotolerant alga *Dunaliella salina*. *Plant Physiol.* 81 92–96. 10.1104/pp.81.1.92

195. Pitman M. G., Läuchli A. (2002). "Global impact of salinity and agricultural ecosystems," in *Salinity: Environment-Plants-Molecules* eds Läuchli R., Lüttge U., editors. (The Netherlands: Kluwer Academic Publishers;) 3–20.

196. Pitman M. G., Läuchli A., Stelzer R. (1981). Ion distribution in roots of barley seedlings measured by electron probe x-ray microanalysis. *Plant Physiol.* 68 673–679. 10.1104/pp.68.3.673

197. Plett D., Safwat G., Gilliham M., Møller I. S., Roy S., Shirley N., et al. (2010). Improved salinity tolerance of rice through cell type-specific expression of AtHKT1;1. *PLoS ONE* 5:e1257110.1371/journal.pone.0012571

198. Popova L. G., Shumkova G. A., Andreev I. M., Balnokin Y. V. (2005). Functional identification of electrogenic Na^+-translocating ATPase in the plasma membrane of the halotolerant microalga *Dunaliella maritima*. *FEBS Lett.* 579 5002–5006. 10.1016/j.febslet.2005.07.087

199. Premkumar L., Greenblatt H. M., Bagashwar U. K., Savchenko T., Gokhman I., Sussman J. L., et al. (2005). 3D structure of a halotolerant algal carbonic anhydrase predicts halotolerance of a mammalian homolog. *Proc. Natl. Acad. Sci. U.S.A.* 102 7493–7498. 10.1073/pnas.0502829102

200. Rains D. W., Epstein E. (1965). Transport of sodium in plant tissue. *Science* 148 1611 10.1126/science.148.3677.1611

201. Ramsey I. S., Delling M., Clapham D. E. (2006). An introduction to TRP channels. *Annu. Rev. Physiol.* 68 619–647. 10.1146/annurev.physiol.68.040204.100431

202. Roberts S. K., Tester M. (1997). A patch clamp study of Na^+transport in maize roots. *J. Exp. Bot.* 48 431–440. 10.1093/jxb/48

203. Rodriguez-Navarro A. (2000). Potassium transport in fungi and plants. *BBA Rev. Biomembr.* 1469 1–30. 10.1016/S0304-4157(99)00013-1

204. Rodríguez-Rosales M. P., Gálvez F. J., Huertas R., Aranda M. N., Baghour M., Cagnac O., et al. (2009). Plant NHX cation/proton antiporters. *Plant Signal. Behav.* 4 265–276. 10.4161/psb.4.4.7919

205. Sassi A., Mieulet D., Khan I., Moreau B., Gaillard I., Sentenac H., et al. (2012). The rice monovalent cation transporter OsHKT2;4: revisited ionic selectivity. *Plant Physiol.* 160 498–510. 10.1104/pp.112.194936

206. Schulze L. M., Britto D. T., Li M., Kronzucker H. J. (2012). A pharmacological analysis of high-affinity sodium transport in barley (*Hordeum vulgare* L.): a $24Na^+/42K^+$ study. *J. Exp. Bot.*63 2479–2489. 10.1093/jxb/err419

207. Serrano R. (1988). Structure and function of proton translocating ATPase in plasma membranes of plants and fungi. *Biochim. Biophys. Acta* 947 1–28. 10.1016/0304-4157(88)90017-2

208. Serrano R. (1996). Salt tolerance in plants and microorganisms: toxicity targets and defense mechanisms. *Int. Rev. Cytol.* 165 1–52. 10.1016/S0074-7696(08)62219-6

209. Shabala L., Cuin T. A., Newman I. A., Shabala S. (2005). Salinity-induced ion flux patterns from the excised roots of*Arabidopsis* sos mutants. *Planta* 222 1041–1050. 10.1007/s00425-005-0074-2

210. Shabala S. (2006). "Non-invasive microelectrode ion flux measurements in plant stress physiology," in *Plant Electrophysiology—Theory and Methods* ed. Volkov A. G., editor. (Berlin: Springer-Verlag;) 35–71.

211. Shabala S. (2009). Salinity and programmed cell death: unravelling mechanisms for ion specific signalling. *J. Exp. Bot.*60 709–712. 10.1093/jxb/erp013

212. Shabala S. (2013). Learning from halophytes: physiological basis and strategies to improve abiotic stress tolerance in crops. *Ann. Bot.* 112 1209–1221. 10.1093/aob/mct205

213. Shabala S., Bose J. (2012). "Application of non-invasive microelectrode flux measurements in plant stress physiology," in*Plant Electrophysiology—Methods and Cell Electrophysiology*ed. Volkov A. G., editor. (Berlin: Springer-Verlag;) 91–126.

214. Shabala S., Demidchik V., Shabala L., Cuin T. A., Smith S. J., Miller A. J., et al. (2006). Extracellular $Ca2^+$ameliorates NaCl-induced K^+ loss from *Arabidopsis* root and leaf cells by controlling plasma membrane K^+-permeable channels. *Plant Physiol.* 141 1653–1665. 10.1104/pp.106.082388

215. Shabala S., Newman I. (2000). Salinity effects on the activity of plasma membrane H^+ and Ca^{2+} transporters in bean leaf mesophyll: masking role of the cell wall. *Ann. Bot.* 85 681–686. 10.1006/anbo.2000.1131

216. Shabala S., Pang J., Zhou M., Shabala L., Cuin T. A., Nick P., et al. (2009). Electrical signalling and cytokinins mediate effects of light and root cutting on ion uptake in intact plants. *Plant Cell Environ.* 32 194–207. 10.1111/j.1365-3040.2008.01914.x

217. Shabala S. N., Lew R. R. (2002). Turgor regulation in osmotically stressed *Arabidopsis* epidermal root cells. Direct support for the role of inorganic ion uptake as revealed by concurrent flux and cell turgor measurements. *Plant Physiol.* 129290–299. 10.1104/pp.020005

218. Shan Q., Wang Y., Li J., Zhang Y., Chen K., Liang Z., et al. (2013). Targeted genome modification of crop plants using a CRISPR-Cas system. *Nat. Biotechnol.* 31 686–688. 10.1038/nbt.2650

219. Shavrukov Y., Bovill J., Afzal I., Hayes J. E., Roy S. J., Tester M., et al. (2013). HVP10 encoding V-PPase is a prime candidate for the barley HvNax3 sodium exclusion gene: evidence from fine mapping and expression analysis. *Planta* 237 1111–1122. 10.1007/s00425-012-1827-3

220. Shi H., Ishitani M., Kim C., Zhu J. K. (2000). The *Arabidopsis thaliana* salt tolerance gene SOS1 encodes a putative Na^+/H^+antiporter. *Proc. Natl. Acad. Sci. U.S.A.* 97 6896–6901. 10.1073/pnas.120170197

221. Shi H., Lee B. H., Wu S. J., Zhu J. K. (2003). Overexpression of a plasma membrane Na^+/H^+ antiporter gene improves salt tolerance in *Arabidopsis thaliana*. *Nat. Biotechnol.* 21 81–85. 10.1038/nbt766

222. Shi H., Quintero F. J., Pardo J. M., Zhu J. K. (2002). The putative plasma membrane Na^+/H^+ antiporter SOS1 controls long-distance Na^+ transport in plants. *Plant Cell* 14 465–477. 10.1105/tpc.010371

223. Silva P., Gerós H. (2009). Regulation by salt of vacuolar H^+-ATPase and H^+-pyrophosphatase activities and Na^+/H^+exchange. *Plant Signal. Behav.* 4 718–726. 10.4161/psb.4.8.9236

224. Skou J. C. (1998). Nobel Lecture. The identification of the sodium pump. *Biosci. Rep.* 18 155–169.

225. Spiller D. G., Wood C. D., Rand D. A., White M. R. (2010). Measurement of single-cell dynamics. *Nature* 465 736–745. 10.1038/nature09232

226. Staal M., Maathuis F. J. M., Elzenga J. T. M., Overbeek J. H. M., Prins H. B. A. (1991). Na^+/H^+ antiport activity in tonoplast vesicles from roots of the salt-tolerant *Plantago maritima* and the salt-sensitive *Plantago media*. *Physiol. Plant.* 82 179–184. 10.1111/j.1399-3054.1991.tb00078.x

227. Subramanyam P., Colecraft H. M. (2015). Ion channel engineering: perspectives and strategies. *J. Mol. Biol.* 427 190–204. 10.1016/j.jmb.2014.09.001

228. Sun J., Chen S. L., Dai S. X., Wang R. G., Li N. Y., Shen X., et al. (2009). Ion flux profiles and plant ion homeostasis control under salt stress. *Plant Signal. Behav.* 4 261–264. 10.4161/psb.4.4.7918

229. Sze H., Li X., Palmgren M. G. (1999). Energization of plant cell membranes by H^+-pumping ATPases. Regulation and biosynthesis. *Plant Cell* 11 677–690.

230. Taji T., Seki M., Satou M., Sakurai T., Kobayashi M., Ishiyama K., et al. (2004). Comparative genomics in salt tolerance between*Arabidopsis* and *Arabidopsis*-related halophyte salt cress using*Arabidopsis* microarray. *Plant Physiol.* 135 1697–1709. 10.1104/pp.104.039909

231. Takahashi F., Tilbrook J., Trittermann C., Berger B., Roy S. J., Seki M., et al. (2015). Comparison of leaf sheath transcriptome profiles with physiological traits of bread wheat cultivars under salinity stress. *PLoS ONE* 10:e013332210.1371/journal.pone.0133322

232. Thiyagarajah M., Fry S. C., Yeo A. R. (1996). In vitro salt tolerance of cell wall enzymes from halophytes and glycophytes.*J. Exp. Bot.* 47 1717–1724. 10.1093/jxb/47.11.1717

233. Tyerman S. D., Beilby M., Whittington J., Juswono U., Newman I., Shabala S. (2001). Oscillations in proton transport revealed from simultaneous measurements of net current and net proton fluxes from isolated root protoplasts: MIFE meets patch-clamp.*Austr. J. Plant Physiol.* 28 591–606. 10.1071/PP01030

234. Tyerman S. D., Skerrett I. M. (1998). Root ion channels and salinity. *Sci. Hortic.* 78 175–235. 10.1016/S0304-4238(98)00194-0

235. Tyerman S. D., Skerrett M., Garrill A., Findlay G. P., Leigh R. A. (1997). Pathways for the permeation of Na^+ and Cl^- into protoplasts derived from the cortex of wheat roots. *J. Exp. Bot.*48 459–480. 10.1093/jxb/48.Special_Issue.459

236. Valencia-Cruz G., Shabala L., Delgado-Enciso I., Shabala S., Bonales-Alatorre E., Pottosin I. I., et al. (2009). Kbg and Kv1.3 channels mediate potassium efflux in the early phase of apoptosis in Jurkat T lymphocytes. *Am. J. Physiol. Cell Physiol.* 297C1544–C1553. 10.1152/ajpcell.00064.2009

237. Verbelen J. P., De Cnodder T., Le J., Vissenberg K., Baluska F. (2006). The root apex of arabidopsis thaliana consists of four distinct zones of growth activities: meristematic zone. *Plant Signal. Behav.* 1 296–304. 10.4161/psb.1.6.3511

238. Volkov V., Amtmann A. (2006). *Thellungiella halophila*, a salt-tolerant relative of *Arabidopsis thaliana*, has specific root ion-channel features supporting K^+/Na^+ homeostasis under salinity stress. *Plant J.* 48 342–353. 10.1111/j.1365-313X.2006.02876.x

239. Volkov V., Boscari A., Clément M., Miller A. J., Amtmann A., Fricke W. (2009). Electrophysiological characterization of pathways for K^+ uptake into growing and non-growing leaf cells of barley. *Plant Cell Environ.* 32 1778–1790. 10.1111/j.1365-3040.2009.02034.x

240. Volkov V., Hachez C., Moshelion M., Draye X., Chaumont F., Fricke W. (2007). Water permeability differs between growing and non-growing barley leaf tissues. *J. Exp. Bot.* 58 377–390. 10.1093/jxb/erl203

241. Volkov V., Wang B., Dominy P. J., Fricke W., Amtmann A. (2004). *Thellungiella halophila*, a salt-tolerant relative of*Arabidopsis thaliana*, possesses effective mechanisms to discriminate between potassium and sodium. *Plant Cell Environ.*27 1–14. 10.1046/j.0016-8025.2003.01116.x

242. Vrbka L., Vondrásek J., Jagoda-Cwiklik B., Vácha R., Jungwirth P. (2006). Quantification and rationalization of the higher affinity of sodium over potassium to protein surfaces. *Proc. Natl. Acad. Sci. U.S.A.* 103 15440–15444. 10.1073/pnas.0606959103

243. Walker D. J., Black C. R., Miller A. J. (1998). The role of cytosolic potassium and pH in the growth of barley roots. *Plant Physiol.* 118 957–964. 10.1104/pp.118.3.957

244. Walker D. J., Smith S. J., Miller A. J. (1995). Simultaneous measurement of intracellular pH and K^+ or NO_3- in barley root cells using triple-barreled, ion-selective microelectrodes. *Plant Physiol.* 108 743–751.

245. Wang B. (2006) *Physiological and Molecular Strategies for Salt Tolerance in Thellungiella halophila, a Close Relative of Arabidopsis thaliana*. Ph.D. thesis, Division of Biochemistry and Molecular Biology, Institute of Biomedical and Life Sciences, University of Glasgow; Glasgow: 235.

246. Wang B., Davenport R. J., Volkov V., Amtmann A. (2006). Low unidirectional sodium influx into root cells restricts net sodium accumulation in *Thellungiella halophila*, a salt-tolerant relative of *Arabidopsis thaliana*. *J. Exp. Bot.* 57 1161–1170. 10.1093/jxb/erj116

247. Waters S., Gilliham M., Hrmova M. (2013). Plant high-affinity potassium (HKT) transporters involved in salinity tolerance: structural insights to probe differences in ion selectivity. *Int. J. Mol. Sci.* 14 7660–7680. 10.3390/ijms14047660

248. Wegner L. H. (2013). Cation selectivity of the plasma membrane of tobacco protoplasts in the electroporated state. *BBA Biomembr.*1828 1973–1981. 10.1016/j.bbamem.2013.04.010

249. Wegner L. H. (2014). The conductance of cellular membranes at supra-physiological voltages. *Bioelectricity* 103 34–38. 10.1016/j.bioelechem.2014.08.005

250. Wegner L. H., Flickinger B., Eing C., Berghöfer T., Hohenberger P., Frey W., et al. (2011). A patch clamp study on the electro-permeabilization of higher plant cells: supra-physiological voltages induce a high-conductance. *BBA Biomembr.* 1808 1728–1736. 10.1016/j.bbamem.2011.01.016

251. Wegner L. H., Frey W., Schönwälder S. (2013). A critical evaluation of whole cell patch clamp studies on electroporation using the voltage sensitive dye ANNINE-6. *Bioelectrochemistry*92 42–46. 10.1016/j.bioelechem.2013.03.002

252. White P. J. (1993). Characterization of a high-conductance, voltage-dependent cation channel from the plasma membrane of rye roots in planar lipid bilayers. *Planta* 191 541–551. 10.1007/BF00195756

253. White P. J., Cooper H. D., Clarkson D. T., Earnshaw M. J., Loughman B. C. (1991). Effects of low temperature on growth and nutrient accumulation in rye (*Secale cereale*) and wheat (*Triticum aestivum*). *Ann. Bot.* 68 23–31.

254. White P. J., Lemtiri-Chlieh F. (1995). Potassium currents across the plasma membrane of protoplasts derived from rye roots: a patch clamp study. *J. Exp. Bot.* 46 497–511. 10.1093/jxb/46.5.497

255. Winter H., Robinson D. G., Heldt H. W. (1994). Subcellular volumes and metabolite concentrations in spinach leaves. *Planta* 193 530–535. 10.1007/BF02411558

256. Wu S. J., Ding L., Zhu J. K. (1996). SOS1, a genetic locus essential for salt tolerance and potassium acquisition. *Plant Cell* 8 617–627. 10.1105/tpc.8.4.617

257. Wyn Jones R. G. A., Pollard A. (1983). "Proteins, enzymes and inorganic ions," in *Encyclopedia of Plant Physiology, New Series* Vol. 15B eds Lauchli A., Person A., editors. (New York: Springer;) 528–562.

258. Xue S., Yao X., Luo W., Jha D., Tester M., Horie T., et al. (2011). AtHKT1;1 mediates Nernstian sodium channel transport properties in *Arabidopsis* root stelar cells. *PLoS ONE* 6:e24725 10.1371/journal.pone.0024725

259. Yamaguchi T., Aharon G. S., Sottosanto J. B., Blumwald E. (2005). Vacuolar Na^+/H^+ antiporter cation selectivity is regulated by calmodulin from within the vacuole in a Ca^{2+}- and pH-dependent manner. *Proc. Natl Acad. Sci. U.S.A.* 102 16107–16112. 10.1073/pnas.0504437102

260. Yamaguchi T., Apse M. P., Shi H., Blumwald E. (2003). Topological analysis of a plant vacuolar Na^+/H^+ antiporter reveals a luminal C terminus that regulates antiporter cation selectivity. *Proc. Natl. Acad. Sci. U.S.A.* 100 12510–12515. 10.1073/pnas.2034966100

261. Yamaguchi T., Hamamoto S., Uozumi N. (2013). Sodium transport system in plant cells. *Front. Plant Sci.* 4:410 10.3389/fpls.2013.00410

262. Yang Q., Chen Z. Z., Zhou X. F., Yin H. B., Li X., Xin X. F., et al. (2009). Overexpression of SOS (Salt Overly Sensitive) genes increases salt tolerance in transgenic *Arabidopsis*. *Mol. Plant* 222–31. 10.1093/mp/ssn058

263. Yue Y., Zhang M., Zhang J., Duan L., Li Z. (2012). SOS1 gene overexpression increased salt tolerance in transgenic tobacco by maintaining a higher K^+/Na^+ ratio. *J. Plant Physiol.* 169 255–261. 10.1016/j.jplph.2011.10.007

264. Zahran H. H., Marín-Manzano M. C., Sánchez-Raya A. J., Bedmar E. J., Venema K., Rodríguez-Rosales M. P. (2007). Effect of salt stress on the expression of NHX-type ion transporters in *Medicago intretexta* and *Melilotus indicus* plants. *Physiol. Plant.* 131 122–130. 10.1111/j.1399-3054.2007.00940.x

265. Zdebik A. A., Zifarelli G., Bergsdorf E. Y., Soliani P., Scheel O., Jentsch T. J., et al. (2008). Determinants of anion-proton coupling in mammalian endosomal CLC proteins. *J. Biol. Chem.* 283 4219–4227. 10.1074/jbc.M708368200

266. Zhang H. X., Blumwald E. (2001). Transgenic salt-tolerant tomato plants accumulate salt in foliage but not in fruit. *Nat. Biotechnol.* 19 765–768. 10.1038/90824

267. Zhao F. Y., Zhang X. J., Li P. H., Zhao Y. X., Zhang H. (2006). Co-expression of the *Suaeda salsa* SsNHX1 and *Arabidopsis* AVP1 confer greater salt tolerance to transgenic rice than the single SsNHX1. *Mol. Breed.* 17 341–353. 10.1007/s11032-006-9005-6

268. Zhao K. F., Fan H., Song J., Sun M. X., Wang B. Z., Zhang S. Q., et al. (2005). Two Na^+ and Cl^- Hyperaccumulators of the Chenopodiaceae. *J. Integr. Plant Biol.* 47 311–318. 10.1111/j.1744-7909.2005.00057.x

269. Zou P., Zhao Y., Douglass A. D., Hochbaum D. R., Brinks D., Werley C. A., et al. (2014). Bright and fast multicoloured voltage reporters via electrochromic FRET. *Nat. Commun.* 5 4625 10.1038/ncomms5625

CHAPTER 11

IMPROVING PHARMACEUTICAL PROTEIN PRODUCTION IN ORYZA SATIVA

Yu-Chieh Kuo [1], Chia-Chun Tan [1,2], Jung-Ting Ku [1], Wei-Cho Hsu [1], Sung-Chieh Su [1], Chung-An Lu [2] and Li-Fen Huang [1]

[1]Graduate School of Biotechnology and Bioengineering, Yuan Ze University, 135 Yuan-Tung Road, Taoyuan 32003, Taiwan

[2]Department of Life Sciences, National Central University, 300, Jhongda Rd., Taoyuan 32001, Taiwan

ABSTRACT

Application of plant expression systems in the production of recombinant proteins has several advantages, such as low maintenance cost, absence of human pathogens, and possession of complex post-translational glycosylation capabilities. Plants have been successfully used to produce recombinant cytokines, vaccines, antibodies, and other proteins, and rice (Oryza sativa) is a potential plant used as recombinant protein expression system. After successful transformation, transgenic rice cells can be either regenerated into whole plants or grown as cell cultures that can be upscaled into bioreactors. This review summarizes recent advances in the production of different recombinant protein produced in rice and describes their production methods as well as methods to improve protein yield and quality. Glycosylation and its impact in plant development and protein production are discussed, and several methods of improving yield and quality that have not been incorporated in rice expression systems are also proposed. Finally, different bioreactor options are explored and their advantages are analyzed.

INTRODUCTION

Production of recombinant protein using plant expression system first dates back to the 1980s when plants' genetic transformability has been confirmed [1]. Soon after, human growth hormone and antibody chains were produced using transgenic tobacco and sunflower [2,3]. In 1997, recombinant avidin was produced commercially using transgenic maize, affirming plants' capabilities to produce recombinant proteins on a large scale [4]. Since then, numerous antibodies, vaccines, cytokines and other proteins have been produced using plant expression systems in the forms of both mature plants and tissue cell cultures [5]. Plants have numerous advantages over other expression systems such as intrinsic safety (plants are free of animal pathogens and viruses, and they can be grown without using animal-derived materials), lower production and capital costs, and glycosylation pattern similar to that of humans [5–7]. Among these advantages, plants' ability to perform complex post-translational modification is particularly important because it allows for the production of complicated glycoproteins with glycosylation, which plays major roles in the bioactivity and stability of certain proteins [8,9]. Rice is a well-studied model plant whose genome has been decoded, and its transformation technology has been very well developed [10,11]. As the third most produced crop in the world, rice is also considered to be hypoallergenic, making it an excellent host for protein production and even administration [12]. Recent reports indicate that rice is a suitable host for recombinant protein due to its competitive yield of certain recombinant proteins, such as human serum albumin,

Citation: Vadim Volkov, Salinity tolerance in plants. Quantitative approach to ion transport starting from halophytes and stepping to genetic and protein engineering for manipulating ion fluxes, doi: 10.3389/fpls.2015.00873.

compared to other plant expression systems such as tobacco cell cultures and potato tubers [13–16]. Suspension cell culture is a widely used method for producing recombinant proteins, and systems upscaled into bioreactors have been utilized commercially [7,17]. Bioreactors have numerous advantages including lower capital and production costs, scalability, safety from viral transmission and other mammalian pathogens [17,18]. They have been widely applied for the production of numerous proteins and are designed to culture and harvest cells in consecutive cycles while continuously renewing nutrient and oxygen within the chamber [19,20]. This review focuses on recombinant proteins that have been successfully produced in rice and summarizes different approaches in improving the yield and quality of the proteins.

RECOMBINANT PHARMACEUTICALS IN RICE

Foreign genes of interest may be incorporated into plant systems through transient transformation such as agroinfiltration, virus infection, and magnifection [21–23]. These methods allow plants to temporarily express recombinant protein within relatively short amount of time compared to stable, long term transformation, although none of these techniques have been widely applied in rice. Alternatively, long term transformation is achieved by inserting the transgene into the plant's genome through methods such as Agrobacterium transfection and particle bombardment [11,24]. Callus cells expressing genes of interest may be grown as suspension cell cultures or regenerated into whole plants [13,16]. Whole plant system has advantages such as large scale production and stable storage means in the form of seeds [16]. In addition, direct oral administration of certain pharmaceutical is possible by using certain parts of the plant, although some limitations still exist [5]. On the other hand, suspension cell cultures also have comparative advantages such as more controlled environment, faster cell growth, and potential to be upscaled into bioreactors [17]. In addition, by utilizing a signal peptide, recombinant proteins produced will be secreted out of the cell into the culture medium, removing the need to extract the protein through cell lysis [25]. Rice has been successfully used to produce human serum albumin, vaccine, antibodies, cytokines, and other pharmaceutical proteins [26]. The characteristics, functions, and yields of protein successfully produced using rice expression systems are summarized (Table 1). Furthermore, the biological activity of the recombinant proteins is also examined. If applicable, the yield rate is compared with that of other plant expression systems.

Table 1: Pharmaceuticals produced in rice and yield comparison to other plants

Product	Plant	Promoter	Localization	Maximum Yield	Reference
Envelope protein of Japanese encephalitis virus (JEV)	Rice	*CaMV 35S*	Leaf	1.9 μg/mg *	[27]
FIP-*fve* fused with the antigen Der p 2 (OsDp2Fve)	Rice	*αAmy8*	Cell culture medium	7.5 mg/L	[28]
Human cytotoxic T-lymphocyte antigen 4-immuno globulin	Rice	*RAmy3D*	Cell culture medium	31.4 mg/L	[29]
Human granulo cyte macrophage-colony stimulating factor (hGM-CSF)	Rice	*Ramy3D*	Cell culture medium	129 mg/L	[30]
Human granulo cyte macrophage-colony stimulating factor (hGM-CSF)	Tobacco	*Gt1/Gt3*	Seed	0.03% TSP **	[31]
Human growth hormone (hGH)	Rice	*RAmy3D*	Cell culture medium	57 mg/L	[32]
Human serum albumin	Rice	*RAmy3D*	Cell culture medium	76.4 mg/L	[13]
Human serum albumin	Tobacco	*CaMV35S*	Cell culture medium	11.88 mg/L	[14]
Human serum albumin	Potato	*B33*	Tuber	0.01 μg/g *	[15]
Human serum albumin	Rice	*Gt13a*	Seed	2.75 g/kg *	[16]
Human α_1-antitrypsin (hAAT)	Rice	*Ramy3D*	Cell culture medium	247 mg/L	[33]
Interferon Gamma (INF-γ)	Rice	*Ubi-1*	Cell culture medium	12 ng/mL	[34]
Interferon Gamma (INF-γ)	Rice	*RAmy3D*	Cell culture medium	17.4 ng/mL	[34]
Interleukin-10 (IL-10)	Rice	*Glub1*	Seed	50 μg/g *	[35]
Interleukin-10 (IL-10)	Tobacco	*CaMV 35S*	Leaf	37.0 μg/g *	[36]
Mouse granulo cyte macrophage-colony stimulating factor (mGM-CSF)	Rice	*RAmy3D*	Cell culture medium	24.6 mg/L	[25]
Mouse granulo cyte macrophage-colony stimulating factor (mGM-CSF)	Tobacco	*RbcS1*	Leaf	19.9 μg/g *	[37]
Cry j I, Cry j II	Rice	*Glub1*	Seed	7 μg/rice grain *	[38]
Der p I	Rice	*Glub1*	Seed	7.5 μg/rice grain *	[39]
Single-chain Fv antibody (ScFvT84.66)	Rice	*Ubi-1*	Leaf	29 μg/g *	[40]
Single-chain Fv antibody (ScFvT84.66)	Rice	*Ubi-1*	Seed	3.8 μg/g *	[41]

Human Serum Albumin

Human serum albumin (HSA) is an unglycosylated, monomeric protein that plays important roles in osmotic regulation and pH buffering as well as acts as a carrier protein for fatty acids, thyroid hormones, and numerous drugs [42]. There are currently several studied platforms for producing this human protein such as Escherichia coli, Saccharomyces cerevisiae, Pichia pastoris, and numerous plant expression systems, such as tobacco cultures and potato tubers [14,15,43–45]. However, due to several limitations, none of these systems have offered a reliable way of providing biologically active HSA in large quantity. Consequently, limited human plasma remains the most important albumin source but is unable to meet increasing demand. Extensive transfusion exposes recipients to body fluid transferred disease such as Hepatitis B, Syphilis, and HIV [46].

One of the first publications involving producing HSA using rice suspension cell system using an inducible promoter αAmy3/RAmy3Dreported a maximum yield of 76.4 mg/L of medium after 4 days of sucrose starvation [13]. The expression system utilized a signal peptide and the HSA, along with other proteins, is secreted into the culture medium. There is more than a 6-fold yield advantage compared to HSA produced using tobacco suspension cell, which was 11.88 mg/L [14]. Alternatively, high yield of recombinant HSA has also been produced in rice endosperm. An endosperm-specific promoter, Gt13a, and its signal peptide were used to direct expressed protein into protein storage vacuoles, and a maximum yield of 2.75 g/kg of brown rice was obtained [16]. This is a significant yield improvement over production of HSA using potato tubers (0.01 g/kg of fresh weight) and well above the threshold for cost-effective industrial production, which was 0.1 g/kg of fresh weight [15]. The structure of HSA plays a critical role in its biological activity, and the most unique characteristics of the protein, the 17 disulfide bonds as well as the only free SH group Cys34, have been crystallized and extensively studied [47]. It has been shown that recombinant HSA (OsrHSA) produced using a rice expression is structurally equivalent to plasma HSA, and all 17 disulfide bonds have been identified on OsrHSA using X-Ray crystallography [16].

Vaccines

In addition to the advantages of protein production using plant expression systems described above, plant-based pharmaceuticals can also be applied through direct oral administration using specific parts of the plant, eliminating the needs for purification [5]. However, there are still challenges in oral delivery of vaccines produced in plant parts. Oral administration requires about 100 folds of vaccine delivered through direct injection [48]. Endotoxins that exist inside plants, such as solanine in potato tubers, also raise safety concerns, as they cannot be easily removed [49]. In cases where further processing, such as cooking, is required, thermolabile proteins will be degraded in the process, further reducing effective yield [49]. Despite the challenges described, rice remains a competitive candidate for producing and delivering recombinant vaccines for two important reasons. First of all, rice does not contain endotoxins harmful to the human body. Second, localization of recombinant protein into seeds provides ideal storage due to low protease activity [50]. Several vaccine antigens have been successfully produced in recombinant rice. Envelope protein of Japanese encephalitis virus (JEV) has been expressed in rice leaves, and the maximum yield was 1.1–1.9 μg/mg of total protein [27]. To examine the biological activity of the recombinant vaccine, mice were immunized using both E. coli derived JEV envelope proteins and recombinant proteins from rice through oral administration. The rice-derived vaccine was able to elicit higher immunoglobin G (IgG) and immunoglobin A (IgA) responses compared to the recombinant protein derived from E. coli [27]. Oral administration of Chlamydophila psittaci antigen, MOMP, fused to the B subunit of E. coli heat-labile enterotoxin (LTB) produced in rice seeds has also successfully induced mouse immunity to mucosal disease [51]. Other vaccines were successfully produced for the intention of oral administration as uncooked rice powder, and biologically activities were examined through animal tests [38,39]. Transgenic rice expressing mouse dominant T cell epitope peptides of Cry j I and Cry j II allergens of Japanese cedar pollen was able to prevent the development of allergen-specific immunoglobin E (IgE) and immunoglobin G (IgG) responses [38]. Alternatively, transgenic rice expressing a fragment (p45–145) of mite allergen (Der p 1) containing immunodominant human and mouse T cell epitopes successfully reduced the serum levels of allergen-specific IgE and IgG [39]. It is worth noting that differences in mammalian and plant glycosylation have caused immunogenic response in both mice and human, indicating that further modification may be required for recombinant protein produced using rice expression systems [52,53].

Antibodies

Antibodies are serum proteins that bind to target molecules with high specificity and are widely used for prevention, detection, and treatment of diseases. Recombinant antibodies are shown to provide immunization against pathogens and are potential answers to disease, especially with increasing microbial resistance towards antibiotics as well as new pathogens being discovered [54]. Currently, recombinant human cytotoxic T-lymphocyte antigen 4-immunoglobulin (hCTLA4Ig) has been successfully produced in rice suspension cells usingαAmy3/RAmy3D promoter with maximum yield of 31.4 mg/L in liquid medium [29]. Another antibody, single-chain Fv antibody (ScFvT84.66) under the control of maize Ubiquitin-1 promoter, has also been expressed in the leaves and calli of transgenic rice, and the yields were 29 μg/g and 3.8 μg/g of fresh weight of leaves and calli, respectively [40,41]. While antibody production in rice has been relatively rare, other antibodies have been successfully produced in other plants. The first recombinant proteins produced in plants were progeny of the cross of two individual transgenic plants, tobacco and sunflower, expressing single immunoglobulin gamma and kappa chains [3]. Antibodies produced in prokaryotic systems often form inclusion bodies, and harsh chemicals must be applied in order to refold the proteins back into their biologically active state [41]. On the other hand, antibodies produced in animal cells are more expensive to maintain and are susceptible to pathogen contamination [55]. In addition to being pathogen-free and capable of correct protein folding, rice also has yield advantages in several other proteins, as mentioned previously. Therefore, it is worth evaluating the possible comparative advantages of producing those antibodies using transgenic rice platform compared to other expression systems.

Cytokines

Cytokines are signaling proteins in intercellular communication and are involved in diverse regulation processes, such as embryogenesis, immune and hematopoietic systems [56]. Due to high production costs, pharmaceutical application of recombinant cytokines is still very limited. Some cytokines that have been successfully produced in rice culture are described below, and their biological activities are examined.

Granulocyte macrophage colony stimulating factor (GM-CSF) is a cytokine used to promote white blood cell proliferation [57]. The first report of human GM-CSF (hGM-CSF) production in rice was in 2003 through suspension cell cultures, and the maximum yield obtained was 129 mg/L [30].

Since then, improvements have been made in rice suspension cell systems producing hGM-CSF by using methods such as humanizing N-Glycan structure and increasing yield by 2–3 folds through reduction of endogenous α-amylase expression, co-expression of proteinase inhibitor, and suppression of cellular cysteine proteinase [8,58–61]. The protein is also produced in other expression systems such as rice seeds and tobacco seeds using rice glutelin promoters Gt1 and Gt3. The maximum yields were 1.3% of total soluble protein in rice seeds and 0.03% of total soluble protein in tobacco seeds [31,62]. Mouse GM-CSF (mGM-CSF) was successfully produced using rice suspension cells, with yield of 24.6 mg/L of medium [25]. The same glycoprotein was also produced in tobacco leaves, and the yield was 19 μg/g of fresh leaves [37]. The biological activities of recombinant human and mouse GM-CSF produced in rice suspension culture has also been confirmed using GM-CSF dependent TF1 cells and murine myeloblastic leukemia NFS-60 cells, respectively [8,25].

Interleukin-10 (IL-10) is a pleotropic cytokine that plays numerous roles in immunoregulation and inflammation [63]. It has been successfully produced in rice seeds using Glub-1 promoter and its signal peptide to localize the protein inside seeds exclusively. The product is determined to be unglycosylated and yield of final purified protein was 2 mg per 40 g of rice used (50 μg/g). The biological activity of recombinant IL-10 was confirmed using mouse bone marrow dendritic cells [35]. The protein was also produced inside tobacco leaves, and the yield was 37.0 μg/g of fresh leaves [36].

INF-γ is a class II interferon responsible for regulating immune response against bacteria and tumor [64]. Production of INF-γ has been performed in rice suspension cells using both constitutive maize ubiquitin promoter and inducible rice αAmy3/RAmy3D promoter, and the biological activity has been confirmed using human A549 cell line against dengue virus. An α-amylase signal peptide was added to both in order to allow for secretion of recombinant INF-γ into the culture medium. Highest yield obtained from the culturing medium of the ubiquitin promoter driven system was 12 ng/mL and 17.4 ng/mL inαAmy3/RAmy3D promoter driven system, and yield found inside the cell was 699.79 ng/g of fresh cell weight for the ubiquitin promoter driven system and 131.6 ng/g of fresh cell weight for theαAmy3/RAmy3D promoter driven system [34].

Other Proteins

Human α_1-antitrypsin (hAAT), a serine trypsin inhibitor, is the most abundant protease inhibitor in human plasma [65]. Its deficiency is known to lead to chronic obstructive pulmonary disease, also known as emphysema [66,67]. Production of human α_1-antitrypsin in recombinant rice suspension cells was first achieved in 2000, with yield of 120 mg/L, cultured in 30 mL of medium with 10% (v/v) of cell density [68]. A 350 mL membrane bioreactor system was used to produce human α_1-antitrypsin, and maximum yield of 247 mg/L was obtained [33]. A larger 5 liter bioreactor was incorporated to producing hAAT, and the maximum yield was 7.3 mg/L [69]. The half-life of recombinant hAAT is shorter than that of plasma hAAT, possibly due to the absence of sialic acid, implying that proper post translational modification will increase the stability of recombinant protein [70].

Human growth hormone (hGH) is responsible for reproduction, growth, and regeneration of certain cells such as muscle and bone cells [71]. It is often prescribed to patients with growth deficiency and is essential for metabolic regulation. Recombinant hGH has been successfully produced in rice suspension cells using αAmy3/RAmy3D as a promoter. The maximum yield is 57 mg/L in a culture medium, and the biological activity of the recombinant protein has been confirmed using Nb2 node lymphoma cells, whose proliferation and growth depend on the presence of hGH [32].

A fungal immunemodulatory protein (FIP-fve) was isolated from golden needle mushroom (Flarnmulina velutipes) and reported to inhibit allergy reactions in mice and regulate Th2 cytokines [72,73]. In addition, it acts as a strong adjuvant that enhanced the immunogenicity of the human papillomavirus type-16 (HPV-16) E7 oncoprotein and conferred increased antitumor effects against the inoculation of HPV-16.E7-expressing TC-1 cells in mice. Recombinant FIP-fve fused with the antigen Der p 2 (OsDp2Fve) has been produced using rice suspension cells under a control of α-amylase (αAmy8) promoter, and the yield was about 7.5 μg/mL (10.5% of total protein) [28].

IMPROVEMENT OF RICE EXPRESSION SYSTEM

Despite its many advantages, production of recombinant proteins using rice expression systems is still limited by yield, ranging from 12 μg/L to 247 mg/L (Table 1). Several strategies targeting various stages of protein expression have been incorporated to improve the yield. In addition, the significance of presence of and difference in glycosylation are discussed, and different modifications, which have been experimented with, are summarized.

Increase Protein Yields

Physical and chemical approaches have been used to improve protein yield. Currently, rice suspension cells are extensively utilized due to their fast cell doubling time (as fast as 1 day) [7,69]. On the other hand, whole plant systems have been very established, and certain organelles, such as seeds and tubers, act as stable storage locations. Thus, selection of expression system should be based on the protein's properties, the expression cassette, and purification concerns. In both systems, further yield improvements can be made by modifying promoters of the expression cassettes, the codons of gene responsible for translating the protein, and modifying genes within the host cells.

Promoters

Different promoters have been chosen based on the properties of the protein produced. In terms of constitutive promoters, the 35S promoter obtained from cauliflower mosaic virus and ubiquitin promoter obtained from maize are most frequently used. The main advantage of constitutive promoters is that they continuously express the gene in all parts of the plant. However, it was demonstrated that spontaneous transgene-silencing phenomena are caused by methylation of the CaMV 35S promoter during transgenic plant growth [74,75]. In addition, there is an increased chance of homology dependent gene silencing (HDGS) occurring. Despite stable integration of 35S:GUSgene into rice genome, β-glucuronidase (GUS) expression was not found in a transgenic rice line due to HDGS [76]. Alternatively, a number of inducible promoters that only activate during specific conditions or tissue-specific promoters have been chosen in the production of numerous recombinant proteins. While gene HDGS has not been identified in inducible systems, inducible systems are also newer than constitutive systems and not as extensively studied. Therefore, the possibility of gene silencing in inducible systems should be explored. Several induced promoters and their characteristics are discussed below.

Recombinant protein systems using steroid-regulated promoters require only trace amount of steroid to activate the expression system. Expression of green florescent protein (GFP) inside rice calli using estrogen receptor fused chimeric transcription factor, XVE, was studied and compared with 35S:GFP expression system [77]. GFP expression levels were similar between the constitutive and inducible systems in the roots, while expression level of the inducible system was lower than the constitutive system in leaves. By using different estradiol concentrations (from 1 to 25 μM), GFP expression level peaked at different times and the green fluorescence emitted longer for cells grown on plates with higher estradiol concentration. GFP expression level remains observable for at least one day after calli were moved to an estradiol-free medium [77]. The relationship between the concentration of estradiol present in the medium and the number of days the bioreactor is operated should be explored, as estradiol degradation will lead to a decrease in protein production. In addition, degradation rate of mRNA of steroid induced system should also be examined in order to determine the ideal time for medium exchange or steroid addition.

Ethanol-induced system expressing distinct reporters, chloramphenicol acetyltransferase (CAT) and GUS, in tomatoes was examined [78]. Garoosi et al. reported that, under constant presence of 0.1% (v/v) ethanol, both CAT and GUS activities were detectable within 4 h [78]. GUS expression level of an alcohol-induced system is comparable to that of 35S constitutive promoter. Ethanol should also be maintained under lower concentration, as volume percentage higher than 1% proved to be toxic for plant tissues [78]. In the case of whole-plant production, consistent alcohol concentration may be hard to control due to its volatility. Finally, continuous drenching of root, even in lower concentration ethanol, may have detrimental effects on plant development.

Metabolite-regulated promoters are triggered in the presence or absence of certain metabolic products. αAmy3/RAmy3D is one of the most widely used metabolite-regulated promoters and is highly expressed during sugar starvation [30]. In a suspension cell system expressing recombinant HSA using αAmy3/RAmy3D promoter, the promoter is activated within 12 h of sugar starvation. Cell viability drops significantly to about 30% within 24 h of sugar starvation. HSA secretion reaches the maximum after four days of sugar starvation, during which cell viability drops nearly to zero. However, once sugar is added into the culture again, cell viability is able to make a full recovery at least for seven cycles, implying that sugar-starvation driven system does not have permanent negative impacts on the cells [13]. Since expression of recombinant protein only takes place under sugar starvation, culture medium exchange must take place in order to maintain cell viability. Lastly, a sugar starvation driven promoter should be sensitive to the absence of sugar so that recombinant proteins can still be produced while cells still have spare resource to produce proteins not essential for its survival.

Production of GUS using the heat-induced small heat shock protein 18.2 (sHSP 18.2) promoter in liquid tobacco root cultures has been performed [79]. Expression level was higher in younger cells and GUS expression level reached maxima when the heat shock treatment is performed at 42 °C for 2 h followed by 27 °C recovery for 24 h. A similar experiment is also performed in transgenic wheat and shows expression throughout the entire plant exposed to heat shock treatment, although the expression level is lower in proximal roots than in root tips [80]. Constant cycles of heat shocks and recovery may cause irreversible effects on cell metabolism, which may lower cell viability. Furthermore, the property of recombinant protein under constant temperature fluctuation should also be studied before using a temperature-regulated expression system. Finally, additional energy is required to perform heating, which should also be taken into consideration during commercialization.

Rice Optimized Codon Usage

The preferred codons vary significantly between different plants, so when the desired foreign gene contains rare codons the host plant, the codon becomes a limiting factor in protein translation [81,82]. This is a major drawback in the goal of mass production of recombinant proteins in plants. To overcome this limitation, a plausible strategy involves increasing translation efficiency through modifications of codons from the original DNA sequence to more suitable ones in the host plant without changing the amino acid sequence [83]. In plant expression systems, expression of Bacillus thuringiensis cryIA(b) gene, primarily used to genetically increase tomatoes resistance towards insects, was increased 100 folds by making 21% changes to the original cryIA(b) DNA sequence [84]. Production of miraculin, a taste-altering glycoprotein, in transgenic tomato is also improved by codon optimization, and yield improved from 196.5 to 286.9 μg/g of fresh weight [81]. Codon modification was also performed in rice to express phytase of Saccyromyces cerevisae, and phytase activity in rice leaves increased from 0.039 U/g of fresh leaves of transgenic rice using unchanged codon to 4.6 U/g using full-length modified codons and 10.6 U/g using truncated codons (U is defined as 1 μmol phosphate released per min at 37 °C) [85]. By combining codon optimization with the usage of KDEL sequence, expression of adioponectin increased from 0.06% to 0.32% of total soluble protein in rice leaves [86].

Endogenous Gene Modification

Modification within the host's genes is also made in order to increase protein productivity through either reduction of protease expression or alleviation of internal resource competition and feedback regulation. The first strategy used involves knocking down genes expressing endogenous secreted protease such as cysteine proteinase [60]. Results here show up to 1.9 fold of human GM-CSF (hGM-CSF) yield improvement, from 150.4 to 289.1 mg/L. In another case studied, expression of endogenous α-amylase has been knocked down through RNAi [59]. Results indicate that the production of hGM-CSF usingαAmy3/RAmy3D promoter has almost doubled from approximately 150 to 280 mg/L. Alternatively, studies have also been performed inNicotiana benthamiana in suppressing transcriptional silencing that is responsible for decreased expression level of recombinant protein. Yield of murine IgG was 10.8 mg per gram of leaf fresh weight for normal transgenic line while yield increased 14-fold to 147.7 mg/g of fresh weight of leaves for transgenic line also containing p19, a virus suppressor found in the tomato bushy stunt virus that is able to interfere with plant's endogenous gene silencing [87,88].

Humanized Glycosylation

One of the biggest challenges for plant expression system remains the differences that exist between mammalian glycosylation and plant glycosylation. In addition to affecting the biological activity of glycoproteins, there are two additional concerns involved with difference in glycosylation. First, human body might recognize plant-specific glycosylation and trigger allergic reactions [89,90]. However, lack of proper glycosylation from normally glycosylated proteins may also induce immune response, as recombinant material produced in E. coli that lacks glycosylation showed increased immunogenicity in patients [91]. Furthermore, glycosylation is also known to improve the stability of plasma glycoprotein such as CC chemokine ligand 2 inside the blood through proper folding and increased conformation stability [92,93].

There are several approaches in modifying the post translational glycosylation system and increasing the similarity between recombinant protein produced in plants and their native forms. One way is by knocking down the expression levels of enzymes, α-1,3-fucosyltransferase and β-1,2-xylosyltransferase, responsible for adding plant specific glycans, fucose and xylose respectively. Current study indicates that knockdown of α-1,3-fucosyltransferase and β-1,2-xylosyltransferase in rice does not have negative effects on the functions of the transgenic suspension cell culture in areas such as cell division, proliferation, and ability to secret proteins out of the cell, while using sucrose as the carbon source [58]. However, it has not been determined whether using alternative carbon sources, such as starch, will affect the growth of the transgenic rice cell line. Rice α-amylase, the major enzyme responsible for breaking down starch, is a glycoprotein [94]. Therefore, its biological activity may be tied to its glycosylation. Thus, it is worth investigating whether incomplete glycosylation caused by RNA interference will affect the biological activity of α-amylase in rice. The physiology of the plants should also be studied and compared to wild type plants. For example, deficiency or absence of α-mannosidase I, located in both the ER and Golgi, has negative impacts such as the formation of aberrant N-glycans, short and swollen roots, and malformation and function loss of cell walls [95]. Knockdown of two xylosyltransferases has been performed inArabidopsis thaliana, and the transgenic plant possessed aberrant root hairs and lacked detectable xylogulcan [96]. These evidences indicate that knockdown or knockout of the gene involved in N-glycosylation may lead to abnormal cell growth and compromise protein production. Consequently, it is suggested that RNA interference should be performed using an inducible system in order to preserve normal plant physiology and express glycosylation change only in proper environment.

Due the importance of high-mannose glycosylation taking place in the ER, another method involves adding the sequence KDEL (Lysine-Aspartic acid-Glutamic acid-Leucine) to the sequence of the recombinant. KDEL is an amino acid sequence capable of performing ER-retention, which prevents the protein from leaving the endoplasmic reticulum (ER) into the Golgi apparatus. Addition of KDEL tag to recombinant protein is rather significant because both plant-specific xylosyltransferase and fucosyltransferase are located inside the Golgi apparatus, and plant-exclusive glycosylation of the recombinant protein would be impossible to perform if the protein is retained inside the ER. Using this method, recombinant human interleukin 10 has been successfully produced from transgenic rice seeds, and the proteins were shown to be unglycosylated and identical to its natural human counterpart, confirming that the addition of the KDEL sequence does not affect the biological activity of the amino acid [35]. In Nicotiana tabacum, cPIPP, a mouse/human chimeric antibody specific for the β-subunit of human chorionic gonadotropin, was produced with a KDEL tag [97]. The glycoprotein localized inside the ER does not contain any of its natural plant-specific β-1,2-xylose and α-1,3-fucose glycans [97]. The stability of recombinant protein retained inside the ER in vivo also does not appear to differ from recombinant protein produced without a KDEL tag, further indicating removal of plant-specific β-1,2-xylose and α-1,3-fucose can be performed to reduce risk of immunogenic responses without compromising protein stability [98]. The biggest disadvantage of adding an ER-retaining tag to the recombinant protein is that cell lysis will be necessary to recover the recombinant protein, as KDEL also prevents its tagged protein from migrating elsewhere, making protein secretion an ineffective strategy. Consequently, additional protein purification steps will be required to isolate protein from other cell content, especially protease from other organelles that will compromise the stability of the recombinant protein.

Genome Stabilization during Rice Cell Culture

While a liquid suspension cell culture significantly improves cell growth rate and allows for recombinant protein secretion through the application of signal peptides, there are still potential concerns involved with the genome integrity of the transgenic plants. The activity level of retrotransposon inside undifferentiated rice suspension cultures is relatively high compared to differentiated cells, partially due to the high speed of replication. Retrotransposons are Class I transposable elements (TE) that are capable of performing reverse transcription using self-encoded reverse transcriptase, and the RNA fragments created will be integrated back into the genome [99,100]. Consequently, the increased number and changing locations of the retrotransposon could have a potential impact on the genomic stability of the host cells, which is a concern in maintaining cell lines expressing recombinant proteins. Thus, the changing level of transposons within cell lines over the course of time is an important indicator. It was observed that in rice cultures containing Tos17 activity, the expression level of the retrotransposon remained high and increased over the course of one year [101].

To preserve genome integrity and limit activity of retrotranspon, it may be advisable to keep unused cells under cryopreservation [102]. Arabidopsis cell lines have been cryogenically frozen at −30 °C or −80 °C and then thawed out, and the cell activity remained similar to unfrozen cells [103]. A similar experiment was also performed in transgenic rice expressing recombinant hCTLA4Ig and cells recovered from cryopreservation banking had cell viability of 88% [102]. These experiments indicate that cryopreservation banking is a useful tool in conserving transgenic rice cell lines under controlled low temperature, and better temperature parameter for both freezing and thawing should be experimented for optimal rice cell viability recovery. It may also be appropriate to store transgenic cell lines capable of producing recombinant proteins in the form of rice grains, as cell division stops at this period. Seeds can be grown into whole plants or induced into cell cultures using auxin such as 2,4-Dichlorophenoxyacetic acid if they are needed for protein production in the future.

LARGE SCALE PRODUCTION

The methods proposed above are used in rice or other plant expression systems to improve average yield per cell. In order to further increase recombinant protein production, plant suspension cell cultures should be upscaled into bioreactors. Large-scale bioreactor producing plant secondary metabolite has already been well established [104]. By using a bioreactor, it is possible for plant production of recombinant protein to increase to a level similar to mammalian expression systems [7]. In 2010, Protalix has developed a high-volume bioreactor (up to 75,000 liters) using carrot as a host organism to produce taliglucerase alfa (http://www.protalix.com/index.asp), a plant-derived enzyme used to treat patients with Gaucher's Disease [105]. The pharmaceutical became the first plant pharmaceutical to be approved by the FDA on 1 May, 2012. Efforts are also made in disposable bioreactors capable of meeting GMP standards [106]. However, there is very limited number of bioreactor production of pharmaceutical using rice cells. While overall protein production increases, average protein production per cell actually may decrease in bioreactors, possibly due to compromised environment and increased endotoxic materials, secondary metabolite, and feedback regulation. For example, production of human α_1-antitrypsin reached 247 mg/L in a 300 mL membrane bioreactor, but dropped down to 7.3 mg/L when upscaled to a 5 liter bioreactor [33,69]. Furthermore, the biggest challenges in bioreactor upscaling are the rheological and hydrodynamic shearing [107], and plant cells are sensitive towards shearing due to inflexible cellulose-based cell wall [7]. Cell aggregate is advantageous in some situations, as increase in sedimentation rate allows media exchange and culture broth recovery become easier to perform. However, increase in cell aggregate makes cells more susceptible to shearing and can cause uneven distribution of oxygen and other nutrients within larger aggregates [108]. A few examples of rice bioreactors are described, and potential new options are proposed below.

Fed-batch reactors involve the addition of nutrient and cells at the start of cultivation or at certain points during the reaction process. The main advantage of the system is its relatively easy implementation. However, the system suffers major drawbacks such as depletion of major nutrients and accumulation of potentially inhibitory metabolites. Furthermore, as the concentration of product increases inside the medium, feedback regulation may take place to limit further production of the recombinant protein. Cell growth also slows down at higher cell density. Cleanup of the system between usages is also very laborious and time consuming. Regardless, batch-fed bioreactors have been successfully utilized to produce hCTLA4Ig and the maximum yield obtained was 76.5 mg/L. A sugar-starvation driven αAmy3/RAmy3D promoter was used here and protein yield was 1.2 times higher than when a constitutive promoter was used, indicating that conditions can be changed to optimize protein production [109]. Perfusion and continuous cultures may also be used to refresh nutrient and remove product as well as metabolites from the bioreactor, and excessive cells can also be removed in order to maintain an ideal growth condition similar to when the bioreactor was first applied.

Current application of rice bioreactor producing recombinant protein with the highest yield utilized a membrane bioreactor [33]. A membrane with 50 kDa molecular weight cut off (MWCO) was suspended inside the bioreactor and used to exchange nutrient and waste liquid. Another gas exchange membrane is located near the bottom of the bioreactor and is used to remove CO_2 and introduce O_2 into the system. The yield of recombinant human alpha-1-antitrypsin in this system was 247 mg/L. The biggest limitation of a membrane based system involves the cleanup of membranes as cell aggregates begin to adsorb on the membrane, as the system usually needs to be paused in order for the cell mass to be removed. While continuous scraping is a viable way of removing debris from the membrane in bioreactors used to treat waste water, care must be taken when applying the same system to plant recombinant protein bioreactors. Disposable membranes are recommended in this case.

In addition to protein yield decrease due to depleting nutrients and accumulating products that may contribute to feedback regulation, long-term bioreactors also have several other challenges, including the time, labor, and cost involved in cleaning the bioreactor containers. Furthermore, loss due to contamination is much more significant when small numbers of large volume bioreactors are used compared to larger numbers of smaller bioreactors. Thus, disposable, smaller volume, bioreactors using plastic containers have become increasingly more popular in animal suspension cells [110]. In addition, there have been studies indicating that protein loss involved in suspension cell cultures may be associated with adsorptive properties of glassware, indicating that material used as bioreactor tank may also play important roles in maximizing protein yield [111]. An air-lift bioreactor using a 5-liter plastic container has been utilized to produce mouse GM-CSF in rice cells [25]. While large volume bioreactors may have advantages in economies of scale, it may be less risky and practical to incorporate smaller bioreactors in large numbers to produce certain proteins.

IMPROVEMENTS IN DOWNSTREAM PROCESSING

Purification cost of plant-derived pharmaceuticals is estimated to account for 80% of total cost [112]. Therefore, reducing the number of purification steps required will greatly reduce the cost associated with protein production. In most of the rice suspension cells system described above, a signal peptide is usually used to guide the protein into the growth medium, eliminating the need for cell lysis. A new idea using the application of oleosin in protein separation has been used in seeds of plants with high oil body contents. Expression system of target protein fused with oleosin has been performed in safflower to produce Apolipoprotein AI (Milano) [113]. Human insulin-like growth factor 1 and insulin have also been produced using oleosin fusion technology inside Arabidopsis [114,115]. While rice is not rich in oil content, it is worth testing whether oleosin fusion protein can be localized in oil-rich section of rice cell or if the oleosin fusion protein can be secreted into the culture medium. Alternatively, if oleosin fusion proteins can be properly secreted out of the cell into the medium using signal peptide to targeting out of the cell into medium, this system may still be useful in purifying recombinant proteins produced in rice.

CONCLUSION AND FUTURE REMARKS

Development of plant-based recombinant protein expression system has greatly improved protein yield in plants. The incorporation of both transient and stable transformation techniques has allowed for either fast recombinant protein expression or the establishment of transgenic cell cultures [5,7]. Whole plant systems are safe, competitive candidates for producing pharmaceuticals and have niche roles in oral administration, although some challenges exist [8,45]. Alternatively, suspension cells can be grown in isolated environments with controlled medium and hormone supply, and cell division time is shortened to 1–1.5 days [69]. Further upscaling to bioreactor allows for even higher protein production. In addition to trying to further improve protein yield and production cost effectiveness, focus should be also placed on humanizing the proteins through glycosylation modification and further improving protein stability and purification processes. Furthermore, as evident by rice's excellent production yield mentioned earlier, it will also be worthwhile to develop and optimize bioreactors for rice. It is projected that four of five top-selling drugs will be therapeutic proteins by 2013 [116], and through further improvement of expression systems and genome stabilization, coupled with existing advantages such as high yield, lack of human pathogens, and complex existing glycosylation systems, rice cells will be an ideal host for producing protein therapeutics.

ACKNOWLEDGMENTS

This work was supported by a grant from the National Science Council of the Republic of China (NSC 99-2313-B-155-001-MY3 to Li-Fen Huang). We thank Shaw S. Wang for his suggestion in this review article.

REFERENCES

1. Bevan M.W., Flavell R.B., Chilton M.D. A chimaeric antibiotic resistance gene as a selectable marker for plant cell transformation. 1983. Biotechnology. 1992;24:367–370.
2. Barta A., Sommergruber K., Thompson D., Hartmuth K., Matzke M.A., Matzke A.J.M. The expression of a napoline synthase human growth hormone chimeric gene in transformed tobacco and sunflower callus tissue. Plant Mol. Biol. 1986;6:347–357.
3. Hiatt A., Cafferkey R., Bowdish K. Production of antibodies in transgenic plants. Nature. 1989;342:76–78.
4. Hood E.E., Witcher D.R., Maddock S., Meyer T., Baszczynski C., Bailey M., Flynn P., Register J., Marshall L., Bon D., et al. Commercial production of avidin from transgenic maize: Characterization of transformant, production, processing, extraction and purification. Mol. Breed. 1997;3:291–306.
5. Obembe O.O., Popoola J.O., Leelavathi S., Reddy S.V. Advances in plant molecular farming. Biotechnol. Adv. 2011;29:210–222.
6. Gomord V., Fitchette A.C., Menu-Bouaouiche L., Saint-Jore-Dupas C., Plasson C., Michaud D., Faye L. Plant-specific glycosylation patterns in the context of therapeutic protein production. Plant Biotechnol. J. 2010;8:564–587.
7. Xu J., Ge X., Dolan M.C. Towards high-yield production of pharmaceutical proteins with plant cell suspension cultures. Biotechnol. Adv. 2011;29:278–299.
8. Kim H.J., Lee D.H., Kim D.K., Han G.B. The glycosylation and in vivo stability of human granulocyte-macrophage colony-stimulating factor produced in rice cells. Biol. Pharm. Bull. 2008;31:290–294.
9. Lienard D., Sourrouille C., Gomord V., Faye L. Pharming and transgenic plants. Biotechnol. Annu. Rev. 2007;13:115–147.
10. IRGSP. The map-based sequence of the rice genome. Nature.2005;436:793–800.
11. Nishimura A., Aichi I., Matsuoka M. A protocol for Agrobacterium-mediated transformation in rice. Nat. Protoc.2006;1:2796–2802.
12. Takaiwa F. Allergen-specific immunotherapy with plant-based oral vaccines. Immunotherapy. 2009;1:517–519.
13. Huang L.F., Liu Y.K., Lu C.A., Hsieh S.L., Yu S.M. Production of human serum albumin by sugar starvation induced promoter and rice cell culture. Transgenic Res. 2005;14:569–581.
14. Sun Q.Y., Ding L.W., Lomonossoff G.P., Sun Y.B., Luo M., Li C.Q., Jiang L., Xu Z.F. Improved expression and purification of recombinant human serum albumin from transgenic tobacco suspension culture. J. Biotechnol. 2011;155:164–172.
15. Farran I., Sanchez-Serrano J.J., Medina J.F., Prieto J., Mingo-Castel A.M. Targeted expression of human serum albumin to potato tubers.Transgenic Res. 2002;11:337–346.
16. He Y., Ning T., Xie T., Qiu Q., Zhang L., Sun Y., Jiang D., Fu K., Yin F., Zhang W., et al. Large-scale production of functional human serum albumin from transgenic rice seeds. Proc. Natl. Acad. Sci. USA.2011;108:19078–19083.
17. Huang T.K., McDonald K.A. Bioreactor systems for in vitroproduction of foreign proteins using plant cell cultures. Biotechnol. Adv. 2012;30:398–409.
18. Kim D.I., Cho G.H., Pedersen H., Chin C.K. A hybrid bioreactor for high density cultivation of plant cell suspensions. Appl. Microbiol. Biotechnol. 1991;34:726–729.
19. Sivakumar G., Medina-Bolivar F., Lay J.O., Jr, Dolan M.C., Condori J., Grubbs S.K., Wright S.M., Baque

M.A., Lee E.J., Paek K.Y. Bioprocess and bioreactor: Next generation technology for production of potential plant-based antidiabetic and antioxidant molecules. Curr. Med. Chem. 2011;18:79–90.

20. Su W.W., Arias R. Continuous plant cell perfusion culture: Bioreactor characterization and secreted enzyme production. J. Biosci. Bioeng. 2003;95:13–20.

21. Kapila J., de Rycke R., van Montagu M., Angenon G. Agrobacterium-mediated transient gene expression system for intact leaves. Plant Sci. 1997;122:101–108.

22. Porta C., Lomonossoff G.P. Viruses as vectors for the expression of foreign sequences in plants. Biotechnol. Genet. Eng. Rev. 2002;19:245–291.

23. Gleba Y., Klimyuk V., Marillonnet S. Magnifection—A new platform for expressing recombinant vaccines in plants. Vaccine.2005;23:2042–2048.

24. Zhao Y., Qian Q., Wang H., Huang D. Hereditary behavior of bar gene cassette is complex in rice mediated by particle bombardment. J. Genet. Genomics. 2007;34:824–835.

25. Liu Y.K., Huang L.F., Ho S.L., Liao C.Y., Liu H.Y., Lai Y.H., Yu S.M., Lu C.A. Production of mouse granulocyte-macrophage colony-stimulating factor by gateway technology and transgenic rice cell culture. Biotechnol. Bioeng. 2012;109:1239–1247.

26. Greenham T., Altosaar I. Molecular strategies to engineer transgenic rice seed compartments for large-scale production of plant-made pharmaceuticals. Methods Mol. Biol. 2013;956:311–326.

27. Wang Y., Deng H., Zhang X., Xiao H., Jiang Y., Song Y., Fang L., Xiao S., Zhen Y., Chen H. Generation and immunogenicity of Japanese encephalitis virus envelope protein expressed in transgenic rice.Biochem. Biophys. Res. Commun. 2009;380:292–297.

28. Su C.F., Kuo I.C., Chen P.W., Huang C.H., Seow S.V., Chua K.Y., Yu S.M. Characterization of an immunomodulatory Der p 2-FIP-fvefusion protein produced in transformed rice suspension cell culture. Transgenic Res. 2012;21:177–192.

29. Lee S.J., Park C.I., Park M.Y., Jung H.S., Ryu W.S., Lim S.M., Tan H.K., Kwon T.H., Yang M.S., Kim D.I. Production and characterization of human CTLA4Ig expressed in transgenic rice cell suspension cultures. Protein Expr. Purif. 2007;51:293–302.

30. Shin Y.J., Hong S.Y., Kwon T.H., Jang Y.S., Yang M.S. High level of expression of recombinant human granulocyte-macrophage colony stimulating factor in transgenic rice cell suspension culture. Biotechnol. Bioeng. 2003;82:778–783.

31. Sardana R.K., Alli Z., Dudani A., Tackaberry E., Panahi M., Narayanan M., Ganz P., Altosaar I. Biological activity of human granulocyte-macrophage colony stimulating factor is maintained in a fusion with seed glutelin peptide. Transgenic Res. 2002;11:521–531.

32. Kim T.G., Baek M.Y., Lee E.K., Kwon T.H., Yang M.S. Expression of human growth hormone in transgenic rice cell suspension culture.Plant Cell. Rep. 2008;27:885–891.

33. McDonald K.A., Hong L.M., Trombly D.M., Xie Q., Jackman A.P. Production of human alpha-1-antitrypsin from transgenic rice cell culture in a membrane bioreactor. Biotechnol. Prog. 2005;21:728–734.

34. Chen T.L., Lin Y.L., Lee Y.L., Yang N.S., Chan M.T. Expression of bioactive human interferon-gamma in transgenic rice cell suspension cultures. Transgenic Res. 2004;13:499–510.

35. Fujiwara Y., Aiki Y., Yang L., Takaiwa F., Kosaka A., Tsuji N.M., Shiraki K., Sekikawa K. Extraction and purification of human interleukin-10 from transgenic rice seeds. Protein Expr. Purif.2010;72:125–130.

36. Bortesi L., Rossato M., Schuster F., Raven N., Stadlmann J., Avesani L., Falorni A., Bazzoni F., Bock R., Schillberg S., et al. Viral and murine interleukin-10 are correctly processed and retain their biological activity when produced in tobacco. BMC Biotechnol.2009;9:22.

37. Gora-Sochacka A., Redkiewicz P., Napiorkowska B., Gaganidze D., Brodzik R., Sirko A. Recombinant mouse granulocyte-macrophage colony-stimulating factor is glycosylated in transgenic tobacco and maintains its biological activity. J. Interferon Cytokine Res.2010;30:135–142.

38. Takagi H., Hiroi T., Yang L., Tada Y., Yuki Y., Takamura K., Ishimitsu R., Kawauchi H., Kiyono H., Takaiwa F. A rice-based edible vaccine expressing multiple T cell epitopes induces oral tolerance for inhibition of Th2-mediated IgE responses. Proc. Natl. Acad. Sci. USA.2005;102:17525–17530.

39. Suzuki K., Kaminuma O., Yang L., Takai T., Mori A., Umezu-Goto M., Ohtomo T., Ohmachi Y., Noda Y., Hirose S., et al. Prevention of allergic asthma by vaccination with transgenic rice seed expressing mite allergen: Induction of allergen-specific oral tolerance without bystander suppression. Plant Biotechnol. J. 2011;9:982–990.

40. Stoger E., Vaquero C., Torres E., Sack M., Nicholson L., Drossard J., Williams S., Keen D., Perrin Y., Christou P., et al. Cereal crops as viable production and storage systems for pharmaceutical ScFv antibodies. Plant. Mol. Biol. 2000;42:583–590.

41. Torres E., Vaquero C., Nicholson L., Sack M., Stoger E., Drossard J., Christou P., Fischer R., Perrin Y. Rice cell culture as an alternative production system for functional diagnostic and therapeutic antibodies.Transgenic Res. 1999;8:441–449.

42. Arroyo V. Human serum albumin: Not just a plasma volume expander. Hepatology. 2009;50:355–357.

43. Lawn R.M., Adelman J., Bock S.C., Franke A.E., Houck C.M., Najarian R.C., Seeburg P.H., Wion K.L. The sequence of human serum albumin cdna and its expression in E. coli. Nucleic Acids Res.1981;9:6103–6114.

44. Quirk A.V., Geisow M.J., Woodrow J.R., Burton S.J., Wood P.C., Sutton A.D., Johnson R.A., Dodsworth N. Production of recombinant human serum albumin from Saccharomyces cerevisiae. Biotechnol. Appl. Biochem. 1989;11:273–287.

45. Kobayashi K., Kuwae S., Ohya T., Ohda T., Ohyama M., Ohi H., Tomomitsu K., Ohmura T. High-level expression of recombinant human serum albumin from the methylotrophic yeast Pichia pastoriswith minimal protease production and activation. J. Biosci. Bioeng.2000;89:55–61.

46. Zhang W., Hu D., Xi Y., Zhang M., Duan G. Spread of HIV in one village in central china with a high prevalence rate of blood-borne aids.Int. J. Infect. Dis. 2006;10:475–480.

47. Sugio S., Kashima A., Mochizuki S., Noda M., Kobayashi K. Crystal structure of human serum albumin at 2.5 a resolution. Protein Eng. 1999;12:439–446.

48. Streatfield S.J., Howard J.A. Plant-based vaccines. Int. J. Parasitol.2003;33:479–493.

49. Sparrow P.A., Irwin J.A., Dale P.J., Twyman R.M., Ma J.K. Pharma-planta: Road testing the developing regulatory guidelines for plant-made pharmaceuticals. Transgenic Res. 2007;16:147–161.

50. Horn M.E., Woodard S.L., Howard J.A. Plant molecular farming: Systems and products. Plant Cell Rep. 2004;22:711–720.

51. Zhang X., Yuan Z., Duan Q., Zhu H., Yu H., Wang Q. Mucosal immunity in mice induced by orally administered transgenic rice.Vaccine. 2009;27:1596–1600.

52. Franconi R., Massa S., Illiano E., Mullar A., Cirilli A., Accardi L., Di Bonito P., Giorgi C., Venuti A. Exploiting the plant secretory pathway to improve the anticancer activity of a plant-derived HPV16 E7 vaccine. Int. J. Immunopathol. Pharmacol. 2006;19:187–197.

53. McCormick A.A., Reddy S., Reinl S.J., Cameron T.I., Czerwinkski D.K., Vojdani F., Hanley K.M., Garger S.J., White E.L., Novak J., et al. Plant-produced idiotype vaccines for the treatment of non-Hodgkin's lymphoma: Safety and immunogenicity in a phase I clinical study. Proc. Natl. Acad. Sci. USA. 2008;105:10131–10136.

54. Berry J.D., Gaudet R.G. Antibodies in infectious diseases: Polyclonals, monoclonals and niche biotechnology. New Biotechnol.2011;28:489–501.

55. Yano A., Takekoshi M., Morita E., Imai S., Nisizawa T., Hanada N. Production of Fab fragment corresponding to surface protein antigen ofStreptococcus mutans serotype c-derived peptide by Escherichia coliand cultured tobacco cells. J. Biosci. Bioeng. 2006;101:251–256.

56. Sirko A., Vanek T., Gora-Sochacka A., Redkiewicz P. Recombinant cytokines from plants. Int. J. Mol. Sci. 2011;12:3536–3552.

57. Metcalf D. The granulocyte-macrophage colony stimulating factors.Cell. 1985;43:5–6.

58. Shin Y.J., Chong Y.J., Yang M.S., Kwon T.H. Production of recombinant human granulocyte macrophage-colony stimulating factor in rice cell suspension culture with a human-like N-glycan structure.Plant. Biotechnol. J. 2011;9:1109–1119.

59. Kim N.S., Kim T.G., Jang Y.S., Shin Y.J., Kwon T.H., Yang M.S. Amylase gene silencing by RNA interference improves recombinant hGM-CSF production in rice suspension culture. Plant. Mol. Biol.2008;68:369–377.

60. Kim T.G., Lee H.J., Jang Y.S., Shin Y.J., Kwon T.H., Yang M.S. Co-expression of proteinase inhibitor enhances recombinant human granulocyte-macrophage colony stimulating factor production in transgenic rice cell suspension culture. Protein Expr. Purif.2008;61:117–121.

61. Kim N.S., Kim T.G., Kim O.H., Ko E.M., Jang Y.S., Jung E.S., Kwon T.H., Yang M.S. Improvement of recombinant hgm-csf production by suppression of cysteine proteinase gene expression using RNA interference in a transgenic rice culture. Plant. Mol. Biol.2008;68:263–275.

62. Sardana R., Dudani A.K., Tackaberry E., Alli Z., Porter S., Rowlandson K., Ganz P., Altosaar I. Biologically active human GM-CSF produced in the seeds of transgenic rice plants. Transgenic Res.2007;16:713–721.

63. Asadullah K., Sterry W., Volk H.D. Interleukin-10 therapy--review of a new approach. Pharmacol Rev. 2003;55:241–269.

64. Schroder K., Hertzog P.J., Ravasi T., Hume D.A. Interferon-gamma: An overview of signals, mechanisms and functions. J. Leukoc. Biol.2004;75:163–189.

65. Blank C.A., Brantly M. Clinical features and molecular characteristics of alpha 1-antitrypsin deficiency. Ann. Allergy.1994;72:105–120. ; quiz 120–102.

66. Fagerhol M.K., Laurell C.B. The polymorphism of "prealbumins" and alpha-1-antitrypsin in human sera. Clin. Chim. Acta. 1967;16:199–203.

67. Travis J., Salvesen G.S. Human plasma proteinase inhibitors. Annu. Rev. Biochem. 1983;52:655–709.

68. Huang J., Sutliff T.D., Wu L., Nandi S., Benge K., Terashima M., Ralston A.H., Drohan W., Huang N., Rodriguez R.L. Expression and purification of functional human alpha-1-antitrypsin from cultured plant cells. Biotechnol. Prog. 2001;17:126–133.

69. Trexler M.M., McDonald K.A., Jackman A.P. Bioreactor production of human alpha(1)-antitrypsin using metabolically regulated plant cell cultures. Biotechnol. Prog. 2002;18:501–508.

70. Lindhout T., Iqbal U., Willis L.M., Reid A.N., Li J., Liu X., Moreno M., Wakarchuk W.W. Site-specific enzymatic polysialylation of therapeutic proteins using bacterial enzymes. Proc. Natl. Acad. Sci. USA. 2011;108:7397–7402.

71. Li C.H., Evans H.M. The isolation of pituitary growth hormone.Science. 1944;99:183–184.

72. Hsieh K.Y., Hsu C.I., Lin J.Y., Tsai C.C., Lin R.H. Oral administration of an edible-mushroom-derived protein inhibits the development of food-allergic reactions in mice. Clin. Exp. Allergy.2003;33:1595–1602.

73. Lee Y.T., Lee S.S., Sun H.L., Lu K.H., Ku M.S., Sheu J.N., Ko J.L., Lue K.H. Effect of the fungal immunomodulatory protein FIP-fve on airway inflammation and cytokine production in mouse asthma model.Cytokine. 2013;61:237–244.

74. Lee G.-H., Sohn S.-H., Park E.-Y., Park Y.-D. Aberrant promoter methylation occurred from multicopy

transgene and SU(VAR)3–9 homolog 9 (SUVH9) gene in transgenic Nicotiana benthamiana. Funct. Plant Biol. 2012;39:764–773.

75. De Neve M., De Buck S., de Wilde C., van Houdt H., Strobbe I., Jacobs A., van Montagu M., Depicker A. Gene silencing results in instability of antibody production in transgenic plants. Mol. Gen. Genet. 1999;260:582–592.

76. Kanno T., Naito S., Shimamoto K. Post-transcriptional gene silencing in cultured rice cells. Plant. Cell. Physiol. 2000;41:321–326.

77. Okuzaki A., Konagaya K., Nanasato Y., Tsuda M., Tabei Y. Estrogen-inducible GFP expression patterns in rice (Oryza sativa L.)Plant. Cell Rep. 2011;30:529–538.

78. Garoosi G.A., Salter M.G., Caddick M.X., Tomsett A.B. Characterization of the ethanol-inducible alc gene expression system in tomato. J. Exp. Bot. 2005;56:1635–1642.

79. Lee K.T., Chen S.C., Chiang B.L., Yamakawa T. Heat-inducible production of beta-glucuronidase in tobacco hairy root cultures. Appl. MicroBiol. Biotechnol. 2007;73:1047–1053.

80. Freeman J., Sparks C.A., West J., Shewry P.R., Jones H.D. Temporal and spatial control of transgene expression using a heat-inducible promoter in transgenic wheat. Plant. Biotechnol. J.2011;9:788–796.

81. Hiwasa-Tanase K., Nyarubona M., Hirai T., Kato K., Ichikawa T., Ezura H. High-level accumulation of recombinant miraculin protein in transgenic tomatoes expressing a synthetic miraculin gene with optimized codon usage terminated by the native miraculin terminator.Plant. Cell Rep. 2011;30:113–124.

82. Murray E.E., Lotzer J., Eberle M. Codon usage in plant genes.Nucleic Acids Res. 1989;17:477–498.

83. Gustafsson C., Govindarajan S., Minshull J. Codon bias and heterologous protein expression. Trends Biotechnol. 2004;22:346–353.

84. Perlak F.J., Fuchs R.L., Dean D.A., McPherson S.L., Fischhoff D.A. Modification of the coding sequence enhances plant expression of insect control protein genes. Proc. Natl. Acad. Sci. USA.1991;88:3324–3328.

85. Hamada A., Yamaguchi K., Ohnishi N., Harada M., Nikumaru S., Honda H. High-level production of yeast (Schwanniomyces occidentalis) phytase in transgenic rice plants by a combination of signal sequence and codon modification of the phytase gene. Plant. Biotechnol. J. 2005;3:43–55.

86. Liu F., Zhao Y.Y., Zheng J.G. High-level expression of modified gene encoding human adiponectin in transgenic rice. Biol. Res.2011;44:369–375.

87. Voinnet O., Rivas S., Mestre P., Baulcombe D. An enhanced transient expression system in plants based on suppression of gene silencing by the p19 protein of tomato bushy stunt virus. Plant. J.2003;33:949–956.

88. Boivin E.B., Lepage E., Matton D.P., De Crescenzo G., Jolicoeur M. Transient expression of antibodies in suspension plant cell suspension cultures is enhanced when co-transformed with the tomato bushy stunt virus p19 viral suppressor of gene silencing. Biotechnol. Prog. 2010;26:1534–1543.

89. Bardor M., Faveeuw C., Fitchette A.C., Gilbert D., Galas L., Trottein F., Faye L., Lerouge P. Immunoreactivity in mammals of two typical plant glyco-epitopes, core alpha(1,3)-fucose and core xylose. Glycobiology. 2003;13:427–434.

90. Jin C., Altmann F., Strasser R., Mach L., Schahs M., Kunert R., Rademacher T., Glossl J., Steinkellner H. A plant-derived human monoclonal antibody induces an anti-carbohydrate immune response in rabbits. Glycobiology. 2008;18:235–241.

91. Almond R.J., Flanagan B.F., Antonopoulos A., Haslam S.M., Dell A., Kimber I., Dearman R.J. Differential immunogenicity and allergenicity of native and recombinant human lactoferrins: Role of glycosylation. Eur. J. Immunol. 2012;43:170–181.

92. Lu D., Yang C., Liu Z. How hydrophobicity and the glycosylation site of glycans affect protein folding and stability: A molecular dynamics simulation. J. Phys. Chem. B. 2012;116:390–400.

93. Ruggiero P., Flati S., Di Cioccio V., Maurizi G., Macchia G., Facchin A., Anacardio R., Maras A., Lucarelli M., Boraschi D. Glycosylation enhances functional stability of the chemotactic cytokine CCL2. Eur. Cytokine Netw. 2003;14:91–96.

94. Miyata S., Akazawa T. Enzymic mechanism of starch breakdown in germinating rice seeds: 12. Biosynthesis of alpha-amylase in relation to protein glycosylation. Plant. Physiol. 1982;70:147–153.

95. Liebminger E., Huttner S., Vavra U., Fischl R., Schoberer J., Grass J., Blaukopf C., Seifert G.J., Altmann F., Mach L., et al. Class I alpha-mannosidases are required for N-glycan processing and root development in Arabidopsis thaliana. Plant. Cell. 2009;21:3850–3867.

96. Cavalier D.M., Lerouxel O., Neumetzler L., Yamauchi K., Reinecke A., Freshour G., Zabotina O.A., Hahn M.G., Burgert I., Pauly M., et al. Disrupting two Arabidopsis thaliana xylosyltransferase genes results in plants deficient in xyloglucan, a major primary cell wall component.Plant. Cell. 2008;20:1519–1537.

97. Sriraman R., Bardor M., Sack M., Vaquero C., Faye L., Fischer R., Finnern R., Lerouge P. Recombinant anti-HCG antibodies retained in the endoplasmic reticulum of transformed plants lack core-xylose and core-alpha(1,3)-fucose residues. Plant. Biotechnol. J. 2004;2:279–287.

98. Triguero A., Cabrera G., Rodriguez M., Soto J., Zamora Y., Perez M., Wormald M.R., Cremata J.A. Differential N-glycosylation of a monoclonal antibody expressed in tobacco leaves with and without endoplasmic reticulum retention signal apparently induces similar in vivo stability in mice. Plant. Biotechnol. J. 2011;9:1120–1130.

99. Langdon T., Seago C., Mende M., Leggett M., Thomas H., Forster J.W., Jones R.N., Jenkins G. Retrotransposon evolution in diverse plant genomes. Genetics. 2000;156:313–325.

100. Grandbastien M.A., Lucas H., Morel J.B., Mhiri C., Vernhettes S., Casacuberta J.M. The expression of the

tobacco Tnt1 retrotransposon is linked to plant defense responses. Genetica. 1997;100:241–252.

101. Lin C., Lin X., Hu L., Yang J., Zhou T., Long L., Xu C., Xing S., Qi B., Dong Y., et al. Dramatic genotypic difference in, and effect of genetic crossing on, tissue culture-induced mobility of retrotransposonTos17 in rice. Plant. Cell Rep. 2012;31:2057–2063.

102. Cho J.S., Hong S.M., Joo S.Y., Yoo J.S., Kim D.I. Cryopreservation of transgenic rice suspension cells producing recombinant hCTLA4Ig. Appl. MicroBiol. Biotechnol. 2007;73:1470–1476.

103. Ogawa Y., Sakurai N., Oikawa A., Kai K., Morishita Y., Mori K., Moriya K., Fujii F., Aoki K., Suzuki H., et al. High-throughput cryopreservation of plant cell cultures for functional genomics. Plant. Cell Physiol. 2012;53:943–952.

104. Hellwig S., Drossard J., Twyman R.M., Fischer R. Plant cell cultures for the production of recombinant proteins. Nat. Biotechnol.2004;22:1415–1422.

105. Hollak C.E. An evidence-based review of the potential benefits of taliglucerase alfa in the treatment of patients with Gaucher disease.Core Evid. 2012;7:15–20.

106. Shih S.M., Doran P.M. Foreign protein production using plant cell and organ cultures: Advantages and limitations. Biotechnol. Adv.2009;27:1036–1042.

107. Huang T.-K., McDonald K.A. Bioreactor engineering for recombinant protein production in plant cell suspension cultures.Biochem. Bioeng. J. 2009;45:168–184.

108. Kieran P.M., Malone D.M., MacLoughlin P.F. Effects of hydrodynamic and interfacial forces on plant cell suspension systems.Adv. Biochem. Eng. Biotechnol. 2000;67:139–177.

109. Park C.-I., Lee S.-J., Kang S.-H., Jung H.-S., Kim D.-I., Lim S.-M. Fed-batch cultivation of transgenic rice cells for the production of hCTLA4Ig using concentrated amino acids. Proc. Biochem.2010;45:67–74.

110. Smelko J.P., Wiltberger K.R., Hickman E.F., Morris B.J., Blackburn T.J., Ryll T. Performance of high intensity fed-batch mammalian cell cultures in disposable bioreactor systems. Biotechnol. Prog. 2011;27:1358–1364.

111. Kwon J.Y., Lee K.H., Cheon S.H., Ryu H.N., Kim S.J., Kim D.I. Adsorptive loss of secreted recombinant proteins in transgenic rice cell suspension cultures. Plant. Cell Rep. 2012;31:551–560.

112. Kusnadi A.R., Nikolov Z.L., Howard J.A. Production of recombinant proteins in transgenic plants: Practical considerations.Biotechnol. Bioeng. 1997;56:473–484.

113. Nykiforuk C.L., Shen Y., Murray E.W., Boothe J.G., Busseuil D., Rheaume E., Tardif J.C., Reid A., Moloney M.M. Expression and recovery of biologically active recombinant apolipoprotein AI(Milano) from transgenic safflower (Carthamus tinctorius) seeds. Plant. Biotechnol. J. 2011;9:250–263.

114. Li W., Li L., Li K., Lin J., Sun X., Tang K. Expression of biologically active human insulin-like growth factor 1 in Arabidopsis thaliana seeds via oleosin fusion technology. Biotechnol. Appl. Biochem. 2011;58:139–146.

115. Nykiforuk C.L., Boothe J.G., Murray E.W., Keon R.G., Goren H.J., Markley N.A., Moloney M.M. Transgenic expression and recovery of biologically active recombinant human insulin fromArabidopsis thaliana seeds. Plant Biotechnol. J. 2006;4:77–85.

116. Goodman M. Market watch: Sales of biologics to show robust growth through to 2013. Nat. Rev. Drug Discov. 2009;8:837.

CHAPTER 12

INSECTICIDAL ACTIVITY OF PLANT LECTINS AND POTENTIAL APPLICATION IN CROP PROTECTION

Maria Lígia R. Macedo [1], Caio F. R. Oliveira [1,2] and Carolina T. Oliveira [1]

[1]Department of Food Technology and Public Health, University of Mato Grosso do Sul, Campo Grande, Mato Grosso do Sul 79070-900, Brazil

[2] Department of Biochemistry and Tissue Biology, University of Campinas, Campinas, Sao Paulo 13083-862, Brazil

ABSTRACT

Lectins constitute a complex group of proteins found in different organisms. These proteins constitute an important field for research, as their structural diversity and affinity for several carbohydrates makes them suitable for numerous biological applications. This review addresses the classification and insecticidal activities of plant lectins, providing an overview of the applicability of these proteins in crop protection. The likely target sites in insect tissues, the mode of action of these proteins, as well as the use of lectins as biotechnological tools for pest control are also described. The use of initial bioassays employing artificial diets has led to the most recent advances in this field, such as plant breeding and the construction of fusion proteins, using lectins for targeting the delivery of toxins and to potentiate expected insecticide effects. Based on the data presented, we emphasize the contribution that plant lectins may make as tools for the development of integrated insect pest control strategies.

INTRODUCTION

More than a century has passed since the first studies regarding plant compounds with hemagglutinating activity that constitute the complex group of proteins currently known as plant lectins. A series of studies have continually implemented novel insights regarding the structure and function of this huge plant protein family. In this review, we present some of these findings and demonstrate the biotechnological potential of plant lectins for pest control, corroborating the practices of the integrated pest management (IPM).

The first description of a compound with hemagglutinating activity was reported in 1888, when Stillmark described a toxic factor extracted from castor beans (*Ricinus communis* L.) with the capability to agglutinate red blood cells from different animals [1]. He attributed both the toxicity and agglutinating activity to this factor, which was named ricin. Later, with the improvement of protein isolation techniques, the first lectin was isolated from *Canavalia ensiformis* seeds, also known as Jack beans. The protein was called concavalin A (ConA) [2] and was the first lectin isolated in crystalline form. Subsequently, the ability of this protein to agglutinate erythrocytes, yeast cells, bacteria and other compounds, such as starch granules, could be extensively examined [3]. For a long period, these proteins were denominated hemagglutinins, due to their ability to agglutinate erythrocytes and other kind of cells. However, it was later discovered that some lectins did not exhibit this agglutinating activity. The term lectin was then coined and has been used since. The word lectin originates from the Latin word *legere*, which means *select*. The current definition for plant lectins proposes the presence of at least one

Citation: Maria Lígia R. Macedo, Caio F. R. Oliveira and Carolina T. Oliveira, Insecticidal Activity of Plant Lectins and Potential Application in Crop Protection, doi:10.3390/molecules20022014.

non-catalytic domain that binds reversibly to a specific mono- or oligosaccharide [4], and this property constitutes the basic criteria for the classification of a protein as a plant lectin, while agglutination is no longer used for such classification criteria. Furthermore, lectins can also demonstrate additional domains, beyond the sugar-binding domains that confer different biological activities [5].

STRUCTURAL CHARACTERISTICS AND CLASSIFICATION OF PLANT LECTINS

Based on their number domains and their characteristics, plant lectins can be divided into four classes. The simplest class, the merolectins, are lectins that possess a single carbohydrate-binding domain. As a result, the merolectins do not present agglutinating activity. Hololectins contain two or multivalent carbohydrates-binding sites. The third class of lectins, the chimerolectins, possess a carbohydrate-binding domain and an additional domain that confers other biological activities. Finally, the superlectins are lectins with two or multivalent carbohydrate domains that are able to recognize structurally unrelated sugars [6].

Currently, the data from genome and transcriptome studies indicate that most of the plant lectin genes have a chimeric domain architecture, reinforcing the idea of a group with bi- or multi-functional biological roles [7]. In addition, a number of lectins from rice, soybean and *Arabidopsis* contain additional domains besides the carbohydrate-binding domain, such as a Kinase domain, suggesting that complex functions for these proteins have been acquired during the evolution of plants [6].

Due to advances in plant biochemistry and molecular biology techniques, over the last two decades, the plant lectins have been classified based on their carbohydrate-binding domains. In 1998, a novel classification system was proposed, grouping the lectins into seven distinct families, considering both sequence similarities and their evolutionary relationships [6]. Since 1998, five novel lectin domains have been identified in plants. At present, plant lectins are classified into 12 different families, with distinct carbohydrate-binding domains. The families are: *Agaricus bisporus* agglutinin homologs,*Amaranthins*, class V chitinase homologs, *Euonymus europaeus* agglutinin family, *Galanthus nivalis*agglutinin family, proteins with hevein domains, jacalins, proteins with legume lectin domains, LysM domain proteins, the *Nicotiana tabacum* agglutinin family and the ricin-B family.

To date, all plant lectins belong to one of these 12 families, with the exception of an exclusively maltose-binding lectin isolated from *Dioscorea batatas* (DB3L), which could not be classified into any known plant lectin family in terms of both structure and sugar specificity. This lectin shows sequence homology to dioscorin B from *Dioscorea alata* with 90% identity and to carbonic anhydrase from*Arabidopsis thaliana* with about 45% identity [8]. How DB3L acquired its hemagglutinating activity remains unknown.

In general, the three-dimensional structure of lectins is composed of a high content of β-sheets with little contribution from α-helixes. The β-sheets are connected by loops forming antiparallel chains. The stability of dimers and tetramers is conferred by hydrophobic interactions, hydrogen bonds and salt links [9]. The lectin carbohydrate-binding sites can be grouped into three overlapping regions. The central region is constituted by a conserved core in which residues interact with metallic ions ($Mn^{2+}$and Ca^{2+}), required for carbohydrate interactions. This core makes little contribution to the lectin's carbohydrate specificity; however, it provides necessary binding energy. Surrounding the core, aromatic residues occupy variable positions in a horseshoe shape. This region is involved in the lectin's monosaccharide specificity. Finally, residues with higher variability are located in the outer zone and are involved in interactions with larger oligosaccharide ligands [9,10].

CONSTITUTIVELY EXPRESSED AND INDUCIBLE LECTINS

Why plants express lectins has intrigued researchers for a long time. Lectins have been detected in hundreds of different plant species and the first studies characterized lectins from mature seeds [9,11]. Vegetative tissues, such as bulbs, tubers, rhizomes, roots, bark, stems, fruits and leaves [5,6] are also rich sources of lectins. Furthermore, lectins possess a broad distribution in plant tissues, and the amounts accumulated in each tissue and species vary in the same proportion.

The role of lectins in plant physiology is unclear. As lectins are often abundantly concentrated (generally 0.1%–10% of the total protein) in seeds or vegetative storage tissues (e.g., bulbs, bark and rhizomes), these proteins are thought to act as storage proteins, representing a source of amino acids for rapid growth and development [6]. The constitutively expressed lectins are also referred to as such "classical lectins" [5]. Accumulating data have revealed that the majority of these lectins accumulate in vacuoles or related organelles, or are secreted into the extracellular compartment [7]. Those lectins that are particularly expressed at high concentrations are also considered to be defense-related proteins against pathogens and herbivorous insects [12,13,14]. These features indicate that plant lectins possess multifunctional roles as supported by the fact that most lectins have a preferential or even exclusive specificity for foreign over plant-specific glycans [15]. The influence of symbiosis with microorganisms has also been studied [16].

In contrast to highly-expressed lectins, another lectin group is expressed under specific conditions and at lower concentrations and are probably involved in further functions and mechanisms currently not fully understood [17]. During the last decade, this new class of plant lectins denominated the "inducible lectins" has been extensively studied. Under normal growth conditions, or in the absence of stress factors, the inducible lectins are not expressed at detectable levels. In contrast to classical lectins, most inducible lectins remain in the nucleus and the cytoplasm of the plant cell [5,18]. The first inducible lectin to be purified and characterized was found in rice seedlings and named *Oryza sativa* agglutinin (Orysata) [19]. Orysata is a mannose-specific lectin belonging to the jacalin-related family. Although Orysata was designated as a lectin only in 2000, ten years earlier Claes *et al.* [20] described Orysata as a salt-inducible protein (SalT). The salT mRNA accumulates rapidly in sheaths and roots of mature plants and seedlings upon salt or drought stress. Later, Moons *et al.* [21] showed that salT expression was also induced by abscisic and jasmonic acid.

In a recent study, Orysata was successfully expressed in transgenic tobacco and studied with respect to its insecticidal activity against three important pest insects: the beet armyworm (*Spodoptera exigua*), the green peach aphid (*Myzus persicae*) and the pea aphid (*Acyrthosiphon pisum*). The latter was fed on an artificial diet containing different concentrations of Orysata. Results demonstrated that Orysata possesses strong insecticidal activity. Larvae of *S. exigua* fed on transformed plants presented higher larval mortality, reduced weight and extension of larval development. Similar to the beet armyworm, *M. persicae* also exhibited higher mortality compared to aphids fed on the wild type plants. Moreover, a low LC_{50} value (79 µg/mL) in the pea aphid (*A. pisum*) bioassay indicates that Orysata has strong negative effects on this insect, compared with other mannose-binding lectins [12].

Studies of changes in wheat (*Triticum aestivum*) gene expression during infestation by Hessian fly (*Mayetiola destructor*) larvae have identified three lectin-like genes expression, suggesting their involvement in the defense response. These genes are designated as *Hfr* (Hessian fly responsive) and promote the expression of a high mannose N-glycan-specific jacalin-like lectin, and two other lectin-like proteins with an amaranthin domain and a chitin-binding hevein domain, named HFR1, HFR2 and HFR3, respectively [22,23,24].

Williams *et al.* [22] reported that infested wheat-resistant plants accumulate higher levels of *Hfr-1*transcripts from one to three days after egg hatching, compared with uninfested or susceptible plants. This period corresponds to the period when first instar larvae inject substances into the plant in an attempt to establish the feeding site. Interestingly, resistant wheat plants accumulate high levels of HFR1 at the larval feeding sites [22]. When infested by *M. destructor*, wheat plants also presented high expression levels of HFR2 and HFR3 [23,25].

In another approach, HFR1 and HFR3, produced as recombinant proteins, showed insecticidal activity towards another major wheat pest, the cereal aphid *Sitobion avenue*, which does not invoke gene-for-gene responses in wheat plants, such as the Hessian fly. Aphids fed on a liquid diet containing HFR1 and HFR3 showed a significant reduction in survival and growth. The authors also observed that HFR3 binds to the midgut region in cereal aphids and is stabilized against degradation by gut enzymes for up to 48 h [26].

Another inducible lectin was purified in 2002 from tobacco (*Nicotiana tabacum*) leaves, and further referred to as *N. tabacum* agglutinin or NICTABA [27]. NICTABA was one of the first nucleocytoplasmic lectins characterized in plants [18]. Its expression is induced in leaf parenchyma cells after treatment with jasmonates, known phytohormones involved in defense responses. Vandenborre *et al.* [28], reported that chewing caterpillars (*Spodoptera littoralis* and *Manduca sexta*) and cell-content-feeding spider mites (*Tetranychus urticae*) also induce NICTABA expression. On the other hand, infestation with phloem-feeding herbivores (*Myzus nicotianae* and *Trialeurodes vaporarorum*) does not result in NICTABA accumulation. These differences suggest that feeding behavior plays an important role in the regulation of NICTABA expression, since phloem feeders cause little or minor damage to plant tissues, in comparison with mites and chewing caterpillars [29].

ANTI-INSECT ACTIVITY AND MODE OF ACTION OF PLANT LECTINS

Of the several biological functions of plant lectins [30], their anti-insect activities have received particular attention [17,31]. Over the last few years, the development of plants resistant to insect attacks has opened a promising field for the use of plant lectins in pest management strategies [32,33]. The use of lectins in transgenic plants has yielded positive results, especially for crops now expressing Cry toxins from *Bacillus thuringiensis* (*Bt*), which demonstrate resistance to sap-sucking insects. Furthermore, the ingestion of lectins in artificial diets or their expression in transgenic plants has been shown to reduce performance in insects belonging to different orders, including Lepidoptera, Coleoptera, and Hemiptera [31]. Although a number of studies have addressed lectin toxicity, the exact mode of action of lectins in providing resistance against insects remains unclear.

The most likely mechanisms underlying the entomotoxic activity of lectins involve interactions with different glycoproteins or glycan structures in insects, which may interfere with a number of physiological processes in these organisms. Since lectins possess at least one carbohydrate-binding domain and different sugar specificities, and considering the variety of glycan structures in the bodies of insects [34,35], possible targets for lectin binding are numerous (Figure 1). Therefore, it is difficult to predict the exact mode of action of each lectin and even more difficult to understand the variability in insect toxicity upon exposure to different plant lectins.

Broadly speaking, antinutritional effects are often observed as a result of lectin ingestion and could affect several biological parameters in insects, such as larval weight, fecundity, pupation and survival. Likewise, an extension in the larval development period and adult emergence is commonly observed [36]. It is important to note that one of the principal characteristics of an insecticidal protein constitutes its resistance to proteolytic degradation in the insect gut [37]. In general, plant lectins present a high resistance to insect digestive enzymes. For example, *Moringa oleifera* lectin (cMoL) is resistant to digestion by *Anagasta kuehniella* proteases for up to 12 h [36] and *Olneya tesota* lectin (PF2) is not degraded by *Zabrotes subfasciatus* digestive enzymes within 24 h [38]. NICTABA shows an even stronger resistance and is not digested for up to three days of incubation with midgut extracts of*S. littoralis* [39].

For some time, efforts have focused on finding more efficient ways to control sap-sucking hemipteran attacks on several important crops [40]. These insects cause dramatic negative impact on plants, especially due to their extremely rapid population growth [41]. Studies of plant lectin effects on Hemipteran pests have provided interesting results regarding their mode of action. The snowdrop lectin (*Galanthus nivalis* agglutinin or GNA) has received particular attention due to its toxic effects against hemipterans and other economically important insect pests. Through immunohistochemical studies, Powell *et al.* [42] showed GNA binding to cell surface carbohydrate moieties in the midgut epithelium of brown rice plant hoppers (*Nilaparvata lugens*). Moreover, GNA caused morphological changes in the epithelium, with disruption of the microvilli brush border region. GNA has also been observed at low levels in the fat bodies, the ovarioles and throughout the hemolymph, possibly causing general systemic effects. A depression in *N. lugens* feeding, as measured by honeydew production, was also observed when insects were fed on transgenic rice plants expressing GNA (2.0% of total soluble protein). In addition, insects had a decreased survival and overall fecundity, besides presenting a retarded development [43].

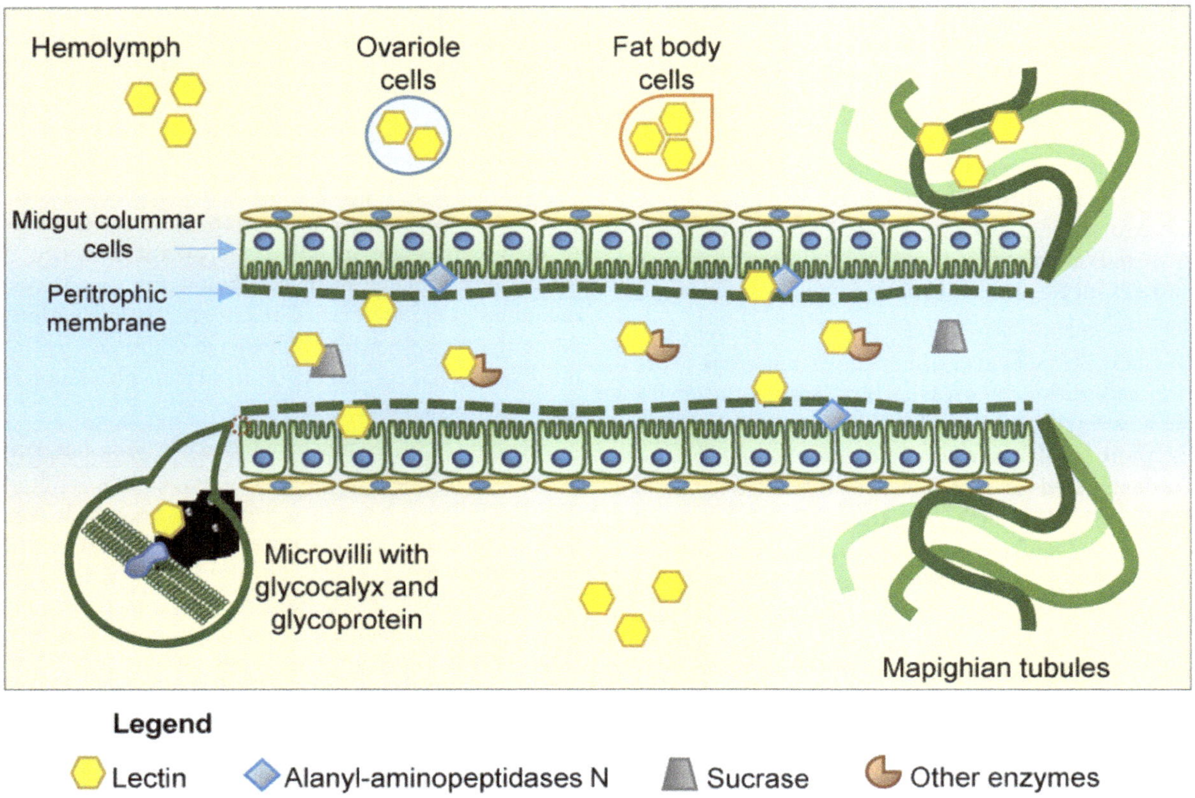

Figure 1: Representative scheme showing the binding sites of plant lectins that have been reported. In addition to the sites displayed in the figure, other receptors are also known: α-amylase, α- and β-glucosidases, trypsin-like enzymes, ferritin, NADH quinone oxidoreductase, vacuolar ATPase; sarcoplasmic reticulum type calcium ATPase, heat shock protein 70, β-subunit ATP synthase and Clathrin heavy chain.

Another lectin that also displays toxicity towards hemipterans is the *Allium sativum* leaf agglutinin (ASAL) [44], a mannose-binding lectin which has been engineered into a variety of transgenic plant species, such as tobacco [45] and Indian mustard (*Brassica juncea*) [46]. Fitches *et al.* [47] demonstrated that the heterodimeric and homodimeric garlic lectins, ASAI and ASAII, produced as recombinant proteins, showed insecticidal effects when fed to pea aphids (*A. pisum*). A dose-dependent effect of these lectins was observed, with survival being most reduced at the highest lectin concentration (2.0 mg/mL). Peptide fingerprinting using MALDI-TOF mass spectrometry identified two abundant membrane-associated aphid gut proteins, alanyl aminopeptidase N and sucrase, as receptors for lectin binding. Alanyl-aminopeptidase N is glycosylated and the carbohydrate side chains contain multiple mannose residues, allowing ASA binding. Furthermore, these proteins accounted for more than 15% of the total gut protein in *A. pisum*, being predominantly associated with the microvillar membrane of the insect midgut [48]. This aminopeptidase also strongly interacts with GNA and ConA. Although binding of the lectin impairs aphid development, the lectin does not affect the insect's aminopeptidase activity, which is thought to be associated with amino acid absorption processes [48]. Several plant lectins have also been shown to interfere with other insect digestive enzyme activities, such as α-amylase [49,50,51], α- and β-glucosidases [52] and trypsin-like enzymes [49,50,53].

Other investigations have identified the binding of GNA and ASAL to important midgut receptors of*N. lugens*, such as ferritin [54] and NADH quinone oxidoreductase [55]. These two mannose-specific lectins (ASAL and GNA) were pyramided into rice lines, which proved to be more resistant to *N. lugens*,*Nephotettix virescens* and *Sogatella furcifera*, resulting in reduced insect survival, fecundity, feeding ability and delayed development [56]. Recently, Roy *et al.* [13] reported the binding of *Colocasia esculenta* tuber agglutinin (CEA) to receptor proteins, identified as vacuolar ATP synthase, sarcoplasmic reticulum type calcium ATPase from *Bemisia tabaci* and heat shock protein 70, ATP synthase β subunit, clathrin heavy chain protein from *Lipaphis erysimi*. The changes resulting from the binding of these lectins probably entail a deficient metabolic functioning, leading to growth retardation, loss of fecundity and death.

The anti-insect activity of plant lectins towards lepidopteran pests has been reported. In a screening study, Machuka *et al.* [57] showed that 16 lectins (from a total of 25 lectins among 15 plant families) had negative effects on the legume pod-borer, *Maruca vitrata*. At least one of the parameters, such as larval survival, weight, feeding ability, pupation, adult emergence and/or fecundity was affected by lectin ingestion. GNA was one of the most active lectins, exhibiting marked effects on all parameters studied.

The negative effects of GNA on other lepidopterans have

also been well documented. Larvae of*Lacanobia oleracea* fed on an artificial diet and on transgenic potatoes expressing GNA presented decreased growth besides longer instar durations [58]. GNA expressed in transgenic sugarcane decreased the performance of *Eorreuma loftini*, presenting reduced larval survival, adult emergence and female fecundity, in the first generation, and reduced pupal weight in the second generation. In contrast, deleterious effects of GNA were not observed on *Diatraea saccharalis*. Moreover, in the first generation, GNA-fed larvae showed an enhanced larval growth, but this was not observed in the second generation [59]. Despite the consistent reports of adverse effects of GNA, such effects may vary with respect to concentration of the lectin and with respect to insect species.

The insecticidal properties of the *Bauhinia monandra* leaf lectin (BmoLL) against the cowpea weevil (*Callosobruchus maculatus*), the Mexican bean weevil (*Z. subfasciatus*) and the Mediterranean flour moth (*A. kuehniella*) have been investigated. For the bruchids, BmoLL exhibited a high mortality (up to 50%) in low concentrations (0.3%–0.5% *w/w*) in comparison with *A. kuehniella*, with no effects on survival when larvae were fed up to 1%. At this concentration, the reduction in *A. kuehniella* larval weight was 40% [51].

Sadeghi *et al.* [60] identified ferritin as a receptor for GNA in the larval midgut of *S. littoralis*, as already mentioned above in *N. lugens*. The authors postulated that part of the insecticidal activity of GNA in insects might be due to binding and interaction with certain proteins in the digestive tract, such as ferritin, in turn interfering in the insects' iron metabolism. In another study on *S. littoralis*, Caccia *et al.* [61] showed that the lectin from *Hippeastrum* hybrid (*Amaryllis*) bulbs, named HHA, interacts with the brush border of midgut cells. FITC-HHA showed that the labeled lectin binds to the cell membrane of columnar cells and is also internalized into columnar cells. This suggests that HHA is able to cross the gut epithelial barrier and pass into the insect hemolymph. In addition, the authors showed that the FITC-HHA internalization mechanism involves clathrin-mediated endocytosis.

Differences in gene expression profiles have been reported in insects fed on a lectin-containing diet. Li *et al.* [62] reported that *Drosophila melanogaster* larvae exposed to a dietary WGA evoked differential expression of 61 transcripts; among these were several chitin-binding transcripts. The authors hypothesized that the midgut responds to the damage caused by WGA to the structure of the PM, producing more chitin for repair or replacement of PM. Other differentially-transcribed gene clusters were associated with digestive enzymes, cytoskeleton organization, detoxification reactions and energy metabolism. Such studies of gene expression analysis may contribute to the better understanding of the broad mode of interaction of plant lectins and insects.

Another possible target site for lectins in the gut region is the peritrophic membrane (PM), since this membrane is mainly composed of chitin-microfibrils and glycoproteins [63]. In contrast, hemipterans do not have a PM. In the case of lepidopterans, a lectin-containing diet can cause deformations in PM formation and disruption of the microvilli. When the European corn borer (*Ostrinia nubilalis*) was fed on a wheat germ agglutinin (WGA)-containing diet, larvae presented hypersecretion of unorganized PM in the anterior midgut as well as disintegration of microvilli, indicating the binding of WGA to the chitin microfibers and several glycosylated matrix proteins [64]. The interaction of WGA may cause voids in the PM, allowing the passage of food particles through the PM and penetration into the microvillar brush border. On the other hand, the same was not observed for *M. sexta*, which presents differences in PM formation and structure, suggesting that these characteristics affect the interaction of WGA with chitin and other components of the PM [65]. WGA also causes abnormalities in the microvilli of*Drosophila melanogaster* [62]. Another example is *Dioscorea batatas* agglutinin (DB1), classified as a mannose-binding lectin, which strongly binds to the *Helicoverpa armigera* larvae brush border and the peritrophic membrane [66].

Recently, in an interesting study, Walski *et al.* [67] evaluated the toxic effects of three lectins that were structurally related but had different molecular sizes. The lectins studied were *Sambucus nigra*agglutinin I and II (SNA-I and SNA-II) and *Rhizoctonia solani* agglutinin (RSA) (a fungus lectin), with molecule sizes of 240, 32 and 15.5 kDa, respectively. *In vitro* experiments with the *Tribolium castaneum* cell line indicated a high cytotoxicity for all three lectins, with SNA-II presenting the most toxic effect, followed by SNA-I and lastly RSA. Curiously, when the lectins were introduced into the larvae diet at 2% (*w/w*), RSA produced the highest toxic effect *in vivo*. Searching for the mechanisms underlying the differences in lectins toxicity, the authors found that RSA passed through the PM more efficiently than SNA-II. Moreover, both lectins were shown to be resistant to proteolysis, in contrast to SNA-I, which is thought to be degraded in the gut. SNA-II presented a partial permeability, which may be attributed to its monomeric form in solution, allowing it to cross the PM, while the oligomers with larger dimensions than the pore membrane are retained in the endoperitrophic space. These findings indicate that the ability to pass through the PM, according to the lectin's molecular size and size of PM pores, is a key feature for lectin toxicity. It is also known that the entomotoxicity of SNA-I and SNA-II against lepidopteran and hemipteran insects is related to the induction of apoptosis via a caspase-dependent pathway in the insect gut [68,69].

BIOTECHNOLOGICAL APPLICATIONS

Insects resistant against single-defense genes, such as those currently used in *Bt*-crops, are overwhelmingly reported [1,5,70]. Thus, the combination of proteins with different modes of action, as well as the correct application of IPM has the potential to improve resistance against insects over the long-term. Based on a series of reports that has shown major effects of plant lectins on insect development, the use of plant lectins in combination with other toxic proteins has been proposed in order to obtain a synergistic effect.

Among the most studied lectins purified from the snowdrop bulbs is *G. nivalis* (Amaryllidaceae). The*G. nivalis* agglutinin, or GNA, is composed of four identical 12 kDa subunits with affinity for α-1,3- or 1,6-linked D-manose residues in carbohydrates [71]. Artificial diet feeding assays have shown the insecticidal effects of GNA for several insects, including sap-sucking insects. Since the major pests for rice are Homopteran, the brown plant hopper (*N. lugens*) and green rice leafhopper (*N. virescens*), transgenic rice plants demonstrating tissue-specific expression of GNA were constructed and evaluated for control of the brown plant hopper [43]. Insects fed on transformed plants exhibited a significant decrease in survival and fecundity, which retarded their development. GNA also exhibited insecticidal effects for the glasshouse potato aphid (*Aulacothum solani*) through transformation of potato plants [72]. GNA expression level of 0.3%–0.4% provoked a significant reduction in both survival and fecundity in glasshouse bioassays.

Transgenic plants expressing GNA were also fed to the white backed plant hopper (*S. furcifera*), showing a remarkable effects on insect survival, fecundity and plant damage [73]. Western blot analysis showed that the GNA expression was 0.3% of the total soluble protein. At this level, a reduction of 90% in survival was recorded for nymphs fed on transgenic plants. Among the survivors, only 25% reached the adult stage and these showed a reduction of 90% in average fecundity. The negative effects of GNA promoted a reduction of 88% in feeding rate, in comparison with insects fed on control plants.

The insecticidal activity and involvement of *N. tabacum* agglutinin (NICTABA) in tobacco defenses have been investigated [39]. In this study, the NICTABA gene was silenced in *N. tabacum* using the RNAi technique. Moreover, the NICTABA gene was expressed in *Nicotiana attenuata*, a species that does not possess genes encoding NICTABA. *S. litorallis* larvae fed on NICTABA-silenced plants showed an increase of 70% in average weight compared with larvae fed on wild-type tobacco. When the larvae were fed with NICTABA-expressing plants, a similar reduction in *S. littoralis* larval weight was recorded (~70%). Bioassays with *M. sexta* larvae also corroborated the reduction of larval weight with the presence of NICTABA in plant tissues. With these bioassays, the authors demonstrated the involvement of NICTABA in the defense of tobacco against insect pests.

Studies evaluating the synergic effect of plant lectins and other proteins have been carried out, as well as studies to explore carbohydrate specificity in order to carry other toxic proteins to active sites. The pea lectin (P-Lec) and the cowpea trypsin inhibitor (CpTI) were expressed in tobacco plants and protection against the tobacco budworm (*Heliothis virescens*) evaluated [74]. In comparison with plants expressing only P-Lec (C^-L^+) or CpTI (C^+L^-), tobacco plants expressing both defense genes (C^+L^+) showed a higher mortality rate and the lowest larval weight. In addition, the leaf damage was markedly reduced, in contrast with control plants or the other treatments.

In 2003, Zhu-Salzman *et al.* [75] constructed a recombinant fusion protein from the soybean cysteine protease inhibitor N (scN) and *Griffonia simplicifolia* lectin II (rGSII). The individual proteins did not present insecticide effects on the cowpea bruchid (*C. maculatus*) development. Subsequently, they compared the insecticide activity of direct protein mixtures when used with a fusion protein, called scN-rGSII. After purification of the fusion protein, the authors demonstrated that both the inhibitory activity and carbohydrate binding were preserved. Bioassays showed that the insecticide activity of scN-rGSII is higher than that of the mixture protein at the same concentration, justifying the use of the fusion protein. At the higher concentrations evaluated (0.67% scN-GSII), the fusion protein induced 100% mortality, while feeding on the separate proteins led to a mortality of 50% on average. The authors suggested that the fusion protein could block the access of scN-degrading proteases, as described in the cowpea bruchid [76], and indicated that different proteins may possess domains with similar or different functions. Through this approach, it is possible to achieve a higher insecticidal activity and control strategy durability, compared to the use of single defense genes.

Some lectins exhibit a strong resistance to gut proteolysis, probably to maintain the protein's carbohydrate affinity. This characteristic can be taken advantage of to explore the delivery of these proteins at other sites, such as the hemolymph. The application of lectins as such protein carriers has been explored for the delivery of a number of compounds, such as hormones and toxins. The potential of these compounds when ingested orally is quite low, since they need to reach the hemolymph to exert their insecticidal effects. Different insecticide hormones have been studied in order to become alternatives for pest management. Due to their low dose and specificity, these compounds may be able to control many physiological and metabolic aspects, such as biosynthesis and muscular controls [77,78]. The injection of *M. sexta* allatostatin (Manse-AS) in *L. oleracea* larvae was found to result in a high mortality rate (up to 80%) while oral feeding did not present any effect [79]. In order to evaluate the potential of GNA as a carrier for Manse-AS into the *L. oleraca* hemolymph by oral ingestion, Fitches and co-workers [80] constructed a GNA/Manse-AS fusion protein. Immunoreactivity assays showed high activity for GNA/Manse-AS fed larvae and size exclusion chromatography demonstrated that this activity was due to a fraction with a molecular weight that was similar to that of the intact fusion protein. In other words, the fusion protein is able to cross the gut epithelium and reach the hemolymph without signals of degradation. This promising result highlights the potential for the use of plant lectins as protein carriers.

Another fusion protein combination was investigated using the biological properties of GNA for the delivery of ButaIT, a toxin from the red scorpion *Mesobuthus tamulus* [81]. The feeding bioassays with ButaIT/GNA were carried out with *S. littoralis*, *Musca domestica* and *T. castaneum*. For *S. littoralis*, ButaIT/GNA-fed larvae showed a mortality of 30% within 24 h. At the end of the experiments, the most interesting result was the marked low weight of the ButaIT/GNA-fed larvae, which was 88% lower in comparison with control-fed larvae. For assays with *M. domestica*, ButaIT/GNA provoked a mortality rate of 35%, 66% and 75% after 24, 48 and 72 h, respectively. *T. castaneum* larvae fed with ButaIT/GNA showed a decrease in average weight of 90% at the end of the experiments and presented a mortality of 50%. Thus, authors demonstrated the application of GNA as a carrier protein for facilitating the delivery of neurotoxins into the circulatory system. Similar approaches are attractive, since the insect hormones and neurotoxins do not offer risks for higher animals and are effective and potentially applicable for pest management.

Another neurotoxin highly insecticidal to a variety of insect pests, the spider-venom peptide ω-hexatoxin-Hv1a (Hv1a), has a limited oral toxicity. Since GNA has been proved to be resistant to proteolytic activity and when ingested it is transported into the hemolymph, Fitches *et al.* [82], investigated the ability of GNA as a carrier for Hv1a. The authors showed that the Hv1a/GNA fusion protein had significant oral activity against the lepidopteran *Mamestra brassicae*. Even more interestingly, the authors presented evidence for binding of orally delivered GNA to the lepidopteran central nervous system, which may facilitate the toxin's action. Recently, another study by Nakasu *et al.* [83], demonstrated that Hv1a/GNA was not toxic to the non-target species, *Apis mellifera*. Although the larvae internalized the Hv1a/GNA toxin, the survival rate was only slightly affected (LD50 > 100 μg bee-1). Likewise, no adverse effects of the fusion protein on honeybee learning and memory were observed.

Through a similar approach, Yang *et al.* [84], demonstrated the efficiency of GNA as a carrier fused to another spider venom peptide, δ-amaurobitoxin-Pl1a. When third instar larvae of *M. brassicae* were fed on the fusion protein (30 μg), 100% larval mortality was noticed within six days. The Pl1a/GNA also caused increased mortality when injected in housefly (*Musca domestica*). *A. pisum* nymphs fed on a Pl1a/GNA-containing diet (1 mg/mL) presented 100% mortality after three days of feeding. The Pl1a/GNA labeled by conjugation with fluorescein demonstrated binding to the aphid gut and persistence in the midgut region even after 48 h.

Recently a new synthetic gene was constructed, comprising a combination of *Bt* toxin Cry1Ac and the*Allium sativum* agglutinin (ASAL) [85]. In bioassays, Cry-ASAL showed enhanced insecticidal activity for *Pectinophora gossypiella* and *H. armigera*, being 8- and 30-fold more toxic for these insects when compared with Cry1Ac feeding. The increase of the entomotoxic effect observed for Cry-ASAL was attributed to its binding to additional receptors in the midgut cells, mediated by the carbohydrate-binding sites of ASAL.

TRI-TROPHIC INTERACTION AND SAFETY ASSESSMENT OF LECTIN-EXPRESSING PLANTS

Many research groups are seeking different plant genes with insecticidal effects for biotechnological purposes. However, the impact of the insertion of a new gene on natural pollinators and the development of beneficial predators has not been investigated with the same frequency. The potential to use lectin-mediated tri-trophic effects for pest control has been studied. However, few results are available and these vary according to the lectin, the target pest and it respective natural predator.

The tri-trophic effect of GNA, when expressed in potato plants, on the two-spot ladybirds (*Adalia bipunctata*), the natural predator of peach-potato aphids (*M. persicae*), has been investigated [86]. The ladybirds were fed with aphids previously fed on three different diets; on control potato plants, on GNA-expressing transgenic potatoes and *Vicia faba* bean plants, considered as optimal for aphid development. Subsequently, a series of physiological parameters were analyzed. No difference in aphid consumption (in terms of weight eaten) by ladybirds was observed between the treatments. However, both egg laying and fertility were significantly reduced in the ladybird group fed with GNA-aphids. Furthermore, the ladybirds fed with GNA-aphids showed a 50% reduction in longevity. The authors suggested that the accumulation of GNA in the insect gut might exert a direct effect on the ladybirds' performance. Interestingly, when the ladybirds were fed with aphids from control plants, the parameters of weight and longevity improved. The authors highlighted the importance of "refuge" areas, at which non-transgenic plants should be planted in a specific arrangement. The adoption of this measure would have two positive impacts: first, it would decrease the selection pressure on the target pest, hindering the breakdown of the plants' genetic resistance. Secondly, the availability of unaffected aphids would corroborate the ladybirds' feeding range.

The effect of insecticide proteins on development of the green lacewing, *Chysoperda carnea*(Neuroptera: Chrysopidae), an insect used in the biological control aphids, has also been studied [87]. The feeding of *C. carnea* larvae with GNA exerted negative effects on larval development and survival. Neither the protease inhibitor SBTI nor the *Bt* toxins, Cry1Ab and Cry1Ac, exerted negative effects on*C. carnea* larval development, while the main effect observed in GNA-treated larvae is reduced fertility in adults insects [88].

Although the predators' performance was affected by the presence of lectin-expressing transgenic plants, similar effects are also observed when the defenses of non-transformed plants are triggered. For example, a seminal work evaluated the effect of tomato defenses elicited by the exogenous application of jasmonic acid (JA) on the performance of both herbivores and parasitoids [89]. The treatment of tomato plants with JA or the insect herbivory stimulates the octadecanoid pathway, culminating in the production of antinutritive compounds such as protease inhibitors and oxidative enzymes [89,90]. A direct relationship has been demonstrated between the induction of plants with JA and increases in parasitism of caterpillar pests. The models used in this study were the endoparasitic wasp *Hyposoter exiguae* and the beet armyworm, *S. exigua*. The JA-treated tomato plants demonstrated a two-fold increase in the number of wasps, in relation to the control plants. The *S. exigua* larvae fed on JA-treated plants demonstrated a reduced growth rate, compared with control larvae, and were parasitized earlier. However, the parasitoids of these larvae also showed impairments in development, such as an extension in larval development and decreased pupal weight.

The impact of transgenic plants on arthropod enemies has been critically reviewed and some provocative results have been obtained [91]. There are some discrepancies in the tests that evaluate the effects of such plants on some insect orders. Additionally, most experimental conditions did not mimic ecological conditions, since on an *ad libitum* diet, constant temperatures, single prey type and no combination of stress factors were used. Finally, small sample sizes were used in these assays, and the variability observed and the statistical analysis methods employed could endanger the conclusions regarding these tests. These findings suggest that current tests should be improved in order to increase the capability of predictions of the potential impact of transgenic crops in the environment.

Another series of tests that has received attention evaluated the biosafety of transgenic plants as a health risk for consumers before marketing. Tests with lectin-expressing plants have been carried out in laboratory animal studies in order to analyze their safety [92]. In these assays, a detailed analysis of compositional data from transgenic and parental plants is made, followed by a clinical study in which Wistar rats are fed during 90 days. In an early study, a rice variety expressing GNA was evaluated for safety [93]. Rats were fed on a diet containing 60% raw rice for 90 days. No adverse effects were reported with regard to clinical, biological, immunological, microbiological and pathological parameters. In a similar study, the immunomodulating effect of PHA-E lectin, expressed in transgenic rice, was compared with rice expressing the *Cry1Ab* toxin from *Bacillus thuringiensis* (*Bt*), a toxin already commercialized in different crops [94]. PHA-E-fed rats presented immunomodulatory effects. An increase in mesenteric lymph node weight was observed, suggesting local effects in the intestine. No immunotoxicological effects were observed for *Bt*-fed rats. The authors concluded the study by reinforcing the importance of evaluating the sensitization of consumers prior to the introduction of new genes to the world market. Complementary studies evaluating the effect of cooked food, such as rice or other seeds, were not made and it is possible that the cooking process may affect the activity of expressed proteins and should be investigated for a global view regarding the production of a new transgenic crop. To date, lectin-expressing crops have been constructed for research proposes only, with no variety being currently commercially available.

CONCLUSIONS AND FUTURE DIRECTIONS

Over the next few years, an increase in world food production is expected. The implementation of agricultural machinery and food technology plays a prominent role in this increase, as well as the amplification of planted land. With the wide food supply, insect attacks will probably increase in the same proportion. Approaches to counter these pests are under study and will be available soon. The use of genes conferring partial resistance against insect pests could provide a more sustainable control strategy [95,96], since both selection pressure on insect resistance and the economic damage in a crop are kept at a low level [97]. Thus, the use of genes related to plant defenses, such as lectin genes, is encouraged. However, a series of factors must be carefully studied, such as, initially, the use of tissue-specific promoters for the insecticide genes, the evaluation of plants in green houses and field trials, tri-trophic effect studies, and the adoption of refuge areas in crops. Secondly, the expression of the insecticide proteins in different tissues, especially those for consumption, the manner of preparation of transgenic food (whether the food is cooked or not), and food safety should be evaluated.

As presented in this review, studies with plant lectins are advanced and, based on these findings, lectin-expressing crops should soon be commercially available. It is important to highlight that the use of insecticide genes is only one of the existing options available for pest control and this technique should be integrated with other practices, such as biological control. The potential of the incorporation of synthetic genes containing lectin domains in plants should be further investigated, since available data are promising. As such, a combination of techniques have the potential to provide alternatives for providing an adequate amount of food to the growing world population.

ACKNOWLEDGMENTS

The authors would like to thank the Brazilian agencies FUNDECT, FINEP, CAPES, CNPq for financial support and student scholarships.

REFERENCES

1. Stillmark, H. Uber Ricin, Eines Giftiges Ferment aus den Samen von *Ricinus communis L.* und Anderson Euphorbiacen. Inaugural Disseration, University of Dorpat, Dorpat, Estonia, 1888.
2. Sumner, J.B. The globulins of the jack bean, *Canavalia ensiformis*. *J. Biol. Chem.* 1919, *37*, 137–142.
3. Sumner, J.B.; Howell, S.F. Identification ofhemagglutinin of jack bean with concanavalin A. *J. Bacteriol.* 1936, *32*, 227.
4. Peumans, W.J.; van Damme, E. Lectins as plant defense proteins. *Plant Physiol.* 1995, *109*, 347.
5. Van Damme, E.J.; Lannoo, N.; Fouquaert, E.; Peumans, W.J. The identification of inducible cytoplasmic/nuclear carbohydrate-binding proteins urges to develop novel concepts about the role of plant lectins. *Glycoconj. J.* 2003, *20*, 449–460.
6. Van Damme, E.J.M.; Peumans, W.J.; Barre, A.; Rouge, P. Plant lectins: A composite of several distinct families of structurally and evolutionary related proteins with diverse biological roles.*Crit. Rev. Plant Sci.* 1998, *17*, 575–692.
7. Van Damme, E.J.M.; Lannoo, N.; Peumans, W.J. Chapter 3 Plant Lectins. In *Advances in Botanical Research*; Jean-Claude, K., Michel, D., Eds.; Academic Press: Waltham, MA, USA, 2008; Volume 48, pp. 107–209.
8. Gaidamashvili, M.; Ohizumi, Y.; Iijima, S.; Takayama, T.; Ogawa, T.; Muramoto, K. Characterization of the yam tuber storage proteins from *Dioscorea batatas* exhibiting unique lectin activities. *J. Biol. Chem.* 2004, *279*, 26028–26035.
9. Sharon, N.; Lis, H. Legume lectins—A large family of homologous proteins. *FASEB J.* 1990, *4*, 3198–3208.
10. Young, N.M.; Oomen, R.P. Analysis of sequence variation among legume lectins: A ring of hypervariable residues forms the perimeter of the carbohydrate-binding site. *J. Mol. Biol.* 1992,*228*, 924–934.
11. Rüdiger, H.; Gabius, H.-J. Plant lectins: Occurrence, biochemistry, functions and applications.*Glycoconj. J.* 2001, *18*, 589–613.
12. Al Atalah, B.; Smagghe, G.; van Damme, E.J.M. Orysata, a jacalin-related lectin from rice, could protect plants against biting-chewing and piercing-sucking insects. *Plant Sci.* 2014, *221–222*, 21–28.
13. Roy, A.; Gupta, S.; Hess, D.; Das, K.P.; Das, S. Binding of insecticidal lectin *Colocasia esculenta*tuber agglutinin (CEA) to midgut receptors of *Bemisia tabaci* and *Lipaphis erysimi* provides clues to its insecticidal potential. *Proteomics* 2014, *14*, 1646–1659.
14. Guo, P.; Wang, Y.; Zhou, X.; Xie, Y.; Wu, H.; Gao, X. Expression of soybean lectin in transgenic tobacco results in enhanced resistance to pathogens and pests. *Plant Sci.* 2013, *211*, 17–22.
15. Peumans, W.J.; Barre, A.; Hao, Q.; Rougé, P.; van Damme, E.J.M. Higher plants developed structurally different motifs to recognize foreign glycans. *Trends Glycosci. Glycotechnol.* 2000,*12*, 83–101.

16. Bohlool, B.; Schmidt, E. Lectins: A possible basis for specificity in the *Rhizobium*—Legume root nodule symbiosis. *Science* 1974, *185*, 269–271.
17. Michiels, K.; van Damme, E.; Smagghe, G. Plant-insect interactions: What can we learn from plant lectins? *Arch. Insect Biochem. Physiol.* 2010, *73*, 193–212.
18. Lannoo, N.; van Damme, E.J.M. Nucleocytoplasmic plant lectins. *Biochim. Biophys. Acta Gen. Subj.* 2010, *1800*, 190–201.
19. Zhang, W.; Peumans, W.J.; Barre, A.; Houles Astoul, C.; Rovira, P.; Rougé, P.; Proost, P.; Truffa-Bachi, P.; Jalali, A.A.H.; van Damme, E.J.M. Isolation and characterization of a jacalin-related mannose-binding lectin from salt-stressed rice (*Oryza sativa*) plants. *Planta* 2000, *210*, 970–978.
20. Claes, B.; Dekeyser, R.; Villarroel, R.; van den Bulcke, M.; Bauw, G.; van Montagu, M.; Caplan, A. Characterization of a rice gene showing organ-specific expression in response to salt stress and drought. *Plant Cell* 1990, *2*, 19–27.
21. Moons, A.; Prinsen, E.; Bauw, G.; van Montagu, M. Antagonistic effects of abscisic acid and jasmonates on salt-stress-inducible transcripts in rice shoots. *Plant Cell* 1997, *9*, 2243–2259.
22. Williams, C.; Collier, C.; Nemacheck, J.; Liang, C.; Cambron, S. A lectin-like wheat gene responds systemically to attempted feeding by avirulent first-instar hessian fly larvae. *J. Chem. Ecol.* 2002, *28*, 1411–1428.
23. Puthoff, D.P.; Sardesai, N.; Subramanyam, S.; Nemacheck, J.A.; Williams, C.E. Hfr-2, a wheat cytolytic toxin-like gene, is up-regulated by virulent hessian fly larval feeding. *Mol. Plant Pathol.*2005, *6*, 411–423.
24. Subramanyam, S.; Smith, D.F.; Clemens, J.C.; Webb, M.A.; Sardesai, N.; Williams, C.E. Functional characterization of HFR1, a high-mannose N-glycan-specific wheat lectin induced by hessian fly larvae. *Plant Physiol.* 2008, *147*, 1412–1426.
25. Giovanini, M.P.; Saltzmann, K.D.; Puthoff, D.P.; Gonzalo, M.; Ohm, H.W.; Williams, C.E. A novel wheat gene encoding a putative chitin-binding lectin is associated with resistance against hessian fly. *Mol. Plant Pathol.* 2007, *8*, 69–82.
26. Pyati, P.; Chellamuthu, A.; Gatehouse, A.M.R.; Fitches, E.; Gatehouse, J.A. Insecticidal activity of wheat hessian fly responsive proteins HFR-1 and HFR-3 towards a non-target wheat pest, cereal aphid (*Sitobion avenae* F.). *J. Insect Physiol.* 2012, *58*, 991–999.
27. Chen, Y.; Peumans, W.J.; Hause, B.; Bras, J.; Kumar, M.; Proost, P.; Barre, A.; Rougé, P.; van Damme, E.J.M. Jasmonate methyl ester induces the synthesis of a cytoplasmic/nuclear chitooligosaccharide-binding lectin in tobacco leaves. *FASEB J.* 2002, *16*, 905–907.
28. Vandenborre, G.; van Damme, E.M.; Smagghe, G. *Nicotiana tabacum* agglutinin expression in response to different biotic challengers. *Arthropod Plant Interact.* 2009, *3*, 193–202.
29. Walling, L.L. Avoiding effective defenses: Strategies employed by phloem-feeding insects. *Plant Physiol.* 2008, *146*, 859–866.
30. Lam, S.; Ng, T. Lectins: Production and practical applications. *Appl. Microbiol. Biotechnol.* 2011,*89*, 45–55.
31. Vandenborre, G.; Smagghe, G.; van Damme, E.J.M. Plant lectins as defense proteins against phytophagous insects. *Phytochemistry* 2011, *72*, 1538–1550.
32. Mohan Babu, R.; Sajeena, A.; Seetharaman, K.; Reddy, M.S. Advances in genetically engineered (transgenic) plants in pest management—An over view. *Crop Prot.* 2003, *22*, 1071–1086.
33. Carlini, C.R.; Grossi-de-Sá, M.F. Plant toxic proteins with insecticidal properties. A review on their potentialities as bioinsecticides. *Toxicon* 2002, *40*, 1515–1539.
34. Vandenborre, G.; Smagghe, G.; Ghesquière, B.; Menschaert, G.; Rao, R.N.; Gevaert, K.; van Damme, E.J.M. Diversity in protein glycosylation among insect species. *PLoS One* 2011, *6*, e16682.
35. Vandenborre, G.; van Damme, E.J.M.; Ghesquière, B.; Menschaert, G.; Hamshou, M.; Rao, R.N.; Gevaert, K.; Smagghe, G. Glycosylation signatures in *Drosophila*: Fishing with lectins. *J. Proteome Res.* 2010, *9*, 3235–3242.
36. Oliveira, C.F.R.; Luz, L.A.; Paiva, P.M.G.; Coelho, L.C.B.B.; Marangoni, S.; Macedo, M.L.R. Evaluation of seed coagulant *Moringa oleifera* lectin (cMoL) as a bioinsecticidal tool with potential for the control of insects. *Process Biochem.* 2011, *46*, 498–504.
37. Felton, G.W. Indigestion is a plant's best defense. *Proc. Natl. Acad. Sci. USA* 2005, *102*, 18771–18772.
38. Lagarda-Diaz, I.; Guzman-Partida, A.M.; Urbano-Hernandez, G.; Ortega-Nieblas, M.M.; Robles-Burgueño, M.R.; Winzerling, J.; Vazquez-Moreno, L. Insecticidal action of PF2 lectin from *Olneya tesota* (Palo Fierro) against *Zabrotes subfasciatus* larvae and midgut glycoconjugate binding. *J. Agric. Food Chem.* 2008, *57*, 689–694.
39. Vandenborre, G.; Groten, K.; Smagghe, G.; Lannoo, N.; Baldwin, I.T.; van Damme, E.J.M.*Nicotiana tabacum* agglutinin is active against Lepidopteran pest insects. *J. Exp. Bot.* 2010, *61*, 1003–1014.
40. Bhatia, V.; Uniyal, P.L.; Bhattacharya, R. Aphid resistance in Brassica crops: Challenges, biotechnological progress and emerging possibilities. *Biotechnol. Adv.* 2011, *29*, 879–888.
41. Goggin, F.L. Plant–aphid interactions: Molecular and ecological perspectives. *Curr. Opin. Plant Biol.* 2007, *10*, 399–408.
42. Powell, K.S.; Spence, J.; Bharathi, M.; Gatehouse, J.A.; Gatehouse, A.M.R. Immunohistochemical and developmental studies to elucidate the mechanism of action of the snowdrop lectin on the rice brown planthopper, *Nilaparvata lugens* (Stal). *J. Insect Physiol.*1998, *44*, 529–539.
43. Rao, K.V.; Rathore, K.S.; Hodges, T.K.; Fu, X.; Stoger, E.; Sudhakar, D.; Williams, S.; Christou, P.; Bharathi, M.; Bown, D.P.; *et al.* Expression of snowdrop lectin

(GNA) in transgenic rice plants confers resistance to rice brown planthopper. *Plant J.* 1998, *15*, 469–477.

44. Bandyopadhyay, S.; Roy, A.; Das, S. Binding of garlic (*Allium sativum*) leaf lectin to the gut receptors of homopteran pests is correlated to its insecticidal activity. *Plant Sci.* 2001, *161*, 1025–1033.
45. Dutta, I.; Saha, P.; Majumder, P.; Sarkar, A.; Chakraborti, D.; Banerjee, S.; Das, S. The efficacy of a novel insecticidal protein, *Allium sativum* leaf lectin (ASAL), against homopteran insects monitored in transgenic tobacco. *Plant Biotechnol. J.* 2005, *3*, 601–611.
46. Dutta, I.; Majumder, P.; Saha, P.; Ray, K.; Das, S. Constitutive and phloem specific expression of *Allium sativum* leaf agglutinin (ASAL) to engineer aphid (*Lipaphis erysimi*) resistance in transgenic indian mustard (*Brassica juncea*). *Plant Sci.* 2005, *169*, 996–1007.
47. Fitches, E.; Wiles, D.; Douglas, A.E.; Hinchliffe, G.; Audsley, N.; Gatehouse, J.A. The insecticidal activity of recombinant garlic lectins towards aphids. *Insect Biochem. Mol. Biol.* 2008, *38*, 905–915.
48. Cristofoletti, P.T.; Mendonça de Sousa, F.A.; Rahbé, Y.; Terra, W.R. Characterization of a membrane-bound aminopeptidase purified from *Acyrthosiphon pisum* midgut cells. *FEBS J.*2006, *273*, 5574–5588.
49. Napoleão, T.H.; Pontual, E.V.; Lima, T.A.; Santos, N.D.L.; Sá, R.A.; Coelho, L.C.B.B.; Navarro, D.M.A.F.; Paiva, P.M.G. Effect of *Myracrodruon urundeuva* leaf lectin on survival and digestive enzymes of *Aedes aegypti* larvae. *Parasitol. Res.* 2012, *110*, 609–616.
50. Agra-Neto, A.C.; Napoleão, T.H.; Pontual, E.V.; Santos, N.D.; Luz Lde, A.; de Oliveira, C.M.; de Melo-Santos, M.A.; Coelho, L.C.; Navarro, D.M.; Paiva, P.M. Effect of *Moringa oleifera* lectins on survival and enzyme activities of *Aedes aegypti* larvae susceptible and resistant to organophosphate. *Parasitol. Res.* 2014, *113*, 175–184.
51. Macedo, M.; das Graças Machado Freire, M.; da Silva, M.; Coelho, L. Insecticidal action of*Bauhinia monandra* leaf lectin (BmoLL) against *Anagasta kuehniella* (Lepidoptera: Pyralidae),*Zabrotes subfasciatus* and *Callosobruchus maculatus* (Coleoptera: Bruchidae). *Comp. Biochem. Physiol. A Comp. Physiol.* 2007, *146*, 486–498.
52. Sprawka, I.; Goławska, S.; Goławski, A.; Chrzanowski, G.; Czerniewicz, P.; Sytykiewicz, H. Entomotoxic action of jackbean lectin (Con A) in bird cherry-oat aphid through the effect on insect enzymes. *J. Plant Interact.* 2013, *9*, 425–433.
53. Coelho, M.B.; Marangoni, S.; Macedo, M.L.R. Insecticidal action of *Annona coriacea* lectin against the flour moth *Anagasta kuehniella* and the rice moth *Corcyra cephalonica* (Lepidoptera: Pyralidae). *Comp. Biochem. Physiol. Part C Toxicol. Pharmcol.* 2007, *146*, 406–414.
54. Du, J.; Foissac, X.; Carss, A.; Gatehouse, A.M.R.; Gatehouse, J.A. Ferritin acts as the most abundant binding protein for snowdrop lectin in the midgut of rice brown planthoppers (*Nilaparvata lugens*). *Insect Biochem. Mol. Biol.* 2000, *30*, 297–305.
55. Bala, A.; Roy, A.; Behura, N.; Hess, D.; Das, S. Insight to the mode of action of *Allium sativum*leaf agglutinin (ASAL) expressing in T3 rice lines on brown planthopper. *Am. J. Plant Sci.* 2013,*4*, 400–407.
56. Bharathi, Y.; Vijaya Kumar, S.; Pasalu, I.C.; Balachandran, S.M.; Reddy, V.D.; Rao, K.V. Pyramided rice lines harbouring *Allium sativum* (ASAL) and *Galanthus nivalis* (GNA) lectin genes impart enhanced resistance against major sap-sucking pests. *J. Biotechnol.* 2011, *152*, 63–71.
57. Machuka, J.; van Damme, E.J.M.; Peumans, W.J.; Jackai, L.E.N. Effect of plant lectins on larval development of the legume pod borer, *Maruca vitrata*. *Entomol. Exp. Appl.* 1999, *93*, 179–187.
58. Gatehouse, J.A.; Gatehouse, A.M.R.; Fitches, E. Effects of snowdrop lectin (GNA) delivered via artificial diet and transgenic plants on the development of tomato moth (*Lacanobia oleracea*) larvae in laboratory and glasshouse trials. *J. Insect Physiol.* 1997, *43*, 727–739.
59. Sétamou, M.; Bernal, J.S.; Legaspi, J.C.; Mirkov, T.E.; Legaspi, B.C., Jr. Evaluation of lectin-expressing transgenic sugarcane against stalkborers (Lepidoptera: Pyralidae): Effects on life history parameters. *J. Econ. Entomol.* 2002, *95*, 469–477.
60. Sadeghi, A.; Smagghe, G.; Proost, P.; van Damme, E.J.M. Ferritin acts as a target site for the snowdrop lectin (GNA) in the midgut of the cotton leafworm *Spodoptera littoralis*. *Insect Sci.*2008, *15*, 513–519.
61. Caccia, S.; van Damme, E.J.M.; de Vos, W.H.; Smagghe, G. Mechanism of entomotoxicity of the plant lectin from *Hippeastrum* hybrid (Amaryllis) in *Spodoptera littoralis* larvae. *J. Insect Physiol.*2012, *58*, 1177–1183.
62. Li, H.M.; Sun, L.; Mittapalli, O.; Muir, W.M.; Xie, J.; Wu, J.; Schemerhorn, B.J.; Sun, W.; Pittendrigh, B.R.; Murdock, L.L. Transcriptional signatures in response to wheat germ agglutinin and starvation in *Drosophila melanogaster* larval midgut. *Insect Mol. Biol.* 2009, *18*, 21–31.
63. Hegedus, D.; Erlandson, M.; Gillott, C.; Toprak, U. New insights into peritrophic matrix synthesis, architecture, and function. *Annu. Rev. Entomol.* 2009, *54*, 285–302.
64. Harper, M.S.; Hopkins, T.L.; Czapla, T.H. Effect of wheat germ agglutinin on formation and structure of the peritrophic membrane in european corn borer (*Ostrinia nubilalis*) larvae. *Tissue Cell* 1998, *30*, 166–176.
65. Hopkins, T.L.; Harper, M.S. Lepidopteran peritrophic membranes and effects of dietary wheat germ agglutinin on their formation and structure. *Arch. Insect Biochem. Physiol.* 2001, *47*, 100–109.
66. Ohizumi, Y.; Gaidamashvili, M.; Ohwada, S.; Matsuda, K.; Kominami, J.; Nakamura-Tsuruta, S.; Hirabayashi, J.; Naganuma, T.; Ogawa, T.; Muramoto, K. Mannose-binding lectin from yam (*Dioscorea batatas*) tubers with insecticidal properties against *Helicoverpa armigera*(Lepidoptera: Noctuidae). *J. Agric. Food Chem.* 2009, *57*, 2896–2902.
67. Walski, T.; van Damme, E.J.M.; Smagghe, G. Penetration through the peritrophic matrix is a key to lectin toxicity

against *Tribolium castaneum*. *J. Insect Physiol.* 2014, *70*, 94–101.

68. Shahidi-Noghabi, S.; Damme, E.J.M.V.; Iga, M.; Smagghe, G. Exposure of insect midgut cells to*Sambucus nigra* L. agglutinins I and II causes cell death via caspase-dependent apoptosis. *J. Insect Physiol.* 2010, *56*, 1101–1107.
69. Shahidi-Noghabi, S.; van Damme, E.J.M.; Mahdian, K.; Smagghe, G. Entomotoxic action of*Sambucus nigra* agglutinin I in *Acyrthosiphon pisum* aphids and *Spodoptera exigua* caterpillars through caspase-3-like-dependent apoptosis. *Arch. Insect Biochem. Physiol.* 2010, *75*, 207–220.
70. Tabashnik, B.E.; Carrière, Y.; Dennehy, T.J.; Morin, S.; Sisterson, M.S.; Roush, R.T.; Shelton, A.M.; Zhao, J.-Z. Insect resistance to transgenic Bt crops: Lessons from the laboratory and field.*J. Econ. Entomol.* 2003, *96*, 1031–1038.
71. Van Damme, E.J.; Allen, A.K.; Peumans, W.J. Isolation and characterization of a lectin with exclusive specificity towards mannose from snowdrop (*Galanthus nivalis*) bulbs. *FEBS Lett.*1987, *215*, 140–144.
72. Down, R.E.; Gatehouse, A.M.R.; Hamilton, W.D.O.; Gatehouse, J.A. Snowdrop lectin inhibits development and decreases fecundity of the glasshouse potato aphid (*Aulacorthum solani*) when administered *in vitro* and via transgenic plants both in laboratory and glasshouse trials. *J. Insect Physiol.* 1996, *42*, 1035–1045.
73. Nagadhara, D.; Ramesh, S.; Pasalu, I.; Rao, Y.K.; Sarma, N.; Reddy, V.; Rao, K. Transgenic rice plants expressing the snowdrop lectin gene (GNA) exhibit high-level resistance to the whitebacked planthopper (*Sogatella furcifera*). *Theor. Appl. Genet.* 2004, *109*, 1399–1405.
74. Boulter, D.; Edwards, G.A.; Gatehouse, A.M.R.; Gatehouse, J.A.; Hilder, V.A. Additive protective effects of different plant-derived insect resistance genes in transgenic tobacco plants. *Crop Prot.*1990, *9*, 351–354.
75. Zhu-Salzman, K.; Ahn, J.E.; Salzman, R.A.; Koiwa, H.; Shade, R.E.; Balfe, S. Fusion of a soybean cysteine protease inhibitor and a legume lectin enhances anti-insect activity synergistically. *Agric. For. Entomol.* 2003, *5*, 317–323.
76. Zhu-Salzman, K.; Koiwa, H.; Salzman, R.; Shade, R.; Ahn, J.E. Cowpea bruchid *Callosobruchus maculatus* uses a three-component strategy to overcome a plant defensive cysteine protease inhibitor. *Insect Mol. Biol.* 2003, *12*, 135–145.
77. Lange, A.B.; Bendena, W.G.; Tobe, S.S. The effect of the thirteen Dip-allatostatins on myogenic and induced contractions of the cockroach (*Diploptera punctata*) hindgut. *J. Insect Physiol.*1995, *41*, 581–588.
78. Weaver, R.J.; Edwards, J.P.; Bendena, W.G.; Tobe, S.S. *Structures, Functions and Occurrences of Insect Allatostatic Peptides*; Seminar Series-Society for Experimental Biology; Cambridge University Press: New York, NY, USA, 1998; pp. 3–32.
79. Audsley, N.; Weaver, R.J.; Edwards, J.P. *In vivo* effects of *Manduca sexta* allatostatin and allatotropin on larvae of the tomato moth, *Lacanobia oleracea*. *Physiol. Entomol.* 2001, *26*, 181–188.
80. Fitches, E.; Audsley, N.; Gatehouse, J.A.; Edwards, J.P. Fusion proteins containing neuropeptides as novel insect contol agents: Snowdrop lectin delivers fused allatostatin to insect haemolymph following oral ingestion. *Insect Biochem. Mol. Biol.* 2002, *32*, 1653–1661.
81. Fitches, E.C.; Bell, H.A.; Powell, M.E.; Back, E.; Sargiotti, C.; Weaver, R.J.; Gatehouse, J.A. Insecticidal activity of scorpion toxin (ButaIT) and snowdrop lectin (GNA) containing fusion proteins towards pest species of different orders. *Pest Manag. Sci.* 2010, *66*, 74–83.
82. Fitches, E.C.; Pyati, P.; King, G.F.; Gatehouse, J.A. Fusion to snowdrop lectin magnifies the oral activity of insecticidal ω-hexatoxin-Hv1a peptide by enabling its delivery to the central nervous system. *PLoS One* 2012, *7*, e39389.
83. Nakasu, E.Y.T.; Williamson, S.M.; Edwards, M.G.; Fitches, E.C.; Gatehouse, J.A.; Wright, G.A.; Gatehouse, A.M.R. Novel biopesticide based on a spider venom peptide shows no adverse effects on honeybees. *Proc. R. Soc. B* 2014, *281*.
84. Yang, S.; Pyati, P.; Fitches, E.; Gatehouse, J.A. A recombinant fusion protein containing a spider toxin specific for the insect voltage-gated sodium ion channel shows oral toxicity towards insects of different orders. *Insect Biochem. Mol. Biol.* 2014, *47*, 1–11.
85. Tajne, S.; Boddupally, D.; Sadumpati, V.; Vudem, D.R.; Khareedu, V.R. Synthetic fusion-protein containing domains of *Bt* Cry1Ac and *Allium sativum* lectin (ASAL) conferred enhanced insecticidal activity against major lepidopteran pests. *J. Biotechnol.* 2014, *171*, 71–75.
86. Birch, A.N.E.; Geoghegan, I.E.; Majerus, M.E.; McNicol, J.W.; Hackett, C.A.; Gatehouse, A.M.; Gatehouse, J.A. Tri-trophic interactions involving pest aphids, predatory 2-spot ladybirds and transgenic potatoes expressing snowdrop lectin for aphid resistance. *Mol. Breed.* 1999, *5*, 75–83.
87. Lawo, N.C.; Romeis, J. Assessing the utilization of a carbohydrate food source and the impact of insecticidal proteins on larvae of the green lacewing, *Chrysoperla carnea*. *Biol. Control* 2008, *44*, 389–398.
88. Li, Y.; Romeis, J. Impact of snowdrop lectin (*Galanthus nivalis* agglutinin; GNA) on adults of the green lacewing, *Chrysoperla carnea*. *J. Insect Physiol.* 2009, *55*, 136–143.
89. Thaler, J.S. Jasmonate-inducible plant defences cause increased parasitism of herbivores. *Nature*1999, *399*, 686–688.
90. Broadway, R.M.; Duffey, S.S.; Pearce, G.; Ryan, C.A. Plant proteinase inhibitors: A defense against herbivorous insects? *Entomol. Exp. Appl.* 1986, *41*, 33–38.
91. Lövei, G.; Arpaia, S. The impact of transgenic plants on natural enemies: A critical review of laboratory studies. *Entomol. Exp. Appl.* 2005, *114*, 1–14.
92. Poulsen, M.; Pedersen, J.W. Assessing biosafety of GM plants containing lectins. *Plant Sci. Rev.*2011, *2010*, 157.
93. Poulsen, M.; Kroghsbo, S.; Schroder, M.; Wilcks, A.; Jacobsen, H.; Miller, A.; Frenzel, T.; Danier, J.;

Rychlik, M.; Shu, Q.; *et al.* A 90-day safety study in wistar rats fed genetically modified rice expressing snowdrop lectin *Galanthus nivalis* (GNA). *Food Chem. Toxicol.* 2007,*45*, 350–363.

94. Kroghsbo, S.; Madsen, C.; Poulsen, M.; Schrøder, M.; Kvist, P.H.; Taylor, M.; Gatehouse, A.; Shu, Q.; Knudsen, I. Immunotoxicological studies of genetically modified rice expressing PHA-E lectin or Bt toxin in wistar rats. *Toxicology* 2008, *245*, 24–34.

95. Gould, F.; Kennedy, G.G.; Johnson, M.T. Effects of natural enemies on the rate of herbivore adaptation to resistant host plants. *Entomol. Exp. Appl.* 1991, *58*, 1–14.

96. Van Emden, H. Host plant–aphidophaga interactions. *Agric. Ecosyst. Environ.* 1995, *52*, 3–11.

97. Poppy, G.M.; Sutherland, J.P. Can biological control benefit from genetically-modified crops? Tritrophic interactions on insect-resistant transgenic plants. *Physiol. Entomol.* 2004, *29*, 257–268.

CHAPTER 13

METABOLOMICS OF GENETICALLY MODIFIED CROPS

Carolina Simó, Clara Ibáñez, Alberto Valdés, Alejandro Cifuentes and Virginia García-Cañas

Laboratory of Foodomics, Institute of Food Science Research (CIAL), Spanish National Research Council (CSIC), Nicolas Cabrera 9, Cantoblanco Campus, Madrid 28049, Spain

ABSTRACT

Metabolomic-based approaches are increasingly applied to analyse genetically modified organisms (GMOs) making it possible to obtain broader and deeper information on the composition of GMOs compared to that obtained from traditional analytical approaches. The combination in metabolomics of advanced analytical methods and bioinformatics tools provides wide chemical compositional data that contributes to corroborate (or not) the substantial equivalence and occurrence of unintended changes resulting from genetic transformation. This review provides insight into recent progress in metabolomics studies on transgenic crops focusing mainly in papers published in the last decade.

INTRODUCTION

The application of genetic engineering is considered one of the leading technological advances in modern biotechnology. The organisms derived from genetic engineering are commonly named genetically modified organisms (GMOs). Since the production of the first genetically modified (GM) plant in 1983, a variety of agronomic traits that include benefits in agronomic productivity and industrial processing have been developed. Among the most relevant traits present in authorized GM crops, tolerance to herbicide and resistance to insects are prevalent worldwide. However, value-added traits such as different micronutrient content, faster ripening, improved feed value, and high levels of antioxidants, have also gained much attention recently [1,2,3].

Despite its important economic potential, authorization and commercialization of GMOs has been always controversial within the scientific community and the public sector. Several aspects of GMOs, including risk assessment, marketing, labeling, and traceability are strictly regulated in the European Union and other countries. In such regulations, the starting point in risk assessment of GMOs relies on the substantial equivalence concept that involves the comparison of the GMO under assessment with traditional varieties. Substantial equivalence concept is based on the assumption that commercialized traditional crops have been consumed for decades and have gained a history of safe use. Therefore, they can be used as comparators for the safety assessment of new GMOs derived from established plant varieties. One of the central safety issues under debate regarding GMOs is the occurrence of unintended changes resulting from genetic transformation. Unintended effects go beyond the primary expected effects of genetic modification, and represent statistically significant differences in a phenotype compared with an appropriate phenotype control [4]. Such unpredictable alterations are considered a significant source of uncertainty

Citation: Carolina Simó, Clara Ibáñez, Alberto Valdés, Alejandro Cifuentes, and Virginia García-Cañas, Metabolomics of Genetically Modified Crops, doi: 10.3390/ijms151018941.

that might have an impact on human health and/or the environment [5].

Substantial equivalence evaluations are commonly approached using targeted analysis of predefined compounds that include natural toxins, macro-, micro-, and anti-nutrients, following recommendations in the Organization of Economic Cooperation and Development (OECD) consensus documents for individual crops [6]. This targeted approach has enabled the identification of unintended effects in some cases; however, its adoption within the substantial equivalence framework has raised several criticisms. Specifically, it has been argued that this targeted approach is biased, and that some unforeseen, unintended effects may escape detection [7]. In response to the bias and uncertainties associated with targeted analysis in comparative compositional evaluation of GMOs, a report by a panel of European Food Safety Agency recommended the development and use of profiling technologies such as omics technologies, with the potential to improve the breadth of comparative analyses [8]. More recently, a panel of experts on risk assessment and management has recommended profiling especially in cases where the most scientifically valid isogenic and conventional comparator would not grow, or not grow as well, under the relevant stress condition [9]. However, certain questions have been raised about the value of molecular profiling for GMO risk assessment [10]. Some arguments against profiling rely on the lack of validated procedures and the difficulty to interpret the differences observed between a certain GMO and its comparator. However, a number of reports demonstrating the suitability and applicability of different profiling approaches for comparative analysis of GMOs suggest good acceptance of these fast-evolving techniques by the scientific community.

Omics technologies are essential tools for understanding the response of organisms to genetic and environmental changes [11]. In this context, metabolomics has the potential to provide new dimensions to GMO analysis, allowing detection of the effects (intended or not) that might take place as a result of genetic engineering application. However, metabolomic analysis faces some challenges since the diversity of metabolites found in plants is by far greater than in other organisms, being the actual size of the plant metabolome unknown [12,13,14,15]. A group of well-established analytical techniques, namely, nuclear magnetic resonance (NMR) and mass spectrometry (MS)-based techniques, are the most commonly used in the vast majority of metabolic profiling and fingerprinting analyses of plants [16,17]. Although NMR requires limited sample preparation, medium to high abundance metabolites will be detected using this technique [18,19]. The field strength improvements in NMR superconducting magnets have increased the spectral resolution and detection sensitivity. Current MS-based approaches yield higher sensitivity than NMR when analyzing complex plant metabolite mixtures. The use of high and ultra-high resolution mass spectrometers greatly improves analytical performance and offers the best combination of selectivity and sensitivity [20,21]. However, to achieve the maximum high-throughput production of metabolic information from the analysis of the largest possible number of plant samples, sample pre-treatment should be reduced to a minimum. Moreover, a variety of novel direct MS-based approaches with great potential for metabolomics have been introduced in last years. An array of direct ionization or desorption/ionization techniques have been developed and combined for this purpose [22,23]. On the other side, the application of metabolomics approaches even to a limited number of samples results in a huge amount of data with its inherent difficulties in making a meaningful interpretation. In the last years, great efforts have been made to apply metabolomics approaches to investigate the compositional equivalence between GMOs and the conventional unmodified organisms. A variety of crops have been studied using mainly MS or NMR-based analytical platforms in combination with several statistical methodologies [4,24]. The most general approach to find meaning in metabolomics datasets involves the application of multivariate analysis methods such as for example, partial squares discriminant analysis (PLS-DA) and principal component analysis (PCA). Multivariate methods allow the identification of the spectral features contributing most to variation or separation for further analysis. PCA is one of the most frequently used unsupervised methods for metabolic fingerprinting and it provides a means to achieve unbiased dimensionality reduction. Unsupervised refers to the modeling being done without user intervention and only on the explanatory variables, leaving any responses optional for later stages in the process. However PCA only reveals group structure when within-group variation is sufficiently less than between group variation. On the other side, PLS-DA often performs more efficiently for the interpretable decomposition than PCA.

This review provides insight into recent progress in metabolomics studies on GM crops focusing mainly in papers published in the last decade (a list is given in Table 1). Below, cutting-edge applications of metabolomics in the context of GMO analysis are highlighted to illustrate its impressive potential.

Table 1: Metabolomic studies on GMOs (genetically modified organisms)

GM Crop	Tissue	Donor Specie	Genetic Modification	Phenotype	Analytical Technique	Reference
Rice	Seed	*B. thuringiensis*	*Cry1Ab*	Insect resistance	FTIR MS, NMR	[25]
	Seed	*B. thuringiensis*	*Cry1Ac, sck*	Insect resistance	GC-FID, GC-EI-Q MS	[26]
	Leaf	*Z. mays*	*C1, R-S*	Flavonoid production	LC-ESI-Q MS, LC-DAD	[27]
	Leaf, seed, root	*O. sativa*	*YK1*	Stress tolerance	CE-ESI-Q MS	[28]
	Seed	*O. sativa*	*RCH10, RAC22, β-Glu, B-RIP*	Antifungal activity	NIRS, GC-EI-Q MS, LC-DAD, ICP-AES	[29]
	Seed	*O. sativa*	*Mod. (Xa23, Xa21* genes)	Insect resistance	GC-EI-Q MS	[30]
	Seed	*B. thuringiensis*	*Cry1Ac, sck*	Insect Resistance	LC-ESI-Q/TOF MS	[31]
	Seed	*E. coli*	*GlgC-TM*	Nutritionally enhanced	LC-ESI-Q MS	[32]
	Seed	*N. tabacum*	*ASA2*	Nutritionally enhanced	LC-ESI-Q MS	[33]
	Seed	*A. tumefaciens*	*Psy-2A-Crtl* *Bar*	Nutritionally enhanced Herbicide tolerance	GC-EI-TOF MS	[34]
	Leaf, seed	*E. coli/O. sativa*	*LysC, dapA/LKR/SDH*	Nutritionally enhanced	LC-FTIR MS, GC-EI-Q MS	[35]
Maize	Grain	*B. thuringiensis*	*Cry1Ab*	Insect resistance	NMR	[36,37]
	Grain	*Z. mays*	*Mod.* (*Rpd3* gene)	Seed development	NMR	[38]
	Grain	*B. thuringiensis*	*Cry1Ab*	Insect resistance	NMR	[39]
	Grain	*B. thuringiensis*	*Cry1Ab*	Insect resistance	CE-ESI-TOF MS	[40]
	Grain	*B. thuringiensis*	*Cry1Ab*	Insect resistance	FT-ICR MS	[41]
	Grain	*B. thuringiensis*	*Bt toxin*	Insect resistance	GC-EI-Q MS	[42]
	Grain	B. thuringiensis	*Cry1Ab*	Insect resistance	GC-EI-Q MS	[43]
	Grain	*B. thuringiensis* *A. tumefaciens*	*Cry1Ab* *CP4 EPSPS*	Insect resistance Herbicide tolerance	GC-EI-Q MS	[44]
	Grain	*Z. mays*	*Mod. (Zmpsy1, Pacrtl, Gllycb, Glbch, ParacrtW* genes)	Nutritionally enhanced	LC-DA, LC-ESI-APCI MS	[45]
	Grain	*B. thuringiensis*	*Bt toxin*	Herbicide tolerance Insect resistance	NMR, GC-EI-Q-MS	[46]
Soybean	Seed	*A. tumefaciens*	*CP4 EPSPS*	Herbicide tolerance	GC-EI-Q MS	[42]
	Seed	*Agrobacterium spp.*	*837ASDIS*	Herbicide tolerance	GC-EI-Q MS	[43]
	Seed	*A. tumefaciens*	*CP4 EPSPS*	Herbicide tolerance	CE-ESI-TOF MS	[47]
	Leaf, EC, seed	*N. tabacum*	*ASA2*	Nutritionally enhanced	GC-EI-Q MS	[48]
	Seed	*A. tumefaciens*	*CP4 EPSPS*	Herbicide tolerance	CE-ESI-TOF MS	[49]
	Seed	*Avena spp*	*Mod.* (*HPPD* gene)	Herbicide tolerance	LC-ESI-Q MS, GC-EI-Q MS	[50]
	Seed	*A. tumefaciens*	*CP4 EPSPS*	Herbicide tolerance	CE-ESI-TOF MS, GC-EI-TOF MS, LC-ESI-Q/TOF MS, ICP MS	[51]
Alfalfa	Stem, leaf	*N. tabacum*	*PAL2*	Nutritionally enhanced	LC-UV	[52]
Pea	Leaf	*S. hygroscopicus*	*Bar*	Herbicide tolerance	NMR	[53]
Wheat	Leaf	*U. maydis*	*Chit/Gluc, RIP, Mod.* (*KP4* gene)	Fungal resistance	LC-DAD, LC-ESI-Q MS	[5]
	Seed	*T. aestivum*	*Glu-A1, Glu-D1*	Nutritionally enhanced	NMR	[54]
Potato	Tuber	*A. pullulans,* *S. tuberosum*	*W2, FK, Mal1, SamDC*	Starch biosynthesis, leaf morphology, ethylene production	GC-EI-Q MS	[6]
	Tuber	*S. tuberosum*	*AGPase, StcPGM, StpPGM*	Altered starch composition	GC-EI-Q MS	[55–57]
	Tuber	*C. scolymus*	*1-SST, 1-FFT*	Inulin synthesis	GC-EI-TOF MS, LC-ESI-Q MS	[58]
	Tuber	*A. tumefaciens*	*Potato virus Y*	Virus resistance	CE-ESI-IT-MS/MS	[59]
	Tuber	*A. thaliana*	*DREB1A*	Stress tolerance	GC-EI-TOF MS, LC-ESI-Q MS	[60]
	Tuber	*A. pullulans*	*W2*	Waxy phenotype	LC-UV, NMR	[61]
	Leaf	*S. cerevisiae*	*TPS1*	Drought resistance	GC-EI-Q MS	[62]
Tomato	Fruit	*A. tumafaciens*	*LBA4404*	Improved texture, mouthfeel, colour	NMR	[63]
	Leaf, fruit	*A. thaliana*	*AtHXK1*	Altered carbohydrate metabolism	GC-EI-Q MS	[64]
	Fruit	*Z. mays*	*LC1, C1*	Increased flavonol content	NMR	[65,66]
	Fruit	*E. coli*	*DXS*	Increased carotenoid content	LC-DAD	[67]
	Fruit	*V. vinifera*	*Stilbene synthase*	Resveratrol synthesis	LC-ESI-Q MS	[68]
	Fruit	*R. dulcifica*	Miraculin	Sweet flavor	GC-EI-TOF MS, LC-ESI-Q/TOF MS, CE-ESI/TOF MS	[69]
Tobacco	Leaf	*E. coli/P. fluorencens*	*Ent/CpmsB*	Salicylic acid producing plants	NMR	[70]
Lettuce	Leaf	*E. coli*	*Asn A*	Growth enhanced	NMR, GC-FID	[71–73]
Cucumber	Fruit	*T. daniellii*	*Thaumatin-II*	Sweet flavor	GC-EI-TOF MS	[74]
	Fruit	*T. daniellii*	*Preprothaumatin-II*	Aroma, sweet flavor	GC-EI-Q/TOF MS	[75]
Raspberry	Fruit	*RBDV*	*Virus movement protein*	Virus resistance	GC-EI-Q MS	[76]
Grapevine	Leaf	*E. coli*	*Adh*	Abiotic stress	GC-EI-Q MS, LC-ESI-IT MS, LC-DAD	[77]
Peppermint	Leaf	*Mentha x piperita*	*Mod.* (*MFS* gene), *DXR*	Essential oils content	GC-FID	[78]
Cabbage	Leaf	*A. tumafaciens*	*Bar*	Herbicide tolerance	LC-DAD, LC-ESI/Q MS	[79]
Papaya	Fruit	*C. papaya*	*Mod.* (*55-1* gene)	Virus resistance	LC-DAD, GC-FID	[80]
	Pulp, Leaf	*C. papaya*	*Mod.* (*rep* gene)	Virus resistance	GC-EI-Q MS, LC-DAD, LC-ESI-Q MS	[81]
Poplar	Cambial region	*P. trichocarpa*	*Mod.* (*hipI-SOD* gene)	Superoxide production	GC-EI-TOF MS, LC-ESI/TOF MS	[82]
Barley	Seed	*B. amyloliquefaciens*	*GluB, ChGP*	Antifungal activity	LC-ESI-IT MS	[83]

AES: Atomic Emission Spectroscopy; APCI: Atmospheric Pressure Chemical Ionization; CE: Capillary Electrophoresis; DAD: Diode Array Detector; EC: Embryogenic culture; EI: Electron Impact; ESI: Electrospray Ionization; FID: Flame Ionization Detector; FTIC: Fourier Transform Infrared Spectroscopy; GC: Gas Chromatography; ICP: Inductively Coupled Plasma; ICR: Ion Cyclotron Resonance; IT: Ion Trap; LC: Liquid Chromatography; *Mod.*: Modification; NIRS: Near Infrared Spectroscopy; Q: Quadrupole; *RBDV*: Raspberry bushy dwarf virus; TOF: Time Of Flight.

METABOLOMICS AND GM (GENETICALLY MODIFIED) CROPS: CASE STUDIES

Rice

Rice (*Oryza sativa* L.) is one of the most important food crops in the world being the main source of calories and protein intake for half of the world population [84,85]. Significant advances have been performed in rice biotechnology in order to solve problems related to disease, insect, pest and abiotic stress (temperature, salt, nutrition, drought, wounding, *etc.*) that cause yield reduction. Metabolomics has proved to be a useful approach in the study of rice. Different technological developments in metabolomics applied to rice have been described [86], including their use in the determination of unexpected and undesirable compounds accumulated in GM rice.

GM rice (transformed with *cry1Ab* gene from *Bacillus thuringinesis*) and its wild type (WT) bred parent line were used as test materials to investigate the suitability of FTIR and NMR for metabolic fingerprinting, in combination with multivariate statistical analysis for sample classification [25]. The overall results indicated the advantage of supervised over unsupervised statistical analysis for classification purposes. The metabolic profiling of three insect-resistant GM rice lines with inserted *sck* (trypsin proteinase inhibitor derived from cowpea) and *cryIAc* transgenes was approached by Zhou *et al.* [26] using GC-FID. In that study, metabolic profiles of wild and GM rice varieties were compared to assess the unintended effects related to the genetic modification. In order to determine the environmental effects on metabolites, wild samples with different sowing dates or sites were analyzed. Results from that study indicated the levels of malic acid, asparagine, sorbitol and gluconic acid differed in rice planted at different locations whereas sucrose, mannitol and glutamic acid levels were the major metabolic differences affected by gene insertion. However, one of the main conclusions of that study was that growing conditions and gene modification induced similar influence on most of metabolites.

Amongst the numerous kinds of flavonoids produced in GM rice (transformed with maize *C1* and *R-S* regulatory genes), dihydroquercetin (taxifolin), dihydroisorhamnetin (3›-*O*-methyl taxifolin) and 3'-*O*-methyl quercetin were the major flavonoids detected using LC [27]. In a different report, Takahashi *et al.* [28] focused on the analysis of GM rice plants overexpressing NADPH-dependent HC-toxin reductase (*YK1*) gene product, which possesses dihydroflavonol-4-reductase (DFR) activity and provides biotic and abiotic stress tolerance. Authors applied CE-MS to analyze polar metabolites involved in glycolysis, tricarboxylic acid cycle and pentose phosphate pathways. The analyses indicated slight changes in the amounts of several metabolites in YK1-overexpressing plant tissues when compared with those of the same rice variety containing the hygromycin-resistant gene (vector alone). For instance, *cis*-aconitate, isocitrate and 2-oxoglutarate levels were higher in leaves, fructose-1,6-bisphosphate and glyceraldehyde-3-phosphate levels were lower in roots, and glutathione levels were significantly increased in calli. Although most of the metabolic changes could not be directly associated with the function of the transgene, authors hypothesized potential links between the elevation of glutathione level in calli, NADPH levels and DFR activity. Jiao *et al.* [29] reported a comprehensive metabolomics approach based on various analytical techniques including NIRS, GC-MS, LC, ICP-AES and chemometrics for the discrimination of three GM rice varieties from their conventional counterparts. The GM rice varieties included: (1) rice with resistance to blast, bearing four antifungal genes, *RCH10*, *RAC22*,*β-1,3-Glu*, and *B-RIP*; (2) rice with resistance to sheath blight, transformed to contain a rice chitinase gene, an alfalfa β-1,3-glucanase gene, and *p35H* containing a hygromycin phosphotransferase gene; and (3) rice with resistance to insects, containing *sck*, trypsin proteinase inhibitor derived from cowpea and *cry1Ac* gene from *Bacillus thuringiensis*. The levels of several amino acids, fatty acids, and vitamins, were altered to different extents in GM rice samples, suggesting that these unintended compositional alterations may be related to the genetic transformation. More recently, the metabolite profiles of a GM rice line C418-Xa21 (with bacterial blight-resistance genes), and two non-GM lines, C418 and C418/Xa23, were investigated using GC-MS [30]. After GC-MS analysis, cluster analysis (CA), PCA and PLS-DA were used to find differences between metabolic fingerprints obtained from different samples. As a conclusion of their study, authors indicated that the GM rice line was substantially equivalent to traditional cultivars apart from the change detected for succinic acid levels that fell outside the boundaries of natural variability observed between the two non-GM varieties. A similar approach, but using a different analytical platform, was adopted by Chang *et al.* [31] to investigate unintended effects of transgenic rice with *cry1Ac* and *sck*genes. In that work, LC-MS-based metabolomics in combination with PCA and PLS-DA were used to find the metabolites that permitted differentiating insect-resistant GM rice from its native counterparts. The authors also considered different sowing dates or locations as source of metabolite variation. Their findings suggested that environmental factors played a greater role than gene insertion for most metabolites. Although results also indicated slight variations in the levels of phytosphingosine, palmitic acid, 5-hydroxy-2-octadenoic acid and three other unidentified metabolites in the GM rice variety, the changes could not be related to the transgene.

The improvement of the nutritional properties in staple foods, such as rice, may have a major impact on the quality of life of the world's population. The over-accumulation of primary or secondary metabolites caused by the introduction of the genetic modification may affect unexpected processes in the plant's physiology through a sequence of several events that may include altered gene expression. Rice has been modified not only to improve agronomic traits but also to enhance its nutritional properties. The effect of the insertion of a cytoplasmic-localized *AGPase* mutant gene from *Escherichia coli* to enhance starch synthesis in rice was evaluated using LC-MS [32]. The studied GM rice lines showed elevated levels of ADP-glucose, accordingly with their higher AGPase activity. The levels of glucose 1-phosphate, UDP-glucose and glucose 6-phosphate were also elevated to the same relative extent in the GM lines compared with the WT rice line. A putative explanation for these changes was the inefficient utilization of ADP-glucose for starch synthesis due to a limitation in their transport into the amyloplast or as substrate by starch synthases. Glucose and fructose levels were also elevated in the GM rice. However, analysis of metabolite ratios showed no significant differences due to genetic manipulation. The same LC-MS analytical platform for metabolite profiling was applied to evaluate the additional effects of the tryptophan-fortified rice line by genetic engineering [33]. No marked effects on the amounts of other major metabolites were described in that work. However, uneven distribution of tryptophan in the plants was described in a time-dependent manner, with the highest level being observed in young developing tissues. Kim *et al.* [34] have recently studied the substantial equivalence between carotenoid fortified GM rice and five conventional rice cultivars (three white and two red grain colors) using GC-MS-based profiling of polar metabolites. It was suggested that GM rice was substantially equivalent to its conventional counterpart since GM rice was indistinguishable from the non-GM counterpart using the proposed non-targeted approach.

The GM-rice lines with increased lysine levels developed by Long *et al.* [35] are representative examples of essential nutrient improvement in crops by genetic engineering. It is noteworthy to mention that lysine is the first limiting essential amino acid in cereal grains (including rice). Authors genetically engineered rice to increase lysine levels following three different strategies: (1) expressing bacterial genes to enhance lysine biosynthesis; (2) using RNA interference of rice genes to down-regulate its catabolism; and (3) combining 1 and 2 to achieve both metabolic effects. The developed GM-rice plants contained free lysine levels increased up to ~12-fold in leaves and ~60-fold in seeds. In this work, the LC-MS technique was highly valuable for investigating the profile of 11 intermediate compounds involved in the Lys metabolic pathway in both rice seeds and leaves. These analyses revealed the existence of lysine catabolism in leaves and different regulatory mechanisms of lysine accumulation between rice leaves and seeds.

Maize

Maize (*Zea mays* L.) is another important crop worldwide for food, animal feed and bioenergy production [87]. GM maize, covering about 25% of total grown maize, has an important place in agriculture. Metabolomics has proved to be useful for predicting important agronomic traits. A first attempt to identify and classify maize seeds obtained from GM maize plants containing the *Cry1Ab* transgene, following a metabolomic fingerprinting approach, was carried out by Manetti *et al.* [36,37]. They demonstrated the capabilities of NMR and multivariate statistical data analysis to classify the maize seeds GM plants and their non-GM counterparts without the need of a complete assignment of the spectra. In another published report, a similar approach was used to study the introduction of the antisense-mediated down-regulation and over-expression of the *Rpd3* gene in the genome of a maize inbred line [38]. Piccioni *et al.* [39] also applied NMR to profile the metabolome of insect-resistant GM maize containing the*cry1A(b)* gene. Using this technique, 40 water-soluble metabolites were identified in all samples; nevertheless, based on the quantitative data, multivariate analysis allowed the discrimination between GM and non-GM samples. The metabolites responsible for such discrimination were ethanol, citric acid, glycine-betaine and trehalose, which showed higher levels in the GM maize samples. A clear link between the alterations in the concentration of these metabolites and the genetic modification was not found.

CE-MS has also proved to be helpful on the detection of the statistically significant differences in the metabolic profiles of varieties of conventional and insect-resistant GM maize [40]. Main differences were observed in the levels of L-carnitine and stachydrine between conventional and GM maize. The potential of combining two MS-based metabolomics approaches (namely, FTIR-MS and CE-MS) and pressurized liquid extraction (PLE), a green extraction technology, was evaluated later by the same research group to study GM maize [41]. Three GM varieties of insect-resistant maize and their corresponding isogenic lines grown under the same field conditions were analyzed. In that work, it was found that some metabolic pathways (amino acid, purine metabolism and folate biosynthesis, among others) were clearly altered in the GM varieties compared with their respective isogenic lines.

The amino acids profiles of insect-resistant GM maize (containing*cry1Ab* gene) and herbicide-tolerant soybean (containing CP4-EPSPS construct) were studied by GC-MS [42]. In that case, fast recovery of amino acids from maize and soybean grains was achieved with supercritical fluid extraction (SFE) with modified CO_2. Following that procedure, various differences were detected in the amino acid profiles in GM maize and GM soybean when compared with their corresponding isogenic non-modified varieties. However, a direct association between the observed changes and the genetic modification could not be found. Following a similar analytical approach by GC-MS, the obtained profile of major and minor fatty acids in insect-resistant GM (*cry1Ab* gene) maize indicated high similarity when compared with its isogenic line grown in the same conditions [43]. On the contrary, the metabolic profiling by GC-MS carried out by Skogerson *et al.* [88] revealed that metabolome content was highly dependent on genotypic variation. More recently, Frank *et al.* [44] used also GC-MS profiling to investigate the impact of genetic modifications of insect-resistant maize (DKC78-15B, TXP 138F) and herbicide-tolerant maize (DKC78-35R)*versus* environmental influences (maize were grown together in different areas). The majority of differences observed were related to environmental factors rather than to the genetic modifications. Among them, location and season were predominant factors on the variability of metabolite profiles (Figure 1). Asiago *et al.* [89] designed a complex experiment to study the potential of metabolomic approaches to elucidate the biological variation in the expression of many metabolites due to environment, genotype, or both. Using GC-MS, 156 and 185 metabolites were measured in grain and forage maize samples, respectively. A similar approach, LC-MS was applied to detect a total of 286 and 857 metabolites in grain and forage samples, respectively [90]. The results indicated that the environment had the highest impact on the relative amounts of metabolites in both grain and forage. Rivera*et al.* [45] developed an ultra-high performance liquid chromatography (UHPLC)-MS method for carotenoid profiling to characterize GM maize lines expressing several carotenogenic genes, obtaining satisfactory results in terms of recoveries (82%–108%), detection limits (0.02–0.07 µg/mL) and repeatability (better than 13% Relative Standard Deviation, RSD) [45].

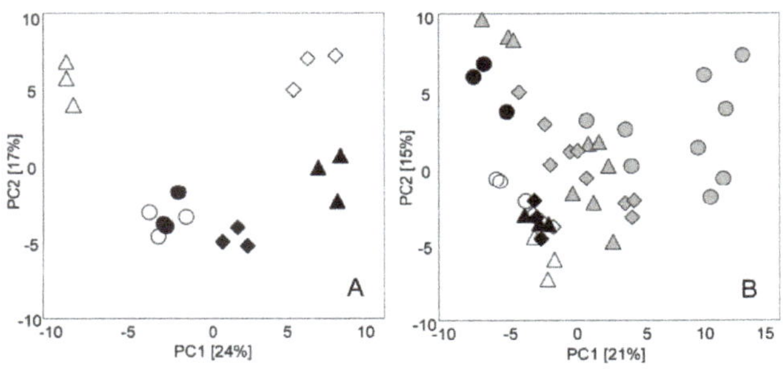

Figure 1: Principal component analysis of GC-MS metabolite profiling data (triplicate analysis of combined fractions I–IV) of Bt maize (Δ, ▲), Roundup Ready maize (◊, ♦), and the near-isogenic counterpart (○, ●) .

In general, studies are focused on acquiring only one omic level. Few attempts have been made to achieve comprehensive omic profiling in GM plants [82]. In this sense, Barros *et al.* [46] published an exploratory study about the use of different omics techniques to study the transcriptome, proteome and metabolome of two GM maize varieties (Bt and RR) for safety evaluation purposes. The effects of environmental conditions (year of harvest, agricultural practices and location) on GM maize were also studied at the three levels of information. Analysis using 1H-NMR and GC-MS platforms for metabolomics, gene expression microarray for transcriptomics and 2-D gel electrophoresis (2-DGE) analysis for proteomics revealed that the environment was shown to cause more variation in the gene, protein and metabolite expression of the maize samples than the different genotypes.

Soybean and Other Legumes

Soybean (*Glycine max*), classified under plant legume, is an important source of vegetable oil and protein. The GM soybean variety tolerant to glyphosate herbicide is one of the most extended GM crops in the world. The first work related to substantial equivalence of GM soybean through a metabolomic approach was carried out by García-Villalba *et al.* [47]. In that work, authors developed a CE-MS-based analytical strategy to compare the metabolic profile of conventional and GM soybean (*CP4 EPSPS* construct) [47]. More than 45 different metabolites, including carboxylic acids, isoflavones, and amino acids were tentatively identified. Among metabolite differences between GM and conventional soybean, most noteworthy differences were found in the concentration of proline, histidine, asparagine and 4-hydroxy-L-threonine. The latter, disappeared in the GM soybean compared to its parental non-GM line. In a comprehensive study, Inaba *et al.* [48] investigated the metabolite profile of herbicide-tolerant GM soybean, expressing an anthranilate synthase (*ASA2*) gene that is characterized by the accumulation of tryptophan in leaves, seeds and embryogenic cultures [48]. Metabolite profiles of different tissues were obtained by GC-MS revealing slight elevation of tyrosine and phenlylalanine levels in the GM soybean line. As authors mentioned in their work, a possible explanation for the elevation of these two amino acids in cells, that have increased levels of tryptophan, could be a feedback insensitive anthranilate synthase due to a mutation or transgene insertion, but further experiments are required to confirm this point. Also, Giuffrida*et al.* [49] studied the herbicide-tolerant GM soybean line using a chiral CE-MS method. Authors observed some quantitative differences in the chiral amino acid profile between the GM soybean and the untransformed genotype. However, further work remains to be done in order to investigate whether these differences could be a direct consequence of the genetic transformation. In a recent work, the issue of substantial equivalence assessment for herbicide-tolerant GM soybean has been addressed [50]. Thus, to study the metabolome within the context of the natural variation, 49 conventional soybean lines and one GM line were analyzed. Using LC-MS and GC-MS, the metabolome of the GM soybean presented no significant deviation from natural variation (represented in Figure 2) with the exception of changes in the targeted engineered pathway.

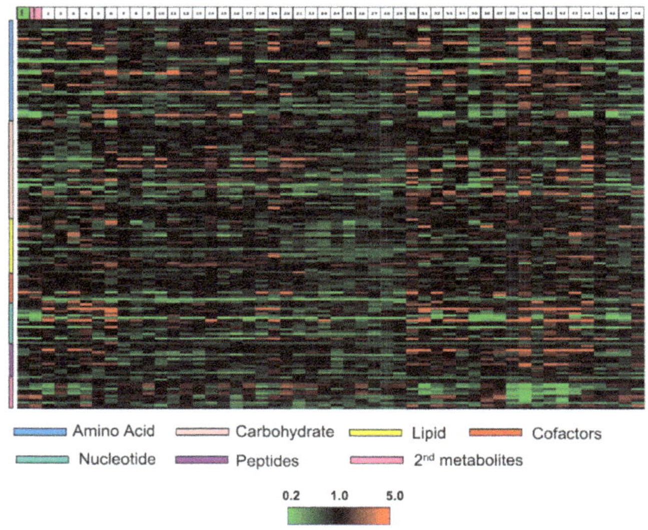

Figure 2: Metabolomic profiles and hierarchical clustering of 169 metabolites across the 49 soybean conventional lines and one GM line. The mean values for 8 biological replications per line were shown. Red and green indicate high and low levels, respectively, .

In a recent paper, Kusano *et al.* [51] have presented the results obtained from a comprehensive metabolomic study on a soybean lineage representing 35 years of breeding and increasing yield potential and three glyphosate-tolerant GM lines. The analytical strategy combined CE-MS, GC-MS, LC-MS and ICP-MS with multivariate analysis to successfully discriminate samples. The differences between the conventional and GM lines were small, indicating that genetic modification is not an important contributor to metabolite variability. Metabolomic data indicated that differences between older and newer soybean varieties provided novel and significant information on the impact of varietal development on biochemical variability, suggesting that safety assessments will need to consider that transgene insertion is not a major source of metabolite variability.

Apart from soybean, alfalfa (*Medicago sativa* L.) is a leading forage legume crop due to its high nutritive value. Several alfalfa lines have been genetically engineered to lower its lignin content with the final goal of improving alfalfa digestibility in ruminants. Thus, Chen *et al.*[52] used LC with UV detection to analyze soluble phenolics, wall-bound phenolics and cell wall lignins from GM alfalfa (with genetically downregulated *O*-methyltransferase genes) samples to study the effects of single gene disruption in the monolignol branch of phenylpropanoid biosynthesis. The results indicated that although the genetic modification decreased lignin biosynthesis, it also affected the levels of cell wall-bound ferulic acid. Pea plant, another well-known legume, has also been object of metabolomics studies. For instance, six independent lines of GM pea, transformed with a plasmid containing five transgenes and a *Ds* transposable element, were studied using NMR [53]. In their study, authors included the analysis of two control groups, the non-transformed pea plant control and the null segregant control from which the transgene has been lost. Multivariate analysis on the basis of PCA and linear discriminant analysis (LDA) failed to provide an acceptable classification of samples. In addition, statistical analysis revealed similar results, suggesting that the null segregant group was significantly different from the wild type, indicating that factors other than the presence of the transgene had significant effects on the metabolite profile.

Wheat

While much of the produced maize and soybeans are directed to feed animals or to be transformed into bioethanol, most wheat is consumed by humans as bread or pasta. Due to the socio-economic impact of this crop, GM wheat varieties have not legally been approved yet. Baker *et al.* [54] investigated the substantial equivalence of a series of GM wheat lines, transformed with *Glu-A1* and *Glu-D1* genes, with improved processing quality. In their study, the authors used NMR and multivariate analysis to compare the metabolite profiles of the GM wheat lines with their corresponding parental lines. The main differences between GM and parent lines were found in maltose and sucrose levels. However, a detailed explanation of the potential link between the observed metabolic changes and the genetic modification was not provided in that work. In the same study, GC-MS was employed for amino acid profiling, observing that differences between the control and GM lines were within the same range as the differences observed due to environmental factors (location and year).

Potato

Potato (*Solanum tuberosum* L.) is another relevant food crop in the world. It is also an alternative source of raw material for bioethanol production [91]. Potato has also been subject of a several metabolomic studies [92]. In a series of works, Roessner *et al.* [55,56,57] have demonstrated the potential of GC-MS for metabolite profiling of GM potato tubers. GC-MS was also the analytical platform of choice for the analysis of key compounds to investigate unintended effects in GM potato lines with altered carbohydrate metabolism (up-regulated or down-regulated fructokinase gene expression) [6]. The statistically significant differences between non-GM and GM lines were not associated with any specific construct. Interestingly, significant differences were also observed between non-transformed tubers and both, tissue culture derived tubers and tubers obtained from transformation with an empty vector, suggesting the occurrence of somaclonal variation. Global analysis of metabolite content using LC-MS and GC-MS was carried out to evaluate the degree of similarity between inulin-producing GM potatoes, transformed with *1-fructosyltransferase 1-SST*) and *fructan:fructan 1-fructosyltransferase 1-FFT*) genes, and conventional cultivars [58]. Multivariate analysis of chromatographic data showed that the most discriminatory ions with a significant impact on genotype separation were predicted to represent fructans of different degree of polymerization (DP), Figure 3.

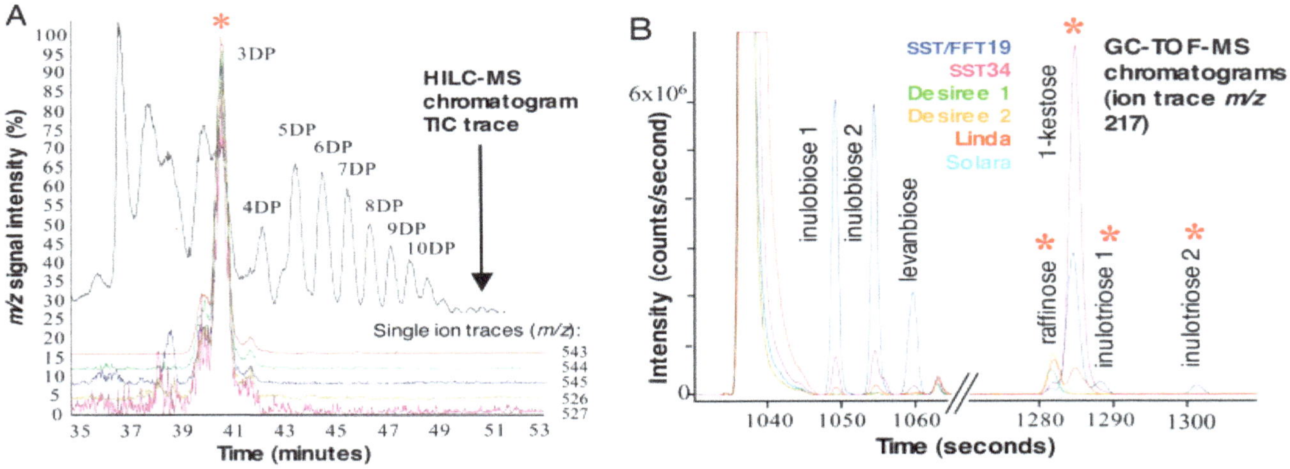

Figure 3: Identification of discriminatory metabolites in GM potato lines, some of them expressing *1-SST* and *1-FFT*(*SST/FFT*), and others expressing *1-SST*(*SST*), by LC-MS and GC-MS. (**A**) Overlaid single-ion chromatograms from LC-MS analysis of top-ranked predicted.

Other relevant transgenic modification was based on the construction of virus resistant potato plants. Tubers were studied by CE-MS in order to determine and compare their glycoalkaloid content with equivalent non-GM potato variety [59]. CE-MS analysis revealed no substantial differences in GM tubers. Apart from targeted changes and the observed large variation in metabolite profile between the conventional cultivars, GM potatoes under study were shown to be substantially equivalent to traditional cultivars. In a recent work, differences between GM lines, containing dehydration response element-binding protein 1A transgene, and the non-GM parent cultivars were investigated [60]. In that work, LC-MS profiling revealed higher levels of the glutathione metabolite, γ-aminobutyric acid and β-cyanoalanine, a byproduct of ethylene biosynthesis, in the GM lines. Combinations of metabolic information provided by different analytical platforms (NMR and LC-UV) have been demonstrated to be largely complementary in terms of metabolites detected in potato tubers [61]. Information obtained from both analytical platforms was combined with the aim to detect unintended effects. Once more, it was found that the largest differences were found not between the GM potatoes and controls but between conventional varieties.

The responses to prolonged drought stress of wild type White Lady and the GM drought-tolerant TPS1 potato lines were studied by Kondrák *et al.* [62] at both transcriptional and metabolic levels. The expression of 57 genes was found to be altered in GM potato leaves compared to that in wild type potato leaves. Substantial increases in the detected proline, inositol and raffinose levels in the leaves of both potato lines seemed to be a general response to drought stress. In general, the biochemical changes detected did not clearly reflect the changes in gene expression. Authors concluded that inositol synthesis was influenced by transcriptional and/or biochemical changes induced, not exclusively by drought, but also by the transgene expression, whereas raffinose and proline synthesis were drought-specific.

Tomato

Tomato (*Lycopersicon esculentum*) is also a major food crop worldwide. The first work on metabolite comparison based on NMR analysis suggested minimal variations between isogenic non-GM and GM lines across sites or seasons (about 95% of all analyzed metabolites presented the same concentration) [63]. A metabolite profiling methodology, based on the combined use of GC-MS, conventional spectrophotometric LC and bioinformatic tools, was employed to study the influence of hexokinase activity on tomato fruit metabolism. To achieve this, different tissues of GM tomato plants overexpressing hexokinase gene product were analyzed at different developmental stages [64]. Their comprehensive analysis revealed some interesting findings regarding the influence of hexokinase on primary metabolism and its dependence on the environmental factors. A NMR study by Le Gall *et al.* [65,66] showed a significant increase of kaempferol glycosides in the flesh of GM tomato over-expressing maize transcription factors *LC* and *C1*. Apart from the significantly increased content of several flavonoid glycosides, the levels of other unrelated metabolites such as citric acid, sucrose, phenylalanine, and trigonelline, among others, were found to be different in GM tomato. However, the reported changes in mean values were relatively minor (less than 3-fold) and within the natural variation that would be observed in a field-grown crop. Long *et al.* [67] used LC to profile carotenoid and phenolic compounds of a panel of tomato lines representing a range of phenotypes of carotenoids or flavonoids levels (wild-type, mutants and various GM tomato lines). The GM lines included the *DXS* up-regulated plants which contained increased carotenoids, and the CRTI line, containing a bacterial (*Erwinia uredovora*) desaturase, which resulted in fruit with elevated (4.0-fold) β-carotene content and lutein levels and reduced (10-fold) phytoene levels. In the same investigation, tomato lines manipulated in cytochrome P450 ferulate 5-hydroxylase to increase the ferulate levels were also included. The study was aimed at investigating how manipulation of carotenoids or flavonoids pathways may affect other secondary metabolism pathways. It could be observed that perturbations to the biosynthesis of flavonoids or carotenoids independently did not affect the overall content of these relevant compounds. LC-MS was also employed for the analysis of GM tomato plants over-expressing grape stilbene synthase genes [68]. The analysis revealed that the genetic modification of the tomato plants originated from different levels of accumulation of four stilbenes (*i.e.*, *trans*-and*cis*-piceid and *trans*- and *cis*-resveratrol), depending on the stages of ripening. Other metabolites (rutin, naringenin, and chlorogenic acid) were suggested to be related to the genetic transformation. More recently, characterization of the GM tomato metabolome was approached by Kusano *et al.* [69]. In that study, two GM tomato varieties over-expressing miraculin glycoprotein and a panel of six traditional tomato cultivars were selected to prove a methodology based on three different metabolomics platforms. More specifically, data from CE-MS, GC-MS and LC-MS were summarized in single consensus datasets for further multivariate analysis. The combination of the three platforms allowed the statistical analysis of datasets containing over 175 unique tentatively identified metabolites and more than 1400 peaks with no or imprecise metabolite annotation. This analytical setup provided metabolite coverage of 85% of the chemical diversity found in the LycoCyc database. Results showed that >92% of the tested peaks in the transgenic lines deviated less from the control line than the accepted limit estimated using the reference panel of traditional cultivars.

Other GM Crops

GM tobacco has also been investigated in metabolomics studies. Choi*et al.* [70] applied 1H NMR and multivariate analysis techniques to differentiate wild type and GM tobacco plants overexpressing salicylate biosynthetic genes. The major compounds contributing to the discrimination were chlorogenic acid, malic acid, glucose and sucrose. Based on literature data, authors discussed potential associations between those altered metabolite levels and the increase of salicylic acid levels in plants; however, none of these changes in primary metabolites were directly related with salicylic acid biosynthesis. Sobolev *et al.* [71,72,73] published a series of papers where they applied NMR to investigate different metabolic aspects of GM lettuce with enhanced growth properties (over-expressing the asparagine synthetase A gene from *E. coli*). Statistical analysis of NMR data demonstrated significant increases in content of short-chain inulin oligosaccharides in GM lettuce leaves compared to those detected in the wild type plant. Interestingly, that was considered an unexpected effect because the transgenes aims at modifying the asparagine level, together with the nitrogen status, rather than the carbohydrate content [71,72,73].

An important application of transgenic modification is focused on the improvement of organoleptic characteristics of food. For instance, a GM cucumber plant has been transformed with the thaumatin-II gene to improve its taste [74]. The GC-MS analysis of five GM cucumber varieties bearing that transgene and its non-modified counterparts revealed that GM lines differ in their metabolic profiles and that those differences could be associated to the transgene integration site. However, the range of some of the observed changes was narrow, and authors classified them as somaclonal effects. Also, common changes in phenylalanine, aspartate, ethanolamine, pipecolate and benzoic acid levels were detected in the GM lines that could not be linked to the genetic modification. Another GM cucumber plant with enhanced aroma properties did not show significant differences when compared with its parental non-GM line by GC-MS [75]. Maolowicki *et al.* [76] studied the impact in certain metabolites in GM raspberry resistant to bushy dwarf virus (RBDV). Volatile compounds of RBDV-resistant raspberry lines were analyzed by GC-MS and compared with the wild-type. Whereas no flavor compounds tested in this study showed any difference between the GM lines and the wild-type raspberry, much larger variations were observed between sites and harvest seasons. The content of phenolic compounds and volatile secondary metabolites was investigated by GC-MS and LC-MS in GM grapevine plants over- and under-expressing alcohol dehydrogenase (*adh*) gene [77]. In that case, the main goal was to study the putative role of alcohol dehydrogenase in plant development and response to stress. Metabolite profiles indicated some differences in the degree of polymerization of proanthocyanidins, as well as increased levels of sucrose, carotenoid- and shikimate-derived volatiles in the GM plant. Supported by literature data, the authors discussed that some link could exist between alcohol dehydrogenase activity and the changes observed in sucrose metabolism. On the other side, the changes observed in secondary metabolites were not directly related with the role of alcohol dehydrogenase in the plant metabolism.

LC-DAD-based metabolic fingerprinting approach was used to evaluate undesirable changes in GM Chinese cabbage containing the *bar* gene [79]. In addition to genetic modification, sample periods (4- and 8-week old plants) were also subjected to study through metabolome analysis. It was observed that the time of samplings affected the metabolome in a higher extent than the genetic modification.

The advantages of using multiple analytical platforms to explore the composition of complex samples become evident. Adopting this strategy, complementary metabolomic information was obtained on the compositional differences between papaya (*Carica papaya* L.) transformed with the replicase gene for resistance to papaya ringspot virus (PRSV) and the non-GM counterpart [80,81]. Profiles of volatile organic compounds, sugar/polyals, organic acids, carotenoids and alkaloids in GM and non-GM papaya were obtained by GC-MS and LC-MS analytical platforms [81]. The metabolite variation between GM and non-GM papaya was slight during the two harvesting times studied in the work. Papayas harvested across different time periods showed a higher degree of compositional variability.

Transcriptome and metabolome analysis on GM barley plants (with different disease-resistance and nutritional traits) and their conventional counterparts grown in field with and without amendment of soil with mycorrhizal (Amykor) was performed by Kogel *et al.* [83]. In their work, a comparative analysis of 72 metabolites obtained in different culture conditions revealed slight differences in the abundance of some of them, as can be seen in Figure 4. The overall conclusion is that cultivar-specific differences in transcriptome and metabolome greatly exceed effects caused by transgene expression.

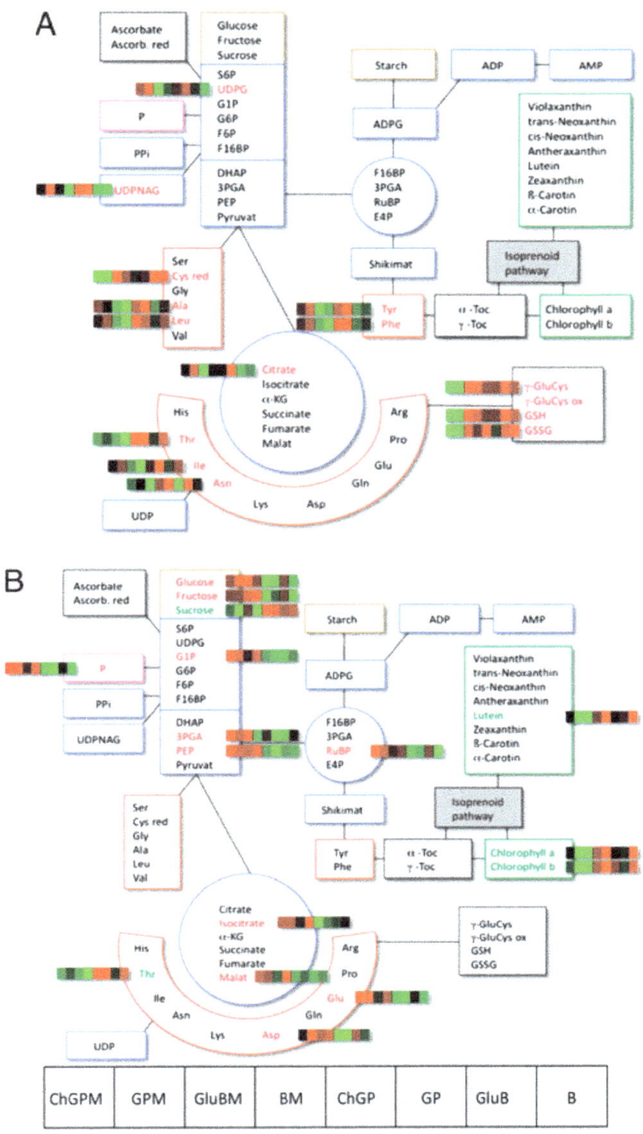

Figure 4: Differentially abundant metabolites in barley leaves. Overview of differentially abundant metabolites from the targeted profiling approach with leaf material from 4-month-old, field-grown barley plants representing the treatments (**A**) cultivar or (**B**) Amykor. ...

CONCLUDING REMARKS

A large number of new transgenic varieties of crops with desired traits are rapidly being introduced in the global market. The application of metabolomics for the safety assessment of GMOs is providing relevant information regarding the associated metabolite alteration as a result of gene modification. Exhaustive, unbiased metabolic profiling or fingerprinting of plants is greatly accelerating in the last years as demonstrated in this review. The chemical complexity of the plant metabolome, as well as the large dynamic range of concentrations are major challenges to be faced by metabolomics. To address these challenges, advances in analytical platforms have played a key role to unravel potential GM effects at the molecular level. Metabolomics studies (e.g., comparing GM crops with their wild type parent lines) are often combined with different culture conditions, from more geographical locations, multiple years, and different growing seasons,*etc.*, in order to investigate the natural metabolome variability. In general, results show that compared with genetic modifications, environmental variations usually produce greater major differences in metabolome composition.

Although the applications of metabolomics have not been validated yet within the regulatory framework for food safety assessment, it can be considered a powerful approach and a key strategy to screen compositional changes increasing the possibility of detecting unintended effects associated with genetic modification in GM plants. However, in order to understand the biological significance and impact of the detected changes, the large amount of data generated in metabolomics studies needs to be processed, integrated and interpreted together with data generated by other high-throughput technologies such as proteomics and transcriptomics as proposed by the new Foodomics strategy [93]. Although combinations of different omics technologies have been applied for the analysis of molecular alterations in GMOs, none of these studies have reported substantial correlations due to the lack of suitable integrative strategies. In this sense, the development of appropriate Systems Biology approaches providing new means to integrate and summarize omics data will help on the applicability of global approaches that can reveal new relationships, which cannot be found otherwise. However, before data integration procedures can be practically applied to GMO analysis, much effort will have to be made to develop effective integrative statistical approaches and appropriate computational frameworks for describing molecular systems and connecting omics databases. In this context, the development and implementation of initiatives to generate tools and databases that allow sharing metabolite data obtained from metabolomics studies will also be important [94].

ACKNOWLEDGMENTS

This work was supported by an AGL2011-29857-C03-01 project (Ministry of Education and Science, Spain). Alberto Valdés thanks the Ministry of Economy and Competitiveness for his FPI pre-doctoral fellowship.

REFERENCES

1. Lönnerdal B. Genetically modified plants for improved trace element nutrition. J. Nutr. 2003;133:1490S–1493S.
2. McGloughlin M.N. Modifying agricultural crops for improved nutrition. New Biotechnol. 2010;30:494–504.
3. Dunwell J.M. Transgenic cereals: Current status and future prospects. J. Cereal Sci. 2014;59:419–434.
4. Cellini F., Chesson A., Colquhoun I., Constable A., Davies H.V., Engel K.H., Gatehouse A.M., Kärenlampi S., Kok E.J., Leguay J.J., et al. Unintended effects and their detection in genetically modified crops.Food Chem. Toxicol. 2004;42:1089–1125.
5. Ioset J.R., Urbaniak B., Ndjoko-Ioset K., Wirth J., Martin F., Gruissem W., Hostettmann K., Sautter C. Flavonoid profiling among wild type and related GM wheat varieties. Plant Mol. Biol.2007;65:645–654.
6. Shepherd L.V.T., McNicol J.W., Razzo R., Taylor M.A., Davies H.V. Assessing the potential for unintended effects in genetically modified potatoes perturbed in metabolic and developmental processes. Targeted analysis of key nutrients and anti-nutrients. Transgenic Res.2006;15:409–425.
7. Millstone E., Brunner E., Mayer S. Beyond "substantial equivalence"Nature. 1999;401:525–526.
8. EFSA. Guidance document of the Scientific Panel on Genetically Modified Organisms for the risk assessment of genetically modified plants and derived food and feed. EFSA J. 2006;99:1–100.
9. Ad Hoc Technical Expert Group (AHTEG) Final report of the ad hoc technical expert group on risk assessment and risk management under the Cartagena protocol on biosafety; In Proceedings of the United Nations Environment Programme Convention for Biodiversity; Ljubljana, Slovenia. 20–23 April 2010; [(accessed on 15 October 2014)]. Available online: http://www.cbd.int/doc/meetings/bs/bsrarm-02/official/bsrarm-02-05-en.pdf.
10. Chassy B.M. Can –omics inform a food safety assessment? Regul. Toxicol. Pharmacol. 2010;58:S62–S70.
11. Valdés A., Simó C., Ibáñez C., García-Cañas V. Foodomics strategies for the analysis of transgenic foods. TrAC Trends Anal. Chem. 2013;52:2–15.
12. Schauer N., Fernie A.R. Plant metabolomics: Towards biological function and mechanism. Trends Plant Sci. 2006;11:508–516.
13. Hall R.D. Plant metabolomics: From holistic hope, to hype, to hot topic. New Phytol. 2006;169:453–468.
14. Harrigan G.G., Martino-Catt S., Glenn K.C. Metabolomics, metabolic diversity and genetic variation in crops. Metabolomics.2007;3:259–272.
15. Dixon R.A., Gang D.R., Charlton A.J., Fiehn O., Kuiper H.A., Reynolds T.L., Tjeerdema R.S., Jeffery E.H., German J.B., Ridley W.P., et al. Applications of metabolomics in agriculture. J. Agric. Food Chem. 2006;54:8984–8994.
16. Seger C., Sturm S. Analytical aspects of plant metabolite profiling platforms: Current standings and future aims. J. Proteome Res.2007;6:480–497.
17. Hegeman A.D. Plant metabolomics-meeting the analytical challenges of comprehensive metabolite analysis. Brief. Funct. Genomics. 2010;9:139–148.
18. Eisenreich W., Bacher A. Advances of high-resolution NMR techniques in the structural and metabolic analysis of plant biochemistry. Phytochemistry. 2007;68:2799–2815.
19. Pan Z., Raftery D. Comparing and combining NMR spectroscopy and mass spectrometry in metabolomics. Anal. Bioanal. Chem.2007;387:525–527.
20. García-Cañas V., Simó C., León C., Ibáñez E., Cifuentes A. MS-based analytical methodologies to characterize genetically modified crops. Mass Spectrom. Rev. 2011;30:396–416.
21. Herrero M., Simó C., García-Cañas V., Ibáñez E., Cifuentes A. Foodomics: MS-based strategies in modern food science and nutrition.Mass Spectrom. Rev. 2012;31:49–69.

22. Ibáñez C., García-Cañas V., Valdés A., Simó C. Novel MS-based approaches and applications in food metabolomics. TrAC Trends Anal. Chem. 2013;52:100–111.

23. Takats z., Wiseman J.M., Cooks R.G. Ambient mass spectrometry using desorption electrospray ionization (DESI): Instrumentation, mechanisms and applications in forensics, chemistry, and biology. J. Mass Spectrom. 2005;40:1261–1275.

24. Hoekenga O.A. Using metabolomics to estimate unintended effects in transgenic crop plants: Problems, promises, and opportunities. J. Biomol. Tech. 2008;19:159–166.

25. Keymanesh K., Darvishi M.H., Sardari S. Metabolome comparison of transgenic and non-transgenic rice by statistical analysis of FTIR and NMR spectra. Rice Sci. 2009;16:119–123.

26. Zhou J., Ma C., Xu H., Yuan K., Lu X., Zhu Z., Wu Y., Xu G. Metabolic profiling of transgenic rice with *cryIAc* and *sck* genes: An evaluation of unintended effects at metabolic level by using GC-FID and GC-MS. J. Chromatogr. B. 2009;877:725–732.

27. Shin Y.M., Park H.J., Yim S.D., Baek N.I., Lee C.H., An G., Woo Y.M. Transgenic rice lines expressing maize *C1* and *R-S* regulatory genes produce various flavonoids in the endosperm. Plant Biol.2006;4:303–315.

28. Takahashi H., Hayashi M., Goto F., Sato S., Soga T., Nishioka T., Tomita M., Kawai-Yamada M., Uchimiya H. Evaluation of metabolic alteration in transgenic rice overexpressing dihydroflavonol-4-reductase. Ann. Bot. 2006;98:819–825.

29. Jiao Z., Si X.X., Li G.K., Zhang Z.M., Xu X.P. Unintended compositional changes in transgenic rice seeds (*Oryza sativa* L.) studied by spectral and chromatographic analysis coupled with chemometrics methods. J. Agric. Food Chem. 2010;58:1746–1754.

30. Wu J., Yu H.C., Dai H.F., Mei W.L., Huang X., Zhu S.F., Peng M. Metabolite profiles of rice cultivars containing bacterial blight-resistant genes are distinctive from susceptible rice. Acta Biochim. Biophys. Sin.2012;44:650–659.

31. Chang Y., Zhao C., Zhu Z., Wu Z., Zhou J., Zhao Y., Lu X., Xu G. Metabolic profiling based on LC/MS to evaluate unintended effects of transgenic rice with *cry1Ac* and *sck* genes. Plant Mol. Biol.2012;78:477–487.

32. Nagai Y.S., Sakulsinghroj C., Edwards G.E., Satoh H., Greene T.W., Blakeslee B., Okita T.W. Control of starch synthesis in cereals: Metabolite analysis of transgenic rice expressing an up-regulated cytoplasmic ADP-Glucose pyrophosphorylase in developing seeds. Cell Physiol. 2009;50:635–643.

33. Matsuda F., Ishihara A., Takanashi K., Morino K., Miyazawa H., Wakasa K., Miyagawa H. Metabolic profiling analysis of genetically modified rice seedlings that overproduce tryptophan reveals the occurrence of its inter-tissue translocation. Plant Biotechnol.2010;27:17–27.

34. Kim J.K., Park S.Y., Lee S.M., Lim S.-H., Kim H.J., Oh S.-D., Yeo Y., Cho H.S., Ha S.H. Unintended polar metabolite profiling of carotenoid-biofortified transgenic rice reveals substantial equivalence to its non-transgenic counterpart. Plant Biotechnol. Rep. 2013;7:121–128.

35. Long X., Liu Q., Chan M., Wang Q., Sun S.S. Metabolic engineering and profiling of rice with increased lysine. Plant Biotechnol. J. 2013;11:490–501.

36. Manetti C., Bianchetti C., Bizzarri M., Casciani L., Castro C., D'Ascenzo G., Delfini M., di Cocco M.E., Lagana A., Miccheli A., et al. NMR-based metabonomic study of transgenic maize.Phytochemistry. 2004;65:3187–3198.

37. Manetti C., Bianchetti C., Casciani L., Castro C., di Cocco M.E., Miccheli A., Motto M., Conti F. A metabonomic study of transgenic maize (*Zea mays*) seeds revealed variations in osmolytes and branched amino acids. J. Exp. Bot. 2006;57:2613–2625.

38. Castro C., Manetti C. A multiway approach to analyze metabonomic data: A study of maize seeds development. Anal. Biochem. 2007;371:194–200.

39. Piccioni F., Capitani D., Zolla L., Mannina L. NMR metabolic profiling of transgenic maize with the *Cry1A(b)* gene. J. Agric. Food Chem. 2009;57:6041–6049.

40. Levandi T., Leon C., Kaljurand M., García-Cañas V., Cifuentes A. Capillary electrophoresis time-of-flight mass spectrometry for comparative metabolomics of transgenic *versus* conventional maize.Anal. Chem. 2008;80:6329–6335.

41. Leon C., Rodriguez-Meizoso I., Lucio M., Garcia-Cañas V., Ibañez E., Schmitt-Kopplin P., Cifuentes A. Metabolomics of transgenic maize combining Fourier transform-ion cyclotron resonance-mass spectrometry, capillary electrophoresis-mass spectrometry and pressurized liquid extraction. J. Chromatogr. A. 2009;1216:7314–7323.

42. Bernal J.L., Nozal M.J., Toribio L., Diego C., Mayo R., Maestre R. Use of supercritical fluid extraction and gas chromatography-mass spectrometry to obtain amino acid profiles from several genetically modified varieties of maize and soybean. J. Chromatogr. A.2008;1192:266–272.

43. Jimenez J.J., Bernal J.L., Nozal M.J., Toribio L., Bernal J. Profile and relative concentrations of fatty acids in corn and soybean seeds from transgenic and isogenic crops. J. Chromatogr. A.2009;1216:7288–7295.

44. Frank T., Röhlig R.M., Davies H.V., Barros E., Engel K. Metabolite profiling of maize kernels—genetic modification *versus* environmental influence. J. Agric. Food Chem. 2012;60:3005–3012.

45. Rivera S.M., Vilaró F., Zhu C., Bai C., Farré G., Christou P., Canela-Garayoa R. Fast quantitative method for the analysis of carotenoids in transgenic maize. J. Agric. Food Chem. 2013;61:5279–5285.

46. Barros E., Lezar S., Anttonen M.J., van Dijk J.P., Röhlig R.M., Kok E.J., Engel K.H. Comparison of two GM maize varieties with a nearisogenic non-GM variety using transcriptomics, proteomics and metabolomics. Plant Biotechnol. J. 2010;8:436–451.

47. García-Villalba R., León C., Dinelli G., Segura-Carretero A., Fernández-Gutiérrez A., Garcia-Cañas V., Cifuentes A. Comparative metabolomic study of transgenic *versus* conventional soybean using capillary electrophoresis-time-of-flight mass spectrometry. J. Chromatogr. A. 2008;1195:164–173.

48. Inaba Y., Brotherton J.E., Ulanov A., Widholm J.M. Expression of a feedback insensitive anthranilate synthase gene from tobacco increases free tryptophan in soybean plants. Plant Cell Rep. 2007;26:1763–1771.

49. Giuffrida A., León C., García-Cañas V., Cucinotta V., Cifuentes A. Modified cyclodextrins for fast and sensitive chiral-capillary electrophoresis-mass spectrometry. Electrophoresis. 2009;30:1734–1742.

50. Clarke J.D., Alexander D.C., Ward D.P., Ryals J.A., Mitchell M.W., Wulff J.E., Guo L. Asessment of genetically modified soybean in relation to natural variation in the soybean seed metabolome. Sci. Rep.2013;3 doi: 10.1038/srep03082.

51. Kusano M., Baxter I., Fukushima A., Oikawa A., Okazaki Y., Nakabayashi R., Bouvrette D.J., Achard F., Jakubowski A.R., Ballam J.M., et al. Assessing metabolomic and chemical diversity of a soybean lineage representing 35 years of breeding. Metabolomics. 2014 doi: 10.1007/s11306-014-0702-6.

52. Chen F., Duran A.L., Blount J.W., Sumner L.W., Dixon R.A. Profiling phenolic metabolites in transgenic alfalfa modified in lignin biosynthesis. Phytochemistry. 2003;64:1013–1021.

53. Charlton A., Allnutt T., Holmes S., Chisholm J., Bean S., Ellis N., Mullineaux P., Oehlschlager S. NMR profiling of transgenic peas. Plant Biotechnol. J. 2004;2:27–35.

54. Baker J.M., Hawkins N.D., Ward J.L., Lovegrove A., Napier J.A., Shewry P.R., Beale M.H. A metabolomic study of substantial equivalence of field-grown genetically modified wheat. Plant Biotechnol. J. 2006;4:381–392.

55. Roessner U., Wagner C., Kopka J., Trethewney R.N., Willmitzer L. Simultaneous analysis of metabolites in potato tuber by gas chromatography-mass spectrometry. Plant J. 2000;23:131–142.

56. Roessner U., Willmitzer L., Fernie A.R. High-Resolution metabolic phenotyping of genetically and environmentally diverse potato tuber systems. Identification of phenocopies. Plant Physiol. 2001;127:749–764.

57. Roessner U., Luedemann A., Brust D., Fiehn O., Linke T., Willmitzer L., Fernie A.R. Metabolic profiling allows comprehensive phenotyping of genetically or environmentally modified plant systems.Plant Cell. 2001;13:11–29.

58. Catchpole G.S., Beckmann M., Enot D.P., Mondhe M., Zywicki B., Taylor J., Hardy N., Smith A., King R.D., Kell D.B., et al. Hierarchical metabolomics demonstrates substantial compositional similarity between genetically modified and conventional potato crops. Proc. Natl. Acad. Sci. USA. 2005;102:14458–14462.

59. Bianco G., Schmitt-Kopplin P., Crescenzi A., Comes S., Kettrup A., Cataldi T.R. Evaluation of glycoalkaloids in tubers of genetically modified virus Y-resistant potato plants (var. Désirée) by non-aqueous capillary electrophoresis coupled with electrospray ionization mass spectrometry (NACE-ESI-MS) Anal. Bioanal. Chem. 2003;375:799–804.

60. Iwaki T., Guo L., Ryals J.A., Yasuda S., Shimazaki T., Kikuchi A., Watanabe K.N., Kasuga M., Yamaguchi-Shinozaki K., Ogawa T., et al. Metabolic profiling of transgenic potato tubers expressing *Arabidopsis*dehydration response element-binding protein 1A (DREB1A) J. Agric. Food Chem. 2013;61:893–900.

61. Defernez M., Gunning Y.M., Parr A.J., Shepherd L.V., Davies H.V., Colquhoun I.J. NMR and HPLC-UV profiling of potatoes with genetic modifications to metabolic pathways. J. Agric. Food Chem.2004;52:6075–6085.

62. Kondrák M., Marincs F., Antal F., Juhász Z., Bánfalvi Z. Effects of yeast trehalose-6-phosphate synthase 1 on gene expression and carbohydrate contents of potato leaves under drought stress conditions.BMC Plant Biol. 2012;12 doi: 10.1186/1471-2229-12-74..

63. Noteborn H.P., Lommen A., van der Jagt R.C., Weseman J.M. Chemical fingerprinting for the evaluation of unintended secondary metabolic changes in transgenic food crops. J. Biotechnol.2000;77:103–114.

64. Roessner-Tunali U., Hegemann B., Lytovchenko A., Carrari F., Bruedigam C., Granot D., Fernie A.R. Metabolic profiling of transgenic tomato plants overexpressing hexokinase reveals that the influence of hexose phosphorylation diminishes during fruit development. Plant Physiol. 2003;133:84–99.

65. Le Gall G., Colquhoun I.J., Davis A.L., Collins G.J., Verhoeyen M.E. Metabolite profiling of tomato (*Lycopersicon esculentum*) using 1H NMR spectroscopy as a tool to detect potential unintended effects following a genetic modification. J. Agric. Food Chem. 2003;51:2447–2456.

66. Le Gall G., DuPont M.S., Mellon F.A., Davis A.L., Collins G.J., Verhoeyen M.E., Colquhoun I.J. Characterization and content of flavonoid glycosides in genetically modified tomato (*Lycopersicon esculentum*) fruits. J. Agric. Food Chem. 2003;51:2438–2446.

67. Long M., Millar D.J., Kimura Y., Donovan G., Rees J., Fraser P.D., Bramley P.M., Bolwell G.P. Metabolite profiling of carotenoid and phenolic pathways in mutant and transgenic lines of tomato: Identification of a high antioxidant fruit line. Phytochemistry.2006;67:1750–1757.

68. Nicoletti I., de Rossi A., Giovinazzo G., Corradini D. Identification and quantification of stilbenes in fruits of transgenic tomato plants (*Lycopersicon esculentum* Mill.) by reversed phase HPLC with photodiode array and mass spectrometry detection. J. Agric. Food Chem. 207;55:3304–3311.

69. Kusano M., Redestig H., Hirai T., Oikawa A., Matsuda F., Fukushima A., Arita M., Watanabe S., Yano M., Hiwasa-Tanase K., et al. Covering chemical diversity of genetically-modified tomatoes using metabolomics

for objective substantial equivalence assessment. PLoS One. 2011;16:e16989.

70. Choi H.K., Choi Y.H., Verberne M., Lefeber A.W., Erkelens C., Verpoorte R. Metabolic fingerprinting of wild type and transgenic tobacco plants by 1H NMR and multivariate analysis technique. Phytochemistry. 2004;65:857–864.

71. Sobolev A.P., Segre A.L., Giannino D., Mariotti D., Nicolodi C., Brosio E., Amato M.E. Strong increase of foliar inulin occurs in transgenic lettuce plants (*Lactuca sativa* L.) overexpressing the*Asparagine Synthetase A* gene from *Escherichia coli*. J. Agric. Food Chem. 2007;55:10827–10831.

72. Sobolev A.P., Capitani D., Giannino D., Nicolodi C., Testone G., Santoro F., Frugis G., Iannelli M.A., Mattoo A.K., Brosio E., et al. NMR-metabolic methodology in the study of GM foods. Nutrients.2010;2:1–15.

73. Sobolev A.P., Testone G., Santoro F., Nicolodi C., Iannelli M.A., Amato M.E., Ianniello A., Brosio E., Giannino D., Mannina L. Quality traits of conventional and transgenic lettuce (*Lactuca sativa* L.) at harvesting by NMR metabolic profiling. J. Agric. Food Chem.2010;58:6928–6936.

74. Tagashira N., Plader W., Filipecki M., Yin Z., Wiśniewska A., Gaj P., Szwacka M., Fiehn O., Hoshi Y., Kondo K., et al. The metabolic profiles of transgenic cucumber lines vary with different chromosomal locations of the transgene. Cell Mol. Biol. Lett. 2005;10:697–710.

75. Zawirska-Wojtasiak R., Gośliński M., Szwacka M., Gajc-Wolska J., Mildner-Szkudlarz S. Aroma evaluation of transgenic, thaumatin II-producing cucumber fruits. J. Food Sci. 2009;74:C204–C210.

76. Malowicki S.M.M., Martin R., Qian M.C. Comparison of sugar, acids, and volatile composition in raspberry bushy dwarf virus-resistant transgenic raspberries and the wild type "meeker" (*Rubus Idaeus* L.) J. Agric. Food Chem. 2008;56:6648–6655.

77. Tesniere C., Torregrosa L., Pradal M., Souquet J.M., Gilles C., Dos Santos K., Chatelet P., Gunata Z. Effects of genetic manipulation of alcohol dehydrogenase levels on the response to stress and the synthesis of secondary metabolites in grapevine leaves. J. Exp. Bot. 2006;57:91–99.

78. Lange B.M., Mahmoud S.S., Wildung M.R., Turner G.W., Davis E.M., Lange I., Bake R.C., Boydston R.A., Croteau R.B. Improving peppermint essential oil yield and composition by metabolic engineering. Proc. Natl. Acad. Sci. USA. 2011;108:16944–16949.

79. Kim J.K., Ryu T.H., Sohn S.I., Kim J.H., Chu S.M., Yu C.Y., Baek H.J. Metabolic fingerprinting study on the substantial equivalence of Genetically Modified (GM) Chinese cabbage to non-gm cabbage. J. Korean Soc. Appl. Biol. 2009;52:186–192.

80. Tripathi S., Suzuki J.Y., Carr J.B., McQuate G.T., Ferreira S.A., Manshardt R.M., Pitz K.Y., Wall M.M., Gonsalves D. Nutritional composition of Rainbow papaya, the first commercialized transgenic fruit crop. J. Food Compos. Anal. 2011;24:140–147.

81. Jiao Z., Deng J.C., Li G.K., Zhang Z.M., Cai Z.W. Study on the compositional differences between transgenic and non-transgenic papaya. J. Food Compos. Anal. 2010;23:640–647.

82. Srivastava V., Obudulu O., Bygdell J., Löfstedt T., Rydén P., Nilsson R., Ahnlund M., Johansson A., Jonsson P., Freyhult E., et al. OnPLS integration of transcriptomic, proteomic and metabolomic data shows multi-level oxidative stress responses in the cambium of transgenic hipI- superoxide dismutase Populus plants. BMC Genomics.2013;14 doi: 10.1186/1471-2164-14-893..

83. Kogel K.H., Voll L.M., Schäfer P., Jansen C., Wu Y., Langen G., Imani J., Hofmann J., Schmiedl A., Sonnewald S., et al. Transcriptome and metabolome profiling of field-grown transgenic barley lack induced differences but show cultivar-specific variances. Proc. Natl. Acad. Sci. USA. 2010;107:6198–6203.

84. Cantrell R.P., Reeves T.G. The rice genome. The cereal of the world's poor takes center stage. Science. 2002;296 doi: 10.1126/science.1070721..

85. Goff S.A., Ricke D., Lan T.H., Presting G., Wang R., Dunn M., Glazebrook J., Sessions A., Oeller P., Varma H., et al. A draft sequence of the rice genome (*Oryza sativa* L. ssp. *japonica*) Science.2002;296:92–100.

86. Oikawa A., Matsuda F., Kusano M., Okazaki Y., Saito K. Rice metabolomics. Rice. 2008;1:63–71.

87. Choudhary V.K., Suresh-Kumar P. Maize production, economics and soil productivity under different organic source of nutrients in eastern himalayan region, India. Int. J. Plant Prod. 2013;7:167–186.

88. Skogerson K., Harrigan G.G., Reynolds T.L., Halls S.C., Ruebelt M., Iandolino A., Pandravada A., Glenn K.C., Fiehn O. Impact of genetics and environment on the metabolite composition of maize grain. J. Agric. Food Chem. 2010;58:3600–3610.

89. Asiago V.M., Hazebroek J., Harp T., Zhong C. Effects of genetics and environment on the metabolome of commercial maize hybrids: A multisite study. J. Agric. Food Chem. 2012;60:11498–11508.

90. Baniasadi H., Vlahakis C., Hazebroek J., Zhong C., Asiago V. Effect of environment and genotype on commercial maize hybrids using LC/MS-based metabolomics. J. Agric. Food Chem.2014;62:1412–1422.

91. Duvernaya W.H., Chinna M.S., Yencho G.C. Hydrolysis and fermentation of sweetpotatoes for production of fermentable sugars and ethanol. Ind. Crop Prod. 2013;42:527–537.

92. Urbanczyk-Wochniak E., Baxter C., Kolbe A., Kopka J., Sweetlove L.J., Fernie A.R. Profiling of diurnal patterns of metabolite and transcript abundance in potato (*Solanum tuberosum*) leaves. Planta.2005;221:891–903.

93. García-Cañas V., Simó C., Herrero M., Ibáñez E., Cifuentes A. Present and future challenges in food analysis: Foodomics. Anal. Chem. 2012;84:10150–10159.

94. Fukushima A., Saito K. Development of the *Arabidopsis*metabolome database. Plant Physiol. 2014 doi: 10.1104/pp.114.240986..

CHAPTER 14

TRANSGENIC PLANTS AS A SUSTAINABLE, TERRESTRIAL SOURCE OF FISH OILS

Johnathan A. Napier,[1] Sarah Usher, 1Richard P. Haslam, 1Noemi Ruiz Lopez, [2]and Olga Sayanova [1]

[1] Department of Biological Chemistry and Crop Protection, Rothamsted Research, Harpenden, UK

[2] Instituto de la Grasa (CSIC), Seville, Spain

ABSTRACT

An alternative, sustainable source of omega-3 long chain polyunsaturated fatty acids is widely recognized as desirable, helping to reduce pressure on current sources (wild capture fisheries) and providing a de novo source of these health beneficial fatty acids. This review will consider the efforts and progress to develop transgenic plants as terrestrial sources of omega-3 fish oils, focusing on recent developments and the possible explanations for advances in the field. We also consider the utility of such a source for use in aquaculture, since this industry is the major consumer of oceanic supplies of omega-3 fish oils.

Citation: Johnathan A. Napier,corresponding author Sarah Usher, Richard P. Haslam, Noemi Ruiz⊠Lopez, and Olga Sayanova, Transgenic plants as a sustainable, terrestrial source of fish oils, doi: 10.1002/ejlt.201400452.

Given the importance of the aquaculture industry in meeting global requirements for healthy foodstuffs, an alternative source of omega-3 fish oils represents a potentially significant breakthrough for this production system.

INTRODUCTION

The aquaculture (fish farming) industry is heavily dependent on fish oil as a commodity ingredient for the formulation of fish feeds—this is based on the nutritional requirements of marine and salmonid species for omega-3 long chain polyunsaturated fatty acids (LC-PUFAs) in their diets, and also to ensure that the final product destined for the consumer contains these health beneficial fish oils 1. Counter-intuitively, most farmed fish species are unable to synthesise omega-3 fish oils such as eicosapentaenoic acid (EPA) and docosahexaenoic acid (DHA), and are entirely dependent on dietary sources 2. Thus, the predominant farmed species such as salmon and trout represent major consumers of oceanically-sourced fish oils (harvested from the so-called reduction fisheries containing species such as anchovies, menhaden, and capelin), and collectively the aquafeed industry consumes in excess of 7 50 000 metric tons of fish oils per annum 3. When this demand is overlaid with the continuing expansion of fish-farming (on average 6% per annum for the last two decades) 4 and the ever-growing human population, there are a number of significant issues. Firstly, fish-farming is the most effective system for producing animal protein for human consumption—the input/output ratios for aquaculture are superior compared to all terrestrial animal production systems 5. Secondly, farmed fish is now near-ubiquitous in many countries—for example, 95% of all salmon sold in the UK is farmed 6. Thirdly, the finished products have well-defined levels of health-beneficial omega-3s 7, though feed inclusion rates are showing a downward trend, most likely as a consequence of price-sensitivity for raw materials such as fish oil 6, 8. Thus, all of these factors need to be considered in the continued attempts to ensure that fish farming remains an economically-viable and environmentally-sustainable means by which to ensure human health and nutrition. However, the continued expansion of the aquaculture industry, and its ability to deliver a nutritionally—important product, are constrained by the sustainable availability of omega-3 fish oils with which to formulate feeds. Reduced wild fish stocks and associated quotas to protect these limit the annual global harvests of fish oils to ~1 million metric tons, of which <20% is used for direct human nutrition 7. Thus, the limited availability of fish oils represents the critical bottleneck in this production system, and one that all major aquafeed companies in this field are trying to overcome. Attempts to replace fish oils in aquafeed diets with conventional vegetable oils (which are devoid ofthe critical fatty acids such as EPA and DHA) result in finished products that lack the health-beneficial omega-3 fish oils and run the risk of losing consumer-confidence in oily fish as a healthy foodstuff 8. It is for all of these reasons that we set out to develop a new, sustainable, source of fish oils via their de novo synthesis in transgenic (genetically modified; GM) plants. This review will consider the approach we have adopted, contrast other alternative systems, and also examine the steps that still need to be taken to ensure that this research is translated into a tangible outcome.

CONTEXT AND PROGRESS TO DATE

The accumulation of LC-PUFA EPA and DHA in the marine environment is predominantly via the biosynthetic activities of unicellular organisms such as microalgae, diatoms, and some bacteria; organisms that form the base of the marine foodweb. In general, most trophic levels above these microbes have limited to zero capacity to make EPA and DHA, most likely as a consequence of their environments being so omega-3 rich and hence providing a continuous dietary source.

Several different biosynthetic pathways for EPA and DHA have been identified in marine microbes, and the biochemical and molecular nature of these have been well described previously 9, 10, 11, but for clarity we will summarize these briefly. The predominant route by which microalgae and diatoms synthesise omega-3 LC-PUFAs is via a series of sequential desaturation and elongation reactions—this is an aerobic process, and can be further sub-divided into two forms—the predominant "Δ6-pathway," in which the first committed step is the introduction of a Δ6-desaturation into a C18 substrate, followed by C2 elongation and further Δ5-desaturation. The much less common route, the so-called alternative or "Δ8-pathway," initiates with the C2 elongation of the C18 substrate, followed by two successive desaturation reactions (Δ8, Δ5) 12. If DHA is also synthesized, this occurs by the C2 elongation of EPA and further (Δ4) desaturation. These two pathways are represented schematically in Fig. Fig.1.1. Multiple genes for all of these activities have been identified in the last decade from a range of different microbial sources.

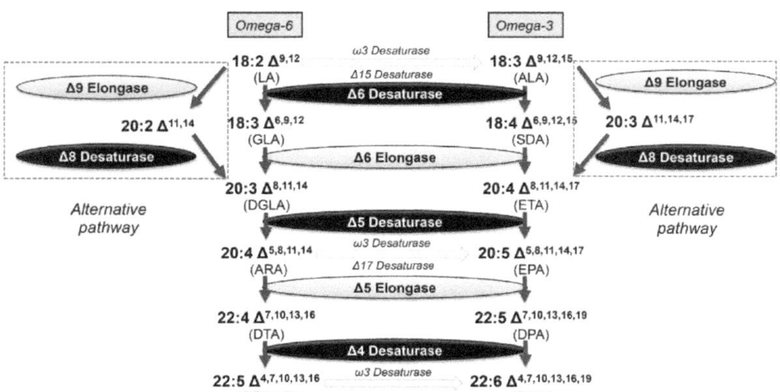

Figure 1: Schematic representation of the aerobic pathways for the biosynthesis of omega-3 long chain polyunsaturated fatty acids. The relevant desaturase and elongase activities, and their associated substrates, are indictated, as is the variant form (the alternative pathway)

An entirely different route favored by some bacteria and a few examples of unicellular marine eukaryotes is the biosynthesis of EPA or DHA via an anaerobic pathway which uses a processive polyketide synthase-like enzymatic system to convert malonyl-CoA to omega-3 LC-PUFAs without any fatty acid intermediates. The genes for this pathway have also been identified and functionally characterized 13.

Collectively, all these different genes and the enzymatic activities they encode, represent a toolkit by which the metabolic engineer/synthetic biologist can attempt to reconstitute the capacity to synthesise EPA and DHA in a heterologous host. However, this is not as straight-forward as it might appear, for a number of reasons. Firstly, in the case of the aerobic pathways, the substrate for the constituent enzymes is pre-existing C18 di- or tri-unsaturated fatty acids. In the case of the anaerobic PKS-like pathway, the substrate is malonyl-CoA, which is a low-abundance primary metabolite. Secondly, the successful reconstitution of the complete pathway requires the coordinated expression and activity of multiple genes, which is non-trivial in many transgenic systems. Thirdly, by definition, the host into which one is introducing the capacity to synthesise EPA and DHA has limited (or no) endogenous capability to make these fatty acids, meaning that native enzymes and pathways will be unfamiliar with these fatty acids—this could result in either selective exclusion of EPA/DHA from desired pathways (such as neutral lipid synthesis) or alternatively, the "trapping" of biosynthetic intermediates in undesired metabolic dead-ends through the inclusive promiscuous activity of house-keeping lipid enzymes. Finally, the underlying biochemical structural and biophysical properties of these enzymes are only just starting to emerge, mainly due to the membrane-associated nature which precludes crystal structure-based studies 14. Thus, although the genes with which to attempt the heterologous expression of omega-3 LC-PUFA biosynthetic pathways have been in hand for over 10 years, the first successes are only now beginning to emerge 15, 16.

In the context of the focus of this review, we will mainly consider the potential of genetically engineered oilseed crop plants to produce EPA and DHA, since it is likely that only such agricultural systems will have the capacity (in terms of scalability) to produce, in an economically viable manner, sufficient omega-3 LC-PUFAs for "high volume/low value" markets such as the aquafeed industry. However, we will also briefly consider attempts to produce these fatty acids in transgenic yeasts and diatoms, and how these contrast with the situation observed in plants.

Engineering the accumulation of EPA and DHA in transgenic plants—first attempts

As mentioned above, examples of all the biosynthetic genes required for the accumulation of EPA and DHA biosynthesis were identified by 2004 9, 17, facilitating their combined introduction into transgenic plants. Prior to that, individual genes from this pathway had been evaluated in plants, notably the Δ6- and Δ5-desaturases—these data gave confidence that such heterologous activities were active in plant cells, pointing the way for subsequent endeavors 18. The first successful (proof-of-concept) demonstration of the feasibility of making EPA in a transgenic plant was published by Qi et al 19, expressing algal components of the alternative pathway in the leaves of Arabidopsis. Interestingly, this not only generated moderate amounts of EPA, but also the C20 omega-6 LC-PUFA arachidonic acid (ARA). Although this study was the first to generate EPA in a transgenic plant, the approach taken was via sequential transformation of individual genes, which was less than optimal (since it means that the three transgenes are genetically unlinked and prone to segregation in subsequent generations). Moreover, although vegetative accumulation of EPA was achieved, this was predominantly in phospholipids, whereas the target lipid species would ideally be neutral storage lipids such as triacylglycerols found in seeds.

The first demonstration of the seed-specific accumulation of EPA was also achieved in the same year by Abbadi et al. 20, who expressed genes of the conventional Δ6-pathway in transgenic linseed, under the control of seed-specific promoters. Analysis of the resulting transgenic plants and their seeds confirmed the presence of low levels of EPA, but also very high levels of C18 Δ6-desaturation products. This lead the authors to hypothesize that this unwanted build-up of a biosynthetic intermediate was as a consequence of poor acyl-exchange between different metabolic pools, and this concept has been further defined as "substrate dichotomy" 18. In this situation, a bottleneck is generated as a result of the differing substrate requirements for the sequential reactions present in the aerobic biosynthesis of LC-PUFAs—specifically, most lower eukaryotic desaturases require their fatty acid substrate to be linked to a glycerolipid backbone, whereas fatty acid elongases require the substrate in the form of acyl-CoAs.

These initial first steps formed the basis for further iterations to drive up the accumulation of target fatty acids (EPA, DHA) and reduce the levels of undesired biosynthetic intermediates (such as the Δ6-desaturation product γ-linolenic acid; GLA). These studies have confirmed first the ability to make significant levels of EPA, albeit with attendant levels of GLA, 21, 22 and then also DHA 21, 23. The critical breakthroughs in optimizing the accumulation of target fatty acids were achieved by: (i) the use of omega-3 specific desaturases, which prevented the accumulation of unwanted omega-6 fatty acids; (ii) the use of acyl-CoA-dependent desaturases for the first committed (Δ6) step on the pathway 24, breaking the substrate-dichotomy bottleneck and reducing the accumulation of GLA 22, 24, 25, 26, 27. As a consequence, it is now technically possible to accumulate fish oil-like levels of omega-3 LC-PUFAs in the seed oils of transgenic plants similar to that found in fish oils, in which EPA and DHA accumulate to ~20% of total fatty acids. A comparison of the fatty acid composition of fish oil, vegetable oil (GM and native), and also algal oil is shown in Fig.

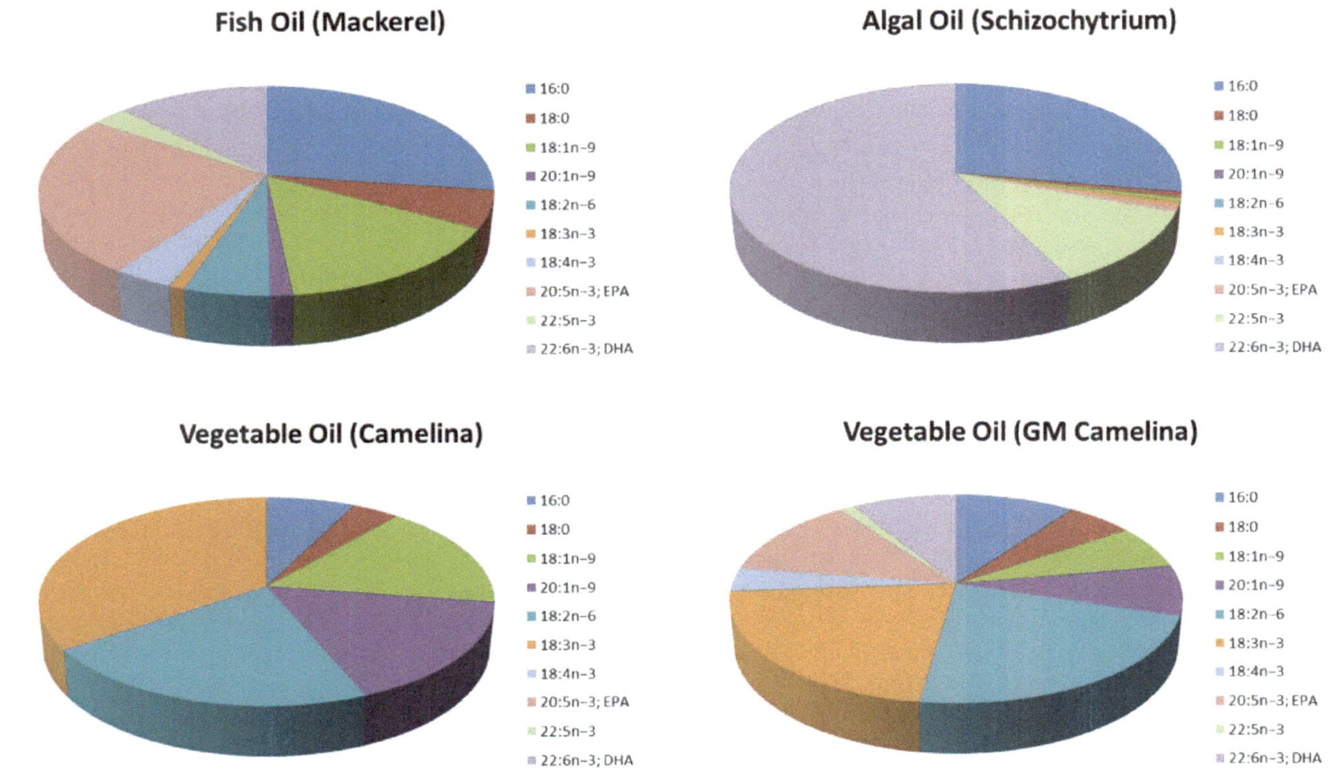

Figure 2: Comparison of fatty acid composition in different sources of omega-3 LC-PUFAs. The major fatty acids present in either fish oil, vegetable oil, GM vegetable oil, or algal oil are presented. Note that non-GM vegetable oil lacks .

Approaches to successful accumulation of target fatty acids

Two discrete paths were adopted to ultimately produce EPA and DHA in the seeds of transgenic plants, though both started from the same position of having a "toolkit" of suitable biosynthetic genes isolated from microalgae and other omega-3 accumulators, validated by heterologous expression in yeast. The first approach, exemplified by Petrie and colleagues at CSIRO (Canberra, Australia), took advantage of *Agrobacterium*-mediated transient expression systems which allow the rapid evaluation of multiple gene combinations (via co-infection) in the leaves of *Nicotiana benthamiana* 28. This allowed the identification of an optimal set of non-native genes which directed the efficient synthesis of EPA and DHA in plants, albeit non-seed tissues. However, by the further co-expression in leaves of a "master regulator" transcription factor *LEC2*, the authors were able to "reprogramme" lipid metabolism in this tissue so that it now more closely resembled that of seeds, with the specific capacity to synthesize seed oil in the form of triacylglycerol. Thus, the host leaf cells under evaluation came to more closely resemble the metabolic context observed in seeds. In addition, an added benefit of this transcription factor-mediated reprogramming was that instead of having to use constitutive promoters to drive the expression of the non-native genes, it was now possible to use seed-specific promoters (since they were now active as a result of the *LEC2* expression), meaning that both regulatory elements and biosynthetic enzymes were validated in a plant host. The resulting study29 allowed for the selection of the optimal cassettes for subsequent stable transformation into plants (discussed below).

The alternative approach, adopted by ourselves and colleagues at the University of Hamburg, Germany was to directly assess and iterate gene combinations and their associated regulatory elements in the seeds of stably transformed transgenic plants 20, 21. The first example of this21, demonstrated the potential of this approach, although the resulting seed oil profile still was suboptimal. Using transgenic Arabidopsis, we subsequently systematically evaluated 22 genes in 9 different combinations, and in 2 different genetic backgrounds 30, 31. For each construct, multiple independent transgenic lines were subject to detailed lipidomic analysis to define the incorporation of non-native fatty acids into different lipid classes, which was then used to inform subsequent iterations. This systematic approach also identified an optimal suite of (different) genes for introduction into crop species 30, 31.

The transition from model to crop

As indicated above, different approaches resulted in the identification of a combination of algal and other lower eukaryotic activities which were capable of directing the synthesis and accumulation of EPA and DHA in model systems such as Arabidopsis or tobacco. However, this was still one step removed from the demonstration of the accumulation of these fatty acids in a suitable oilseed crop plant. Given the significant volumes of fish oils being annually removed from the oceans, it is only via the use of agricultural crops (and the scalability that brings) can one hope to address the challenges associated with generating a potential substitute.

To date, two oilseed crop species have been identified as potential hosts for the omega-3 LC-PUFA biosynthetic trait—canola, a cultivar of rapeseed (*Brassica napus* L.) and (*Camelina sativa*). In the case of canola, there are no peer-reviewed data yet in the public domain to allow an accurate assessment. However, for Camelina, both the Rothamsted and CSIRO groups have demonstrated the suitability of this crop to act as an efficient host for the accumulation of EPA and DHA, although interestingly, the fatty acid profile obtained varies between the two studies. In the first publically available data 32, we built on our iterative studies in Arabidopsis to test two optimized constructs in transgenic Camelina—one directing the synthesis of just EPA, and one making both EPA and DHA. Based on the total fatty acid composition of seeds from stably transformed lines, the average level of EPA alone was 24%, and for EPA and DHA, 11 and 8%, respectively32. Very low levels of undesired C18 biosynthetic intermediates such as GLA and stearidonic acid were observed. Interestingly, for both the EPA and EPA+DHA lines, the endogenous levels of α-linolenic acid (ALA) were markedly decreased (presumptively since this fatty acid serves as direct precursor for the synthesis of EPA), whereas linoleic acid was effectively unchanged, most likely due to the transgene presence of a FAD2 Δ12-desaturase. Surprisingly, levels of oleic acid were also strongly reduced for both constructs. Overall, the total level of target omega-3 LC-PUFAs was 19% (EPA+DHA), though this was found to be greater than 25% in some single seeds.

In the CSIRO study 33, a construct (GA7) previously evaluated in Arabidopsis (Petrie et al., 2012) was introduced into Camelina, and the performance compared with two additional constructs (mod-F, mod-G) in which minor changes were made. Significant levels of DHA (up to 12.4% for mod-F) were obtained for all three constructs, though the levels of EPA were markedly low (maximally 3.3% for mod-F). In contrast to the results observed by Ruiz-Lopez et al., 32, levels of ALA were only moderately reduced, whereas LA was down from 18.1 to 8.4% (mod-F), whereas accumulation of the C18 Δ6-desaturation product stearidonic acid was noticeably higher. The observed differences between the various constructs almost certainly resulted from the contribution of a number of different factors, including but not limited to: (i) precise expression of seed-specific promoters; (ii) superior examples of the same enzyme activity; (iii) construct design and orientation of elements; and (iv) site of integration in the Camelina genome. Perhaps the most striking and relevant difference between the two studies is the variation in the level of EPA—although the levels of DHA are broadly comparable (>10%), Petrie et al. 33 reported only 0.8–3.3% EPA for the three constructs, in contrast to the >11% observed by Ruiz-Lopez et al. 32. The most likely explanation for this is the use of a highly active Δ5-elongase in the former study, which delivers increased levels of DHA at the expense of EPA.

One additional interesting observation from these studies relates to the apparent utility (or not) of using a model system to predict the subsequent outcome in crop systems. Based on both sets of results obtained in Camelina, it appears as though previous studies in Arabidopsis underestimated the levels of target fatty acids obtainable in the crop, though obviously these earlier studies formed an important part of the proof-of-concept. For example, in the earlier Arabidopsis study by Ruiz-Lopez et al. 30, 31, maximal levels of EPA observed were ~8%, whereas in Camelina this was 24% 32. Similarly, Petrie et al. 23 reported an average of 5.5% DHA in T2 Arabidopsis seeds, compared with 9.6% DHA in T2 Camelina seeds 33. Bearing in mind that in both studies, the same identical constructs (EPA-A5.1; GA7, respectively) were introduced into the two different hosts, these data would imply that, at least for the omega-3 LC-PUFA trait, Arabidopsis may underestimate the outcome in more complex (genomically and biochemically) crops. In that respect, it may be that in addition to the four influencing factors described above, an additional consideration should be the "metabolic context" into which the heterologous pathway is superimposed 34. Since the transgene-derived activities need to biochemically integrate with endogenous lipid metabolism to reconstitute the efficient synthesis of EPA and DHA, it may be beneficial to have multiple copies of such native genes, each with perhaps minor differences in expression pattern, substrate-specificity, or efficiency. Thus, working in a hexaploid crop such as Camelina provides more opportunities (as a result of increased variation via homeologues) for the smoother incorporation of the heterologous pathway, and hence, higher synthesis and accumulation of target fatty acids.

Utility and application of novel plant oils

The availability of these modified Camelina oils is a very recent development, so there is currently only limited data on their properties and applications of them. However, two studies are of note. Firstly, Mansour et al. 35 examined the detailed composition of Camelina oil derived from the GA7 construct, looking at both glycerolipid composition and also sterols. Interestingly, the maximum level of DHA present in TAG (the dominant neutral lipid) was 6.8%, and basal levels of EPA (0.4%). There was significantly less DHA (1.6%) in phospholipids, though the residual protein meal also contained DHA (5.4%) though the overall lipid content in this fraction was very low (0.3% of starting seed weight). Although this preliminary study did not indicate which generation or event for the GA7 construct was examined, it is noteworthy that the levels of DHA in TAG are somewhat reduced compared to the original study (6.8% vs. 9.6%; respectively) 33, 35. There was no significant difference between the GA7 line and wildtype Camelina with regards to levels or profiles of sterols, indicating that the presence of non-native omega-3 LC-PUFAs did not alter this pathway.

More recently, the first study on the application of these novel oils has been published—Betancor et al. 36 used the EPA-containing Camelina oil described by Ruiz-Lopez et al. 32 to determine if this oil could substitute for fish oil in a salmon feeding trial. In this study, juvenile (post-smolt) Atlantic salmon were fed diets containing fish oil, EPA-containing Camelina oil, or wild type Camelina oil. All three diets were isocalorific, and also contained low levels of DHA (via the deliberate use of fish meal as a protein source) to ensure no deficiency for this fatty acid. The fish were fed the different diets for seven weeks, after which they had all doubled in weight and displayed no indicators for ill-health. Fatty acid analysis of the flesh and the liver indicated the expected accumulation of EPA, and in the latter tissue, some conversion to DHA 36. Transcriptomic analyses of these tissues also indicated the up-regulation of genes involved in omega-LC-PUFA metabolism, though both Camelina treatments (WT, GM) induced transcriptome signatures more similar to each other than the fish oil treatment—this most likely reflects the fact that the GM oil still contains fatty acids normally associated with vegetable oils, but not observed in fish oils (such as linoleic acid), and also the common presence of (phyto) sterols, which are known to induce the up-regulation of the cholesterol pathway in fish 37. In that respect, the EPA-containing GM Camelina oil can be considered a "hybrid" oil, representing an in vivo blend of plant and marine fatty acids (see also Fig. Fig.22 for graphical representation). Given that most aquafeed diet formulations are now a mixture of fish oil and vegetable oil, this GM Camelina oil is well-suited for direct use in aquaculture. Moreover, this study, for the first time, demonstrates the utility of using novel oils derived from a GM plant as a safe and effective replacement for fish oils.

Alternative production systems for the synthesis of omega-3 LC-PUFAs

Currently, two additional approaches also contribute to the de novo production of omega-3 LC-PUFAs, via either the fermentation of a GM yeast strain, or via the cultivation of native strains of microalgae, and as such, may have the capacity to augment piscine sources of these fatty acids. In the latter case of algae cultivation such production systems use a diverse range of different algal strains, and play an important role in providing suitable nutrients to the earlier stages of the life cycle of various species used in fish-farming, though the actual levels of omega-3 LC-PUFAs produced by this process are very modest (<1% of annual fish oil production rates). Currently, there are no examples of genetically modified algal strains being used in this way, though recently it has been reported that a GM strain of *Phaeodactylum tricornutum* had been produced which had increased levels of DHA 38.

In the case of microbial fermentation, work from DuPont demonstrated the feasibility of using the oleaginous yeast *Yarrowia lipolytica* as a heterologous host for the production of EPA. This tour-de-force of metabolic engineering introduced components of the alternative pathway into the yeast to enable it synthesize e EPA, with very high levels (up to 15% dry weight) of this fatty acid being obtained. Serendipitously, part of this high level of accumulation arose as a result of the insertional mutagenesis (via integration of the transgenes) of a gene (PXA10) involved in peroxisomal biogenesis, resulting in decreased levels of fatty acid catabolism by beta-oxidation 39. The utility and applications of this source of EPA are discussed in a recent review 40.

CONCLUSIONS

As reviewed above, it is clear that the promise of a plant-based source of omega-3 LC-PUFAs has now been deomonstrated, at least in the case of Camelina. However, there still remain a number of hurdles to clear before such a crop becomes a commercial reality and contributes to the pressing need for an alternative, sustainable source of omega-3 LC-PUFAs. Given the progressive and forward-looking nature of the aquafeed industry, it is hoped that such novel ingredients can, subject to appropriate regulatory and safety approval, become incorporated into a wide range of fish diets, and in turn, relieve pressure on the oceanic stocks of the reduction fisheries. Since Camelina as an oilseed crop can easily yield 0.75 ton of oil/ha, then a GM oil containing similar levels of EPA and DHA to that found in fish oils could make a significant contribution to off-setting oceanic sources. For example, 2 00 000 ha of GM Camelina could produce 1 50 000 MT of oil which could serve as a direct replacement for fish oils in aquafeed, representing 15% of the global oceanic harvest of these oils. And although 2 00 000 ha might seem like a large area, in agricultural terms, it is a quite modest scale—for comparison, the current annual Canadian sowing of related oilseeds such as canola is in excess of 7 million ha. Thus, the proposed 2 00 000 ha of Camelina represents only a small (<3%) area of land currently given over to vegetable oil production in one country, yet could significantly increase the de novo synthesis of omega-3 LC-PUFAs.

Converting this hypothetical concept into reality will require a number of different factors to be dealt with, including seeking and obtaining regulatory approval to grow a novel GM crop, better understanding of the agronomy and processing of Camelina, use and formulation of the novel oil by end-users in the feed sector, and ultimately acceptance by the consumer. None of these steps are trivial or should be taken for granted, and will most likely require significant resources and organizational focus. However, given the increasing demand for omega-3 LC-PUFAs, and the demonstrated examples of "proof-of-concept" described in this article, it is hoped that a terrestrial, GM oilseed source of EPA and DHA might be available by the end of this decade.

REFERENCES

1. Saravanan P., Davidson N. C., Schmidt E. B., Calder P. C.,Cardiovascular effects of marine omega-3 fatty acids. Lancet 2010,376, 540–550.
2. Tacon A. G., Metian M., Global overview on the use of fish meal and fish oil in industrially compounded aquafe eds trends and future prospects. Aquaculture 2008, 285, 146–158.
3. Hixson S. M., Fish nutrition and current issues in aquaculture: The balance in providing safe and nutritious seafood, in an environmentally sustainable manner. J. Aquac. Res. Devel. 2014, 5, 1–10.
4. Bostock J., McAndrew B., Richards R., Jauncey K., et al.,Aquaculture: Global status and trends. Philos. Trans. R. Soc. Lond. B2010, 365, 2897–2912.
5. Salem N., Eggersdorfer M., Is the world supply of omega-3 fatty acids adequate for optimal human health?. Curr. Opin. Clin. Nutr. Metab. Care 2015, 18, 147–154.
6. Tacon A. G., Metian M., Fish matters: Importance of aquatic foods in human nutrition and global food supply. Rev. Fisheries Sci. 2013, 21, 22–38.
7. Henriques J., Dick J. R., Tocher D. R., Bell J. G., Nutritional quality of salmon products available from major retailers in the UK: Content and composition of n-3 long-chain PUFA. Br. J. Nutr. 2014, 112, 964–975.
8. Nichols P. D., Glencross B., Petrie J. R., Singh S. P., Readily available sources of long-chain omega-3 oils: Is farmed Australian seafood a better source of the good oil than wild-caught seafood?Nutrients 2014, 6, 1063–1079.
9. Sayanova O. V., Napier J. A., Eicosapentaenoic acid: Biosynthetic routes and the potential for synthesis in

transgenic plants.Phytochemical 2004, 65, 147–158.

10. Venegas-Calerón M., Sayanova O., Napier J. A., An alternative to fish oils: Metabolic engineering of oil-seed crops to produce omega-3 long chain polyunsaturated fatty acids. Prog. Lipid Res. 2010, 49, 108–119.
11. Petrie J. R., Singh S. P., Expanding the docosahexaenoic acid food web for sustainable production: Engineering lower plant pathways into higher plants. AoB Plants 2011, plr 011.
12. Ruiz-Lopez N, Haslam RP, Usher S, Napier JA, Sayanova O., An alternative pathway for the effective production of the omega -3 long-chain polyunsaturates EPA and ETA in transgenic oilseeds. Plant Biotechnol. J. 2015, Jan 30 DOI: 10.1111/pbi.12328
13. Metz J. G., Roessler P., Facciotti D., Levering C., et al., Production of polyunsaturated fatty acids by polyketide synthases in both prokaryotes and eukaryotes. Science 2001, 293, 290–293.
14. McCartney A. W., Dyer J. M., Dhanoa P. K., Kim P. K., et al.,Membrane-bound fatty acid desaturases are inserted co-translationally into the ER and contain different ER retrieval motifs at their carboxy termini. Plant J. 2004, 37, 156–173.
15. Cahoon E. B., Shockey J. M., Dietrich C. R., Gidda S. K., et al.,Engineering oilseeds for sustainable production of industrial and nutritional feedstocks: Solving bottlenecks in fatty acid flux. Curr. Opin. Plant Biol. 2007, 3, 236–244.
16. Ruiz-Lopez N., Usher S., Sayanova O. V., Napier J. A., Haslam R. P., Modifying the lipid content and composition of plant seeds: Engineering the production of LC-PUFA. Appl. Microbiol. Biotechnol.2015, 99, 143–154.
17. Kinney A. J., Metabolic engineering in plants for human health and nutrition. Curr. Opin. Biotechnol. 2006, 17, 130–138.
18. Napier J. A., The production of unusual fatty acids in transgenic plants. Annu. Rev. Plant Biol. 2007, 58, 295–319.
19. Qi B., Fraser T., Mugford S., Dobson G., et al., Production of very long chain polyunsaturated ω-3 and ω-6 fatty acids in plants. Nat. Biotech. 2004, 22, 739–745.
20. Abbadi A., Domergue F., Bauer J., Napier J. A., et al., Biosynthesis of very-long-chain polyunsaturated fatty acids in transgenic oilseeds: Constraints on their accumulation. Plant Cell. 2004, 16, 2734–2748.
21. Wu G., Truksa M., Datla N., Vrinten P., et al., Stepwise engineering to produce high yields of very long-chain polyunsaturated fatty acids in plants. Nat. Biotechnol. 2005, 23, 1013–1017.
22. Cheng B., Wu G., Vrinten P., Falk K., et al., Towards the production of high levels of eicosapentaenoic acid in transgenic plants: The effects of different host species, genes and promoters. Transgenic Res. 2010,19, 221–229.
23. Petrie J. R., Shrestha P., Zhou X. R., Mansour M. P., et al.,Metabolic engineering plant seeds with fish oil-like levels of DHA.PLoS ONE 2012, 7, 49165.
24. Domergue F., Abbadi A., Zähringer U., Moreau H., Heinz E., In vivo characterization of the first acyl-CoA D6-desaturase from a member of the plant kingdom, the microalga Ostreococcus tauri.Biochem. J. 2005, 389, 483–490.
25. Hoffmann M., Wagner M., Abbadi A., Fulda M., Feussner I.,Metabolic engineering of ω-3 very long chain polyunsaturated fatty acid production by an exclusively acyl-CoA-dependent pathway. J. Biol. Chem. 2008, 283, 22352–22362.
26. Petrie J. R., Shrestha P., Mansour M. P., Nichols P. D., et al.,Metabolic engineering of omega-3 long-chain polyunsaturated fatty acids in plants using an acyl-CoA D6-desaturase with omega3-preference from the marine microalga *Micromonas pusilla* . Metab. Eng. 2010, 12, 233–240.
27. Sayanova O., Ruiz-Lopez N., Haslam R. P., Napier J. A., The role of Δ6-desaturase acyl-carrier specificity in the efficient synthesis of long-chain polyunsaturated fatty acids in transgenic plants. Plant Biotechnol. J. 2012, 10, 195–206.
28. Wood C. C., Petrie J. R., Shrestha P., Mansour M. P., et al., A leaf-based assay using interchangeable design principles to rapidly assemble multistep recombinant pathways. Plant Biotechnol. J. 2009, 7, 914–924.
29. Petrie J. R., Shrestha P., Liu Q., Mansour M. P., et al., Rapid expression of transgenes driven by seed-specific constructs in leaf tissue: DHA production. Plant Meth. 2010, 6, 8.
30. Ruiz-López N., Haslam R. P., Venegas-Calerón M., Li T., et al.,Enhancing the accumulation of omega-3 long chain polyunsaturated fatty acids in transgenic *Arabidopsis thaliana* via iterative metabolic engineering and genetic crossing. Transgenic Res. 2012, 21, 1233–1243.
31. Ruiz-Lopez N., Haslam R. P., Usher S. L., Napier J. A., Sayanova O., Reconstitution of EPA and DHA biosynthesis in Arabidopsis: iterative metabolic engineering for the synthesis of n-3 LC-PUFAs in transgenic plants. Metab. Eng. 2013, 17, 30–41.
32. Ruiz-Lopez N., Haslam R. P., Napier J. A., Sayanova O., Successful high-level accumulation of fish oil omega-3 long-chain polyunsaturated fatty acids in a transgenic oilseed crop. Plant J. 2014, 77, 198–208.
33. Petrie J. R., Shrestha P., Belide S., Kennedy Y., et al., Metabolic engineering *Camelina sativa* with fish oil-like levels of DHA. PLoS ONE 2014, 9, e85061.
34. Napier J. A., Haslam R. P., Beaudoin F., Cahoon E. B.,Understanding and manipulating plant lipid composition: Metabolic engineering leads the way. Curr. Opin. Plant Biol. 2014, 19, 68–75.
35. Mansour M. P., Shrestha P., Belide S., Petrie J. R., et al.,Characterization of oilseed lipids from "DHA-producing Camelina sativa": A new transformed land plant containing long-chain omega-3 oils. Nutrients 2014, 6, 776–789.
36. Betancor M. B., Sprague M., Usher S., Sayanova O., et al., A nutritionally-enhanced oil from transgenic *Camelina sativa* effectively replaces fish oil as a source of eicosapentaenoic acid for fish. Sci. Rep.2015, 5, 8104.

37. Leaver M. J., Villeneuve L. A., Obach A., Jensen L., et al.,Functional genomics reveals increases in cholesterol biosynthetic genes and highly unsaturated fatty acid biosynthesis after dietary substitution of fish oil with vegetable oils in Atlantic salmon (Salmo salar). BMC Genomics 2008, 24, 299.

38. Hamilton M. L., Haslam R. P., Napier J. A., Sayanova O.,Metabolic engineering of *Phaeodactylum tricornutum* for the enhanced accumulation of omega-3 long chain polyunsaturated fatty acids.Metab. Eng. 2014, 22, 3–9.

39. Xue Z., Sharpe P. L., Hong S. P., Yadav N. S., et al., Production of omega-3 eicosapentaenoic acid by metabolic engineering of *Yarrowia lipolytica* . Nat. Biotechnol. 2013, 31, 734–740.

40. Xie D., Jackson E. N., Zhu Q., Sustainable source of omega-3 eicosapentaenoic acid from metabolically engineered *Yarrowia lipolytica*: From fundamental research to commercial production. Appl. Microbiol. Biotechnol. 2015, 99, 1599–1610.

CHAPTER 15

PLASTID GENOMICS IN HORTICULTURAL SPECIES: IMPORTANCE AND APPLICATIONS FOR PLANT POPULATION GENETICS, EVOLUTION, AND BIOTECHNOLOGY

Marcelo Rogalski,[1] Leila do Nascimento Vieira,[2] Hugo P. Fraga,[2] and Miguel P. Guerra[2]

[1] Laboratório de Fisiologia Molecular de Plantas, Departamento de Biologia Vegetal, Universidade Federal de Viçosa, Viçosa, Brazil

[2] Laboratório de Fisiologia do Desenvolvimento e Genética Vegetal, Programa de Pós-graduação em Recursos Genéticos Vegetais, Centro de Ciências Agrárias, Universidade Federal de Santa Catarina, Florianópolis, Brazil

ABSTRACT

During the evolution of the eukaryotic cell, plastids, and mitochondria arose from an endosymbiotic process, which determined the presence of three genetic compartments into the incipient plant cell. After that, these three genetic materials from host and symbiont suffered several rearrangements, bringing on a complex interaction between nuclear and organellar gene products. Nowadays, plastids harbor a small genome with ~130 genes in a 100–220 kb sequence in higher plants. Plastid genes are mostly highly conserved between plant species, being useful for phylogenetic analysis in higher taxa. However, intergenic spacers have a relatively higher mutation rate and are important markers to phylogeographical and plant population genetics analyses. The predominant uniparental inheritance of plastids is like a highly desirable feature for phylogeny studies. Moreover, the gene content and genome rearrangements are efficient tools to capture and understand evolutionary events between different plant species. Currently, genetic engineering of the plastid genome (plastome) offers a number of attractive advantages as high-level of foreign protein expression, marker gene excision, gene expression in operon and transgene containment because of maternal inheritance of plastid genome in most crops. Therefore, plastid genome can be used for adding new characteristics related to synthesis of metabolic compounds, biopharmaceutical, and tolerance to biotic and abiotic stresses. Here, we describe the importance and applications of plastid genome as tools for genetic and evolutionary studies, and plastid transformation focusing on increasing the performance of horticultural species in the field.

INTRODUCTION

The existence of plastids represents one of the principal features that distinguish plant from other eukaryotic cells. Except for some gametic cells, plastids are assumed to be present as one of several different types in all living cells of higher plants, which show its essentiality for cell viability (Zhang et al., 2003; Kuroiwa, 2010; Nagata, 2010). These different plastid types have specific characteristics and functions, i.e., proplastids (present in meristematic regions of the plant); chloroplasts (chlorophyll-containg plastids specialized in photosynthesis); chromoplasts (colored plastids able to store high amounts of carotenoids present in petals of flowers and fruits); amyloplasts (mainly present in storage tissues such as tubers and seed endosperm); elaioplasts (lipid-storing plastids); leucoplasts (pigment-less plastids present mainly in root cells); and etioplasts (achlorophyllous plastids present in cotyledons of dark-grown angiosperm seedlings; Lopez-Juez and Pyke, 2005; Egea et al., 2010; Pyke, 2011; Bock, 2014; Osteryoung and Pyke, 2014). Moreover, plastids are involved in other

Citation: Marcelo Rogalski, Leila do Nascimento Vieira, Hugo P. Fraga, and Miguel P. Guerra, Plastid genomics in horticultural species: importance and applications for plant population genetics, evolution, and biotechnology, doi: 10.3389/fpls.2015.00586.

essential cellular processes such as lipid, hormone, amino acid, and phytochrome biosynthesis as well as nitrate and sulfate assimilation (Tetlow et al., 2004; Waters et al., 2004; Aldridge et al., 2005; Rogalski and Carrer, 2011; Galili and Amir, 2013; Galili et al., 2014).

At the beginning of the last century, a non-Mendelian inheritance of leaf variegation in *Mirabilis jalapa* and *Pelargonium zonale* was proposed, suggesting the plastids would contain their own genome (Baur, 1909, 1910; Correns, 1909; Hagemann, 2000, 2002; Greiner et al., 2011). This hypothesis was confirmed with the discovery of plastid DNA (Chun et al., 1963; Sager and Ishida, 1963; Tewari and Wildman, 1966). Today we know that the plastid genome (plastome) size of photosynthetically active seed plants varies between 120 and 220 kb in a circularly mapping genome, encoding 120–130 genes. The plastome is commonly mapped as a single circular molecule, however, it shows a high dynamic structure (i.e., linear molecules, branched complexes, and circular molecules) and ploydy level in each chloroplast (Bendich, 2004). Thus, inside a single cell, the plastome may occur at high copy number, with up to thousands of genome copies. Mesophyll cells of higher plants can contain 700–2000 copies of plastome, which depend on the developmental stage of the leaves and the plant species (Golczyk et al., 2014). These multiple copies are packed together in large nucleoprotein bodies, the plastid nucleoids (Golczyk et al., 2014; Krupinska et al., 2014; Powikrowska et al., 2014). Generally, the plastid DNA in photosynthetic active plant tissues (i.e., chloroplasts) forms up to 10–20% of total cellular DNA content (Bendich, 1987; Bock, 2001; Golczyk et al., 2014).

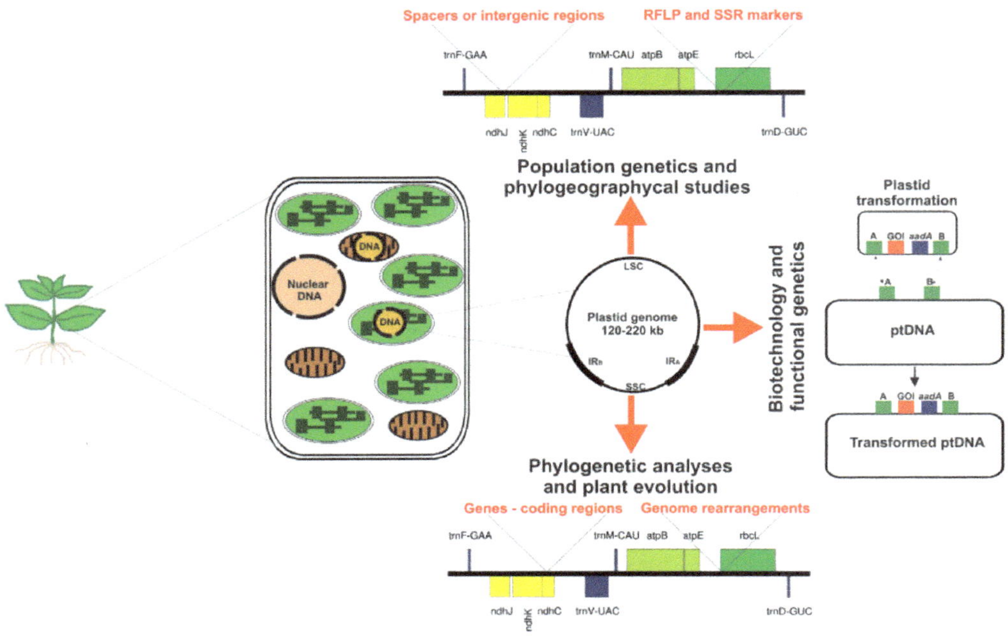

Figure 1: Illustration of a plant cell shows the genetic material into the three cellular compartments. Different sequences of plastid DNA are used for several applications as population genetics and phylogeographycal studies (intergenic spacers, RFLP and SSR molecular ...

Although the evolutionary forces that gave rise to the characteristic diversity of sizes, rearrangements, structure, and compactness of contemporary plastomes are poorly understood, nowadays the plastome has been used as basis for analyses of phylogeny and evolution (Leebens-Mack et al., 2005; Jansen et al., 2007; Parks et al., 2009;Moore et al., 2010; Crosby and Smith, 2012; Vieira et al., 2014a), population genetics (Angioi et al., 2009b; Nock et al., 2011; Yang et al., 2013; Wheeler et al., 2014), plastid gene transfer to nucleus (Huang et al., 2003, 2005; Stegemann et al., 2003; Bock, 2006; Stegemann and Bock, 2006), exchange of plastome between different species (Stegemann et al., 2012; Fuentes et al., 2014), plant speciation (Greiner et al., 2011), functional genomics (Svab et al., 1990; Svab and Maliga, 1993), and plastid gene expression machinery (Ruf et al., 1997; Hager et al., 1999; Drescher et al., 2000; Shikanai et al., 2001; Maliga, 2004;Kode et al., 2005; Rogalski et al., 2006, 2008; Alkatib et al., 2012). In addition to basic research, plastome studies may be focused in plastome transformation for biotechnological applications, i.e., adding new agronomic traits, manipulation of metabolic pathways, enhanced pest resistance, increase of biomass and production of enzyme for biofuel industry, and molecular farming in species related to agriculture and horticulture (Maliga and Bock, 2011; Verma et al., 2013; Wang et al., 2013; Bock, 2014; Shenoy et al., 2014; Shil et al., 2014; Zhang et al., 2015). All of these plastid/plastome applications are summarized in the**Figure** .

Here, we review recent progress in plastid genomics in horticultural species. We focus on plastid evolution, gene content, size, inheritance, genomic structure, and rearrangements. We present information about the plastid genome of horticultural species and the current use of this information for different areas. We also briefly highlight the application the plastid genome information on genetic diversity and divergence within natural plant populations, evolution and the importance of plastid genomic for biotechnological use.

PLASTID ORIGIN AND EVOLUTION

The evolutionary history of plastids is based on the endosymbiotic theory which posits that plastids and mitochondria originated from an engulfment of free-living eubacteria over a billion years ago, an α-proteobacteria and a cyanobacterium ancestor, respectively, giving rise the present-day plant cell (Timmis et al., 2004; Reyes-Prieto et al., 2007; Bock and Timmis, 2008; Gould et al., 2008; Archibald, 2009;Kleine et al., 2009; Keeling, 2013; Zimorski et al., 2014). The main evidence of the origin of organelles via the endosymbiotic theory is the molecular, genetic, physiological, and biochemical similarities to prokaryotic cells of ancestors (Zimorski et al., 2014). From these symbionts, the eukaryotic cell acquired the novel biochemistry as oxidative phosphorylation and photosynthesis (Timmis et al., 2004;Bock and Timmis, 2008; Green, 2011). The acquisition of organelles was one of the most important evolutionary processes, given that the association between host and symbionts resulted in a cell with three compartments containing genetic information: the nucleus, mitochondria, and plastids. The combination of three genomes, or host and symbiont genetic compartments, was followed by a dramatic reorganization of the genomes with loss of dispensable genes from organelles, elimination of common genetic information, transfer of genes from organelles to the nucleus, import of products of these transferred genes into the organelles and a complex interaction between nuclear and organellar gene products with the acquisition of new gene functions (Martin et al., 1998; Dyall et al., 2004; Timmis et al., 2004;Bock, 2006; Bock and Timmis, 2008; Zimorski et al., 2014). As a consequence, the size of organelle genomes was drastically reduced during the evolution of the plant cell (Bock, 2007, 2014; Jansen and Ruhlman, 2012).

Nowadays, plastids retain a small prokaryotic chromosome containing no more than 200 protein coding genes (Glöckner et al., 2000; Zimorski et al., 2014) from more than 3200 present in their cyanobacterial ancestor (Kaneko et al., 1996). Even containing a reduced genome and small number of protein-coding genes, plastids harbor thousands of proteins (Meisinger et al., 2008; Kleine et al., 2009), which means that the plastid proteome do not reflect its genome. Most of the genes present into the symbiont genome reside now in the nucleus, where they became functional and their products (i.e., proteins) continue to have their original function in the plastids. Some genes have similarly migrated to the nucleus, however, acquiring a new function which is not related to the prokaryotic ancestor (Martin et al., 2002; Rousseau-Gueutin et al., 2012). Proteomic and genomic analyses suggest that approximately 93–99% of the proteins present in plastids are encoded in the nucleus (Richly et al., 2003; Richly and Leister, 2004; Meisinger et al., 2008; Kleine et al., 2009).

Some experiments *in vivo* were carried out to recapitulate the movement of plastid DNA to the nucleus by using of tobacco transplastomic plants (Huang et al., 2003, 2005; Stegemann et al., 2003;Bock and Timmis, 2008). These experiments suggest that, during the evolution, organellar DNA have been constantly transferred to the nucleus and regularly incorporated into chromosomes. Few experiments have been done to show how a plastid gene becomes functional in the nucleus (Stegemann and Bock, 2006) and the stability of gene expression after nuclear insertion (Sheppard and Timmis, 2009). These different experimental approaches using transplastomic plants showed that the gene transfer from plastid genome to the nucleus is an ongoing process and occurs at a surprisingly high frequency (Huang et al., 2003, 2005; Stegemann et al., 2003; Bock, 2006;Stegemann and Bock, 2006; Bock and Timmis, 2008).

Following the reduction of plastome size, gene content and expression capacity during the evolution, some angiosperm species acquired new lifestyle as parasite plants as examples *Epifagus virginiana* and different species of the *Cuscuta* genus (Wolfe et al., 1992; Lohan and Wolfe, 1998; McNeal et al., 2007). This adaptation to the new life has resulted in an attenuation of the plastome and plastid gene expression machinery and, consequently, a high dependency on the host plant. These alterations in the plastome include loss of photosynthesis-related genes, deletion of a meaningful part of the genetic information and impaired photosynthesis capacity in some species (Funk et al., 2007;Gould et al., 2008; Tsai and Manos, 2010). Except in parasite plants where photosynthesis is dispensable, the plastome sequence and gene content of different species of higher plants are highly conserved, and some experimental evidences *in vivo* have suggested that plastid gene expression is essential for cell survival and development (Rogalski et al., 2006, 2008; Alkatib et al., 2012; Tiller and Bock, 2014). However, a different situation was observed in the high-throughput sequencing and transcriptomic analyses of *Polytomella* spp., a free-living non-photosynthetic green algae closely related to the model organism*Chlamydomonas reinhardtii* (Smith and Lee, 2014). For this species, data analyses revealed no plastid genome-derived reads and, although*Polytomella* spp. has plastids, they do not contain a genome and seems to be only a cellular compartment needed for cellular metabolism (Smith and Lee, 2014).

Regulation of Gene Expression in Plastids

Plastid gene expression involves not only the activation of a set of plastid genes required for plastid biogenesis and photosynthesis, but also the modulation of gene expression during chloroplast development and in response to different environmental factors (Pfannschmidt, 2003;Pfannschmidt et al., 2003). The plastid gene expression must rely on the nucleus for most of their structural proteins and regulatory factors, and a complex signaling pathways are involved, demonstrating the interdependence and need for coordination of gene expression between these cellular genetic compartments (Bräutigam et al., 2007; Greiner et al., 2011). All steps of plastid gene expression are dependent on nuclear gene expression given that nuclear gene products (i.e., proteins) are required for transcription, processing, translation, post-translation modification, and turnover of plastid proteins (Marín-Navarro et al., 2007; Wobbe et al., 2008; Berry et al., 2013; Chi et al., 2013; Small et al., 2013; Petrillo et al., 2014; Ramundo and Rochaix, 2014).

This complex interaction between nuclear genome and organellar genome (i.e., plastome) plays a crucial role in the plant cell controlling the entire metabolism. Moreover, organellar and nuclear genomes constitute a tightly integrated functional unit that co-evolves. This integration between cellular genomes (e.g., plastid genome and nuclear genome) is involved in speciation processes where the lack of functional interaction between genomes results in reproductive barriers between populations (Greiner et al., 2011). Fail in the interaction between plastid genome and nuclear genome can induce genome incompatibilities affecting phenotypically the progenies and resulting in hybrid bleaching, hybrid variegation, or disturbance of the sexual phase (Bogdanova, 2007; Greiner et al., 2008, 2015; Bogdanova et al., 2009;Greiner and Bock, 2013), which affect directly the survival of the plants on natural environment. The sequencing of plastid genome, because of its small size and relatively low number of genes, is a valuable and essential tool to investigate the cause of these incompatibilities (Greiner et al., 2008, 2011; Besnard et al., 2011; Bogdanova et al., 2012; Greiner and Bock, 2013).

Plant Population Genetic Studies in Horticultural Species based on Plastid Genomes

Plastid genomes, unlike most nuclear chromosomes, are typically uniparentally inherited. For sexually reproducing species with male and female gametes, maternal plastid inheritance is the norm (Zhang and Sodmergen, 2010), although it was indicated that about 20% of angiosperms exhibits the potential for paternal plastid transmission (Corriveau and Coleman, 1988; Zhang et al., 2003; Zhang and Sodmergen, 2010; Schneider et al., 2015). Studies have identified diverse species with paternal (mainly conifers) or biparental modes of plastid inheritance (Crosby and Smith, 2012). This uniparental mode of inheritance allows the generation of inferences about the relative contributions of seed and pollen flow to the genetic structure of natural populations by comparing nuclear and plastid markers (Provan et al., 2001; Roullier et al., 2011; Delplancke et al., 2012; Khadivi-Khub et al., 2013).

Effective genetic population size is a parameter influenced by the mode of inheritance. The haploid nature of chloroplast genome is related to its reduced genetic variation. Since the effective population size of a haploid genome is 1/4 in dioecious plants and 1/2 in monoecious plants of the nuclear genome, coalescence times and time to fixation of chloroplast DNA haplotypes within a population are shorter than in diploid genomes (Small et al., 2004). Moreover, different plastid genes evolve at different rates, allowing measuring evolutionary distance at many taxonomy levels (Palmer, 1985; Shaw et al., 2007). This low evolving rate along with the absence of recombination, uniparentally inherited nature in most plant species perceived in plastid genome may greatly facilitate the use of plastid DNA markers in plant population genetic studies (Palmer, 1985; Powell et al., 1995; Provan et al., 2001).

In the 80's, the use of phylogenetic studies based on plastid genomes began to show promising results (Palmer, 1985). The *rbcL* gene was widely sequenced from many plant taxa, generating a suitable database for plant phylogenetic studies at family level and higher taxa (Palmer et al., 1988; Hasebe et al., 1992; Brunsfeld et al., 1994; Setoguchi et al., 1998). However, in some cases *rbcL* gene and other coding regions proved to be highly conserved searching the answer of questions between closely related genera (Gielly and Taberlet, 1994). Since the non-coding regions are likely to evolve faster than coding regions (Gielly and Taberlet, 1994), the analysis of non-coding regions of plastid DNA (i.e., introns and intergenic spacers) was a strategy applied to clarify the relationships at lower taxonomic levels, see Figure. This strategy has solved some of the questions in the context of phylogenetic studies, and later many unexplored non-coding regions of the plastid genome proved promising to bring even more additional information to this line of study (Shaw et al., 2005, 2007). For instance, the pairwise sequence divergence across genes, introns, and spacers in*Helianthus* (Asteraceae) and *Lactuca* (Asteraceae) has resulted in the discovery of fast-evolving DNA sequences for use in species-level phylogenetics (Timme et al., 2007).

The plastid markers, restriction fragment length polymorphisms (RFLPs), began to be used for evolution (Palmer et al., 1983), phylogenetic (Jansen and Palmer, 1987; Sandbrink et al., 1989), and plastid diversity analyses (Ogihara and Tsunewaki, 1988; Dally and Second, 1990). This method has the disadvantage of requiring reasonable amounts plant material, digestible, and nuclear DNA-free plastid DNA, associated with laborious experimental procedures of the Southern hybridization-based RFLP method (Sandbrink et al., 1989;Provan et al., 2001).

In the following decade, Powell et al. (1995) reported the presence of simple nucleotide repeats in plastids, exhibiting length variation and polymorphism in higher levels than those of plastid RFLPs. This marker became a widely used plastid marker, known as chloroplast simple sequence repeats (cpSSRs), consisting of repetitive DNA sequences in tandemly repeated motifs of six base pairs (bp) or less, which have aroused considerable interest due to their ability to generate highly informative DNA markers (Provan et al., 2001). Even though in chloroplasts genomes the occurrence of di-, tri-, tetra-, penta-, and hexanucleotide repeats is less common (George et al., 2015). These regions may be used for both intraspecific and interspecific variability analyses, with practical value for monitoring gene flow, population differentiation and cytoplasmic diversity (Powell et al., 1995).

The development and application of these plastid molecular markers was demonstrated by Angioi et al. (2009a), who developed a useful set of cpSSR markers to study the genetic diversity of *Phaseolus* spp. and other legumes. These markers could discriminate among the genera, and among and within the species of the *Phaseolus* genus. Shortly after, these set of markers were applied to characterized a *Phaseolus vulgaris*collection from Italy, clarifying the origin of the Sardinian (Italy) bean germplasm by comparing local accessions with commercial and Americas varieties (Angioi et al., 2009b). These data generated important information to elucidate the colonization process of *P. vulgaris* in Europe and to define an appropriate management of the local genetic resources, particularly for breeding purposes.

Similarly, Khadivi-Khub et al. (2013) characterized the genetic and phylogenetic relationships of eight wild *Prunus L.* subgen. *Cerasus*species naturally growing in Iran and three commercial species based on nuclear and cpSSR. These markers were able to discriminate all species analyzed, with high level of polymorphism detected, indicating high inter- and intraspecific genetic variation. A close correlation was observed between intraspecific variation and geographical distribution, providing bases for conservation suggestion for these native populations of wild *Cerasus* germplasm and for future breeding activity. By using the same strategy of combining nuclear and cpSSR markers, but for different purposes, Delplancke et al. (2012)investigated spontaneous gene flow among wild and domesticated*Prunus*. Two key almond tree species were selected, the cultivated*Prunus dulcis* and the wild relative *Prunus orientalis*. They identified high genetic diversity levels in both species along with substantial and symmetric gene flow between the domesticated and the wild species. This crop-to-wild gene flow study highlights the importance of use of*ad hoc* transgene containment strategies for this species before the introduction of genetically modified cultivars.

The cpSSRs can also be applied to elucidate evolutionary questions in high economic interesting species and with intriguing domestication processes, such as the evolutionary history of wheat species. The characterization of a large set of accessions of *Triticum spp.*, provided very strong evidence that neither *Triticum urartu* nor *Aegilops tauschii*was the maternal and thus cytoplasmic donor for polyploid wheats cultivated today (Leigh et al., 2013).

In sweet potato landraces, nuclear and cpSSR markers combined also allowed the origin and dispersal investigation, providing bases to suggest at least two independent domestications processes for these species, in Central/ Caribbean America and in the north-western part of South America. The comparison of nuclear and chloroplast data also suggests that exchanges of clones and sexual reproduction were both important processes in landrace diversification in this clonally propagated crop. These analyses provided useful tools for rationalizing the conservation and use of sweet potato germplasm collections (Roullier et al., 2011).

Another relevant area of cpDNA markers application is phylogeographical analysis, e.g., in following colonization after the ice age. The fact that chloroplasts present mainly uniparental inheritance means that they show a clearer geographical structure than nuclear markers, notably in wind-pollinated species (Petit et al., 2005). In this way, glacial refugia have been identified for several tree species, such as *Quercus petraea*, *Quercus pubescens*, *Fagus sylvatica* (Petit et al., 2002), and *Populus nigra* (Cottrell et al., 2005).

PLASTID GENOME IN HORTICULTURAL SPECIES

In general, land plant chloroplast genomes are mostly conserved and contain basically two groups of genes. The first group comprises components for the photosynthetic machinery – photosystem I (PSI), photosystem II (PSII), the cytochrome b6/f complex and the ATP synthase. The second group includes the genes required for the genetic system of plastids – subunits of an RNA polymerase, rRNAs, tRNAs, and ribosomal proteins. The tobacco plastid genome, for example, consists of 155,943 bp and contains a pair of inverted repeat regions (IRA and IRB) separated by a small (SSC) and a large (LSC) single copy region. It contains 115 genes, 79 protein-encoding genes and 35 encoding stable RNA species (Shinozaki et al., 1986;Wakasugi et al., 1998; Yukawa et al., 2005). This plastid genome organization is highly conserved in angiosperms, with very few exceptions (Guo et al., 2007; Hansen et al., 2007; Tangphatsornruang et al., 2010; Do et al., 2014; Gurdon and Maliga, 2014). In gymnosperms, the loss of the large IR has been reported in several species, mainly in conifers. Some authors believed that a pair of large IR could stabilize the plastid genome against major structural rearrangements (Palmer and Thompson, 1982; Strauss et al., 1988;Hirao et al., 2008). However, recently Sabir et al. (2014) found evidences that the loss of the IR in legumes is not the major driving force behind the genomic upheaval, and hypothesized that other factors, such as the extent and location of repetitive DNA, may be more important in destabilizing these genomes.

Table 1: List of genes category, group, and names commonly identified in plant plastid genomes

Category of genes	Group of gene	Name of gene					
Self-replication	Ribosomal RNA genes	*rrn16*	*rrn23*	*rrn5*	*rrn4.5*		
	Transfer RNA genes	*trnA* -UGC	*trnC* -GCA	*trnD* -GUC	*trnE* -UUC	*trnF* -GAA	*trnfM* -CAU
		trnG -UCC	*trnG* -GCC	*trnH* -GUG	*trnI* -CAU	*trnI* -GAU	*trnK* -UUU
		trnL -CAA	*trnL* -UAA	*trnL* -UAG	*trnM* -CAU	*trnN* -GUU	*trnP* -GGG
		trnP -UGG	*trnQ* -UUG	*trnR* -ACG	*trnR* -UCU	*trnR* -CCG	*trnS* -GCU
		trnS -UGA	*trnS* -GGA	*trnT* -UGU	*trnT* -GGU	*trnV* -GAC	*trnV* -UAC
		trnW -CCA	*trnY* -GUA				
	Small subunit of ribosome	*rps2*	*rps3*	*rps4*	*rps7*	*rps8*	*rps11*
		rps12	*rps14*	*rps15*	*rps16*	*rps18*	*rps19*
	Large subunit of ribosome	*rpl2*	*rpl14*	*rpl16*	*rpl20*	*rpl22*	*rpl23*
		rpl32	*rpl33*	*rpl36*			
	DNA-dependent RNA polymerase	*rpoA*	*rpoB*	*rpoC1*	*rpoC2*		
	Translational initiation factor	*infA*					
Genes for photosynthesis	Subunits of photosystem I (PSI)	*psaA*	*psaB*	*psaC*	*psaI*	*psaJ*	*psaM*
		ycf3	*ycf4*				
	Subunits of photosystem II (PSII)	*psbA*	*psbB*	*psbC*	*psbD*	*psbE*	*psbF*
		psbH	*psbI*	*psbJ*	*psbK*	*psbL*	*psbM*
		psbN	*psbT*	*psbZ*			
	Subunits of cytochrome	*petA*	*petB*	*petD*	*petG*	*petL*	*petN*
	Subunits of ATP synthase	*atpA*	*atpB*	*atpE*	*atpF*	*atpH*	*atpI*
	Large subunit of Rubisco	*rbcl*					
	Chlorophyll biosynthesis	*chlB*	*chlL*	*chlN*			
	Subunits of NADH dehydrogenase	*ndhA*	*ndhB*	*ndhC*	*ndhD*	*ndhE*	*ndhF*
		ndhG	*ndhH*	*ndhI*	*ndhJ*	*ndhK*	
Other genes	Maturase	*matK*					
	Envelope membrane protein	*cemA*					
	Subunit of acetyl-CoA	*accD*					
	C-type cytochrome synthesis gene	*ccsA*					
	Protease	*clpP*					
	Component of TIC complex	*ycf1*					
Genes of unknown function	Conserved open reading frames	*ycf2*	*ycf12*				
Pseudogenes		*ycf68*	*ycf15*				

We used four phylogenetically distant species to unify in a single table the gene content of plastid genomes. Small changes may occur according to species. The species used as reference were Prunus persica (eudicotyledon), Elaeis guineensis (monocotyledon), Picea abies (conifer Clade I), and Taxus mairei (Conifer Clade II).

In order to unify in a single table a list of genes category, group, and names commonly identified in plant plastid genomes, we arbitrarily choose four representative horticultural species from very distant taxa, two angiosperms: *Prunus persica* (eudicotyledon) and *Elaeis guineensis* (monocotyledon), and two gymnosperms: *Picea abies*(conifer Clade I) and *Taxus mairei* . Although plastid genome shows a high conservative gene content, small changes may occur according to the species, such as complete gene losses, or presence of pseudogenes.

The beginning of the complete plastid genome sequencing was in the 1980s, with the almost simultaneously sequence release of *Nicotiana tabacum* (Shinozaki et al., 1986) and *Marchantia polymorpha* (Ohyama et al., 1986). The plastid genome sequencing, especially in tobacco, together with the development of plastid transformation for this species have allowed the investigation of the function of several plastid genes (Svab et al., 1990; Svab and Maliga, 1993). Consequently, during the next years until the present, plastid genes and gene expression machinery have been extensively studied (Ruf et al., 1997; Hager et al., 1999; Drescher et al., 2000; Shikanai et al., 2001; Maliga, 2004; Kode et al., 2005; Rogalski et al., 2006, 2008; Alkatib et al., 2012).

In the last decades, several research groups around the world have centered its efforts on the sequencing of plastid genomes of various taxonomic groups. Today, more than 600 land plant species has its plastid genome sequence available in Genbank web page (www.ncbi.nlm.nih.gov/genomes/GenomesGroup.cgi?taxid=2759&opt=plastid). In this review, we highlight species of high interest to horticulture, as tomato (NC_007898; Daniell et al., 2006), potato (NC_008096; Chung et al., 2006), lettuce (NC_007578; Timme et al., 2007), spinach (NC_002202; Schmitz-Linneweber et al., 2001), onion (NC_024813), carrot (NC_008325; Ruhlman et al., 2006); ornamental species, as orchids, i.e., *Phalaenopsis aphrodite*(NC_007499; Chang et al., 2006), *Cymbidium aloifolium* (NC_021429;Yang et al., 2013), and *Cattleya crispata* (NC_026568; da Rocha Perini et al., 2015), *Lilium* (NC_026787), *Magnolia kwangsiensis*(NC_015892; Kuang et al., 2011); fruit crops, as strawberry (NC_015206; Shulaev et al., 2011), peach (NC_014697; Jansen et al., 2011), orange (NC_008334; Bausher et al., 2006), banana (HF677508;Martin et al., 2013); medicinal species, as *Camellia grandibracteata*(NC_024659; Huang et al., 2014), *Salvia* (NC_020431; Qian et al., 2013), *Artemisia frigida* (NC_020607; Liu et al., 2013); and forestry species, as *Eucalyptus aromaphloia* (NC_022396; Bayly et al., 2013),*Pinus contorta* (NC_011153; Cronn et al., 2008), *Picea abies*(NC_021456; Nystedt et al., 2013).

The beginning of plastid genome sequencing involved cloning of chloroplast DNA into plasmid vectors, followed by selection of chloroplast DNA-containing clones, and then sequencing the clones in traditional Sanger-based sequencers using both plasmid and chloroplast-specific primers (Jansen et al., 2005). With the emergence of pyrosequencing, more specifically with the Genome Sequencer 20 (GS 20) system (Roche, Basel, Switzerland), to clone template DNA into bacterial vectors became no more necessary, and genome sequence could be obtained in a single 5-h run with a few days of template preparation (Moore et al., 2006).

Shortly after, Cronn et al. (2008) PCR-amplified eight *Pinus* plastid genomes and adapted multiplex sequencing-by-synthesis (MSBS) to simultaneously sequence multiple plastid genomes using the Illumina Genome Analyzer (Illumina Inc., San Diego, CA, USA). The use of the PCR-based methods to amplify overlapping fragments from conserved gene loci in plastid genomes is time consuming and can be more difficult to implement considering that gene organization differs among plants (Atherton et al., 2010). Atherton et al. (2010) demonstrated a suitable alternative approach, isolating chloroplasts and then using the capacity of high-throughput sequencer Illumina Genome Analyzer II to obtain purified and complete plastid sequences. This technique allowed the obtainment of reads sequence easy to assemble for building the complete plastid genome map.

With the advances of next-generation sequencing, it is becoming increasingly faster and cost-effective to sequence and assemble plastid genomes. The isolation of chloroplast DNA is a facilitator in the sequencing data assembly (Vieira et al., 2014b), but the capacity of current sequencing technologies allows effective analysis of the chloroplast genome sequence by sequencing total DNA (Henry et al., 2014). Using this approach, the chloroplast insertions in the nuclear genome can be distinguished by their much lower copy number, and the short-read sequences from plastid genome are easy discriminated from nuclear reads by alignment with a reference plastome (Henry et al., 2014). Thus, depending on the available framework, nowadays plastid genome sequence may be realized from amplification of chloroplast DNA using long range PCR in species that chloroplast isolation is more challenged and hard to be reached.

Thereby, the complete genome sequencing in Fabaceae family allowed the comparison in two horticultural species of high economic potential,*Glycine max* and *P. vulgaris* with the considered outstanding model for genome research *Medicago truncatula*. All the three legumes present very similar gene content and order, and lack the *rpl22* gene (Saski et al., 2005; Guo et al., 2007). However, the *rps16* is an intron-containing and functional gene in *G. max*, a pseudogene in *P. vulgaris* and absent in *M. truncatula* (Saski et al., 2005; Guo et al., 2007). *M. truncatula*also differ by missing one copy of the IR (Saski et al., 2005). Studies point out that the presence of small repeats of *psbA* and *rbcL* in legumes that have lost one copy of the IR indicate that this loss has only occurred once during the evolutionary history of legumes (Cai et al., 2008; Gurdon and Maliga, 2014). *P. vulgaris* differs from the others by containing an additional pseudogene, *rpl33*. Interestingly, *P. vulgaris*chloroplast genome show higher evolutionary rates on genomic and gene levels than *G. max*, which is believed to be a consequence of pressure from both mutation and natural selection (Guo et al., 2007).

In Rosids, a large monophyletic clade of Angiosperms, comprising 17 orders, many of them containing species with high economic interest, several plastid genomes were sequenced (Hu et al., 2011; Jansen et al., 2011; Rodríguez-Moreno et al., 2011; Njuguna et al., 2013). These plastid genome sequences enabled the identification of a common gene lost in Passifloraceae and Fagaceae, the *rpl22* (Jansen et al., 2011). In*Passiflora sp.*, *Castaneae* sp., and *Quercus sp.* the *rpl22* was present in the chloroplast genome as a pseudogene, and in *Castanea* sp. and*Quercus* sp. it was identified a complete copy of this gene in the nuclear genome, characterizing a functional gene transfer from plastid to nucleus (Jansen et al., 2011). As described above, some species from Fabaceae family also lacks *rpl22*. These results together allowed Jansen et al. (2011) to suggest that these *rpl22* gene transfers occurred approximately 56–58, 34–37, and 26–27 Ma for the Fabaceae, Fagaceae, and Passifloraceae, respectively (Jansen et al., 2011).

Comparisons of chloroplast genome organization between *Solanum lycopersicum* and *Solanum bulbocastanum* showed that, at gene order, these genomes are identical, and this conservation extends to more distantly related genera (tobacco and *Atropa*) of Solanaceae (Daniell et al., 2006). These authors also analyzed repeated sequences in Solanaceae chloroplast genomes, revealing 42 groups of repeats shared among various members of the family. In addition, 37 of these 42 repeats are found in all four genomes examined, occurring in the same location, either in genes, introns or within intergenic spacers, suggesting a high level of conservation of repeat structure. In the same way, Chung et al. (2006) reported that the complete sequence of*Solanum tuberosum* chloroplast genome revealed extensive similarity to six Solanaceae species in terms of the gene content and structure, suggesting a common chloroplast evolutionary lineage within Solanaceae.

Plastid Biotechnology of Horticultural Crops

The plastid genome genetic engineering of crop plants is an attractive platform for biotechnologists to increase characteristics of interest for agriculture and horticulture (Clarke and Daniell, 2011; Maliga and Bock, 2011; Rogalski and Carrer, 2011; Hanson et al., 2013; Bock, 2014). This technology offers several exceptional features and advantages when compared with nuclear transformation, among which can be included high transgene expression levels with accumulation of foreign proteins up to >70% of the total soluble cellular protein (Oey et al., 2009; Ruhlman et al., 2010), capacity for multigene stacking in operons in a single genetic transformation event (Quesada-Vargas et al., 2005; Lu et al., 2013; Bock, 2014), precise transgene integration via homologous recombination (Cerutti et al., 1992), absence of epigenetic effects or gene silencing (Bock, 2001, 2013; Maliga, 2004) and exclusion of transgenes transmission by pollen due to maternal inheritance of plastids in most angiosperms (Daniell, 2007; Ruf et al., 2007; Svab and Maliga, 2007). The plastid transformation vector design and the transgene insertion via two homologous recombination events into the plastid genome are illustrated in the.

Other desirable, but not exclusive, feature is the possibility of efficient elimination of the selection marker gene via Cre-lox site-specific recombination (Lutz et al., 2006), φC31 phage site-specific integrase (Kittiwongwattana et al., 2007), serine recombinase Bxb1(Shao et al., 2014) and/or use of direct repeats for gene excision via homologous recombination (Dufourmantel et al., 2007). This is an exceptional advantage because it allows the production of transgenic plants without the insertion of antibiotic resistance genes, eliminate any possibility of antibiotic resistance gene flow to neighboring crop fields or to crop wild relatives growing near the transgenic crops. Moreover, it permits the recycling of selectable marker genes, which can be reused in a new genetic transformation event in the same transgenic plant (Carrer et al., 1993; Svab and Maliga, 1993; Corneille et al., 2001; Barone et al., 2009; Li et al., 2011).

After the successful plastid transformation of the first higher plant species, *N. tabacum* (Svab et al., 1990; Svab and Maliga, 1993), several aspects of plastid transformation were studied and optimized to increase the potential of transplastomic technology for biotechnological aspects such as crop improvement (Jin et al., 2011, 2012; Lu et al., 2013), herbicide, insect, and diseases resistance (Lutz et al., 2001; Ye et al., 2001; Dufourmantel et al., 2005; Bock, 2007; Kiani et al., 2013; Zhang et al., 2015), abiotic and biotic stresses (Kumar et al., 2004; Jin et al., 2011; Bansal et al., 2012; Chen et al., 2014), metabolic engineering (Hasunuma et al., 2008; Apel and Bock, 2009; Lu et al., 2013), phytoremediation (Yun et al., 2009; Ruiz and Daniell, 2009; Ruiz et al., 2011), bioreactors (Maliga and Bock, 2011; Bock, 2013, 2014), vaccines and biopharmaceuticals (Daniell et al., 2009; Clarke et al., 2013; Kwon et al., 2013), enzymes (Petersen and Bock, 2011; Jin et al., 2012; Verma et al., 2013), biomass and raw material for industry (Verma and Daniell, 2007; Agrawal et al., 2011; Jin et al., 2011; Verma et al., 2013). This research have also focused on plastid gene expression in different plastid types in the same plant (Kahlau and Bock, 2008;Valkov et al., 2009), regulatory elements to be used in different plastid types (i.e., chloroplasts, chromoplasts and amyloplasts; Zhang et al., 2012; Caroca et al., 2013) and foreign protein stability in plastids (Apel et al., 2010; Elghabi et al., 2011; De Marchis et al., 2012).

Although tobacco (*N. tabacum*) is the species transformed with the highest efficiency, commercial use of this technology for cultivar improvement is totally dependent on spread of the technology to plant species of agriculture and horticultural interest. Probably, the wide spread of plastid technology to several horticultural crops is dependent on the development of a highly efficient tissue culture system (via organogenesis or somatic embryogenesis), which is observed in tobacco as model species for plastid transformation (Díaz and Koop, 2014;Maliga and Tungsuchat-Huang, 2014; Staub, 2014). Currently, several horticultural species have been efficiently transformed, including tomato (Ruf et al., 2001; Nugent et al., 2005; Ruf and Bock, 2014), lettuce (Lelivelt et al., 2005; Kanamoto et al., 2006; Ruhlman, 2014), potato (Sidorov et al., 1999; Nguyen et al., 2005; Scotti et al., 2011;Valkov et al., 2014), carrot (Kumar et al., 2004), eggplant (Singh et al., 2010; Bansal and Singh, 2014), cabbage (Liu et al., 2007, 2008; Tseng et al., 2014), cauliflower (Nugent et al., 2006) and sugar beet (De Marchis et al., 2009; De Marchis and Bellucci, 2014). Among the horticultural species mentioned above, lettuce, tomato and potato are the most studied species regarding the gene expression and biotechnological applications; lettuce is a model species for edible leaf chloroplasts (Boyhan and Daniell, 2011; Maldaner et al., 2013; Yabuta et al., 2013), tomato and potato are model species for edible organs as fruits and tubers containing chromoplasts (Wurbs et al., 2007; Apel and Bock, 2009; Lu et al., 2013) and amyloplasts (Valkov et al., 2011;Segretin et al., 2012), respectively.

Lettuce, as a model of edible tissue containing chloroplasts, plastid type with the elevated ploidy and highest gene expression (Kahlau and Bock, 2008; Valkov et al., 2009; Zhang et al., 2012; Caroca et al., 2013;Bock, 2014), is currently the target species for expression of antigens, pharmaceutical proteins and vaccines (Boyhan and Daniell, 2011;Maldaner et al., 2013), and also metabolic engineering (Yabuta et al., 2013). The first example of the use of lettuce plastid genome to produce proteins of pharmaceutical interest was made by Boyhan and Daniell (2011), who observed in old lettuce leaves the accumulation of proinsulin up to 53% of total leaf protein. The same study showed that the accumulation was stable even in senescent and dried lettuce leaves, facilitating their processing and storage in the field. This genetic engineering strategy can reduce significantly the costs and facilitate oral delivery of plant-derived pharmaceutical compounds using edible plant leaves (Boyhan and Daniell, 2011). Recently, another study showed the efficient and stable production of the tetra-epitope peptide antigen from E protein of dengue virus in lettuce transplastomic plants (Maldaner et al., 2013). The tetra-epitope peptide expressed in lettuce plastid genomes shows to be efficient to use as antigen in diagnostic assays demonstrating an overall sensitivity of 71.7% and specificity of 100% (Maldaner et al., 2013). Besides to the pharmaceutical area, lettuce chloroplasts were also used to manipulate the metabolic pathway of the tocochromanol (vitamin E) by expression of the enzymes tocopherol cyclase, γ-tocopherol methyltransferase, or both in an operon (Yabuta et al., 2013). The expression of the different genes, alone or combined, resulted in an increase of total tocochromanol content in transplastomic plants, which indicate that chloroplast genetic engineering can be successful used to improve vitamin E quality and quantity in a plant green edible tissue (Yabuta et al., 2013).

The application of plastid transformation technology in tomato was target to metabolic engineering of plastid pigments. The first successful example showed the feasibility to engineer a nutritionally important metabolic human nutrient in non-green plastids. Apel and Bock (2009)overexpressed the enzyme lycopene β-cyclase from the daffodil (*Narcissus pseudonarcissus*) and observed an increase up to 50% in provitamin A content in tomato fruits (an important antioxidant and essential vitamin for human nutrition), which changed the color from red to orange due to the conversion of lycopene into β-carotene. Another example in tomato chloroplasts and chromoplasts was the increase of tocochromanol, which provides tocopherols and tocotrienols (vitamin E), in a complex and successful transcription and translation strategy of a multigene operon containing three genes related to tocochromanol biosynthesis (Lu et al., 2013). The tomato transplastomic plants showed an increase of up to 10-fold in total tocochromanol accumulation (Lu et al., 2013).

Potato contains edible tubers, which have amyloplasts, plastids related to starch accumulation as the plant energetic reserve. Potato is by far the most important non-cereal source of starch and carbohydrates for human nutrition and is the most consumed species in many countries around the world. The first transplastomic events in potato were obtained by Sidorov et al. (1999) and Nguyen et al. (2005) by expression of the resistant marker gene, *aadA*, and the green fluorescent protein (*Gfp*), however, it was a limited method due to the low transformation frequencies and low transgene expression in tubers of potato transplastomic plants. Later, by optimizing of the selection/regeneration procedure, using of new transformation vectors and new regulatory sequences for transgene expression in leaves and tubers, Valkov et al. (2011) confirmed general differences in expression patterns in the two organs containing different plastids leaves (chloroplasts) and tubers (amyloplasts). Although expression in tubers was generally low, it reached up to 0.02% of total soluble protein in comparison with 4% of total protein soluble in potato chloroplasts. In the same year the efficiency of plastid transformation was improved by using of new target regions for insertion of transgenes in the potato plastid genome (Scotti et al., 2011), but this report did not mention about the accumulation of foreign proteins.

Cabbage, as lettuce, represents a plant species with edible leaves containing chloroplasts. The plastid transformation of cabbage was reached by Liu et al. (2007), who expressed the resistant marker gene,*aadA*, and the reporter gene, *uidA*. The study demonstrated a transformation efficiency ranging from 2.7 to 3.3% and a successful accumulation of β-glucuronidase protein in transformed cabbage between 3.2 and 5.2% of total soluble protein. After the development of an efficient plastid transformation in this species, Liu et al. (2008)changed the constructs to express the *cry1Ab* gene targeting to the resistance to *Plutella xylostella* in two cabbage varieties. The *cry1Ab*gene codifies *Bacillus thuringiensis* Cry1Ab delta-endotoxin (Jabeen et al., 2010). The expression of Cry1Ab protein was detected in the range of 4.8–11.1% of total soluble protein in mature leaves of transplastomic plants of the two varieties. This study demonstrated that transplastomic plants displayed significantly higher resistance to *Plutella xylostella*and induces 100% insect mortality after 7 days (Liu et al., 2008).

The only report of carrot plastid transformation was focused on salt tolerance by overexpression of betaine aldehyde dehydrogenase (Kumar et al., 2004). The betaine aldehyde dehydrogenase enzyme activity in carrot transplastomic cells was enhanced eightfold, which accumulated about 50-fold more betaine than cells of control plants. Transplastomic carrot plants grew in the presence of high concentrations of up to 400 mM of NaCl, which is the highest level of salt tolerance reported so far among genetically modified crop plants (Kumar et al., 2004). In this study, it was also observed that the accumulation levels of betaine aldehyde dehydrogenase show a variation dependent on plastid type. The betaine aldehyde dehydrogenase expression reached 74.8% in edible parts (roots), containing chromoplasts, an inferior value compared to leaves (100%), a mainly chloroplasts-containing tissue. This study showed the potential of plastid genome engineering technology to increase salt tolerance in a horticultural crop given that salinity affects drastically and negatively crop productivity and quality (Kumar et al., 2004).

The plastid transformation technology was recently developed for other three horticultural crops as follows: eggplant (Singh et al., 2010; Bansal and Singh, 2014), cauliflower (Nugent et al., 2006) and sugar beet (De Marchis et al., 2009; De Marchis and Bellucci, 2014). These studies did not focus on characteristics of interest for horticulture or agriculture, notwithstanding the plastid genome transformation was developed for them. Although these species have an important economic role in several countries and plastid transformation have the potential to add new traits in order to increase the performance in the field, plastome manipulation have many opportunities in different areas of biotechnology and remains to be done in these species and several others.

Plastid genome sequencing of the target species is an essential tool for correct integration of the transgenes into the plastid genome given that plastid genomes of higher plants are extremely gene-dense and are complexly regulated by operons separated by short intergenic spacer region, which have to be maintained intact given that any disruption can affect the expression of several genes (Shinozaki et al., 1986;Wakasugi et al., 2001; Krech et al., 2012; Bock, 2013). The plastid genome sequencing is also important to identify and characterize endogenous regulatory regions such as promoters, 5′ e 3′ untranslated regions to optimize transgene expression (Ruhlman et al., 2010; Maliga and Bock, 2011; Bock, 2013, 2014). Furthermore, the characterization of endogenous regulatory sequences from plastid genome sequences and transgene expression in edible plant organs containing different plastid types (e.g., leaves, fruits, and tubers) will facilitate the expression of new metabolic pathways and transgenes for the production of healthy nutritional compounds, biopharmaceutical compounds, agriculture useful traits and biomass and raw material for biofuel and chemical industry (Kahlau and Bock, 2008; Valkov et al., 2009; Caroca et al., 2013).

CONCLUDING REMARKS

Plastid genomes are highly conserved with very low rates of substitutions when compared to nuclear genomes. Plastid genes, non-coding regions, RFLP and SSR markers have been frequently used to measure the evolutionary distance at many plant taxonomy levels. This markers are also very helpful for phylogeographical and plant population genetics analyses, as seed and pollen flow studies to the characterize population structure, population differentiation and cytoplasmic diversity. However, the limited number of plastid genome sequences for some species, families and genera restrings the quality and efficacy of this kind of analyses. Nowadays, the increasing number of whole plastid genomes are being used for phylogenetic analyses and have proven to be effective tools to resolve evolutionary relationships and genetic diversity or divergence in plant populations, especially at lower taxonomic levels, which limited sequence variation is available. Plastid genome is also an important tool to analyze genetic distance and plant speciation given that it is possible to relate plastid haplotypes with morphological characteristics in natural population as observed in the *Oenothera* genus.

The interesting features of plastid compartment and genome, the exceptional advantages of plastid genome engineering and crescent necessity of horticultural crops for human consumption as food, raw material for industry and cost reduction for production of biopharmaceutical compounds, makes the plastid transformation a potential tool to manipulate different species for industry and food purposes. The rapidly growing number of plastid genomes available in the organelle genome resource database can be used to generate high efficient plastid transformation vectors, since sequences of genes, intergenic regions and regulatory elements are crucial information for design of efficient plastid transformation strategies.

Moreover, the improvement of tissue culture system for horticultural crops would help to spread this technology to several species which plastid transformation was not reached at the moment. The regeneration capacity of the tissues is still the bottleneck for a large number of species, given the fact that tobacco has become the model species for plastid transformation due to its high capacity for *in vitro* regeneration.

Due to the high potential and environment-friendly characteristics of plastid engineering, the knowledge acquired during the last two decades about this technology, and the enormous field to be explored in horticultural crops, plastid genomic and transformation constitute a high valuable tool to add new traits and increase the marker value of commercial crops. Moreover, plastid transformation is already safer than nuclear transformation due to exceptionally maternal inheritance of plastids in most angiosperms and lack of dissemination of transgenes via pollen, avoiding contamination of natural germoplasm resources. In addition, horticultural crops can be maintained in closed greenhouse worldwide by using of soil-containing pots or hydroponic systems which can enhance security of transgenic plants, without transgene flux, for several commercial applications.

Finally, plastid genome sequencing is an essential tool for several applications related to plant science. The first knowledge about plastid genome was the starting point to elucidate many processes related to plastid gene function, expression machinery, evolution and transfer of genes to other genetic cellular compartments as mitochondria and the nucleus. This gain of knowledge in last three decades, from the first plastid genome sequenced to present day, makes the plastid genome the best studied genetic compartment of the plant cell. The improvement of chloroplast isolation and the evolution of technology of genome sequencing will make plastid genome sequencing routine in many laboratories and will certainly contribute to unveil several unknown questions about plant cell genetic of families/species that no information about plastid genome is available.

CONFLICT OF INTEREST STATEMENT

The authors declare that the research was conducted in the absence of any commercial or financial relationships that could be construed as a potential conflict of interest.

REFERENCES

1. Agrawal P., Verma D., Daniell H. (2011). Expression of*Trichoderma reesei* β-mannanase in tobacco chloroplasts and its utilization in lignocellulosic woody biomass hydrolysis. *PLoS ONE* 6:e29302 10.1371/journal.pone.0029302
2. Aldridge C., Maple J., Moller S. G. (2005). The molecular biology of plastid division in higher plants. *J. Exp. Bot.* 56 1061–1077. 10.1093/jxb/eri118
3. Alkatib S., Scharff L. B., Rogalski M., Fleischmann T. T., Matthes A., Seeger S., et al. (2012). The conributions of wobbling and superwobbling to the reading of the genetic code.*PLoS Genet.* 8:e1003076 10.1371/journal.pgen.1003076
4. Angioi S. A., Desiderio F., Rau D., Bitocchi E., Attene G., Papa R. (2009a). Development and use of chloroplast microsatellites in*Phaseolus* spp. and other legumes. *Plant Biol. (Stuttg).* 11 598–612. 10.1111/j.1438-8677.2008.00143.x
5. Angioi S. A., Rau D., Rodriguez M., Logozzo G., Desiderio F., Papa R., et al. (2009b). Nuclear and chloroplast microsatellite diversity in *Phaseolus vulgaris* L. from Sardinia (Italy). *Mol. Breeding* 23 413–429. 10.1007/s11032-008-9245-8

6. Apel W., Bock R. (2009). Enhancement of carotenoid biosynthesis in transplastomic tomatoes by induced lycopene-to-provitamin A conversion. *Plant Physiol.* 151 59–66. 10.1104/pp.109.140533
7. Apel W., Schulze W. X., Bock R. (2010). Identification of protein stability determinants in chloroplasts. *Plant J.* 63 636–650. 10.1111/j.1365-313X.2010.04268.x
8. Archibald J. M. (2009). The puzzle of plastid evolution. *Curr. Biol.* 19 81–88. 10.1016/j.cub.2008.11.067
9. Atherton R. A., McComish B. J., Shepherd L. D., Berry L. A., Albert N. W., Lockhart P. J. (2010). Whole genome sequencing of enriched chloroplast DNA using the Illumina GAII platform.*Plant Methods.* 6 22 10.1186/1746-4811-6-22
10. Bansal K. C., Singh A. K. (2014). Plastid transformation in eggplant. *Methods Mol. Biol.* 1132 305–316. 10.1007/978-1-62703-995-6_19
11. Bansal K. C., Singh A. K., Wani S. H. (2012). Plastid transformation for abiotic stress tolerance in plants. *Methods Mol. Biol.* 913 351–358. 10.1007/978-1-61779-986-0_23
12. Barone P., Zhang X. H., Widholm J. M. (2009). Tobacco plastid transformation using the feedback-insensitive anthranilate synthase asubunit of tobacco (ASA2) as a new selectable marker.*J. Exp. Bot.* 60 3195–3202. 10.1093/jxb/erp160
13. BaurE.(1909).DasWesenunddieErblichkeitsverhältnisse der "Varietates albomarginatae hort." von *Pelargonium zonale. Z. Indukt. Abstamm. Ver.* 1 330–351.
14. Baur E. (1910). Untersuchungen über die vererbung von chromatophorenmerkmalen bei melandrium, antirrhinum und aquilegia. *Z. Indukt. Abstamm. Ver.* 4 81–102.
15. Bausher M. G., Singh N. D., Lee S., Jansen R. K., Daniell H. (2006). The complete chloroplast genome sequence of *Citrus sinensis* (L.) Osbeck var 'Ridge Pineapple': organization and phylogenetic relationships to other angiosperms. *BMC Plant Biol.*6:21 10.1186/1471-2229-6-21
16. Bayly M. J., Rigault P., Spokevicius A., Ladiges P. Y., Ades P. K., Anderson C., et al. (2013). Chloroplast genome analysis of Australian eucalypts – *Eucalyptus*, *Corymbia*, *Angophora*,*Allosyncarpia* and *Stockwellia* (Myrtaceae). *Mol. Phylogenet. Evol.* 69 704–716. 10.1016/j.ympev.2013.07.006
17. Bendich A. J. (1987). Why do chloroplasts and mitochondria contain so many copies of their genome? *Bioessays* 6 279–282. 10.1002/bies.950060608
18. Bendich A. J. (2004). Circular chloroplast chromosomes: the grand illusion. *Plant Cell* 16 1661–1666. 10.1105/tpc.160771
19. Berry J. O., Yerramsetty P., Zielinski A. M., Mure C. M. (2013).Photosynthetic gene expression in higher plants. *Photosynth. Res.*117 91–120. 10.1007/s11120-013-9880-8
20. Besnard G., Hernández P., Khadari B., Dorado G., Savolainen V. (2011). Genomic profiling of plastid DNA variation in the Mediterranean olive tree. *BMC Plant Biol.* 11:80 10.1186/1471-2229-11-80
21. Bock R. (2001). Transgenic plastids in basic research and plant biotechnology. *J. Mol. Biol.* 312 425–438. 10.1006/jmbi.2001.4960
22. Bock R. (2006). Extranuclear inheritance: gene transfer out of plastids. *Prog. Bot.* 67 75–100. 10.1007/3-540-27998-9_4
23. Bock R. (2007). Plastid biotechnology: prospects for herbicide and insect resistance, metabolic engineering and molecular farming. *Curr. Opin. Biotechnol.* 18 100–106. 10.1016/j.copbio.2006.12.001
24. Bock R. (2013). Strategies for metabolic pathway engineering with multiple transgenes. *Plant Mol. Biol.* 83 21–31. 10.1007/s11103-013-0045-0
25. Bock R. (2014). Engineering plastid genomes: methods, tools, and applications in basic research and biotechnology. *Annu. Rev. Plant Biol.* 66 211–241. 10.1146/annurev-arplant-050213-040212
26. Bock R., Timmis J. N. (2008). Reconstructing evolution: gene transfer from plastids to the nucleus. *Bioessays* 30 556–566. 10.1002/bies.20761
27. Bogdanova V. S. (2007). Inheritance of organelle DNA markers in a pea cross associated with nuclear-cytoplasmic incompatibility. *Theor. Appl. Genet.* 114 333–339. 10.1007/s00122-006-0436-6
28. Bogdanova V. S., Galieva E. R., Kosterin O. E. (2009). Genetic analysis of nuclear-cytoplasmic incompatibility in pea associated with cytoplasm of an accession of wild subspecies *Pisum sativum*subsp. elatius (Bieb.) Schmahl. *Theor. Appl. Genet.* 118 801–809. 10.1007/s00122-008-0940-y
29. Bogdanova V. S., Galieva E. R., Yadrikhinskiy A. K., Kosterin O. E. (2012). Inheritance and genetic mapping of two nuclear genes involved in nuclear-cytoplasmic incompatibility in peas (*Pisum sativum* L.). *Theor. Appl. Genet.* 124 1503–1512. 10.1007/s00122-012-1804-z
30. Boyhan D., Daniell H. (2011). Low-cost production of proinsulin in tobacco and lettuce chloroplasts for injectable or oral delivery of functional insulin and C-peptide. *Plant Biotechnol. J.* 9 585–598. 10.1111/j.1467-7652.2010.00582.x
31. Bräutigam K., Dietzel L., Pfannschmidt T. (2007). "Plastid-nucleus communication: anterograde and retrograde signalling in the development and function of plastids," in *Cell and Molecular Biology of Plastids*, ed. Bock R., editor. (Göteborg: Göteborg University;), 409–455.
32. Brunsfeld S. J., Soltis P. S., Soltis D. E., Gadek P. A., Quinn C. J., Strenge D. D., et al. (1994). Phylogenetic relationships among the genera of Taxodiaceae and Cupressaceae: evidence from rbcL sequences. *Syst. Bot.* 19 253–262. 10.2307/2419600
33. Cai Z., Guisinger M., Kim H. G., Ruck E., Blazier J. C., McMurtry V., et al. (2008). Extensive reorganization of the plastid genome of *Trifolium subterraneum* (Fabaceae) is associated with numerous repeated

sequences and novel DNA insertions. *J. Mol. Evol.* 67 696–704. 10.1007/s00239-008-9180-7

34. Caroca R., Howell K. A., Hasse C., Ruf S., Bock R. (2013).Design of chimeric expression elements that confer high-level gene activity in chromoplasts. *Plant J.* 73 368–379. 10.1111/tpj.12031

35. Carrer H., Hockenberry T. N., Svab Z., Maliga P. (1993). Kanamycin resistance as a selectable marker for plastid transformation in tobacco. *Mol. Gen. Genet.* 241 49–56. 10.1007/BF00280200

36. Cerutti H., Osman M., Grandoni P., Jagendorf A. T. (1992). A homolog of *Escherichia coli* RecA protein in plastids of higher plants. *Proc. Natl. Acad. Sci. U.S.A.* 89 8068–8072. 10.1073/pnas.89.17.8068

37. Chang C. C., Lin H. C., Lin I. P., Chow T. Y., Chen H. H., Chen W. H., et al. (2006). The chloroplast genome of *Phalaenopsis aphrodite* (Orchidaceae): comparative analysis of evolutionary rate with that of grasses and its phylogenetic implications. *Mol. Biol. Evol.* 23 279–291. 10.1093/molbev/msj029

38. Chen P. J., Senthilkumar R., Jane W. N., He Y., Tian Z., Yeh K. W. (2014). Transplastomic *Nicotiana benthamiana* plants expressing multiple defence genes encoding protease inhibitors and chitinase display broad-spectrum resistance against insects, pathogens and abioticstresses. *Plant Biotechnol. J.* 12 503–515. 10.1111/pbi.12157

39. Chi W., Sun X., Zhang L. (2013). Intracellular signaling from plastid to nucleus. *Annu. Rev. Plant Biol.* 64 559–582. 10.1146/annurev-arplant-050312-120147

40. Chun E. H., Vaughan M. H., Rich A. (1963). The isolation and characterization of DNA associated with chloroplast preparations. *J. Mol. Biol.* 7 130–141. 10.1016/S0022-2836(63)80042-X

41. Chung H., Jung J. D., Park H., Kim J., Cha H. W., Min S. R., et al. (2006). The complete chloroplast genome sequences of*Solanum tuberosum* and comparative analysis with Solanaceae species identified the presence of a 241-bp deletion in cultivated potato chloroplast DNA sequence. *Plant Cell Rep.* 25 1369–1379. 10.1007/s00299-006-0196-4

42. Clarke J. L., Daniell H. (2011). Plastid biotechnology for crop production: present status and future perspectives. *Plant Mol. Biol.* 76 211–220. 10.1007/s11103-011-9767-z

43. Clarke J. L., Waheed M. T., Lössl A. G., Martinussen I., Daniell H. (2013). How can plant genetic engineering contribute to cost-effective fish vaccine development for promoting sustainable aquaculture? *Plant Mol. Biol.* 83 33–40. 10.1007/s11103-013-0081-9

44. Corneille S., Lutz K., Svab Z., Maliga P. (2001). Efficient elimination of selectable marker genes from the plastid genome by the CRE-lox site-specific recombination system. *Plant J.* 27171–178. 10.1046/j.1365-313x.2001.01068.x

45. Correns C. (1909). Vererbungsversuche mit blass(gelb) grünen und buntblättrigen Sippen bei *Mirabilis jalapa, Urtica pilulifera*und *Lunaria annua*. *Z. Indukt. Abstamm. Ver.* 1 291–329.

46. Corriveau J. L., Coleman A. W. (1988). Rapid screening method to detect potential biparental inheritance of plastid DNA and results for over 200 angiosperm species. *Am. J. Bot.* 75 1443–1458. 10.2307/2444695

47. Cottrell J. E., Krystufek V., Tabbener H. E., Milner A. D., Connolly T., Sing L., et al. (2005). Postglacial migration of*Populus nigra* L.: lessons learnt from chloroplast DNA. *For. Ecol. Manage.* 206 71–90. 10.1016/j.foreco.2004.10.052

48. Cronn R., Liston A., Parks M., Gernandt D. S., Shen R., Mockler T. (2008). Multiplex sequencing of plant chloroplast genomes using Solexa sequencing-by-synthesis technology. *Nucleic Acids Res.* 36:e122 10.1093/nar/gkn502

49. Crosby K., Smith D. R. (2012). Does the mode of plastid inheritance influence plastid genome architecture? *PLoS ONE*.7:e46260 10.1371/journal.pone.0046260

50. da Rocha Perini V., Leles B., Furtado C., Prosdocimi F. (2015).Complete chloroplast genome of the orchid *Cattleya crispata*(Orchidaceae: Laeliinae), a *Neotropical rupiculous* species.*Mitochondrial DNA* [Epub ahead of print].10.3109/19401736.2014.1003850

51. Dally A. M., Second G. (1990). Chloroplast DNA diversity in wild and cultivated species of rice (Genus *Oryza*, section Oryza). Cladistic-mutation and genetic-distance analysis. *Theor. Appl. Genet.* 80 209–222. 10.1007/BF00224389

52. Daniell H. (2007). Transgene containment by maternal inheritance: effective or elusive? *Proc. Natl. Acad. Sci. U.S.A.*104 6879–6880. 10.1073/pnas.0702219104

53. Daniell H., Lee S. B., Grevich J., Saski C., Quesada-Vargas T., Guda C., et al. (2006). Complete chloroplast genome sequences of *Solanum bulbocastanum*, *Solanum lycopersicum* and comparative analyses with other Solanaceae genomes. *Theor. Appl. Genet.* 112 1503–1518. 10.1007/s00122-006-0254-x

54. Daniell H., Singh N. D., Mason H., Streatfield S. J. (2009). Plant-made vaccine antigens and biopharmaceuticals. *Trends Plant Sci.*14 669–679. 10.1016/j.tplants.2009.09.009

55. De Marchis F., Bellucci M. (2014). Plastid transformation in sugar beet: *Beta vulgaris*. *Methods Mol. Biol.* 1132 367–373. 10.1007/978-1-62703-995-6_24

56. De Marchis F., Pompa A., Bellucci M. (2012). Plastid proteostasis and heterologous protein accumulation in transplastomic plants. *Plant Physiol.* 160 571–581. 10.1104/pp.112.203778

57. De Marchis F., Wang Y., Stevanato P., Arcioni S., Bellucci M. (2009). Genetic transformation of the sugar beet plastome.*Transgenic Res.* 18 17–30. 10.1007/s11248-008-9193-4

58. Delplancke M., Alvarez N., Espíndola A., Joly H., Benoit L., Brouck E., et al. (2012). Gene flow among wild and domesticated almond species: insights from chloroplast and nuclear markers.*Evol. Appl.* 5 317–329. 10.1111/j.1752-4571.2011.00223.x

59. Díaz A. H., Koop H. U. (2014). Nicotiana tabacum: PEG-mediated plastid transformation. *Methods Mol. Biol.* 1132 165–175. 10.1007/978-1-62703-995-6_9

60. Do H. D., Kim J. S., Kim J. H. (2014). A trnI_CAU triplication event in the complete chloroplast genome of Paris verticillata M.Bieb. (Melanthiaceae, Liliales). *Genome Biol. Evol.* 6 1699–1706. 10.1093/gbe/evu138

61. Drescher A., Ruf S., Calsa T., Carrer H., Bock R. (2000). The two largest chloroplast genome-encoded open reading frames of higher plants are essential genes. *Plant J.* 22 97–104. 10.1046/j.1365-313x.2000.00722.x

62. Dufourmantel N., Dubald M., Matringe M., Canard H., Garcon F., Job C., et al. (2007). Generation and characterization of soybean and marker-free tobacco plastid transformants over-expressing a bacterial 4-hydroxyphenylpyruvate dioxygenase which provides strong herbicide tolerance. *Plant Biotechnol. J.* 5118–133. 10.1111/j.1467-7652.2006.00226.x

63. Dufourmantel N., Tissot G., Goutorbe F., Garçon F., Muhr C., Jansen S., et al. (2005). Generation and analysis of soybean plastid transformants expressing *Bacillus thuringiensis* Cry1Ab protoxin. *Plant Mol. Biol.* 58 659–668. 10.1007/s11103-005-7405-3

64. Dyall S. D., Brown M. T., Johnson P. J. (2004). Ancient invasions: from endosymbionts to organelles. *Science* 304 253–257. 10.1126/science.1094884

65. Egea I., Barsan C., Bian W., Purgatto E., Latche A., Chervin C., et al. (2010). Chromoplast differentiation: current status and perspectives. *Plant Cell Physiol.* 51 1601–1611. 10.1093/pcp/pcq136

66. Elghabi Z., Karcher D., Zhou F., Ruf S., Bock R. (2011). Optimization of the expression of the HIV fusion inhibitor cyanovirin-N from the tobacco plastid genome. *Plant Biotechnol. J.* 9 599–608. 10.1111/j.1467-7652.2011.00598.x

67. Fuentes I., Stegemann S., Golczyk H., Karcher D., Bock R. (2014). Horizontal genome transfer as an asexual path to the formation of new species. *Nature* 511 232–235. 10.1038/nature13291

68. Funk H. T., Berg S., Krupinska K., Maier U. G., Krause K. (2007). Complete DNA sequences of the plastid genomes of two parasitic flowering plant species, *Cuscuta reflexa* and *Cuscuta gronovii*. *BMC Plant Biol.* 7:45 10.1186/1471-2229-7-45

69. Galili G., Amir R. (2013). Fortifying plants with the essential amino acids lysine and methionine to improve nutritional quality.*Plant Biotechnol. J.* 11 211–222. 10.1111/pbi.12025

70. Galili G., Avin-Wittenberg T., Angelovici R., Fernie A. R. (2014). The role of photosynthesis and amino acid metabolism in the energy status during seed development. *Front. Plant Sci.*5:447 10.3389/fpls.2014.00447

71. George B., Bhatt B. S., Awasthi M., George B., Singh A. K. (2015). Comparative analysis of microsatellites in chloroplast genomes of lower and higher plants. *Curr. Genet.* [Epub ahead of print].10.1007/s00294-015-0495-9

72. Gielly L., Taberlet P. (1994). The use of chloroplast DNA to resolve plant phylogenies: noncoding versus rbcL sequences.*Mol. Biol. Evol.* 11 769–777.

73. Glöckner G., Rosenthal A., Valentin K. (2000). The structure and gene repertoire of an ancient red algal plastid genome. *J. Mol. Evol.* 51 382–390. 10.1007/s002390010101

74. Golczyk H., Greiner S., Wanner G., Weihe A., Bock R., Borner T., et al. (2014). Chloroplast DNA in mature and senescing leaves: a reappraisal. *Plant Cell* 26 847–854. 10.1105/tpc.113.117465

75. Gould S. B., Waller R. F., McFadden G. I. (2008). Plastid evolution. *Annu. Rev. Plant Biol.* 59 491–517. 10.1146/annurev.arplant.59.032607.092915

76. Green B. R. (2011). Chloroplast genomes of photosynthetic eukaryotes. *Plant J.* 66 34–44. 10.1111/j.1365-313X.2011.04541.x

77. Greiner S., Bock R. (2013). Tuning a menage a trois: co-evolution and co-adaptation of nuclear and organellar genomes in plants. *Bioessays* 35 354–365. 10.1002/bies.201200137

78. Greiner S., Rauwolf U., Meurer J., Herrmann R. G. (2011). The role of plastids in plant speciation. *Mol. Ecol.* 20 671–691. 10.1111/j.1365-294X.2010.04984.x

79. Greiner S., Sobanski J., Bock R. (2015). Why are most organelle genomes transmitted maternally? *Bioessays* 37 80–94. 10.1002/bies.201400110

80. Greiner S., Wang X., Rauwolf U., Silber M. V., Mayer K., Meurer J., et al. (2008). The complete nucleotide sequences of the five genetically distinct plastid genomes of *Oenothera*, subsection *Oenothera*: I. sequence evaluation and plastome evolution. *Nucleic Acids Res.* 36 2366–2378. 10.1093/nar/gkn081

81. Guo X., Castillo-Ramírez S., González V., Bustos P., Fernández-Vázquez J. L., Santamaría R. I., et al. (2007). Rapid evolutionary change of common bean (*Phaseolus vulgaris* L) plastome, and the genomic diversification of legume chloroplasts. *BMC Genomics* 8:228 10.1186/1471-2164-8-228

82. Gurdon C., Maliga P. (2014). Two distinct plastid genome configurations and unprecedented intraspecies length variation in the accD coding region in *Medicago truncatula*. *DNA Res.* 4417–427. 10.1093/dnares/dsu007

83. Hagemann R. (2000). Erwin Baur or Carl Correns: who really created the theory of plastid inheritance? *J. Hered.* 91 435–440. 10.1093/jhered/91.6.435

84. Hagemann R. (2002). Milestones in plastid genetics of higher plants. *Prog. Bot.* 63 1–51. 10.1007/978-3-642-56276-1_1

85. Hager M., Biehler K., Illerhaus J., Ruf S., Bock R. (1999).Targeted inactivation of the smallest plastid genome-encoded open reading frame reveals a novel and essential subunit of the cytochrome b(6)f complex. *EMBO J.* 18 5834–5842. 10.1093/emboj/18.21.5834

86. Hansen A. K., Escobar L. K., Gilbert L. E., Jansen R. K. (2007).Paternal, maternal, and biparental inheritance of the chloroplast genome in *Passiflora* (Passifloraceae): implications for phylogenetic studies. *Am. J. Bot.* 94 42–46. 10.3732/ajb.94.1.42

87. Hanson M. R., Gray B. N., Ahner B. A. (2013). Chloroplast transformation for engineering of photosynthesis. *J. Exp. Bot.* 64731–742. 10.1093/jxb/ers325

88. Hasebe M., Kofuji R., Ito M., Kato M., Iwatsuki K., Ueda K. (1992). Phylogeny of gymnosperms inferred from rbcL gene sequences. *Bot. Mag. Tokyo* 105 673–679. 10.1007/BF02489441

89. Hasunuma T., Miyazawa S. I., Yoshimura S., Shinzaki Y., Tomizawa K. I., Shindo K., et al. (2008). Biosynthesis of astaxanthin in tobacco leaves by transplastomic engineering.*Plant J.* 55 857–868. 10.1111/j.1365-313X.2008.03559.x

90. Henry R. J., Rice N., Edwards M., Nock C. J. (2014). "Next-Generation technologies to determine plastid genome sequences," in *Chloroplast Biotechnology*, ed. Maliga P., editor. (New York, NY: Humana Press;), 39–46. 10.1007/978-1-62703-995-6_2

91. Hirao T., Watanabe A., Kurita M., Kondo T., Takata K. (2008).Complete nucleotide sequence of the *Cryptomeria japonica* D. Don. chloroplast genome and comparative chloroplast genomics: diversified genomic structure of coniferous species. *BMC Plant Biol.* 8:70 10.1186/1471-2229-8-70

92. Hu Z. Y., Hua W., Huang S. M., Wang H. Z. (2011). Complete chloroplast genome sequence of rapeseed (*Brassica napus* L.) and its evolutionary implications. *Genet. Resour. Crop Evol.* 58875–887. 10.1007/s10722-010-9626-9

93. Huang C. Y., Ayliffe M. A., Timmis J. N. (2003). Direct measurement of the transfer rate of chloroplast DNA into the nucleus. *Nature* 422 72–76. 10.1038/nature01435

94. Huang C. Y., Grunheit N., Ahmadinejad N., Timmis J. N., Martin W. (2005). Mutational decay and age of chloroplast and mitochondrial genomes transferred recently to angiosperm nuclear chromosomes. *Plant Physiol.* 138 1723–1733. 10.1104/pp.105.060327

95. Huang H., Shi C., Liu Y., Mao S. Y., Gao L. Z. (2014). Thirteen*Camellia* chloroplast genome sequences determined by high-throughput sequencing: genome structure and phylogenetic relationships. *BMC Evol. Biol.* 14:151 10.1186/1471-2148-14-151

96. Jabeen R., Khan M. S., Zafar Y., Anjum T. (2010). Codon optimization of cry1Ab gene for hyper expression in plant organelles. *Mol. Biol. Rep.* 37 1011–1017. 10.1007/s11033-009-9802-1

97. Jansen R. K., Cai Z., Raubeson L. A., Daniell H., Depamphilis C. W., Leebens-Mack J., et al. (2007). Analysis of 81 genes from 64 plastid genomes resolves relationships in angiosperms and identifies genome-scale evolutionary patterns. *Proc. Natl. Acad. Sci. U.S.A.* 104 19369–19374. 10.1073/pnas.0709121104

98. Jansen R. K., Palmer J. D. (1987). A chloroplast DNA inversion marks an ancient evolutionary split in the sunflower family (Asteraceae). *Proc. Natl. Acad. Sci. U.S.A.* 84 5818–5822. 10.1073/pnas.84.16.5818

99. Jansen R. K., Raubeson L. A., Boore J. L., dePamphilis C. W., Chumley T. W., Haberle R. C., et al. (2005). Methods for obtaining and analyzing whole chloroplast genome sequences.*Methods Enzymol.* 395 348–384. 10.1016/S0076-6879(05)95020-9

100. Jansen R. K., Ruhlman T. A. (2012). "Plastid genomes of seed plants," in *Genomics of Chloroplasts and Mitochondria*, eds Bock R., Knoop V., editors. (Dordrecht: Springer;), 103–126. 10.1007/978-94-007-2920-9_5

101. Jansen R. K., Saski C., Lee S. B., Hansen A. K., Daniell H. (2011). Complete plastid genome sequences of three rosids (*Castanea*, *Prunus*, *Theobroma*): evidence for at least two independent transfers of rpl22 to the nucleus. *Mol. Biol. Evol.* 28835–847. 10.1093/molbev/msq261

102. Jin S., Kanagaraj A., Verma D., Lange T., Daniell H. (2011).Release of hormones from conjugates: chloroplast expression of beta-glucosidase results in elevated phytohormone levels associated with significant increase in biomass and protection from aphids or whiteflies conferred by sucrose esters. *Plant Physiol.* 155 222–235. 10.1104/pp.110.160754

103. Jin S., Zhang X., Daniell H. (2012). Pinellia ternata agglutinin expression in chloroplasts confers broad spectrum resistance against aphid, whitefly, Lepidopteran insects, bacterial and viral pathogens. *Plant Biotechnol. J.* 10 313–327. 10.1111/j.1467-7652.2011.00663.x

104. Kahlau S., Bock R. (2008). Plastid transcriptomics and translatomics of tomato fruit development and chloroplast-to-chromoplast differentiation: chromoplast gene expression largely serves the production of a single protein. *Plant Cell* 20 856–874. 10.1105/tpc.107.055202

105. Kanamoto H., Yamashita A., Asao H., Okumura S., Takase H., Hattori M., et al. (2006). Efficient and stable transformation of*Lactuca sativa* L. cv. Cisco (lettuce) plastids. *Transgenic Res.* 15205–217. 10.1007/s11248-005-3997-2

106. Kaneko T., Sato S., Kotani H., Tanaka A., Asamizu E., Nakamura Y., et al. (1996). Sequence analysis of the genome of the unicellular cyanobacterium *Synechocystis* sp. strain PCC6803. II. Sequence determination of the entire genome and assignment of potential protein-coding regions. *DNA Res.* 3 109–136. 10.1093/dnares/3.3.109

107. Keeling P. J. (2013). The number, speed, and impact of plastid endosymbioses in eukaryotic evolution. *Annu. Rev. Plant Biol.* 64583–607. 10.1146/annurev-arplant-050312-120144

108. Khadivi-Khub A., Zamani Z., Fattahi R., Wünsch A. (2013).Genetic variation in wild Prunus L. subgen. Cerasus germplasm from Iran characterized by nuclear and chloroplast SSR markers.*Trees* 28 471–485. 10.1007/s00468-013-0964-z

109. Kiani S., Mohamed B. B., Shehzad K., Jamal A., Shahid M. N., Shahid A. A., et al. (2013). Chloroplast-targeted expression of recombinant crystal-protein gene in cotton: an unconventional combat with resistant pests. *J. Biotechnol.* 166 88–96. 10.1016/j.jbiotec.2013.04.011

110. Kittiwongwattana C., Lutz K., Clark M., Maliga P. (2007).Plastid marker gene excision by the phiC31 phage site-specific recombinase. *Plant Mol. Biol.* 64 137–143. 10.1007/s11103-007-9140-4

111. Kleine T., Maier U. G., Leister D. (2009). DNA transfer from organelles to the nucleus: the idiosyncratic genetics of endosymbiosis. *Annu. Rev. Plant Biol.* 60 115–138. 10.1146/annurev.arplant.043008.092119

112. Kode V., Mudd E. A., Iamtham S., Day A. (2005). The tobacco plastid accD gene is essential and is required for leaf development. *Plant J.* 44 237–244. 10.1111/j.1365-313X.2005.02533.x

113. Krech K., Ruf S., Masduki F. F., Thiele W., Bednarczyk D., Albus C. A., et al. (2012). The plastid genome-encoded Ycf4 protein functions as a nonessential assembly factor for photosystem I in higher plants. *Plant Physiol.* 159 579–591. 10.1104/pp.112.196642

114. Krupinska K., Oetke S., Desel C., Mulisch M., Schafer A., Hollmann J., et al. (2014). WHIRLY1 is a major organizer of chloroplast nucleoids. *Front. Plant Sci.* 5:3210.3389/fpls.2014.00432

115. Kuang D. Y., Wu H., Wang Y. L., Gao L. M., Zhang S. Z., Lu L. (2011). Complete chloroplast genome sequence of Magnolia kwangsiensis (Magnoliaceae): implication for DNA barcoding and population genetics. *Genome* 54 663–673. 10.1139/G11-026

116. Kumar S., Dhingra A., Daniell H. (2004). Plastid-expressed betaine aldehyde dehydrogenase gene in carrot cultured cells, roots, and leaves confers enhanced salt tolerance. *Plant Physiol.*136 2843–2854. 10.1104/pp.104.045187

117. Kuroiwa T. (2010). 100 years since the discovery of non-Mendelian plastid phenotypes. *J. Plant Res.* 123 125–129. 10.1007/s10265-009-0283-z

118. Kwon K. C., Verma D., Jin S., Singh N. D., Daniell H. (2013).Release of proteins from intact chloroplasts induced by reactive oxygen species during biotic and abiotic stress. *PLoS ONE*8:e67106 10.1371/journal.pone.0067106

119. Leebens-Mack J., Raubeson L. A., Cui L., Kuehl J. V., Fourcade M. H., Chumley T. W., et al. (2005). Identifying the basal angiosperm node in chloroplast genome phylogenies: sampling one's way out of the Felsenstein zone. *Mol. Biol. Evol.* 22 1948–1963. 10.1093/molbev/msi191

120. Leigh F. J., Mackay I., Oliveira H. R., Gosman N. E., Horsnell R. A., Jones H., et al. (2013). Using diversity of the chloroplast genome to examine evolutionary history of wheat species. *Genet. Resour. Crop Evol.* 60 1831–1842. 10.1007/s10722-013-9957-4

121. Lelivelt C. L., McCabe M. S., Newell C. A., Desnoo C. B., van Dun K. M., Birch-Machin I., et al. (2005). Stable plastid transformation in lettuce (*Lactuca sativa* L.). *Plant Mol. Biol.* 58763–774. 10.1007/s11103-005-7704-8

122. Li W., Ruf S., Bock R. (2011). Chlorophenicol acetyltransferase as slectable marker for platid transformation. *Plant Mol. Biol.* 76443–451. 10.1007/s11103-010-9678-4

123. Liu C. W., Lin C. C., Chen J. J., Tseng M. J. (2007). Stable chloroplast transformation in cabbage (*Brassica oleracea* L. var.*capitata* L.) by particle bombardment. *Plant Cell Rep.* 26 1733–1744. 10.1007/s00299-007-0374-z

124. Liu C. W., Lin C. C., Yiu J. C., Chen J. J., Tseng M. J. (2008).Expression of a *Bacillus thuringiensis* toxin (cry1Ab) gene in cabbage (*Brassica oleracea* L. var. *capitata* L.) chloroplasts confers high insecticidal efficacy against *Plutella xylostella.Theor. Appl. Genet.* 117 75–88. 10.1007/s00122-008-0754-y

125. Liu Y., Huo N., Dong L., Wang Y., Zhang S., Young H. A., et al. (2013). Complete chloroplast genome sequences of Mongolia medicine *Artemisia frigida* and phylogenetic relationships with other plants. *PLoS ONE* 8:e57533 10.1371/journal.pone.0057533

126. Lohan A. J., Wolfe K. H. (1998). A subset of conserved tRNA genes in plastid DNA of nongreen plants. *Genetics* 150 425–433.

127. Lopez-Juez E., Pyke K. A. (2005). Plastids unleashed: their development and their integration in plant development. *Int. J. Dev. Biol.* 49 557–577. 10.1387/ijdb.051997el

128. Lu Y., Rijzaani H., Karcher D., Ruf S., Bock R. (2013). Efficient metabolic pathway engineering in transgenic tobacco and tomato plastids with synthetic multigene operons. *Proc. Natl. Acad. Sci. U.S.A.* 110 623–632. 10.1073/pnas.1216898110

129. Lutz K. A., Bosacchi M. H., Maliga P. (2006). Plastid marker-gene excision by transiently expressed CRE recombinase. *Plant J.* 45 447–456. 10.1111/j.1365-313X.2005.02608.x

130. Lutz K. A., Knapp J. E., Maliga P. (2001). Expression of bar in the plastid genome confers herbicide resistance. *Plant Physiol.*125 1585–1590. 10.1104/pp.125.4.1585

131. Maldaner F. R., Aragão F. J., dos Santos F. B., Franco O. L., Lima M. R. Q., Resende R. O., et al. (2013). Dengue virus tetra-epitope peptide expressed in lettuce chloroplasts for potential use in dengue diagnosis. *Appl. Microbiol. Biotechnol.* 97 5721–5729. 10.1007/s00253-013-4918-6

132. Maliga P. (2004). Plastid transformation in higher plants. *Annu. Rev. Plant Biol.* 55 289–313. 10.1146/annurev.arplant.55.031903.141633

133. Maliga P., Bock R. (2011). Plastid biotechnology: food, fuel, and medicine for the 21st century. *Plant Physiol.* 155 1501–1510. 10.1104/pp.110.170969

134. Maliga P., Tungsuchat-Huang T. (2014). Plastid transformation in*Nicotiana tabacum* and *Nicotiana sylvestris* by biolistic DNA delivery to leaves. *Methods Mol. Biol.* 1132 147–163. 10.1007/978-1-62703-995-6_8

135. Marín-Navarro J., Manuell A. L., Wu J. P., Mayfield S. (2007).Chloroplast translation regulation. *Photosynth. Res.* 94 359–374. 10.1007/s11120-007-9183-z

136. Martin G., Baurens F. C., Cardi C., Aury J. M., D'Hont A. (2013). The complete chloroplast genome of banana (*Musa acuminata*, Zingiberales): insight into plastid monocotyledon evolution. *PLoS ONE* 8:e67350 10.1371/journal.pone.0067350

137. Martin W., Rujan T., Richly E., Hansen A., Cornelsen S., Lins T., et al. (2002). Evolutionary analysis of *Arabidopsis*, cyanobacterial, and chloroplast genomes reveals plastid phylogeny and thousands of cyanobacterial genes in the nucleus.*Proc. Natl. Acad. Sci. U.S.A.* 99 12246–12251. 10.1073/pnas.182432999

138. Martin W., Stoebe B., Goremykin V., Hapsmann S., Hasegawa M., Kowallik K. V. (1998). Gene transfer to the nucleus and the evolution of chloroplasts. *Nature* 393 162–165. 10.1038/30234

139. McNeal J. R., Kuehl J. V., Boore J. L., de Pamphilis C. W. (2007). Complete plastid genome sequences suggest strong selection for retention of photosynthetic genes in the parasitic plant genus Cuscuta. *BMC Plant Biol.* 7:57 10.1186/1471-2229-7-57

140. Meisinger C., Sickmann A., Pfanner N. (2008). The mitochondrial proteome: from inventory to function. *Cell* 13422–24. 10.1016/j.cell.2008.06.043

141. Moore M. J., Dhingra A., Soltis P. S., Shaw R., Farmerie W. G., Folta K. M., et al. (2006). Rapid and accurate pyrosequencing of angiosperm plastid genomes. *BMC Plant Biol.* 6:1710.1186/1471-2229-6-17

142. Moore M. J., Soltis P. S., Bell C. D., Burleigh J. G., Soltis D. E. (2010). Phylogenetic analysis of 83 plastid genes further resolves the early diversification of eudicots. *Proc. Natl. Acad. Sci. U.S.A.*107 4623–4628. 10.1073/pnas.0907801107

143. Nagata N. (2010). Mechanisms for independent cytoplasmic inheritance of mitochondria and plastids in angiosperms. *J. Plant Res.* 123 193–199. 10.1007/s10265-009-0293-x

144. Nguyen T. T., Nugent G., Cardi T., Dix P. J. (2005). Generation of homoplasmic plastid transformants of a commercial cultivar of potato (*Solanum tuberosum* L.). *Plant Sci.* 168 1495–1500. 10.1016/j.plantsci.2005.01.023

145. Njuguna W., Liston A., Cronn R., Ashman T. L., Bassil N. (2013). Insights into phylogeny, sex function and age of *Fragaria*based on whole chloroplast genome sequencing. *Mol. Phylogenet. Evol.* 66 17–29. 10.1016/j.ympev.2012.08.026

146. Nock C. J., Waters D. L., Edwards M. A., Bowen S. G., Rice N., Cordeiro G. M., et al. (2011). Chloroplast genome sequences from total DNA for plant identification. *Plant Biotechnol. J.* 9328–333. 10.1111/j.1467-7652.2010.00558.x

147. Nugent G. D., Coyne S., Nguyen T. T., Kavanagh T. A., Dix P. J. (2006). Nuclear and plastid transformation of *Brassica oleracea*var. botrytis (cauliflower) using PEG-mediated uptake of DNA into protoplasts. *Plant Sci.* 170 135–142. 10.1016/j.plantsci.2005.08.020

148. Nugent G. D., Ten Have M., van der Gulik A., Dix P. J., Uijtewaal B. A., Mordhorst A. P. (2005). Plastid transformants of tomato selected using mutations affecting ribosome structure.*Plant Cell Rep.* 24 341–349. 10.1007/s00299-005-0930-3

149. Nystedt B., Street N. R., Wetterbom A., Zuccolo A., Lin Y. C., Scofield D. G., et al. (2013). The Norway spruce genome sequence and conifer genome evolution. *Nature* 497 579–584. 10.1038/nature12211

150. Oey M., Lohse M., Scharff L. B., Kreikemeyer B., Bock R. (2009). Plastid production of protein antibiotics against pneumonia via a new strategy for high-level expression of antimicrobial proteins. *Proc. Natl. Acad. Sci. U.S.A.* 106 6579–6584. 10.1073/pnas.0813146106

151. Ogihara Y., Tsunewaki K. (1988). Diversity and evolution of chloroplast DNA in *Triticum* and *Aegilops* as revealed by restriction fragment analysis. *Theor. Appl. Genet.* 76 321–332. 10.1007/BF00265331

152. Ohyama K., Fukuzawa H., Kohchi T., Shirai H., Sano T., Sano S., et al. (1986). Chloroplast gene organization deduced from complete sequence of liverwort *Marchantia polymorpha*chloroplast DNA. *Nature* 322 572–574. 10.1038/322572a0

153. Osteryoung K. W., Pyke K. A. (2014). Division and dynamic morphology of plastids. *Annu. Rev. Plant Biol.* 65 443–472. 10.1146/annurev-arplant-050213-035748

154. Palmer J. D. (1985). Chloroplast DNA and molecular phylogeny.*BioEssays* 2 263–267. 10.1002/bies.950020607

155. Palmer J. D., Jansen R. K., Michaels H., Manhart J., Chase M. (1988). Chloroplast DNA variation and plant phylogeny. *Ann. Mo. Bot. Gard.* 75 1180–1206. 10.2307/2399279

156. Palmer J. D., Shields C. R., Cohen D. B., Orton T. J. (1983).Chloroplast DNA evolution and the origin of amphidiploid*Brassica* species. *Theor. Appl. Genet.* 65 181–189. 10.1007/BF00308062

157. Palmer J. D., Thompson W. F. (1982). Chloroplast DNA rearrangements are more frequent when a large inverted repeat sequence is lost. *Cell* 29 537–550. 10.1016/0092-8674(82)90170-2

158. Parks M., Cronn R., Liston A. (2009). Increasing phylogenetic resolution at low taxonomic levels using massively parallel sequencing of chloroplast genomes. *BMC Biol.* 7:8410.1186/1741-7007-7-84

159. Petersen K., Bock R. (2011). High-level expression of a suite of thermostable cell wall-degrading enzymes from the chloroplast genome. *Plant Mol. Biol.* 76 311–321. 10.1007/s11103-011-9742-8

160. Petit R. J., Brewer S., Bordács S., Burg K., Cheddadi R., Coart E., et al. (2002). Identification of refugia and post-glacial colonisation routes of European white oaks based on chloroplast DNA and fossil pollen evidence. *For. Ecol. Manage.* 156 49–74. 10.1016/S0378-1127(01)00634-X

161. Petit R. J., Duminil J., Fineschi S., Hampe A., Salvini D., Vendramin G. G. (2005). Invited review: comparative organization of chloroplast, mitochondrial and nuclear diversity in plant populations. *Mol. Ecol.* 14 689–701. 10.1111/j.1365-294X.2004.02410.x

162. Petrillo E., Godoy Herz M. A., Fuchs A., Reifer D., Fuller J., Yanovsky M. J., et al. (2014). A chloroplast retrograde signal regulates nuclear alternative splicing. *Science* 344 427–430. 10.1126/science.1250322

163. Pfannschmidt T. (2003). Chloroplast redox signals: how photosynthesis controls its own genes. *Trends Plant Sci.* 8 33–41. 10.1016/S1360-1385(02)00005-5

164. Pfannschmidt T., Schutze K., Fey V., Sherameti I., Oelmuller R. (2003). Chloroplast redox control of nuclear gene expression–a new class of plastid signals in interorganellar communication.*Antioxid. Redox Signal.* 5 95–101. 10.1089/152308603321223586

165. Powell W., Morgantet M., Andre C., McNicol J. W., Machray G. C., Doyle J. J., et al. (1995). Hypervariable microsatellites provide a general source of polymorphic DNA markers for the chloroplast genome. *Curr. Biol.* 5 1023–1029. 10.1016/S0960-9822(95)00206-5

166. Powikrowska M., Oetke S., Jensen P. E., Krupinska K. (2014).Dynamic composition, shaping and organization of plastid nucleoids. *Front. Plant Sci.* 5:424 10.3389/fpls.2014.00424

167. Provan J., Powell W., Hollingsworth P. M. (2001). Chloroplast microsatellites: new tools for studies in plant ecology and evolution. *Trends Ecol. Evol.* 16 142–147. 10.1016/S0169-5347(00)02097-8

168. Pyke K. A. (2011). Analysis of plastid number, size, and distribution in Arabidopsis plants by light and fluorescence microscopy. *Methods Mol. Biol.* 774 19–32. 10.1007/978-1-61779-234-2_2

169. Qian J., Song J., Gao H., Zhu Y., Xu J., Pang X., et al. (2013).The complete chloroplast genome sequence of the medicinal plant Salvia miltiorrhiza. *PLoS ONE* 8:e5760710.1371/journal.pone.0057607

170. Quesada-Vargas T., Ruiz O. N., Daniell H. (2005). Characterization of heterologous multigene operons in transgenic chloroplasts: transcription, processing, and translation. *Plant Physiol.* 138 1746–1762. 10.1104/pp.105.063040

171. Ramundo S., Rochaix J. D. (2014). Chloroplast unfolded protein response, a new plastid stress signaling pathway? *Plant Signal. Behav.* 9:e972874 10.4161/15592316.2014.972874

172. Reyes-Prieto A., Weber A. P., Bhattacharya D. (2007). The origin and establishment of the plastid in algae and plants. *Annu. Rev. Genet.* 41 147–168. 10.1146/annurev.genet.41.110306.130134

173. Richly E., Chinnery P. F., Leister D. (2003). Evolutionary diversification of mitochondrial proteomes: implications for human disease. *Trends Genet.* 19 356–362. 10.1016/S0168-9525(03)00137-9

174. Richly E., Leister D. (2004). An improved prediction of chloroplast proteins reveals diversities and commonalities in the chloroplast proteomes of *Arabidopsis* and rice. *Gene* 329 11–16. 10.1016/j.gene.2004.01.008

175. Rodríguez-Moreno L., González V. M., Benjak A., Martí M. C., Puigdomènech P., Aranda M. A., et al. (2011). Determination of the melon chloroplast and mitochondrial genome sequences reveals that the largest reported mitochondrial genome in plants contains a significant amount of DNA having a nuclear origin.*BMC Genomics* 12:424 10.1186/1471-2164-12-424

176. Rogalski M., Carrer H. (2011). Engineering plastid fatty acid biosynthesis to improve food quality and biofuel production in higher plants. *Plant Biotechnol. J.* 9 554–564. 10.1111/j.1467-7652.2011.00621.x

177. Rogalski M., Karcher D., Bock R. (2008). Superwobbling facilitates translation with reduced tRNA sets. *Nat. Struct. Mol. Biol.* 15 192–198. 10.1038/nsmb.1370

178. Rogalski M., Ruf S., Bock R. (2006). Tobacco plastid ribosomal protein S18 is essential for cell survival. *Nucleic Acids Res.* 344537–4545. 10.1093/nar/gkl634

179. Roullier C., Rossel G., Tay D., McKey D., Lebot V. (2011).Combining chloroplast and nuclear microsatellites to investigate origin and dispersal of New World sweet potato landraces. *Mol. Ecol.* 20 3963–3977. 10.1111/j.1365-294X.2011.05229.x

180. Rousseau-Gueutin M., Lloyd A. H., Sheppard A. E., Timmis J. N. (2012). "Gene transfer to the nucleus," in *Organelle Genetics: Evolution of Organelle Genomes and Gene Expression*, ed. Bullerwell C. E., editor. (New York, NY: Springer;), 147–171. 10.1007/978-3-642-22380-8_7

181. Ruf S., Bock R. (2014). Plastid transformation in tomato.*Methods Mol. Biol.* 1132 265–276. 10.1007/978-1-62703-995-6_16

182. Ruf S., Hermann M., Berger I. J., Carrer H., Bock R. (2001).Stable genetic transformation of tomato plastids and expression of a foreign protein in fruit. *Nat. Biotechnol.* 19 870–875. 10.1038/nbt0901-870

183. Ruf S., Karcher D., Bock R. (2007). Determining the transgene containment level provided by chloroplast transformation. *Proc. Natl. Acad. Sci. U.S.A.* 104 6998–7002. 10.1073/pnas.0700008104

184. Ruf S., Kossel H., Bock R. (1997). Targeted inactivation of a tobacco intron-containing open reading frame reveals a novel chloroplast-encoded photosystem I-related gene. *J. Cell Biol.* 13995–102. 10.1083/jcb.139.1.95

185. Ruhlman T. A. (2014). Plastid transformation in lettuce (*Lactuca sativa* L.) by biolistic DNA delivery. *Methods Mol. Biol.* 1132331–343. 10.1007/978-1-62703-995-6_21

186. Ruhlman T., Lee S. B., Jansen R. K., Hostetler J. B., Tallon L. J., Town C. D., et al. (2006). Complete plastid genome sequence of*Daucus carota*: implications for biotechnology and phylogeny of angiosperms. *BMC Genomics* 7:222 10.1186/1471-2164-7-222

187. Ruhlman T., Verma D., Samson N., Daniell H. (2010). The role of heterologous chloroplast sequence elements in transgene integration and expression. *Plant Physiol.* 152 2088–2104. 10.1104/pp.109.152017

188. Ruiz O. N., Alvarez D., Torres C., Roman L., Daniell H. (2011).Metallothionein expression in chloroplasts enhances mercury accumulation and phytoremediation capability. *Plant Biotechnol. J.* 9 609–617. 10.1111/j.1467-7652.2011.00616.x

189. Ruiz O. N., Daniell H. (2009). Genetic engineering to enhance mercury phytoremediation. *Curr. Opin. Biotechnol.* 20 213–219. 10.1016/j.copbio.2009.02.010

190. Sabir J., Schwarz E., Ellison N., Zhang J., Baeshen N. A., Mutwakil M., et al. (2014). Evolutionary and biotechnology implications of plastid genome variation in the inverted-repeat-lacking clade of legumes. *Plant Biotech. J.* 12 743–754. 10.1111/pbi.12179

191. Sager R., Ishida M. R. (1963). Chloroplast DNA in chlamydomonas. *Proc. Natl. Acad. Sci. U.S.A.* 50 725–730. 10.1073/pnas.50.4.725

192. Sandbrink J. M., Vellekoop P., Van Ham R., Van Brederode J. (1989). A method for evolutionary

studies on RFLP of chloroplast DNA, applicable to a range of plant species.*Biochem. Syst. Ecol.* 17 45–49. 10.1016/0305-1978(89)90041-0

193. Saski C., Lee S. B., Daniell H., Wood T. C., Tomkins J., Kim H. G., et al. (2005). Complete chloroplast genome sequence of Glycine max and comparative analyses with other legume genomes. *Plant Mol. Biol.* 59 309–322. 10.1007/s11103-005-8882-0
194. Schmitz-Linneweber C., Maier R. M., Alcaraz J. P., Cottet A., Herrmann R. G., Mache R. (2001). The plastid chromosome of spinach (*Spinacia oleracea*): complete nucleotide sequence and gene organization. *Plant Mol. Biol.* 45 307–315. 10.1023/A:1006478403810
195. Schneider A., Stelljes C., Adams C., Kirchner S., Burkhard G., Jarzombski S. (2015). Low frequency paternal transmission of plastid genes in Brassicaceae. *Transgenic Res.* 24 267–277. 10.1007/s11248-014-9842-8
196. Scotti N., Valkov V. T., Cardi T. (2011). Improvement of plastid transformation efficiency in potato by using vectors with homologous flanking sequences. *GM Crops.* 2 89–91. 10.4161/gmcr.2.2.17504
197. Segretin M. E., Lentz E. M., Wirth S. A., Morgenfeld M. M., Bravo-Almonacid F. F. (2012). Transformation of *Solanum tuberosum* plastids allows high expression levels of β-glucuronidase both in leaves and microtubers developed in vitro.*Planta* 235 807–818. 10.1007/s00425-011-1541-6
198. Setoguchi H., Osawa T. A., Pintaud J. C., Jaffré T., Veillon J. M. (1998). Phylogenetic relationships within Araucariaceae based on rbcL gene sequences. *Am. J. Bot.* 85 1507–1516. 10.2307/2446478
199. Shao M., Kumar S., Thomson J. G. (2014). Precise excision of plastid DNA by the large serine recombinase Bxb1. *Plant Biotechnol. J.* 12 322–329. 10.1111/pbi.12139
200. Shaw J., Lickey E. B., Beck J. T., Farmer S. B., Liu W., Miller J., et al. (2005). The tortoise and the hare II: relative utility of 21 noncoding chloroplast DNA sequences for phylogenetic analysis.*Am. J. Bot.* 92 142–166. 10.3732/ajb.94.3.275
201. Shaw J., Lickey E. B., Schilling E. E., Small R. L. (2007). Comparison of whole chloroplast genome sequences to choose noncoding regions for phylogenetic studies in angiosperms: the tortoise and the hare III. *Am. J. Bot.* 94 275–288. 10.3732/ajb.94.3.275
202. Shenoy V., Kwon K. C., Rathinasabapathy A., Lin S., Jin G., Song C., et al. (2014). Oral delivery of Angiotensin-converting enzyme 2 and Angiotensin-(1-7) bioencapsulated in plant cells attenuates pulmonary hypertension. *Hypertension* 64 1248–1259. 10.1161/HYPERTENSIONAHA.114.03871
203. Sheppard A. E., Timmis J. N. (2009). Instability of plastid DNA in the nuclear genome. *PLoS Genet.* 5:810.1371/journal.pgen.1000323
204. Shikanai T., Shimizu K., Ueda K., Nishimura Y., Kuroiwa T., Hashimoto T. (2001). The chloroplast clpP gene, encoding a proteolytic subunit of ATP-dependent protease, is indispensable for chloroplast development in tobacco. *Plant Cell Physiol.* 42264–273. 10.1093/pcp/pce031
205. Shil P. K., Kwon K. C., Zhu P., Verma A., Daniell H., Li Q. (2014). Oral delivery of ACE2/Ang-(1-7) bioencapsulated in plant cells protects against experimental uveitis and autoimmune uveoretinitis. *Mol. Ther.* 22 2069–2082. 10.1038/mt.2014.179
206. Shinozaki K., Ohme M., Tanaka M., Wakasugi T., Hayashida N., Matsubayashi T., et al. (1986). The complete nucleotide sequence of the tobacco chloroplast genome: its gene organization and expression. *EMBO J.* 5 2043–2049. 10.1007/BF02669253
207. Shulaev V., Sargent D. J., Crowhurst R. N., Mockler T. C., Folkerts O., Delcher A. L., et al. (2011). The genome of woodland strawberry (*Fragaria vesca*). *Nat. Genet.* 43 109–116. 10.1038/ng.740
208. Sidorov V. A., Kasten D., Pang S. Z., Hajdukiewicz P. T., Staub J. M., Nehra N. S. (1999). Technical advance: stable chloroplast transformation in potato: use of green fluorescent protein as a plastid marker. *Plant J.* 19 209–216. 10.1046/j.1365-313X.1999.00508.x
209. Singh A. K., Verma S. S., Bansal K. C. (2010). Plastid transformation in eggplant (*Solanum melongena* L.). *Transgenic Res.* 19 113–119. 10.1007/s11248-009-9290-z
210. Small I. D., Rackham O., Filipovska A. (2013). Organelle transcriptomes: products of a deconstructed genome. *Curr. Opin. Microbiol.* 16 652–658. 10.1016/j.mib.2013.07.011
211. Small R., Cronn R. C., Wendel J. F. (2004). L. A. S. JOHNSON REVIEW No. 2. Use of nuclear genes for phylogeny reconstruction in plants. *Aust. Syst. Bot.* 17 145–170. 10.1071/SB03015
212. Smith D. R., Lee R. W. (2014). A plastid without a genome: evidence from the nonphotosynthetic green algal genus polytomella. *Plant Physiol.* 164 1812–1819. 10.1104/pp.113.233718
213. Staub J. M. (2014). Plastid transformation of tobacco suspension cell cultures. *Methods Mol. Biol.* 1132 1771–1785. 10.1007/978-1-62703-995-6_10
214. Stegemann S., Bock R. (2006). Experimental reconstruction of functional gene transfer from the tobacco plastid genome to the nucleus. *Plant Cell* 18 2869–2878. 10.1105/tpc.106.046466
215. Stegemann S., Hartmann S., Ruf S., Bock R. (2003). High-frequency gene transfer from the chloroplast genome to the nucleus. *Proc. Natl. Acad. Sci. U.S.A.* 100 8828–8833. 10.1073/pnas.1430924100
216. Stegemann S., Keuthe M., Greiner S., Bock R. (2012). Horizontal transfer of chloroplast genomes between plant species. *Proc. Natl. Acad. Sci. U.S.A.* 109 2434–2438. 10.1073/pnas.1114076109
217. Strauss S. H., Palmer J. D., Howe G. T., Doerksen A. H. (1988).Chloroplast genomes of two conifers lack a large inverted repeat and are extensively rearranged. *Proc. Natl. Acad. Sci. U.S.A.* 853898–3902. 10.1073/pnas.85.11.3898

218. Svab Z., Hajdukiewicz P., Maliga P. (1990). Stable transformation of plastids in higher plants. *Proc. Natl. Acad. Sci. U.S.A.* 87 8526–8530. 10.1073/pnas.87.21.8526

219. Svab Z., Maliga P. (1993). High-frequency plastid transformation in tobacco by selection for a chimeric aadA gene. *Proc. Natl. Acad. Sci. U.S.A.* 90 913–917. 10.1073/pnas.90.3.913

220. Svab Z., Maliga P. (2007). Exceptional transmission of plastids and mitochondria from the transplastomic pollen parent and its impact on transgene containment. *Proc. Natl. Acad. Sci. U.S.A.*104 7003–7008. 10.1073/pnas.0700063104

221. Tangphatsornruang S., Sangsrakru D., Chanprasert J., Uthaipaisanwong P., Yoocha T., Jomchai N., et al. (2010). The chloroplast genome sequence of mungbean (*Vigna radiata*) determined by high-throughput pyrosequencing: structural organization and phylogenetic relationships. *DNA Res.* 17 11–22. 10.1093/dnares/dsp025

222. Tetlow I. J., Morell M. K., Emes M. J. (2004). Recent developments in understanding the regulation of starch metabolism in higher plants. *J. Exp. Bot.* 55 2131–2145. 10.1093/jxb/erh248

223. Tewari K. K., Wildman S. G. (1966). Chloroplast DNA from tobacco leaves. *Science* 153 1269–1271. 10.1126/science.153.3741.1269

224. Tiller N., Bock R. (2014). The translational apparatus of plastids and its role in plant development. *Mol. Plant.* 7 1105–1120. 10.1093/mp/ssu022

225. Timme R. E., Kuehl J. V., Boore J. L., Jansen R. K. (2007). A comparative analysis of the *Lactuca* and *Helianthus* (Asteraceae) plastid genomes: identification of divergent regions and categorization of shared repeats. *Am. J. Bot.* 94 302–312. 10.3732/ajb.94.3.302

226. Timmis J. N., Ayliffe M. A., Huang C. Y., Martin W. (2004).Endosymbiotic gene transfer: organelle genomes forge eukaryotic chromosomes. *Nat. Rev. Genet.* 5 123–135. 10.1038/nrg1271

227. Tsai Y. H., Manos P. S. (2010). Host density drives the postglacial migration of the tree parasite, Epifagus virginiana.*Proc. Natl. Acad. Sci. U.S.A.* 107 17035–17040. 10.1073/pnas.1006225107

228. Tseng M. J., Yang M. T., Chu W. R., Liu C. W. (2014). Plastid transformation in cabbage (*Brassica oleracea* L. var. capitata L.) by the biolistic process. *Methods Mol. Biol.* 1132 355–366. 10.1007/978-1-62703-995-6_23

229. Valkov V. T., Gargano D., Manna C., Formisano G., Dix P. J., Gray J. C., et al. (2011). High efficiency plastid transformation in potato and regulation of transgene expression in leaves and tubers by alternative 5' and 3' regulatory sequences. *Transgenic Res.* 20137–151. 10.1007/s11248-010-9402-9

230. Valkov V. T., Gargano D., Scotti N., Cardi T. (2014). Plastid transformation in potato: *Solanum tuberosum*. *Methods Mol. Biol.*1132 295–303. 10.1007/978-1-62703-995-6_18

231. Valkov V. T., Scotti N., Kahlau S., Maclean D., Grillo S., Gray J. C., et al. (2009). Genome-wide analysis of plastid gene expression in potato leaf chloroplasts and tuber amyloplasts: transcriptional and posttranscriptional control. *Plant Physiol.* 1502030–2044. 10.1104/pp.109.140483

232. Verma D., Daniell H. (2007). Chloroplast vector systems for biotechnology applications. *Plant Physiol.* 145 1129–1143. 10.1104/pp.107.106690

233. Verma D., Jin S., Kanagaraj A., Singh N. D., Daniel J., Kolattukudy P. E., et al. (2013). Expression of fungal cutinase and swollenin in tobacco chloroplasts reveals novel enzyme functions and/or substrates. *PLoS ONE* 8:e5718710.1371/journal.pone.0057187

234. Vieira L. N., Faoro H., Rogalski M., Fraga H. P. F., Cardoso R. L., Souza E. M., et al. (2014a). The complete chloroplast genome sequence of *Podocarpus lambertii*: genome structure, evolutionary aspects, gene content and SSR detection. *PLoS ONE*9:e90618 10.1371/journal.pone.0090618

235. Vieira L. N., Faoro H., Fraga H. P. F., Rogalski M., Souza E. M., Pedrosa F. O., et al. (2014b). An improved protocol for intact chloroplasts and cpDNA isolation in conifers. *PLoS ONE*9:e84792 10.1371/journal.pone.0084792

236. Wakasugi T., Tsudzuki J., Ito S., Nakashima K., Tsudzuki T. (1998). Loss of all ndh genes as determined by sequencing the entire chloroplast genome of the black pine Pinus thunbergii.*Proc. Natl. Acad. Sci. U.S.A.* 91 9794–9798. 10.1073/pnas.91.21.9794

237. Wakasugi T., Tsudzuki T., Sugiura M. (2001). The genomics of land plant chloroplasts: gene content and alteration of genomic information by RNA editing. *Photosynth. Res.* 70 107–118. 10.1023/A:1013892009589

238. Wang X., Sherman A., Liao G., Leong K. W., Daniell H., Terhorst C., et al. (2013). Mechanism of oral tolerance induction to therapeutic proteins. *Adv. Drug Deliv. Rev.* 65 759–773. 10.1016/j.addr.2012.10.013

239. Waters M. T., Fray R. G., Pyke K. A. (2004). Stromule formation is dependent upon plastid size, plastid differentiation status and the density of plastids within the cell. *Plant J.* 39 655–667. 10.1111/j.1365-313X.2004.02164.x

240. Wheeler G. L., Dorman H. E., Buchanan A., Challagundla L., Wallace L. E. (2014). A review of the prevalence, utility, and caveats of using chloroplast simple sequence repeats for studies of plant biology. *Appl. Plant Sci.* 2:140005910.3732/apps.1400059

241. Wobbe L., Schwarz C., Nickelsen J., Kruse O. (2008). Translational control of photosynthetic gene expression in phototrophic eukaryotes. *Physiol. Plant.* 133 507–515. 10.1111/j.1399-3054.2008.01091.x

242. Wolfe K. H., Morden C. W., Palmer J. D. (1992). Small single-copy region of plastid DNA in the non-photosynthetic angiosperm *Epifagus virginiana* contains only two genes. Differences among dicots, monocots and bryophytes in gene organization at a non-bioenergetic locus. *J. Mol. Biol.* 223 95–104. 10.1016/0022-2836(92)90718-Y

243. Wu C. S., Chaw S. M. (2013). Highly rearranged and size-variable chloroplast genomes in conifer II clade (cupressophytes): evolution towards shorter intergenic

spacers.*Plant Biotechnol. J.* 12 344–353. 10.1111/pbi.12141

244. Wurbs D., Ruf S., Bock R. (2007). Contained metabolic engineering in tomatoes by expression of carotenoid biosynthesis genes from the plastid genome. *Plant J.* 49 276–288. 10.1111/j.1365-313X.2006.02960.x

245. Yabuta Y., Tanaka H., Yoshimura S., Suzuki A., Tamoi M., Maruta T., et al. (2013). Improvement of vitamin E quality and quantity in tobacco and lettuce by chloroplast genetic engineering. *Transgenic Res.* 22 391–402. 10.1007/s11248-012-9656-5

246. Yang J. B., Tang M., Li H. T., Zhang Z. R., Li D. Z. (2013). Complete chloroplast genome of the genus *Cymbidium*: lights into the species identification, phylogenetic implications and population genetic analyses. *BMC Evol. Biol.* 13:8410.1186/1471-2148-13-84

247. Ye G.-N., Hajdukiewicz P. T. J., Broyles D., Rodriguez D., Xu C. W., Nehra N., et al. (2001). Plastid-expressed 5-enolpyruvylshikimate-3-phosphate synthase genes provide high level glyphosate tolerance in tobacco. *Plant J.* 25 261–270. 10.1046/j.1365-313x.2001.00958.x

248. Yi X., Gao L., Wang B., Su Y., Wang T. (2013). The complete chloroplast genome sequence of *Cephalotaxus oliveri*(Cephalotaxaceae): evolutionary comparison of cephalotaxus chloroplast DNAs and insights into the loss of inverted repeat copies in gymnosperms. *Genome Biol. Evol.* 5 688–698. 10.1093/gbe/evt042

249. Yukawa M., Tsudzuki T., Sugiura M. (2005). The 2005 version of the chloroplast DNA sequence from tobacco (*Nicotiana tabacum*). *Plant Mol. Biol. Rep.* 23 359–365. 10.1007/BF02788884

250. Yun C. S., Hasegawa H., Nanamiya H., Terakawa T., Tozawa Y. (2009). Novel bacterial N-acetyltransferase gene for herbicide detoxification in land plants and selection maker in plant transformation. *Biosci. Biotechnol. Biochem.* 73 1000–1006. 10.1271/bbb.80777

251. Zhang J., Khan S. A., Hasse C., Ruf S., Heckel D. G., Bock R. (2015). Pest control. Full crop protection from an insect pest by expression of long double-stranded RNAs in plastids. *Science*347 991–994. 10.1126/science.1261680

252. Zhang J., Ruf S., Hasse C., Childs L., Scharff L. B., Bock R. (2012). Identification of cis-elements conferring high levels of gene expression in non-green plastids. *Plant J.* 72 115–128. 10.1111/j.1365-313X.2012.05065.x

253. Zhang Q., Liu Y., Sodmergen (2003). Examination of the cytoplasmic DNA in male reproductive cells to determine the potential for cytoplasmic inheritance in 295 angiosperm species.*Plant Cell Physiol.* 44 941–951. 10.1093/pcp/pcg121

254. Zhang Q., Sodmergen (2010). Why does biparental plastid inheritance revive in angiosperms? *J. Plant Res.* 123 201–206. 10.1007/s10265-009-0291-z

255. Zimorski V., Ku C., Martin W. F., Gould S. B. (2014). Endosymbiotic theory for organelle origins. *Curr. Opin. Microbiol.* 22 38–48. 10.1016/j.mib.2014.09.008

CHAPTER 16

SURVEYING THE POTENTIAL OF SECRETED ANTIMICROBIAL PEPTIDES TO ENHANCE PLANT DISEASE RESISTANCE

Susan Breen[1], Peter S. Solomon[1], Frank Bedon[2, 3] and Delphine Vincent [2]

[1] Plant Sciences Division, Research School of Biology, The Australian National University, Canberra, ACT, Australia

[2] Department of Economic Development, AgriBio, Bundoora, VIC, Australia

3 AgriBio, La Trobe University, Bundoora, VIC, Australia

ABSTRACT

Antimicrobial peptides (AMPs) are natural products found across diverse taxa as part of the innate immune system against pathogen attacks. Some AMPs are synthesized through the canonical gene expression machinery and are called ribosomal AMPs. Other AMPs are assembled by modular enzymes generating nonribosomal AMPs and harbor unusual structural diversity. Plants synthesize an array of AMPs, yet are still subject to many pathogen invasions. Crop breeding programs struggle to release new cultivars in which complete disease resistance is achieved, and usually such resistance becomes quickly overcome by the targeted pathogens which have a shorter generation time. AMPs could offer a solution by exploring not only plant-derived AMPs, related or unrelated to the crop of interest, but also non-plant AMPs produced by bacteria, fungi, oomycetes or animals. This review highlights some promising candidates within the plant kingdom and elsewhere, and offers some perspectives on how to identify and validate their bioactivities. Technological advances, particularly in mass spectrometry (MS) and nuclear magnetic resonance (NMR), have been instrumental in identifying and elucidating the structure of novel AMPs, especially nonribosomal peptides which cannot be identified through genomics approaches. The majority of non-plant AMPs showing potential for plant disease immunity are often tested using in vitro assays. The greatest challenge remains the functional validation of candidate AMPs in plants through transgenic experiments, particularly introducing nonribosomal AMPs into crops.

INTRODUCTION

Peptides with antimicrobial activities, or antimicrobial peptides (AMPs), have emerged as key components of the innate immune system in almost all living organisms since their first discovery from culture supernatant of the soil bacteria *Bacillus brevis* more than seven decades ago (Dubos, 1939a,b). AMPs are naturally synthesized low molecular mass products [up to 100 amino acids (AAs)] that act against microbial pathogens. AMPs are structurally and biochemically highly diverse but typically include positively charged AAs and hydrophobic or hydrophilic moieties facilitating more or less their aqueous solubility and interaction with the negatively charged parts of the phospholipidic microbial cell membranes (reviewed in Montesinos, 2007; Mousa and Raizada, 2015; Wang et al., 2015). As a consequence of their diversity, specific AMPs are more effective in interacting and disrupting targeted microbial membranes, such as fungal pathogens for example (Marx,2004; Hegedüs and Marx, 2013; Vincent and Bedon, 2013; van der Weerden et al., 2013).

Citation: Susan Breen Peter S. Solomon Frank Bedon and Delphine Vincent, Surveying the potential of secreted antimicrobial peptides to enhance plant disease resistance, doi: 10.3389/fpls.2015.00900.

AMPs can be classified as either ribosomal or nonribosomal according to their mode of synthesis by the cells. Ribosomal peptides are gene-encoded peptides usually resulting from cleavage of a pro-protein, while nonribosomal peptides are assembled by multimodular enzymes called NonRibosomal Peptide Synthetases (NRPS). These NRPSs are usually organized in one operon, for bacteria, or in gene clusters for eukaryotes, and they synthesize one peptide per gene cluster or operon. NPRSs can generate macrocyclic peptides with an unusual structural diversity achieved through the assembly of not only the 20 canonical AAs, but also D-configured- and β-AAs, methylated, glycosylated and phosphorylated residues, heterocyclic elements and even fatty acid (FA) chains (Marahiel, 2009). Cyclic lipopeptides containing FAs are examples of such nonribosomal peptides (for review Lee and Kim,2015; Patel et al., 2015). Ribosomal peptides are difficult to predict *in silico* from transcriptomic and genomic sequencing projects due to their small size and high diversity. There are also no known generic cleavage sites that could indicate a potential peptide. Tools are available allowing*in silico* genome mining, as was demonstrated for *Bacillus sp.* (Aleti et al., 2015). The most comprehensive AMPs database to date, named ADAM, is publically available and currently contains 7007 unique peptide sequences and 759 structures (http://bioinformatics.cs.ntou.edu.tw/ADAM/links.html) (Lee et al.,2015), yet some of the nonribosomal AMPs listed in this review are missing. Other AMP databases exist (reviewed in Holaskova et al.,2015). Methods other than genomics are therefore advantageous, particularly mass spectra (MS)-based proteomics strategies as they allow the direct identification of the AMPs and their isoforms. MS analysis of secreted peptides generates reliable and useful information such as the molecular weight, the AA sequence, as well as the length of the FA chain in the case of lipopeptides (Zhao et al., 2014). Furthermore, MS imaging technology has emerged as a powerful tool to not only identify novel AMPs but also locate them *in situ* (Debois et al., 2013). Nuclear magnetic resonance (NMR) analyses are complementary because they yield structural elucidation such as cyclic structures (Wäspi et al., 1998; Sammer et al., 2009).

Plant protection and resistance against pathogens have been traditionally and are still currently addressed with chemical uses and breeding programs. Nevertheless, the possibilities arising from the study of AMPs will contribute to control the plant pathogens whose virulence is driven by perpetual adaptation through mutation. The aim of this review is to discuss naturally synthesized and secreted AMPs offering biocontrol potential for crop species. In this review, we chose to focus on well-studied AMPs representative of the main taxa such as bacteria, fungi, animals, and plants. The first section covers the plant-secreted peptides induced as a defense mechanism following pathogen attack or herbivory. The last section focuses on peptides secreted by non-plant species (bacteria, fungi and animals) that are subject to microbial challenges, and the potential of these peptides to help plant species fight fungal diseases by applying a biocontrol approach. Figure illustrates AMPs described in the review with their mode-of-action where known. Figure outlines a workflow to isolate, identify and validate AMPs bearing biocontrol potential in plant immunity. We discuss successful cases of increased pathogen resistance in transgenic plants and their potential for increased crop yield. Additional information on transgenic plants expressing AMPs toward improved disease resistance can be found in the reviews from Ramadevi et al. (2011) and Holaskova et al. (2015). Not covered in this review is a promising strategy for synthetic peptides production through combinatorial chemistry which offers an alternative approach to developing AMPs for agricultural applications (López-García et al.,2002; Choi and Moon, 2009; Rebollar et al., 2014).

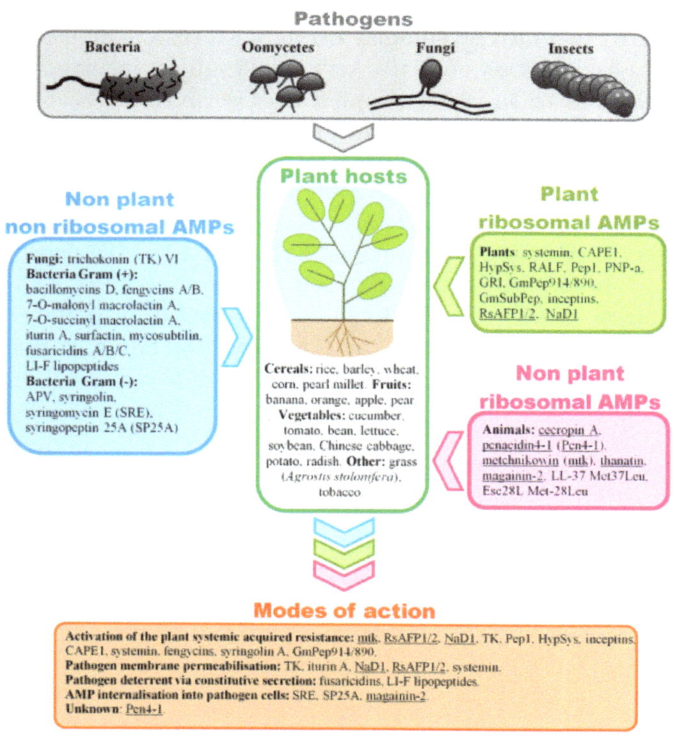

Figure 1: Summary of the modes of action of non-plant and plant related AMPs involved in host resistance against plant pathogens reviewed here. A number of pests (pathogens and herbivores) can attack a crop which deploys an arsenal of defense to stay healthy. Yet .

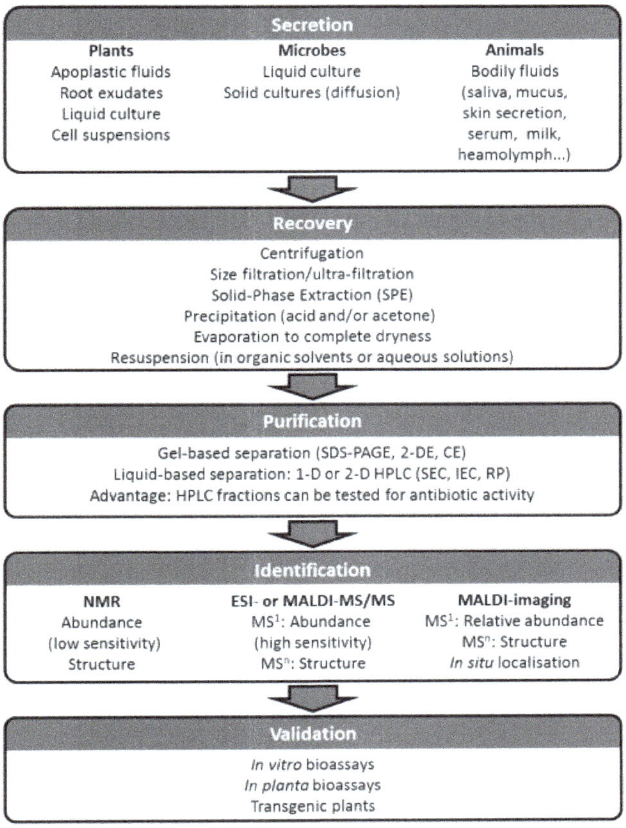

Figure 2: Diagram for the identification and validation of AMPs bearing biocontrol potential in plant immunity. Secreted AMPs can be recovered from any inter- or extra-cellular space and from any organism. Being dilute, secreted AMPs will need to be concentrated.

Secreted Plant Ribosomal Peptides for Signaling and Defense

The advances in sequencing technology over the last few decades have led to an increasing number of plant genomes being widely available. The first sequenced plant genome was *Arabidopsis thaliana* (Intitative,2000). Analysis of this genome revealed that plants encode many more predicted peptide transporters and receptors than was expected based on animal systems (Intitative, 2000; Shiu and Bleecker, 2003). This implies that many of the plant peptides involved in plant growth and development, cell-to-cell communication and defense responses have yet to be identified. Plant AMPs are also a rich source of plant defense compounds which can be grouped based on their structure. The 8 main classes in plant AMPs are cyclotides, lipid transfer proteins, defensins, thionins, snakins, hevein-like, vicillin-like, and knottins (Goyal and Mattoo, 2014). This section will focus on known plant endogenous peptides which are secreted upon recognition of microbial colonization of the plant and aid in plant defense signaling. Some of these have been comprehensively reviewed (Ryan and Pearce, 2003; Boller, 2005; Farrokhi et al., 2008; Yamaguchi and Huffaker, 2011) but updated information will be presented here. The defensins will be looked at in detail as they have been successfully taken through to field trials recently. Other AMPs showing induction of resistance in transgenic plants have also been described previously (Oard and Enright, 2006; Muramoto et al., 2012; Verma et al., 2012; Zhu et al., 2012). These other classes of AMPs with increased resistance have been well reviewed (Goyal and Mattoo, 2014). Where possible the receptors detecting these endogenous peptides will also be discussed. Table summarizes the plant secreted peptides discussed in this review; AMP precursors and receptors, where known, are indicated.

Table 1: Plant ribosomal AMPs

Peptide name	AA length	Precursor protein	Origin species	Target pests	Peptide receptor	References
Systemin	18	Prosystemin	Tomato, Potato, Pepper, and Nightshade	Wounding insects	BRI1 binds systemin	Pearce et al., 1991
					Another unidentified receptor initiates signaling	
HypSys	15–20	preproHypSys	Tobacco, Tomato, Petunia, Sweet potato, Black nightshade, Potato	Wounding insects	Not known	Pearce et al., 2001
Pep1	23	PROPEP1	*A. thaliana*, Tomato, Potato, Maize	*Pythium irregular; Pythium dissotocum*	PEPR1 and PEPR2	Huffaker et al., 2006
GmPep914	8	GmPROPEP914	Soybean	Not known	Not known	Yamaguchi et al., 2011
GmPep890	8	GmPROPEP890	Soybean	Not known	Not known	Yamaguchi et al., 2011
GmSubPep	12	Glyma18g48580	Soybean	Not known	Not known	Pearce et al., 2010
CAPE1	11	PR-1b (P14a)	Tomato	Bacteria and insects	Not known	Chen et al., 2014
RsAFP1	44	RsAFP1	Radish	Pathogenic fungi and yeast	Not known	Terras et al., 1992
RsAFP2	36	RsAFP2	Radish	Pathogenic fungi and yeast	Not known	Terras et al., 1992
NaD1	47	NaD1 precursor	Ornamental tobacco (*Nicotiana alata*)	Pathogenic fungi	Not known	van der Weerden et al., 2008
Inceptins	Nov-13	Chloroplastic ATP synthase γ-subunit	Cowpea (*Vigna unguiculata*)	Wounding insects	Not known	Schmelz et al., 2006

PLANT SIGNALING PEPTIDES

Systemin and its Receptor: The First Peptide Acknowledged as a Plant Hormone

Systemin is the most well-known secreted plant peptide involved in activation of defense signaling and it was the first peptide to be acknowledged as a plant hormone (Ryan and Pearce, 2003). Systemin is an 18-AA peptide which is derived from a 200-AA precursor protein called prosystemin (Pearce et al., 1991; McGurl and Ryan, 1992; Farrokhi et al., 2008; Table). The isolation of systemin was achieved by feeding cut *Solanum lycopersicum* (tomato) stems with a few microliters of tomato leaf juice, causing the plants to produce proteinase inhibitor proteins, including systemin (Pearce et al., 1991; Ryan and Pearce, 1998). Prosystemin localizes in the cytoplasm of phloem parenchyma cells, however upon wounding or death of these cells, prosystemin is processed to systemin, probably by proteinases. This allows systemin to diffuse into the apoplast and be detected by receptors on the plasma membrane of mesophyll cells (Boller, 2005). Systemin induces protease inhibitor production within local tissue in order to suppress wounding by insect proteases. However, it also leads to a systemic wound response by activating the jasmonic acid (JA) signaling pathway (Stratmann, 2003; Farrokhi et al., 2008). The production of this peptide appears to be under diversifying selection which suggests that some herbivores may have evolved defenses against systemin signaling (Boller, 2005). Prosystemin has only been found in species of the Solaneae a subtribe of the Solanaceae family e.g., tomato, potato, pepper and nightshade but not in tobacco, another Solanaceae (Scheer and Ryan, 2002; Ryan and Pearce, 2003). Since systemin has not been identified in tobacco this plant also shows no alkalization response to systemin treatment (0.0025–25 nM) (Ryan and Pearce, 2003). Therefore, it seems systemin is restricted to a few species and is not found globally in plants. This suggests that for the Solaneae plants, pre-treatment with systemin could reduce the wounding effects of insects in the field. Systemin pre-treatment also has the potential to reduce other infections, for example virus and bacterial infections transmitted by insects.

Over a decade after the first report of systemin its tomato cell surface receptor was identified (Scheer and Ryan, 2002). SlSR160 is a 160 kDa receptor belonging to the family of leucine rich repeat receptor-like kinases (LRR-RLKs), and like classical LRR-RLKs, it contains an extracellular leucine rich repeat domain, a transmembrane domain and an intracellular kinase domain. In cell culture experiments, systemin detection by SlSR160 activates a complex signaling pathway which results in activation of a mitogen-activated protein kinase (MAPK), rapid alkalinization of the extracellular medium via blockage of a proton pump in the cell membrane, the activation of phospholipases along with activation of phytodienoic acid and jasmonic acid (JA) for defense signaling (Felix and Boller, 1995; Schaller and Oecking, 1999; Ryan and Pearce, 2003). It was later discovered that the systemin receptor, SlSR160, and the brassinosteroid receptor from *A. thaliana*, AtBRI1, are homologs which contain all the same domains, the most conserved of which have 83–90% similarity (Montoya, 2002; Scheer and Ryan, 2002). BRI1 is the cell surface receptor for the brassinosteroid pathway which detects the hormone brassinolide (BL), a regulator of plant growth and development. This shows dual function for this receptor in both plant defense and maintenance of growth and development. In recent years however, the dual function of BRI1/SR160 has come into question. Initially dual function was demonstrated by the expression of tomato SR160/BRI1 in tobacco which allowed the transgenic plants to respond to systemin treatment (Scheer et al., 2003). A mutant tomato *Slbri1* gene called *cu-3*, was used to further examine the dual function of SR160/BRI1. The tomato*cu-3* plants showed a Brassinosteroid (BR) deficient growth phenotype i.e., stunted growth and curled leaves, however these plants also could not induce systemin signaling (Scheer et al., 2003). Confirming this receptor has a dual role in plant defense as well as in growth and development (Montoya, 2002; Scheer et al., 2003). However, more recently it was shown that the reduced responsiveness of *Slcu3* plants to systemin was due to the stunted nature of these plants and *Slcu3* cell cultures responded the same as the wild type (WT) to treatment with systemin (Holton et al., 2007). These results have led to some controversies in this field and a more recent study has sought to try and understand these conflicting results. Malinowski et al. (2009) showed by expression of tomato BRI1 in tobacco cells that BRI1 binds to systemin however the systemin signaling did not show any increase by the overexpression of tomato BRI1. This report confirms results previously described by Holton et al. (2007), that silencing of SlBRI1 in tomato results in a *bri1* phenotype but did not affect systemin response (Malinowski et al., 2009). These results imply that BRI1 can bind systemin but it is not the ligand that initiates intracellular signaling (Malinowski et al., 2009). Therefore, there could be another ligand for systemin yet to be identified. The newest data also explains why *A. thaliana* which contains BRI1 cannot sense systemin treatment or activate downstream defense signals. In *A. thaliana* there is some redundancy within the BRI1 gene family as three other cell surface receptors have been identified with high sequence similarity, BRI1-like (BRL1, BRL2, and BRL3). It was shown that BRL1 and BRL3 can complement BRI1 but BRL2 cannot, BRL2 is required for provascular differentiation in leaves (Clay and Nelson, 2002; Caño-Delgado et al.,2004). Given there is a gene family in *A. thaliana*, it is highly likely that this is true in Solanacea species. Therefore, it could be that one of these other receptors is required for systemin signal transduction.

HypSys: The First Peptides Identified in Tobacco

HypSys's are another group of well-known wound-induced peptides which show functional relatedness to systemin and contain multiple hydroxyproline residues. These polypeptide hormones were first identified in tobacco using cell suspension assays treated with a crude peptide fraction from tobacco leaves which resulted in the alkalinization of medium. Two 18-AA polypeptide hormones were identified in tobacco and found to be derived from each end of a 165-AA polyprotein hormone precursor, pro-TobSys-A (Pearce et al., 2001; Table). These two peptides were named tobacco hydroxyproline-rich systemin (TobHypSys) I and II (Ryan and Pearce, 2003). Although HypSys's may have similar downstream functions to systemin, it appears the physical properties and the processing of these two peptides differ greatly. The synthesis of HypSys potentially involves the secretory pathway given the presence of carbohydrate residues and hydroxyproline residues, while in contrast systemin contains no hydroxyprolines and no carbohydrate residues (Pearce et al., 2001). HypSys's have also been identified in a wider variety of plants than systemin including tomato (TomHypSys I, II and II), petunia (PhHypSys I, II, and III) sweet potato, black nightshade and potato (Pearce and Ryan, 2003; Pearce et al., 2007, 2009; Chen et al.,2008; Bhattacharya et al., 2013). The TomHypSys are 20, 18, and 15-AAs in length, respectively, while the PhHypSys I, II, and III are 19, 20, and 18-AAs in length, respectively (Pearce and Ryan, 2003; Pearce et al., 2007). In these identified precursor proteins for the HypSys peptides it was noted that a 30 base pair (bp) sequence around the peptidase splice site was conserved in all, this conservation and the high sequence homology over the whole protein is allowing faster and easier identification of these peptides therefore avoiding long troublesome purification from leaf extracts (Bhattacharya et al., 2013).

The 146-AA precursor protein of the three TomHypSys peptides localizes to the cell wall matrix in the phloem parenchyma cells of tomato leaves (Narváez-Vásquez et al., 2005). The tomato, tobacco and black nightshade HypSys's all induce the production of protease inhibitors when supplied through cut petioles. Furthermore, the tomato preproHypSys gene is induced upon wounding of leaves and upon systemin or JA treatment of the leaves (Pearce et al., 2001, 2009; Pearce and Ryan, 2003). In contrast, the petunia HypSys's showed no induction of protease inhibitors against herbivore attacks, however they did induce the expression of *defensin 1* (Pearce et al., 2007). *Defensin 1*is a gene associated with innate immunity in plants, suggesting that the petunia HypSys's still appear to function in plant defense signaling pathways. The petunia HypSys's are functionally more similar to the AtPep family of peptides than to the other HypSys's implying that there may be some functional diversity within peptide families specific to some plants (Bhattacharya et al., 2013). The StHypSys's are functionally comparable to the tobacco and tomato HypSys peptide, however they appear to have additional functions as they also activate enzymes to protect against oxidative stress and free-radical generation (Bhattacharya et al., 2013). These are both outcomes of herbivore attack and pathogen infection therefore the StHypSys peptides appear to act as defense elicitors against both insects and pathogens (Bhattacharya et al., 2013). Although the experiments were not done it could be expected that pre-treatment of plants with HypSys prior to insect attack or pathogen infection would result in reduced symptoms as the defense pathways are primed.

Pep1 and its Receptor: The Model Plant Arabidopsis Gives up its First Peptide

Pep1 from *A. thaliana* is another peptide which is associated with the activation of the plant defense system (Huffaker et al., 2006). This peptide consists of 23 AA residues and is derived from the C-terminus of a 92-AA precursor protein called PROPEP1 (Huffaker et al., 2006; Farrokhi et al., 2008; Table). AtPROPEP1 shows low level expression in all tissue types but increased expression is evident when plants are wounded or treated with methyl jasmonate (MeJA) or ethlylene (ET) (Huffaker et al., 2006). This suggests a role for this protein in plant defense. Plants treated through cut petioles with 20 nM AtPep1, showed increased expression of *Plant Defensin 1.2* (*PDF1.2*), *Pathogenesis-related Protein 1* (*PR-1*) and *PROPEP1* along with the production of H_2O_2 (Huffaker et al., 2006; Huffaker and Ryan, 2007). The production of H_2O_2 and *PDF1.2* are both associated with innate immunity in plants while *PR-1* is known to be induced during Pathogen-Associated Molecular Patterns (PAMP) Triggered Immunity (PTI) however its function remains unknown. Huffaker et al. (2006) and Huffaker and Ryan (2007) used mutants for the JA and ET pathways, to show that AtPep1 functions upstream of the JA/ET pathways, as expression of *PDF1.2* and *PR-1*, triggered by the peptides, was blocked. Transgenic plants constitutively overexpressing*AtPROPEP1* showed an increase in root mass compared to WT plants. When these overexpressing plants were infected with the root pathogen*Pythium irregular* the increase in root growth compared to infected WT plants was still apparent however, the aerial parts of the plants show no difference in growth or disease symptoms between the transgenic and WT. This implies that overexpression of *PROPEP1* provides a growth advantage to the roots even in the presence of a pathogen (Huffaker et al., 2006). This growth advantage looks to be a by-product of*PROPEP1* overexpression as the aerial components of the plants do not show increased growth implying that nutrient acquisition has not increased. Six annotated homologs of the precursor protein have been identified in *A. thaliana*, two of which (PROPEP2 and 3) show increased expression when plants are infected with bacterial, fungal and oomycete pathogens (Huffaker et al., 2006). The expression of *PDF1.2*and *PR-1* was increased in plants treated with AtPep1 (Huffaker et al.,2006). All of these results indicate that *PROPEP1-3* has a role in a feedback loop for defense signaling that is activated in the presence of pathogens and also increases root development (Huffaker and Ryan,2007).

Orthologs of these *A. thaliana* genes have since been found in other plant species suggesting an important and ancient role for these peptides. Important crop species are among those to be investigated for the presence of Pep peptides. Recently an ortholog of *AtPROPEP1* was identified in *Solanum lycopersicum*, *SlPROPEP*, which has high AA identity (96%) with the C-terminal region of the *S. tuberosum* gene (Trivilin et al., 2014). Silencing of the tomato ortholog increased the susceptibility of plants to *Pythium dissotocum* but also decreased the expression of key defense response genes e.g., *PR-1, PR-5, ERF1, LOX-D*, and *DEF2* (Trivilin et al., 2014). These results showed that SlPROPEP plays a role in resistance against *P. dissotocum*. Huffaker et al. (2011) identified and characterized an ortholog of *AtPROPEP1* in maize. Similar to *AtPROPEP1, ZmPROPEP1* is induced in response to JA or peptide treatment, but also fungal infection. In maize plants treated with ZmPep1 it was noted that the expression of several defense genes were up-regulated, *endochitinase A, PR-4, PRms, SerPIN*, and*Benzoxazineless1* (Huffaker et al., 2011). Interestingly, this work also showed that pre-treatment of maize leaves with 25 pmol of ZmPep1 increased resistance of leaves to *Cochliobolis heterostrophus* and pre-treatment of stalks with 5 nmol of ZmPep1 increased resistance to*Colletotrichum graminicola* (Huffaker et al., 2011). The work described above clearly shows a benefit to the plants when pre-treated with these peptides in reducing pathogenicity. Therefore, these peptides may be useful for crop plants in areas that are known to have high pathogenicity instead of the use of fungicides or pesticides by either foliar application or with the generation of transgenic crop plants.

Two cell surface receptors for AtPep1, PEPR1 and PEPR2, have been identified in *A. thaliana* and demonstrated to be membrane associated leucine rich repeat (LRR) receptor kinases (Yamaguchi et al., 2006,2010). These two receptors are 76% similar at the protein level and both showed induced expression upon wounding, also following exogenous application of Pep1 peptide and MeJA (Yamaguchi et al.,2010). PEPR1 was confirmed as recognizing Pep1 resulting in the activation of plant defense responses (Yamaguchi et al., 2006). Gain of function was also observed when *AtPEPR1* was overexpressed in transgenic tobacco cell lines resulting in the rapid alkalinization of the media upon treatment with AtPep1 (Yamaguchi et al., 2006). Like previously described experiments, pre-treatment of *A. thaliana* with exogenous Pep1 resulted in increased resistance to *Pseudomonas syringae pv. tomato DC3000* (Yamaguchi et al., 2010). PEPR1 and PEPR2 show limited redundancy as single mutants of each showed reduced resistance to *P. syringae pv. tomato* while a double mutant is susceptible to *P. syringae pv. tomato* (Yamaguchi et al., 2010). This work has elegantly identified the cell surface receptor of a secreted plant peptide which induces defense responses against pathogen invasion.

Soybean Peptides: Soybean is Potentially the Next High Peptide Yielding Plant

Whilst tomato and tobacco have traditionally been used for peptide discovery, soybean is rapidly becoming the new dicot model as a source of new peptides. Recently many new peptide families have been identified in soybean, GmPep914 and GmPep890 are two recently identified peptides shown to cause the rapid alkalinization of cell culture medium (Yamaguchi et al., 2011). These peptides consist of 8 AA residues and are derived from the C-terminal end of 52-AA precursor proteins GmPROPEP914 and GmPROPEP890 (Yamaguchi et al., 2011; Table). Treatment of soybean leaves with GmPep914 and GmPep890 induced the expression of their precursor proteins as well as defense genes including *CYP93A1*, a cytochrome P450 gene,*chitinaseb1-1* and *Gmachs1* (*Glycine max chalcone synthase1*) which is involved in phytoalexin production (Yamaguchi et al., 2011). GMPep914 is unrelated to any of the previously identified peptides with a known function in plant defense and it is also the smallest peptide found to have such a role (Yamaguchi et al., 2011).

GmSubPep was also recently identified in soybean. The sequence of this 12-AA peptide was found within an extracellular protein family called subtilisin-like protease (subtilase), specifically in the gene*Glyma18g48580* (Pearce et al., 2010; Table). This peptide appears unique for soybean as systemin is for tomato due to the GmSubPep sequence located in the protease-associated (PA) domain of*Glyma18g48580* which is a unique region of the subtilases within legumes (Pearce et al., 2010). Also when this peptide was tested on tomato, tobacco, *Arabidopsis* and corn cell cultures no alkalinization was observed (Pearce et al., 2010). The receptor of Glyma18g48580 was not induced upon wounding or treatment with MeJA, however treatment of cell cultures with the peptide resulted in an increase in some defense genes (*Cyp93A1, Chib-1b, and PDR12;* Pearce et al.,2010). Glyma18g48580 is predicted to be an apoplastic protein of unknown function and it has been speculated that this protein comes into contact with non-self molecules, due to its apoplastic localization. This in turn could induce cleavage of the peptide resulting in activation of defense signaling pathways. No work has been done so far with these soybean peptides to determine if they are able to increase plant resistance to pathogens or insects but given that defense genes are up-regulated it is conceivable that this is the case.

CAPE1: Unraveling the Potential Function of PR-1

The CAPE1 peptide, was recently isolated from tomato and has been shown to be a Damage-Associated Molecular Pattern (DAMP) elicitor (Chen et al., 2014). CAPE1 was identified using endogenous peptide mixtures extracted from stressed and unstressed plants, these extracts were analyzed by nanoflow ultrahigh performance liquid chromatography mass spectrometry (nanoUHPLC-MS). CAPE1 is an 11-AA peptide derived from the C-terminus of a 159-AA precursor protein called PR-1b (also known as P14a; Chen et al., 2014; Table). Expression analysis showed that CAPE1 was induced by wounding and that treatment of tomato plants with the CAPE1 peptide resulted in the induction of defense hormones such as JA and salicylic acid (SA) as well as increased expression of several pathogen-related marker genes, *PR-2, PR-7* and *PR1b* (Chen et al., 2014). However, CAPE1 treated plants did not induce PTI-responsive genes, suggesting that CAPE1 may induce systemic resistance rather than PTI (Chen et al., 2014). The authors also showed that pre-treatment of tomato leaves with CAPE1 prior to infection with *P. syringae pv. tomato* resulted in reduced disease symptoms in the absence of Hypersensitive Response (HR) (Chen et al., 2014). CAPE1 also showed an anti-herbivore activity; *Spodoptera litura* larvae feeding experiments using CAPE1 pre-treated tomato leaves resulted in a 20% reduction in size and weight (Chen et al.,2014). This data demonstrates that CAPE1 has antibacterial and anti-herbivore activity. Curiously anti-fungal activity was not reported on. Comparing these pre-treatment results to those discussed above for the ZmPep1 peptide (Huffaker et al., 2011) shows that these peptides appear to prime plants for pathogen resistance.

Sequence analysis has shown that the CAPE1 peptide is conserved in many flowering plants. The authors also suggest that three AAs immediately in front of the peptide may be the cleavage site (CNYx) and this could act as a conserved motif that could indicate a bio-active peptide in other species (Chen et al., 2014). Pathogenesis-related protein 1 from *A. thaliana* (AtPR1) contains the CAPE1 and potential bio-active motifs (CNYx. PxGNxxxxPY) but has the least protein similarity to tomato PR-1b. However, this CAPE1 peptide from AtPR1 was shown to increase immunity of *A. thaliana* to *P. syringae pv. tomato* indicating that this peptide functions similarly to the CAPE1 from tomato PR-1b (Chen et al., 2014).

The authors suggest a motif that could be used to search other RNAseq or genome database to investigate whether the motif is used in other peptide cleavage sites, showing a downstream application of this technique. As for the plant resistance increase that is observed when the plants are pre-treated with CAPE1 it gives further information about the PR1 protein which has always been associated with plant defense but a function of this protein has been elusive. Since PR1 is present in most plant species this could be a good biocontrol agent against pathogens and shows an ancient and probably conserved role for this protein.

Inceptins: Insect Feeding aids Plant Defense

Inceptins fall into a small category of plant peptides as they are processed by enzymes from the invading organism, not by the plant itself. Inceptins are 11–13 AA acidic peptides which contain disulphide bridges and are derived from the chloroplastic ATP synthase γ-subunit (Schmelz et al., 2006; Table). These peptides were identified from the saliva of larval fall armyworms (*Spodoptera frugiperda*) that were feeding on cowpea (*Vigna unguiculata*) plants. Inceptin peptides are cleaved from the chloroplastic ATP synthase γ-subunit during feeding of the larvae on the leaves; the peptide in the saliva of *S. frugiperda* is then detected by the plant, resulting in the activation of defense related genes (Schmelz et al., 2006). The perception of inceptin by cowpea plants results in activation of the volatiles homoterpene (E)-4,8-dimethyl-1,3,7-nonatriene (DMNT) and cinnamic acid which are known to attract natural insect enemies but also results in induction of JA and SA hormones (Schmelz et al., 2006, 2007).

Defensins: Peptides Endogenous to all Plant Species

Plant defensins are produced in all plant species and are either constitutively expressed in storage and reproductive organs or can be induced during biotic or abiotic stress (Vriens et al., 2014). These peptides are known as cationic peptides and are usually 45–54 AAs in length. These peptides typically harbor a cysteine-stabilized αβ-motif (CSαβ) which has an α-helix and a triple-stranded antiparallel β-sheet which is stabilized by four disulphide bridges (Vriens et al., 2014).

There are typically two groups of plant defensins based on the proproteins. Both groups contain a signal peptide which targets the protein to the endoplasmic reticulum where the mature protein is folded and enters the secretory pathway. One of these groups contains an additional prodomain on the C-terminus which is cleaved during processing through the secretory pathway (Vriens et al., 2014). Some well-known plant defensins are RsAFP1 and RsAFP2 from the seeds of radish (Terras et al., 1992; Table), MsDef1 and MtDef4 from the seeds of *Medicago* species (Gao et al., 2000; Ramamoorthy et al., 2007; Sagaram et al., 2013), NaD1 from the flowers of tobacco (Lay et al.,2003a,b) and Psd1 from seeds of pea pods (Almeida et al., 2000). Each of these peptides has antifungal activity but so far very few defensins have been found with antibacterial activity. In recent years the use of transgenics has been employed to investigate some defensins for their ability to protect plants from fungal infection or insect wounding. Transgenic rice plants containing a plant defensin from *Brassica rapa*, BrD1, showed increased resistance to the insect brown plant hopper (Choi et al., 2009). Transgenic tomato plants containing MsDef1 showed increased seedling resistance to *F. oxysporum* f. sp. *Lycopersici*(Abdallah et al., 2010). While transgenic tobacco and potato plants containing the defensin NmDef01 from *N. megalosiphon* showed increased resistance to the oomycete *Phytophthora infestans* in both glasshouse and field trials (Portieles et al., 2010). The mode-of-action and pathogenicity experiments of two defensins, RsAFPs and NaD1, will be discussed here in more detail.

Raphunus sativus antifungal protein 1 (RsAFP1) and RsAFP2 were first isolated from radish seeds by use of ammonium sulfate fractionation and anion-exchange chromatography and were found to produce 5 kDa peptides which are assembled and stabilized by disulfide bridges (Terras et al., 1992). The RsAFP1 peptide is 44 AAs in length and RsAFP2 is 36 AAs in length (Terras et al., 1992). Within the first 36 AAs of these peptides there is a high amount of conservation with only 2 AAs difference. These changes though are significant and result in a higher positive charge for RsAFP2 compared to RsAFP1. RsAFP2 has a low half maximal inhibitory concentration (IC_{50}) range of 0.4–25 pg/ml when tested against plant pathogenic fungi while RsAFP1 has a larger IC_{50} range of 0.3–100 pg/ml. The antifungal activity of RsAFP2 is also more efficacious compared to RsAFP1 in the presence of salts (Terras et al., 1992). RsAFP2 interacts with fungal glucosylceramide (GlcCer) found in the membrane and cell wall of fungal cells (Thevissen et al., 2004). GlcCer is usually associated with other components which form lipid rafts in fungal membranes and cell walls (Vriens et al., 2014). The binding to GlcCer results in activation of MAP kinase and cell wall integrity signaling pathways, the production of Reactive Oxygen Species (ROS), induction of ion fluxes and activation of caspases, resulting in abnormal hyphal growth and division (De Samblanx et al., 1997; Navarro-García et al., 2005; Aerts et al., 2009; Vriens et al., 2014). The ability of RsAFP2 to inhibit fungal infection of whole plants was tested with the generation of s transgenic wheat constitutively expressing RsAFP2 (Li et al., 2011). Compared to untransformed control plants, the transgenic RsAFP2 plants showed increased resistance to *Fusarium graminearum* and*Rhizoctonia cerealis* in glasshouse experiments over several generations and in field trials for T_4 and T_5 generations (Li et al.,2011). The authors also noted that this resistance was heritable (Li et al., 2011). Another well investigated defensin is *Nicotiana alata*Defensin 1 (NaD1; Table). NaD1 is one of the rarer plant defensins as it was isolated from the flower of ornamental tobacco (*Nicotiana alata*) and not the seed, where most defensins have been found. NaD1 is only active against filamentous fungi and has no effect against bacteria, yeast or human cell lines (van der Weerden et al., 2008). The fungal membrane ligand for NaD1 has recently been identified as phospholipid phosphatidylinositol 4,5-bisphosphate (PIP2; Poon et al., 2014). These authors demonstrated that NaD1 oligomers interact with the head group of two PIP2 molecules and that this complex is required for membrane permeabilization and subsequent cell death (Poon et al., 2014). If the NaD1 oligomer structure is disrupted then antifungal activity is also lost (van der Weerden et al., 2008; Poon et al., 2014). Once the fungal cell wall has been disrupted NaD1 has been shown to form an aperture of 14–23 Å in size through which it may enter the cytoplasm (van der Weerden et al., 2008). It has been shown that NaD1 is able to localize to the cytoplasm of hyphal cells and these cells subsequently show granulation of the hyphal cytoplasm and cell death (van der Weerden et al., 2008). This indicates that defensins may not just target the cell membrane, but also have intracellular functions. Once NaD1 is inside the cytoplasm of hyphal cells, ROS production occurred within these cells suggesting that cell death is occurring.

There is still more to be learnt about the mode-of-action of NaD1 but this defensin has great potential as a natural anti-fungal in plants other than tobacco. This has been elegantly demonstrated by Gaspar et al. (2014), where they generated transgenic homozygous cotton lines constitutively overexpressing NaD1. These lines were tested in greenhouse assays for resistance to *F. oxysporum* f. sp. *vasinfectum*(*Fov*) with one line being chosen for field trials in soil naturally infected with *Fov* and *Verticillium dahlia* (Gaspar et al., 2014). These greenhouse assays and field trials of NaD1 expressing transgenic plants showed an increase in resistance and cotton production compared to the non-transgenic parental line when *Fov* and *V. dahlia* are present in the soil (Gaspar et al., 2014). The NaD1-expressing plants also showed no detrimental agronomic properties compared to the control non-transgenic parental lines. This is a good example of a defensin taken from one plant species and used as a biocontrol agent in an economically important crops species against soil borne fungi.

All of the peptides discussed above have shown a direct correlation with the induction of the plant defense pathways to protect against herbivores and/or pathogens. In some cases experimental data is available to show a decrease in infection of plants when pre-treated with these peptides. This is a small number of plant defense peptides and without doubt there are many more to be identified. As yet undiscovered peptides may be species specific while others could be ancient defense signals active in all higher plants.

AMPs Secreted by Non-Plant Organisms, and Their Potential for Plant Pathogen Resistance Engineering

Whilst plants have proven (and will continue to do so) a rich source of antimicrobial peptides, non-plant organisms are also subject to pathogen attack and synthesize AMPs as a defense mechanism. These non-plant AMPs can in turn be either sprayed on crops or artificially introduced into crop species to engineer pathogen resistance using transgenic methods. Tables 2 and 3 summarizes the non-plant AMPs discussed in this review, detailing their class and molecular weight, the analytical methods employed for their identification in the publications cited here, the organisms naturally secreting them, the targeted pathogens and plant hosts.

Table 2: Non-plant nonribosomal AMPs

Peptide name per taxa of origin	AA length	Mr	Peptide analysis	Origin species	Target pathogens	Plant host	References
BACTERIUM GRAM –							
2-amino-3-(oxirane-2,3-dicarboxamido)-propanoyl-valine (APV)	3	317	LC-ESI-MS/MS, NMR	*Pantoea agglomerans* 48b/90 (Pa48b)	*Agrobacterium tumefaciens, Erwinia amylovora, Pseudomonas syringae pathovars, Serratia marcescens, Bacillus subtilis*	None tested	Sammer et al., 2009
2-amino-3-(oxirane-2,3-dicarboxamido)-propanoyl-valine (APV)	3	317	Not in this study	*Pantoea agglomerans* 48b/90 (Pa48b)	*Pseudomonas syringae* pv. *glycinea*	Soybean	Sammer et al., 2012
syringolin		493	FAB-MS, ESI-MS, NMR	*Pseudomonas syringae* pv. *syringa*	*Pyricularia oryzae*	Rice	Wäspi et al., 1998
Syringolin A			Not in this study	*Pseudomonas syringae* pv. *syringa*	*Blumeria graminis* f sp. *tritici*	Wheat	Wäspi et al., 2001
Syringomycin E (SRE)		1225	HPLC-ESI-Q-MS	*Pseudomonas syringae* pv. *syringae* strain B359	*Fusarium oxysporum, Verticillium dahliae, Botrytis cinerea, Penicillium expansum, Phytophthora infestans, and Rhodotorula pilimanae*	Apple	Fogliano et al., 2002
Syringopeptin 25A (SP25A)		2401	HPLC-ESI-Q-MS	*Pseudomonas syringae* pv. *syringae* strain B359	*Fusarium oxysporum, Verticillium dahliae, Botrytis cinerea, Penicillium expansum, Phytophthora infestans, and Rhodotorula pilimanae*	Apple	Fogliano et al., 2002
BACTERIUM GRAM +							
Bacillomycins D		1073, 1059, 1045	HPLC-ESI-MS/MS	*Bacillus amyloliquefaciens* Q-426	*Fusarium oxysporum* f. sp. *spinaciae*	None tested	Zhao et al., 2014
Fengycins A		1464, 1450	HPLC-ESI-MS/MS	*Bacillus amyloliquefaciens* Q-426	*Fusarium oxysporum* f. sp. *spinaciae*	None tested	Zhao et al., 2014
Fengycins B		1506, 1478	HPLC-ESI-MS/MS	*Bacillus amyloliquefaciens* Q-426	*Fusarium oxysporum* f. sp. *spinaciae*	None tested	Zhao et al., 2014
7-O-malonyl macrolactin A		488	LC-ESI-Q-Trap-Ms/MS	*Bacillus amyloliquefaciens* strain NJN-6	*Fusarium oxysporum* f. sp. *cubense, Ralstonia solanacearum*	Banana	Yuan et al., 2012
7-O-succinyl macrolactin A		502	LC-ESI-Q-Trap-Ms/MS	*Bacillus amyloliquefaciens* strain NJN-6	*Fusarium oxysporum* f. sp. *cubense, Ralstonia solanacearum*	Banana	Yuan et al., 2012
Bacillomycins D	7	1031, 1045	LC-ESI-Q-Trap-Ms/MS	*Bacillus amyloliquefaciens* strain NJN-6	*Fusarium oxysporum* f. sp. *cubense, Ralstonia solanacearum*	Banana	Yuan et al., 2012
Fengycin			Not in this study	*Bacillus amyloliquefaciens* strain PPCB004	*Alternaria citri, Botryosphaeria* sp., *Colletotrichum gloeosporioides, Fusicoccum aromaticum, Lasiodiplodia theobromae, Penicillium crustosum, and Phomopsis perse*	Orange	Arrebola et al., 2010
Iturin A			Not in this study	*Bacillus amyloliquefaciens* strain PPCB004	*Alternaria citri, Botryosphaeria* sp., *Colletotrichum gloeosporioides, Fusicoccum aromaticum, Lasiodiplodia theobromae, Penicillium crustosum, and Phomopsis perse*	Orange	Arrebola et al., 2010

Surfactin		Not in this study	*Bacillus amyloliquefaciens* strain PPCB004	*Alternaria citri*, *Botryosphaeria* sp., *Colletotrichum gloeosporioides*, *Fusicoccum aromaticum*, *Lasiodiplodia theobromae*, *Penicillium crustosum*, and *Phomopsis perse*	Orange	Arrebola et al., 2010
Fengycins A (C15–C17)	1464, 1477	MALDI-TOF MS	*Bacillus atrophaeus* CAB-1	*Botrytis cinerea*, *Sphaerotheca fuliginea*	Cucumber	Zhang et al., 2013
Mycosubtilin		Not in this study	*Bacillus subtilis*	*Bremia lactucae*	Lettuce	Deravel et al., 2014
Surfactin		Not in this study	*Bacillus subtilis*	*Bremia lactucae*	Lettuce	Deravel et al., 2014
Fengycins	1435, 1505	HPLC-ESI-MS	*Bacillus subtilis* strain M4	*Pythium ultimum*, *Botrytis cinerea*	Bean, apple	Ongena et al., 2005
Iturin		HPLC-ESI-MS	*Bacillus subtilis* strain M4	*Pythium ultimum*, *Botrytis cinerea*	Bean, apple	Ongena et al., 2005
Surfactin		HPLC-ESI-MS	*Bacillus subtilis* strain M4	*Pythium ultimum*, *Botrytis cinerea*	Bean, apple	Ongena et al., 2005
Fusaricidins A, B, and C	884, 898, 948	MALDI FT-ICR MS imaging	*Paenibacillus polymyxa* strain Pp56	*Rhodotorula aurantica*, *Botrytis cinerea*, *Fusarium oxysporum*, *Cladosporium cucumerinum*, *Phythium aphanidermatum*, *Pseudomonas syringae*	None tested	Debois et al., 2013
LI-F lipopeptides	912, 926	MALDI FT-ICR MS imaging	*Paenibacillus polymyxa* strain Pp56	*Rhodotorula aurantica*, *Botrytis cinerea*, *Fusarium oxysporum*, *Cladosporium cucumerinum*, *Phythium aphanidermatum*, *Pseudomonas syringae*	None tested	Debois et al., 2013
FUNGI						
Trichokonin VI (TK VI)		Not in this study	*Trichoderma pseudokoningii* strain SMF2	*Fusarium oxysporum*	None tested	Shi et al., 2012
Trichokonin (TK)		Not in this study	*Trichoderma pseudokoningii* strain SMF2	*Pectobacterium carotovorum* subsp. *carotovorum*	Chinese cabbage	Li et al., 2014

Table 3: Animal ribosomal AMPs

Peptide name	AA length	Mr	Peptide analysis	Origin Species	Target pathogens	Plant host	References
Penaeidin4-1 (Pen4-1)	47		Not in this study	*Litopenaeus setiferus* (shrimp)	*Sclerotinia homoecarpa*, *Rhizoctonia solani*	*Agrostis stolonifera*	Zhou et al., 2011
Metchnikowin (Mtk isoforms A and B)	26	3025, 3045	HPLC-ESI-MS, Edman	*Drosophila melanogaster*	*Micrococcus luteus*, *Neurospora crassa*	None tested	Levashina et al., 1995
Metchnikowin (Mtk isoforms A and B)	26		Not in this study	*Drosophila melanogaster*	*Fusarium graminearum*	Barley	Rahnamaeian et al., 2009
Metchnikowin (Mtk isoforms A and B)	26		Not in this study	*Drosophila melanogaster*	*Blumeria graminis* f. sp. *hordei*	Barley	Rahnamaeian and Vilcinskas, 2012
Thanatin	21	2436	2-D HPLC (CXC-RP)-ESI-MS	*Podisus maculiventris* (stinkbug)	*Magnaporthe oryzae*	Rice	Imamura et al., 2010
Thanatin			Immunoblots	*Podisus maculiventris* (stinkbug)	*Aspergillus flavus*	Maize	Schubert et al., 2015
Cecropin A	37	4000	Immunoblots, SDS-PAGE shotgun	*Hyalophora cecropia* (moth)	*Fusarium verticillioides*, *Dickeya dadantii*	Rice	Bundó et al., 2014
Magainin-2 (mag)	23		Not in this study	*Xenopus laevis* (frog)	*Sclerospora graminicola*	Pearl millet	Ramadevi et al., 2014
Cathelicidin (LL-37 Met37Leu)	37	4000	Immunoblots	*Homo sapiens* (Human)	*Pectobacterium carotovorum* subsp. *carotovorum*, *Fusarium oxysporum* f. sp. *Lycopersici*, *Colletotrichum higginsianum*, *Rhizoctonia solani*	Chinese cabbage	Jung et al., 2012
Esculentin-1 (Esc28L Met-28Leu)	46	5513 (with SP)	Immunoblots, LC-MALDI-TOF	*Rana esculenta* (frog)	*Pseudomonas syringae* pv. *Tabaci*, *Pseudomonas aeruginosa*, *Phytophthora nicotianae*	Tobacco	Ponti et al., 2003

Nonribosomal AMPs Identified from Microbes and Tested by Exogenous Application

Biological control of plant pests through the use of natural antagonistic microorganisms producing a vast array of AMPs has emerged as a promising alternative to reduce the use of chemicalpesticides.CycliclipopeptidesarenonribosomalAMPs predominately produced by*Bacillus* and *Paenibacillus* spp. and increasing numbers of lipopeptides are being isolated from *Pseudomonas* spp. as well (Patel et al., 2015). The following section describes the most favorable AMPs for crop disease resistance in such bacteria as well as in fungi.

Syringolin, Syringomycin, and Syringopeptin from *Pseudomonas* sp. Bacteria

Syringolin A is a peptide secreted by the Gram-negative bacterium *P. syringae* pv. *syringae*. Syringolin A has a ring structure composed of 5-methyl-4-amino-2-hexenoic acid and 3,4-dehydrolysine (Wäspi et al.,1998; Table). The α-amino group is joined by a peptide bond to a valine linked to another valine via a urea moiety (Wäspi et al., 1998). Syringolin A was recovered from liquid cultures by centrifugation, filtering, ultra-filtration, followed by gel filtration chromatography. The application of HPLC-purified syringolin onto *Oryza sativa* (rice) detached leaves induced resistance toward the fungal plant pathogen*Pyricularia oryza* (rice blast disease). Plate assays showed that syringolin A did not directly affect the growth of *P. oryzae*. Furthermore, no visible phytotoxic effects were observed when applied on detached rice leaves at the highest concentration (0.05 mM). Gene expression analysis of detached leaves sprayed with syringolin A solution and subsequently inoculated with *P. oryzae* indicated the increased transcript abundance of defense genes (including *Pir7b*, *Pir2*, *Pir2*, and *Rir1* mRNAs). Consequently, it was proposed that syringolin A elicits rice defense responses through acquired resistance rather than being directly anti-fungal (Wäspi et al., 1998). Subsequently, syringolin A was trialed in other pathosystems. *Triticum aestivum* (wheat) detached leaves were sprayed with a syringolin A solution 2 days prior to inoculation with *Blumeria graminis* f. sp *tritici* (powdery mildew; Wäspi et al., 2001). As previously observed with rice blast disease, wheat leaves exposed to syringolin A prior to fungal infection appeared more resistant. Curative effects of syringolin A were also reported in this study with whole wheat plants first infected with powdery mildew and then sprayed with a syringolin A solution. Fungal colonization of wheat tissues was arrested if syringolin A was exogenously applied 2 days post-inoculation. This study suggests that the mode-of-action of syringolin A either targets the host cells in a way that maintains host›s hypersensitivity or reverses the suppression of host defense imposed by the pathogen (Wäspi et al., 2001).

Cyclic lipodepsipeptides (LDPs) include two other AMPs secreted by*P. syringae* pv. *syringae*, syringomycin E (SRE) and syringopeptin 25A (SP25A) (Table). SRE is formed by nona-peptide lactones acylated with a long-chain 3-hydroxy FA. SP25A is composed of long and highly hydrophobic peptide chain, with a polar lactonized penta- or octa-peptide moiety at the C terminus. It has been speculated that pathogen cell walls would present a natural barrier which impairs peptide anti-microbial activity, therefore making pest cell wall porous would enhance AMPs action (Fogliano et al., 2002). LDPs were extracted from centrifuged liquid cultures followed by acetone precipitation and further fractionated using reverse-phase (RP) HPLC. SRE and SP25A were identified by MS. SRE and SP25A, together with cell wall degrading enzymes (CWDEs) from *Trichoderma atroviride*, were incorporated into *in vitro* assays performed on liquid cultures of *F. oxysporum, V. dahliae, Botrytis cinerea, Penicillium expansum, Phytophthora infestans*, and *Rhodotorula pilimanae* (Fogliano et al.,2002; Table). These analyses revealed that while SRE or SP25A alone did not inhibit fungal growth, in the presence of endochitinase and/or glucanase SRE and SP25A prevented spore germination, thus unraveling synergism between LDPs and CWDEs. Furthermore, postharvest assays performed on wounded apple fruits first inoculated with *B. cinerea*, then with *P. syringae*, and/or with *T. atroviride*indicated that the least number of *B. cinerea* related-wounds were observed when both *P. syringae* and *T. atroviride* were combined (Fogliano et al., 2002). This study thus provided support for a synergistic interaction between *P. syringae* LDPs and *T. atroviride*CWDEs against a variety of fungi, thereby supporting the hypothesis that, in a biocontrol strategy, the efficacy of AMPs in plant pathogen resistance would be improved if pathogen cell walls were also made vulnerable (Fogliano et al., 2002). This experiment however has not validated which LDPs and CWDEs are responsible for the synergism observed *in vivo*.

Fengycins, Iturins, Surfactins, Bacillomycins, Macrolactin, and Mycosubtilin from *Bacillus* sp. Bacteria

Potent antifungal lipopeptides such as surfactins, iturins and fengycins are produced by some strains of the Gram-positive bacterium *Bacillus subtilis* such as strain M4 which are beneficial rhizobacteria (Ongena et al., 2005; Table). *B. subtilis* strain M4 was also found to secrete various fengycin homologs, with the great majority of them harboring a C16 or C17 FA chain (Ongena et al., 2005). An enriched lipopeptide extract was obtained by Solid Phase Extraction (SPE) of the crude cell-free culture broth of M4 strain eluted using methanol, followed by HPLC separations and ElectroSpray Ionization (ESI)-MS analyses. *In vitro* assays showed that the lipopeptide-enriched supernatant had a strong inhibitory effect on the growth of *F. oxysporum, Pythium ultimum, Rhizoctonia solani, Rhizopus sp.*, and *B. cinerea*. The antimicrobial potential of these peptides has also been described *in planta*. Post-harvest assays on immature apple fruits preconditioned with M4 supernatant or the lipopeptide-enriched supernatant showed increased pathogen resistance, mostly attributable to fengycins (Ongena et al., 2005). Similarly, bean seedlings with roots pre-treated with M4 supernatant prior to *B. cinerea* leaf inoculation displayed reduced disease symptoms. *B. subtilis* M4 thus shows great potential as biocontrol agent to better manage soilborne, foliar and post-harvest diseases. The lipopeptide modes-of-action, in particular that of fengycins, rely not only on the direct inhibition of the plant pathogen on infected plant organs, but also on an indirect interaction mediated through the host plant via systemic resistance induction. Direct interaction was demonstrated through the disease control provided by treatment of fruits with lipopeptide-enriched supernatant and by *in situ*detection of fengycins in inhibitory amounts. Indirect interaction was mediated by the production of phenolic compounds involved in or derived from the defense-related phenylpropanoid metabolism upon pathogen attack (Ongena et al., 2005).

More recently the same set of secreted lipopeptides, fengycins, iturins, and surfactins, along with bacillomycin, was detected in the Gram-positive bacterium *Bacillus amyloliquefaciens* strain PPCB004. Bacteria were cultured in liquid medium; lipopeptides were extracted using n-butanol, followed by complete evaporation and resuspension in methanol. Growth assays showed that these peptides had inhibitory properties on the mycelial growth and/or spore germination of various post-harvest fungal pathogens, including *Alternaria citri, Botryosphaeria sp., Colletotrichum gloeosporioides, Fusicoccum aromaticum, Lasiodiplodia theobromae, Penicillium crustosum*, and*Phomopsis perse* (Arrebola et al., 2010; Table). In this study, iturin A demonstrated the strongest inhibitory effect. Post-harvest assays performed on orange fruits inoculated with *A. citri* and *C. gloeosporioides* displayed less disease incidence when treated with *B. amyloliquefaciens* strain PPCB004 either before or after inoculation. The mode-of-action for iturin A has been proposed to disrupt the fungal cytoplasmic membrane, creating transmembrane channels, resulting in the release of vital ions such as K^+, thus preventing spore germination and impairing mycelium development (Arrebola et al., 2010).

The lipopeptides bacillomycin D, fengycins A, and B are secreted by *B. amyloliquefaciens* strain Q-426 (Zhao et al., 2014; Table). They were recovered from cell-free supernatant of liquid cultures, further filtered and acid precipitated, prior to drying and methanol resuspension. Using two-dimensional (2-D) HPLC separation and tandem MS (MS/MS) analyses their molecular structure was elucidated. The peptide moiety comprised of 7–10 AA residues arranged in a cyclic structure, while the lipid moiety was composed of a chain of 14–17 FAs. Fengycin A purified from strain Q-426 disrupted the germination of spores from the fungal pathogen *F. oxysporum* f. sp. *spinaciae* in a dose dependent manner, with complete inhibition above 50 ug/mL. Fengycin A also modified hyphal growth, albeit without affecting cell membrane permeability (Zhao et al., 2014).

Along with iturin A, isoforms of bacillomycin D and of the macrolactin family, macrolactin A, 7-O-malonyl macrolactin A, and 7-O-succinyl macrolactin A were identified by HPLC-MS/MS analyses from the supernatant of centrifuged liquid cultures of *B. amyloliquefaciens* strain NJN-6 (Yuan et al., 2011, 2012; Table). This strain was isolated from the root system of a healthy banana plant. Plate assays showed that both purified bacillomycin D isoforms inhibited the hyphal growth of *F.*

oxysporum f. sp. *cubense*, a banana fungal pathogen. Similarly, all three macrolactins inhibited the growth of Gram-negative pathogenic bacteria*Ralstonia solanacearum*. The authors demonstrate that the macrolactin activities are maintained after 1 month at room temperature (Yuan et al.,2012); such shelf-life information is valuable for an in-field application strategy. Performing tests on plant tissues infected with these pathogens would have further validated the antifungal activity of these AMPs.

Fengycin A was also secreted by the Gram-positive bacterium *Bacillus atrophaeus* strain CAB-1 and showed antagonistic effect against airborne plant fungal pathogens *B. cinerea* and *Sphaerotheca fuliginea*. Two groups of lipopeptides were identified in strain CAB-1 by MS analyses: fengycins and unknown lipopeptides. CAB-1's lipopeptide extracts obtained by centrifugation of the culture broth followed by acid precipitation yielded no detectable iturin or surfactin compounds (Zhang et al., 2013). The fengycins produced by strain CAB-1 were a mixture of isoforms with various acyl side chain lengths from C15 to C17 (Table). The C16 isoform of fengycin A was secreted in greater abundance than C15 and C17 isoforms and therefore would contribute more to growth inhibition of cucumber powdery mildew, *S. fuliginea*, and tomato gray mold, *B. cinerea* (Zhang et al., 2013).

Mycosubtilin is a nonribosomal cyclic lipopeptide. Surfactin and mycosubtilin were purified from *B. subtilis* strains BBG131 and BBG125, respectively. Surfactin was produced through an integrated process in a bubbleless membrane bioreactor while mycosubtilin was produced using an overflowing fed-batch process. Subsequent steps of ultrafiltration, diafiltration, evaporation and freeze-dried were required for purification (Deravel et al., 2014). Surfactin and mycosubtilin combined were reported to have dose-dependent synergetic antibiotic effects on lettuce leaves sprayed with the peptide solutions prior to inoculation with its fungal obligate pathogen *Bremia lactucae* (Deravel et al., 2014; Table 2). This data shows that combining several AMPs opens even greater possibilities in terms of biocontrol strategies, yet it also poses challenges as to designing experiments assessing various AMP combinations and engineering plants resistant to microbial pathogens.

Fusaricidins and LI-F Lipopeptides from *Paenibacillus Polymyxa*bacterium

Fusaricidins and closely related LI-F lipopeptides are synthesized by*Paenibacillus polymyxa* and display inhibitory activities against plant pathogens (Table). These nonribosomal peptides consist of a guanidinylated β-hydroxy FA linked to a cyclic hexapeptide including four D configured-AA residues (Debois et al., 2013). Growth inhibition effects were also observed against a variety of microbes: *Rhodotorula aurantica* (yeast), *F. oxysporum, B. cinerea, Cladosporium cucumerinum* (fungi), *Phythium aphanidermatum* (oomycetes), and *P. syringae* (Gram negative bacterium). These AMPs were identified using a matrix-assisted laser desorption ionization (MALDI) imaging approach. By using a clever experimental design involving the insertion of a sterilized MALDI glass slide coated with indium tin oxide at the bottom of the Petri dish covered with a gelified sterile nutrient medium (Debois et al., 2013). *P. polymyxa* was streaked over the glass slide and*F. oxysporum* was inoculated nearby the slide and incubated for 11 days. The glass slide was removed from the Petri dish and completely dried under vacuum prior to MALDI imaging analysis. The authors were able to obtain spectra and fragmentation patterns of the compounds released by strain Pp56 and responsible for the inhibition of*F. oxysporum* mycelial development (Debois et al., 2013). The antagonistic interaction between fusaricidins and LI-F lipopeptides and*F. oxysporum* could thus be visualized by acquiring MS spectra of the different antibiotic compounds exhibiting distinct localizations along the slide. A time-course analysis revealed the early secretion of fusaricidin B, a mixture of LI-F05b/06b/08a, and LI-F08b, away from*F. oxysporum* hyphae, thus suggesting that their production was not triggered by the presence of the fungus. These antibiotics would be readily secreted and would accumulate in toxic amount outside *P. polymyxa* cells to deter any potential pathogen attack. Indeed their distribution patterns visualized by MALDI imaging coincided with the*F. oxysporum* hyphal inhibition zone observed on the culture plates (Debois et al., 2013).

2-amino-3-(Oxirane-2,3-Dicarboxamido)-Propanoyl-Valine (APV) from*Pantoea Agglomerans* Bacterium

APV was HPLC purified from a polar extract from liquid broth supernatant, analyzed using ESI-MS/MS and NMR, and identified as the main antibiotic compound of *Pantoea agglomerans* strain 48b/90 (Pa48b; Sammer et al., 2009). Pa48b is a Gram-negative bacterium that was isolated from soybean leaves which showed limited fire blight disease symptoms caused by the bacterium *Erwinia amylovora*. Using plate assays, APV inhibitory effect was tested against various microbial phytopathogen species, including *Agrobacterium tumefaciens, E. amylovora*, several *P. syringae* pathovars, *Serratia marcescens*, and *B. subtilis* (Table). APV successfully inhibited the growth of pathogens on the minimum synthetic medium in a dose dependent manner, but not on the complex medium, likely due to the presence in the latter of *N*-acetylglucosamine which would compensate for APV inhibitory effects (Sammer et al., 2009). This result suggests that the antagonistic effect of APV and *N*-acetylglucosamine on phytopathogen colonization depends on nutrient availability. In a follow-up study, the *APV*biosynthesis gene cluster was analyzed, and located onto a megaplasmid in Pa48b (Sammer et al., 2012). In this cluster, two genes are likely to be involved in APV biosynthesis regulation, whose transcription is tightly coordinated with translation to avoid precursor cytotoxicity. *In silico* sequence analysis of the *APV* gene cluster revealed a 99% identity to the diaminopropionate-peptide biosynthesis cluster of *P. agglomerans* strain CU0119. One of the genes of this cluster, *DdaI*, is predicted to be a transmembrane efflux pump mediating self-resistance to antibiotics (Sammer et al., 2012). *In planta*assays were then undertaken in which soybean leaves were first inoculated with *P. syringae* pv. *glycinea* prior to applying an APV solution onto the infected wounds. Consistent with the previous study, APV inhibitory effect was dose-dependent, yet only led to minor decrease of the disease symptoms. Therefore, APV was not confirmed to be the key antibiotic factor in the antagonism (Sammer et al., 2012).

Trichokonins (TK) from *Trichoderma Pseudokoningii* Fungus

The trichokonin (TK) family is composed of three major peptaibols, TKs VI, VII and VIII, produced by the ascomycota fungus *Trichoderma pseudokoningii* strain SMF2. Peptaibols are characterized by the presence of an unusual AA, a-aminoisobutyric acid, a C-terminal-hydroxylated and a N-terminal-acetylated AA. Because of their linear and amphipathic nature, peptaibols can form voltage-dependent ion channels in lipid bilayer membranes. TKs were obtained from solid state fermentation of *T. pseudokoningii* strain SMF2, followed by gel filtration of the crude extract and preparative HPLC separation. *In vitro*plate assays showed that TK VI exhibited antimicrobial activities against various pathogenic fungi and oomycetes, *Ascochyta citrullina, B. cinerea, F. oxysporum, Phytophthora parasitica*, and *V. dahlia* (Shi et al., 2012, Table). Toxicity assays using *F. oxysporum* protoplasts treated with TK VI showed the appearance of ROS, fragmentation of nuclear DNA, along with a change of fungal membrane permeability and disintegration of subcellular structures. The antimicrobial efficacy of TK was also tested on Chinese cabbage (*Brassica rapa*) leaves inoculated with the bacterial pathogen, *Pectobacterium carotovorum*subsp. *carotovorum* (Li et al., 2014). As observed by Shi et al. (2012), TK treatment led to an increase in the production of ROS, along with increased activities of pathogenesis-related proteins, and the activation of SA signaling pathway in the cabbage host. These data show that the TK family can induce cell death of pathogenic fungi and oomycetes as well as induce the activation of plant defense pathways.

Animal Ribosomal AMPs Validated in Transgenic Plants

Unlike nonribosomal AMPs, ribosomal AMPs offer the great advantage to be synthetised by genes that can be manipulated and inserted into a plant of interest. Only a few ribosomal AMPs underwent functional validation *in planta* for pest resistance and they all originate from animal species; they are illustrated in this section.

Penaeidin 4-1 from Shrimp

The antimicrobial peptide Penaeidin 4-1 (Pen4-1) was isolated from Atlantic white shrimp (*Litopenaeus setiferus*) under pathogen challenge. Pen4-1 is composed of 47 AAs including six cysteine residues forming three disulfide bridges (Cuthbertson et al., 2004; Table). Pen4-1 was purified using affinity chromatography as follows: pooled *L. setiferus* haemocyte extracts were concentrated using a SPE and applied to an affinity resin containing a Pen4-1-specific antibody. Pen4-1 can inhibit multiple plant pathogenic fungal species, such as *B. cinerea, Penicillium crustosum*, and *F. oxysporum*. Penaeidins harbor a unique two-domain structure, a proline rich N-terminal domain (PRD) and a cysteine-rich domain (CRD) with a stable alpha-helical structure, which might have contributed to its broad range of microbial targets, primarily Gram-positive bacteria and fungi. Transgenic lines of a commercial creeping bentgrass (*Agrostis stolonifera*) displayed enhanced resistance to the fungal pathogens *Sclerotinia homoecarpa* and *R. solani* as a result of Pen4-1 ectopic expression (Zhou et al., 2011; Table). Testing Pen4-1 into other plant hosts of *S. homoecarpa* and *R. solani* would confirm its potential as a biocontrol agent.

Metchnikowin (mtk), Thanatin, and Cecropin a from Insects

Metchnikowin (mtk) is an immune-inducible peptide synthesized in the fat body of*Drosophila melanogaster* as a 52-AA pre-pro-peptide upon microbial challenges (Levashina et al., 1995; Table). *In vitro* assays demonstrated that mtk inhibits the growth of drosophila pathogens, the Gram-positive bacterium *Micrococcus luteus* and the ascomycete fungus *Neurospora crassa*. The mtk AA sequence was determined by use of the Edman sequencing method as a proline-rich 26-AA peptide. MS analyses revealed two mtk isoforms of 3025 and 3045 Da, respectively, due to AA substitution (Rahnamaeian et al., 2009; Table). Low concentrations of mtk inhibited the *in vitro* growth of the pathogenic fungi *Fusarium graminearum* and *Fusarium culmorum*. Transgenic barley plants expressing the *D. melanogaster mtk* gene in its 52-AA pre-propeptide form under the control of the inducible*mannopine synthase* (*mas*) gene promoter were produced. Mas promoter's induction was triggered by wounding, plant growth hormones, as well as fungal infection. Mtk was successfully processed into its mature form and targeted to the apoplastic compartment in transgenic plants. When inoculated with *F. graminearum*, transgenic plants displayed higher frequencies of typical defense responses such as HR of attacked cells and the development of callose deposition at the cell wall underneath attempted penetration sites. This enhanced plant innate immune response was substantiated by the up-regulation of Pathogenesis-Related genes *PR-1* and *PR-5* (Rahnamaeian et al., 2009). The activation of systemic acquired resistance (SAR) was confirmed at the transcript level in a more recent follow-up study in which the mtk barley transgenic plants were infected with *B. graminis* f. sp. *hordei* and subjected to RT-PCR analyses (Rahnamaeian and Vilcinskas, 2012; Table). When the transgenic barley plants were infected with *F. graminearum*, mtk treatment impeded the development of functional haustorium, whose formation is crucial for commencement of biotrophic interaction.

Thanatin is produced by the stinkbug *Podisus maculiventri*; it consists of 21 AA residues (2.4 kDa) which form an internal disulphide bond important for its antimicrobial activity (Imamura et al., 2010; Table). A recombinant *thanatin* gene under the constitutive control of*cauliflower mosaic virus 35S* (*CaMV35S*) gene promoter was introduced into rice plants which were then exposed to blast disease (*Magnaporthe oryzae*). Thanatin was extracted from crude extracts of the transformants, purified using 2-D HPLC (cation exchange chromatography followed by RP chromatography) and its identity confirmed by MS. Although blast disease symptoms were observed on both the transgenic lines and the WT plants, diseased areas on the transformants were significantly smaller than those on the WT plants. While transgenic rice plants were not fully protected against *M. oryzae*, they had acquired partial resistance (Imamura et al., 2010). Very recently, *thanatin* gene was introduced into maize and controlled by the*ubiquitin-1* promoter which targets the expressed AMP to the plant apoplastic space (Schubert et al., 2015; Table). Transgenic maize plants were grown until ears were fully developed and the mature ears were then exposed to the pathogenic fungus *Aspergillus flavus*. The expression of *thanatin* in maize transformants led to a significant reduction in fungal biomass of *A. flavus* relative to that of WT plants (Schubert et al., 2015). The mode-of-action of thanatin in plants has yet to be fully understood.

Cecropin AMPs were first isolated from the haemolymph of the moth*Hyalophora cecropia*; cecropin A (37 AAs, 4 kDa; Table) exhibits a rapid, potent and long-lasting lytic activity against prominent bacterial and fungal phytopathogens (Bundó et al., 2014). Rice transgenic plants were obtained in which *cecropin A* expression was targeted to the seed endosperms by using the tissue-specific promoters *Glutelin B1* or*Glutelin B4*. Seeds from rice transgenic plants exhibited resistance to infection by fungal (*Fusarium verticillioides*) and bacterial (*Dickeya dadantii*) pathogens (Bundó et al., 2014). The mode-of-action of cecropin A in rice seeds remains to be investigated.

Magainin-2 (MAG) and Esculentin-1 (Esc28L) from Frogs

Magainins are ribosomal AMPs produced in the skin of *Xenopus* frogs. Growth assays in the presence of the 23-AA magainin-2 showed that the peptide inhibited the growth of various Gram-positive and Gram-negative bacteria, a few fungi and protozoa (Zasloff, 1987). The *mag* gene encoding for the MAGAININ-2 protein was ectopically over-expressed in pearl millet (*Pennisetum glaucum*) under the control of the constitutive *CaMV 35S* gene promoter and challenged with three strains of downy mildew (*Sclerospora graminicola* viz. Sg 384, Sg 445, and Sg 492; Ramadevi et al., 2014; Table). While some pearl millet transgenic lines exhibited a slight decrease in disease incidence 7 days post inoculation relative to the controls, none demonstrated full resistance (Ramadevi et al., 2014). Performing statistical analyses however would have helped test the significance of these results. The authors attribute the minor response of their genetically modified plants to the complexity of cell wall and cell membrane components of the oomycete pathogen, which are similar in composition and structure to those of host plant cells. Perhaps combining MAGAININ-2 with cell wall degrading enzymes, as tested by Fogliano et al. (2002) whom combined SRE or SP25A peptides with endochitinase and/or glucanase, would prove an efficient strategy.

Esculentins are highly potent AMPs of 46 AAs exclusively found in the skin secretion of the frog *Rana esculenta*. Esc28L is a variant of esculentin-1b artificially created by substituting the Methionine residue at position 28 with a Leucine and an additional Methionine in position 47 (Ponti et al., 1999). The idea of using variant peptides to increase plant pathogen resistance while targeting them to the extracellular space was first explored by Ponti et al. (1999, 2003; Table). Extracellular targeting not only eliminates potential toxicity of the variant AMP toward host cells, while permitting direct contact with pathogens growing and multiplying in the extracellular space. In a follow-up study, Esc28L was fused to the Signal Peptide (SP) sequence of*Phaseolus vulgaris* EndopolyGalacturonase-Inhibiting Protein (PGIP) to target Esc28L to the secretory pathway. Transgenic tobacco (*Nicotiana tabacum*) lines were produced in which the constitutive expression of Esc28L conferred enhanced resistance to the bacterial pathogens of tobacco, *P. syringae* pv. *tabaci* and *Pseudomonas aeruginosa*, as well as against the fungal pathogen *Phytophthora nicotianae*, and moreover demonstrated insecticidal effect against drosophila (Ponti et al., 2003).

Cathelicidin LL-37 Variant (Met37Leu) and Archaic Wallaby Antimicrobial (WAM) Peptide from Mammals

The human cathelicidin antimicrobial protein hCAP18 is synthesized in neutrophils as a preproprotein which comprises a conserved cathelin prodomain, and a non-conserved C-terminal peptide. The latter is enzymatically cleaved after secretion forming LL-37, a 37-AA functional antimicrobial peptide. Inspired by the work of Ponti et al. (2003), a mutated variant of LL-37 was created by substituting the Methionine residue at position 37 with a Leucine (LL-37 Met37Leu) and was fused to the SP sequence of PGIP to elicit secretion (Jung et al., 2012; Table). LL-37 Met37Leu was then overexpressed in Chinese cabbage plants. Transgenic lines were subsequently inoculated with various bacterial (P. carotovorum subsp. carotovorum) and fungal (F. oxysporum f. sp. lycopersici, Colletotrichum higginsianum, R. solani) pathogens. The transgenic plants displayed enhanced pathogen resistance with decreased disease symptoms relative to the controls (Jung et al., 2012). No mode-of-action was proposed in this study but since LL-37 Met37Leu is targeted to the extracellular space of the transgenic plants, it could be assumed that a direct interaction occurs between the peptide and the pathogen.

Mammal cathelicidins as a whole are worth investigating for plant engineering programs aiming at improving pathogen resistance. Fourteen, twelve, and eight divergent cathelicidin genes were identified in the wallaby, possum and platypus genomes, respectively. Of these, the proteins WAM 1 and WAM2, and Platypus AntiMicrobial (PAM) 1 and PAM2, were tested against various bacterial and yeast pathogens, and shown to be much more potent than LL-37 (Wang et al., 2011). A phylogenetic approach was then used to design an archaic WAM predicted to have originated 59 million years ago and ancestral to the major clade of marsupial AMPs including the modern WAM1 and WAM2 (Wang et al., 2011). In theory, it should be more difficult for the modern pathogens to overcome plant resistance mediated by an archaic AMPs.

Conclusions and Perspectives

Secreted peptides with antimicrobial activities are proving useful as biocontrol agents in agriculture in order to increase crop yields by minimizing quantity and quality losses due to pathogenic diseases. Many of the peptides discussed above have been used to create transgenic plants which showed increased resistance to pathogens in *in vivo* laboratory based experiments, for example Pep1, systemin, mtk, and Met37Leu.

Recently, field trials of transgenic cotton plants expressing the tobacco (*Nicotiana alata*) peptide NaD1, which targets PIP2 in the fungal membrane in order to cause membrane permeabilization, showed increase resistance to fungal pathogens (Gaspar et al., 2014). This peptide was previously shown to have antifungal activity against a variety of filamentous fungi *in vitro* and this translated well to *in vivo*field trials using transgenic cotton plants against *F. oxysporum* f. sp.*vasinfectum* and *V. dahlia* (Lay et al., 2003a; Gaspar et al., 2014). Among the peptides reviewed here many have shown some form of inhibition of pathogenic fungi, bacteria and insects. However, in many cases this inhibition was shown *in vitro* by addition of the peptide to the culture medium or *in vivo* by spraying of plant leaves or addition of the peptide to cut petioles or in soil; however all of these approaches are small scale, laboratory based and mainly use model plants (i.e., non crop species). For the majority of these peptides no work has been conducted to test the effect of priming plants on normal growth and development in crop species. In the case of the NaD1 transgenic cotton it was noted that no detrimental agronomic properties were observed in field trials whilst in contrast transgenic potato overexpressing the DF2 defensin altered plant development (Stotz et al., 2009). It should be noted that priming could potentially negatively impact crop performance as energy may be diverted from growth and development of the plant to the defense system. However, this was not the case for the NaD1 transgenic cotton, therefore only more field trials will be able to determine if this is a genuine concern. Moreover, without incorporation of the peptide into the genome as in the case of NaD1 transgenic cotton, crops would potentially have to be sprayed with the peptide of choice as they are with fungicides and pesticides at the moment. The subsequent cost for production and application should be economically assessed in comparison with the current chemical formulation used in fields, along with the impact on the environment.

The discovery of new valuable secreted peptides is associated with the advancing technology of MS, in particular in combination with HPLC separation, which is able to yield high throughput identification and structural characterization. The approach taken by Chen et al. (2014), for identification of the CAPE1 peptide, is an efficient way to use proteomics and MS to identify novel ribosomal and nonribosomal peptides. This technique allows a comparison between different stress conditions to investigate when peptides are induced, e.g., healthy plants vs. wounded, infected etc. Not only does this approach identify peptides but it also gives sequence information for the peptides which could help with subsequent identification of the proprotein. As data-rich as LC-MS experiments can be, they usually require time-consuming, labor-intensive extraction and separation methods. Furthermore, LC-MS does not typically provide information on the localization of such compounds. Therefore, MALDI imaging technology offers a promising alternative as it not only allows the identification of AMPs but also their *in situ* tissue localization. MALDI imaging has successfully been applied *in vitro* (Debois et al., 2013). Coupled with traditional histology, MS imaging informs on cellular localization with a resolution down to 5 μm of not only proteins, and peptides like AMPs, but also metabolites such as lipids, sugars, in a multiplex fashion within the very same tissue section simultaneously (Aichler and Walch, 2015). Such MS imaging methods would be greatly advantageous if directly applied on diseased plants organs, as compounds co-localizing with visible symptoms would be potential targets for the discovery of novel AMPs.

A crucial step in validating the efficacy of AMPs toward plant disease resistance involves introducing such AMPs into the crop of interest and exposing the transformed crop to its pathogens. To our knowledge, such transgenic experiments have only been attempted with ribosomal or artificially designed AMPs. This arises from the fact that transgenic plants are produced by introducing a foreign piece of DNA which then goes through the transcription and translation machinery to synthesize the AMP. Creating transgenic plants by introducing nonribosomal AMPs, which by definition do not undergo ribosomal synthesis, remains a complete challenge. Perhaps an alternative would be to introduce the enzymes responsible for the synthesis of these nonribosomal AMPs, the so-called NRPS, into the crop in order to acquire disease resistance. Using a synthetic biology approach it has been shown that NRPSs can be genetically engineered to improve the antibacterial properties of a lipopeptide and expand its spectrum against human pathogens (Nguyen et al., 2010). We could not find any reports in the literature related to plant resistance against pathogens. However, the link between NRPSs from biocontrol agents and host crops was recently established between the mycoparasite and facultative root symbiont *Trichoderma virens* and maize plants (Mukherjee et al.,2012). The analysis of the loss-of-function mutants of *T. virens* revealed that a hybrid enzyme polyketide synthase (PKS)/NRPS, Tex13, was involved in up-regulation of the defense gene *phenylalanine ammonia-lyase* (*pal)* in maize; Tex13 was more than 40-fold induced during interactions of *T. virens* with maize roots (Mukherjee et al., 2012).

In the evolutionary arms race between plant hosts and pathogens, usually the pathogen wins due to shorter generation times which make them more dynamic than crops, thereby rapidly overcoming plant immunity. Therefore, any tactic that would give the plant host the upper hand would be of immense interest to agricultural programs. One strategy could exploit the fact that different antibiotic peptides can act in synergy, therefore introducing several AMPs in crops would broaden the spectrum of plant disease resistance. Along with the AMPs, enzymes targeting microbial cell wall could also be introduced to the crop of interest to facilitate the entry of secreted AMPs across pathogen physical barriers. Another strategy would be exploring ancient and extinct peptides as archaic AMPs might be more effective than the modern AMPs found in living creatures because microbial pathogens have not been exposed to them for millions of years. Therefore, the modern pathogen would not have developed resistance against the archaic peptide. Engineering crops expressing such archaic AMPs would not only help achieve disease resistance but also slow down becoming overcome by the targeted pathogens (Wang et al., 2011). A last promising strategy is to use *de novo*-designed synthetic peptides. It is out of the scope of this review which focuses on naturally occurring AMPs, yet special mention should be made of two recent reports in which crops transformed with genes coding for synthetic peptides displayed acquired pathogen resistance (Nadal et al., 2012; Zeitler et al., 2013).

ACKNOWLEDGMENTS

The authors would like to thank Prof. Ben Cocks for fruitful feedback.

REFERENCES

1. Abdallah N. A., Shah D., Abbas D., Madkour M. (2010). Stable integration and expression of a plant defensin in tomato confers resistance

to *Fusarium* wilt. GM Crops. 1, 344–350. 10.4161/gmcr.1.5.15091

2. Aerts A. M., Carmona-Gutierrez D., Lefevre S., Govaert G., François I. E. J. A., Madeo F., et al. . (2009). The antifungal plant defensin RsAFP2 from radish induces apoptosis in a metacaspase independent way in *Candida albicans*. FEBS Lett. 583, 2513–2516. 10.1016/j.febslet.2009.07.004
3. Aichler M., Walch A. (2015). MALDI Imaging mass spectrometry: current frontiers and perspectives in pathology research and practice. Lab. Invest. 95, 422–431. 10.1038/labinvest.2014.156
4. Aleti G., Sessitsch A., Brader G. (2015). Genome mining: prediction of lipopeptides and polyketides from *Bacillus* and related Firmicutes. Comput. Struct. Biotechnol. J. 13, 192–203. 10.1016/j.csbj.2015.03.003
5. Almeida M. S., Cabral K. M., Zingali R. B., Kurtenbach E. (2000). Characterization of two novel defence peptides from pea (*Pisum sativum*) seeds. Arch. Biochem. Biophys. 378, 278–286. 10.1006/abbi.2000.1824
6. Arrebola E., Jacobs R., Korsten L. (2010). Iturin A is the principal inhibitor in the biocontrol activity of *Bacillus amyloliquefaciens* PPCB004 against postharvest fungal pathogens. J. Appl. Microbiol. 108, 386–395. 10.1111/j.1365-2672.2009.04438.x
7. Bhattacharya R., Koramutla M. K., Negi M., Pearce G., Ryan C. A. (2013). Hydroxyproline-rich glycopeptide signals in potato elicit signalling associated with defence against insects and pathogens. Plant Sci. 207, 88–97. 10.1016/j.plantsci.2013.03.002
8. Boller T. (2005). Peptide signalling in plant development and self/non-self perception. Curr. Opin. Cell Biol. 17, 116–122. 10.1016/j.ceb.2005.02.007
9. Bundó M., Montesinos L., Izquierdo E., Campo S., Mieulet D., Guiderdoni E., et al. . (2014). Production of cecropin A antimicrobial peptide in rice seed endosperm. BMC Plant Biol.14:102. 10.1186/1471-2229-14-102
10. Caño-Delgado A., Yin Y., Yu C., Vafeados D., Mora-García S., Cheng J. C., et al. . (2004). BRL1 and BRL3 are novel brassinosteroid receptors that function in vascular differentiation in *Arabidopsis*. Development 131, 5341–5351. 10.1242/dev.01403
11. Chen Y.-C., Siems W. F., Pearce G., Ryan C. A. (2008). Six peptide wound signals derived from a single precursor protein in*Ipomoea batatas* leaves activate the expression of the defence gene sporamin. J. Biol. Chem. 283, 11469–11476. 10.1074/jbc.M709002200
12. Chen Y.-L., Lee C.-Y., Cheng K.-T., Chang W.-H., Huang R.-N., Nam H. G., et al. . (2014). Quantitative peptidomics study reveals that a wound-induced peptide from PR-1 regulates immune signaling in tomato. Plant Cell. 26, 4135–4148. 10.1105/tpc.114.131185
13. Choi J., Moon E. (2009). Identification of novel bioactive hexapeptides against phytopathogenic bacteria through rapid screening of a synthetic combinatorial library. J. Microbiol. Biotechnol. 19, 792–802. 10.4014/jmb.0809.497
14. Choi M.-S., Kim Y.-H., Park H.-M., Seo B.-Y., Jung J.-K., Kim S.-T., et al. . (2009). Expression of BrD1, a plant defensin from*Brassica rapa*, confers resistance against brown planthopper (*Nilaparvata lugens*) in transgenic rices. Mol. Cells 28, 131–137. 10.1007/s10059-009-0117-9
15. Clay N. K., Nelson T. (2002). VH1, a provascular cell-specific receptor kinase that influences leaf cell patterns in Arabidopsis.Plant Cell 14, 2707–2722. 10.1105/tpc.005884
16. Cuthbertson B., Büllesbach E., Fievet J., Bachère E., Gross P. (2004). A new class (penaeidin class 4) of antimicrobial peptides from the Atlantic white shrimp (*Litopenaeus setiferus*) exhibits target specificity and an independent proline-richdomain function. Biochem. J. 381, 79–86. 10.1042/BJ20040330
17. Debois D., Ongena M., Cawoy H., De Pauw E. (2013). MALDI-FTICR MS imaging as a powerful tool to identify *Paenibacillus*antibiotics involved in the inhibition of plant pathogens. J. Am. Soc. Mass Spectrom. 24, 1202–1213. 10.1007/s13361-013-0620-2
18. Deravel J., Lemière S., Coutte F., Krier F., Van Hese N., Béchet M., et al. . (2014). Mycosubtilin and surfactin are efficient, low ecotoxicity molecules for the biocontrol of lettuce downy mildew. Appl. Microbiol. Biotechnol. 98, 6255–6264. 10.1007/s00253-014-5663-1
19. De Samblanx G. W., Goderis I. J., Thevissen K., Raemaekers R., Fant F., Borremans F., et al. . (1997). Mutational analysis of a plant defensin from radish (*Raphanus sativus L.*) reveals two adjacent sites important for antifungal activity. J. Biol. Chem.272, 1171–1179. 10.1074/jbc.272.2.1171
20. Dubos R. J. (1939a). Studies on a bactericidal agent extracted from a soil bacillus: I. Preparation of the agent. Its activity *in vitro*. J. Exp. Med. 70, 1–10. 10.1084/jem.70.1.1
21. Dubos R. J. (1939b). Studies on a bactericidal agent extracted from a soil bacillus: II. Protective effect of the bactericidal agent against experimental pneumococcus infection in mice. J. Exp. Med. 70, 11–17. 10.1084/jem.70.1.11
22. Farrokhi N., Whitelegge J. P., Brusslan J. A. (2008). Plant peptides and peptidomics. Plant Biotechnol. J. 6, 105–134. 10.1111/j.1467-7652.2007.00315.x
23. Felix G., Boller T. (1995). Systemin induces rapid ion fluxes and ethylene biosynthesis in *Lycopersicon peruvianum* cells. Plant J.7, 381–389. 10.1046/j.1365-313X.1995.7030381.x
24. Fogliano V., Ballio A., Gallo M., Woo S., Scala F., Lorito M. (2002). *Pseudomonas* lipodepsipeptides and fungal cell wall-degrading enzymes act synergistically in biological control. Mol. Plant Microbe Interact. 15, 323–333. 10.1094/MPMI.2002.15.4.323
25. Gao A. G., Hakimi S. M., Mittanck C. A., Wu Y., Woerner B. M., Stark D. M., et al. . (2000). Fungal pathogen protection in potato by expression of a plant defensin peptide. Nat. Biotechnol. 18, 1307–1310. 10.1038/82436

26. Gaspar Y. M., McKenna J. A., McGinness B. S., Hinch J., Poon S., Connelly A. A., et al. . (2014). Field resistance to *Fusarium oxysporum* and *Verticillium dahliae* in transgenic cotton expressing the plant defensin *NaD1*. J. Exp. Bot. 65, 1541–1550. 10.1093/jxb/eru021

27. Goyal R. K., Mattoo A. K. (2014). Multitasking antimicrobial peptides in plant development and host defence against biotic/abiotic stress. Plant Sci. 228, 135–149. 10.1016/j.plantsci.2014.05.012

28. Hegedüs N., Marx F. (2013). Antifungal proteins: more than antimicrobials? Fungal Biol. Rev. 26, 132–145. 10.1016/j.fbr.2012.07.002

29. Holaskova E., Galuszka P., Frebort I., Oz M. T. (2015).Antimicrobial peptide production and plant-based expression systems for medical and agricultural biotechnology. Biotechnol. Adv. 33, 1005–1023. 10.1016/j.biotechadv.2015.03.007

30. Holton N., Caño-Delgado A., Harrison K., Montoya T., Chory J., Bishop G. J. (2007). Tomato BRASSINOSTEROID INSENSITIVE1 is required for systemin-induced root elongation in *Solanum pimpinellifolium* but is not essential for wound signaling. Plant Cell 19, 1709–1717. 10.1105/tpc.106.047795

31. Huffaker A., Dafoe N. J., Schmelz E. A. (2011). ZmPep1, an ortholog of *Arabidopsis* elicitor peptide 1, regulates maize innate immunity and enhances disease resistance. Plant Physiol. 155, 1325–1338. 10.1104/pp.110.166710

32. Huffaker A., Pearce G., Ryan C. A. (2006). An endogenous peptide signal in *Arabidopsis* activates components of the innate immune response. Proc. Natl. Acad. Sci. U.S.A. 103, 10098–10103. 10.1073/pnas.0603727103

33. Huffaker A., Ryan C. A. (2007). Endogenous peptide defence signals in *Arabidopsis* differentially amplify signaling for the innate immune response. Proc. Natl. Acad. Sci. U.S.A. 104, 10732–10736. 10.1073/pnas.0703343104

34. Imamura T., Yasuda M., Kusano H., Nakashita H., Ohno Y., Kamakura T., et al. . (2010). Acquired resistance to the rice blast in transgenic rice accumulating the antimicrobial peptide thanatin. Transgenic Res. 19, 415–424. 10.1007/s11248-009-9320-x

35. Intitative T. A. G. (2000). Analysis of the genome sequence of the flowering plant *Arabidopsis thaliana*. Nature 408, 796–815. 10.1038/35048692

36. Jung Y. J., Lee S. Y., Moon Y. S., Kang K. K. (2012). Enhanced resistance to bacterial and fungal pathogens by overexpression of a human cathelicidin antimicrobial peptide (hCAP18/LL-37) in Chinese cabbage. Plant Biotechnol. Rep. 6, 39–46. 10.1007/s11816-011-0193-0

37. Lay F. T., Brugliera F., Anderson M. A. (2003a). Isolation and properties of floral defensins from ornamental tobacco and petunia. Plant Physiol. 131, 1283–1293. 10.1104/pp.102.016626

38. Lay F. T., Schirra H. J., Scanlon M. J., Anderson M. A., Craik D. J. (2003b). The three-dimensional solution structure of NaD1, a new floral defensin from *Nicotiana alata* and its application to a homology model of the crop defense protein alfAFP. J. Mol. Biol. 325, 175–188. 10.1016/S0022-2836(02)01103-8

39. Lee D. W., Kim B. S. (2015). Antimicrobial cyclic peptides for plant disease control. Plant Pathol. J. 31, 1–11. 10.5423/PPJ.RW.08.2014.0074

40. Lee H. T., Lee C. C., Yang J. R., Lai J. Z., Chang K. Y. (2015). A large-scale structural classification of antimicrobial peptides.Biomed Res. Int. 2015:475062. 10.1155/2015/475062

41. Levashina E. A., Ohresser S., Bulet P., Reichhart J. M., Hetru C., Hoffmann J. A. (1995). Metchnikowin, a novel immune-inducible proline-rich peptide from *Drosophila* with antibacterial and antifungal properties. Eur. J. Biochem. 233, 694–700. 10.1111/j.1432-1033.1995.694_2.x

42. Li H. Y., Luo Y., Zhang X. S., Shi W. L., Gong Z. T., Shi M., et al. . (2014). Trichokonins from *Trichoderma pseudokoningii*SMF2 induce resistance against Gram-negative *Pectobacterium carotovorum* subsp. carotovorum in Chinese cabbage. FEMS Microbiol. Lett. 354, 75–82. 10.1111/1574-6968.12427

43. Li Z., Zhou M., Zhang Z., Ren L., Du L., Zhang B., et al. . (2011). Expression of a radish defensin in transgenic wheat confers increased resistance to *Fusarium graminearum* and*Rhizoctonia cerealis*. Funct. Integr. Genomics 11, 63–70. 10.1007/s10142-011-0211-x

44. López-García B., Pérez-Payá E., Marcos J. F. (2002). Identification of novel hexapeptides bioactive against phytopathogenic fungi through screening of a synthetic peptide combinatorial library. Appl. Environ. Microbiol. 68, 2453–2460. 10.1128/AEM.68.5.2453-2460.2002

45. Malinowski R., Higgins R., Luo Y., Piper L., Nazir A., Bajwa V. S., et al. . (2009). The tomato brassinosteroid receptor BRI1 increases binding of systemin to tobacco plasma membranes, but is not involved in systemin signaling. Plant Mol. Biol. 70, 603–616. 10.1007/s11103-009-9494-x

46. Marahiel M. A. (2009). Working outside the protein-synthesis rules: insights into nonribosomal peptide synthesis. J. Pept. Sci.15, 799–807. 10.1002/psc.1183

47. Marx F. (2004). Small, basic antifungal proteins secreted from filamentous ascomycetes: a comparative study regarding expression, structure, function and potential application. Appl. Microbiol. Biotechnol. 65, 133–142. 10.1007/s00253-004-1600-z

48. McGurl B., Ryan C. A. (1992). The organization of the prosystemin gene. Plant Mol. Biol. 20, 405–409. 10.1007/BF00040600

49. Montesinos E. (2007). Antimicrobial peptides and plant disease control. FEMS Microbiol. Lett. 270, 1–11. 10.1111/j.1574-6968.2007.00683.x

50. Montoya T. (2002). Cloning the tomato Curl3 gene highlights the putative dual role of the Leucine-Rich Repeat receptor kinase tBRI1/SR160 in plant steroid hormone and peptide hormone signaling. Plant Cell. 14, 3163–3176. 10.1105/tpc.006379

51. Mousa W. K., Raizada M. N. (2015). Biodiversity of genes encoding anti-microbial traits within plant associated microbes.Front. Plant Sci. 6:231. 10.3389/fpls.2015.00231

52. Mukherjee P. K., Buensanteai N., Moran-Diez M. E., Druzhinina I. S., Kenerley C. M. (2012). Functional analysis of non-ribosomal peptide synthetases (NRPSs) in *Trichoderma virens*reveals a polyketide synthase (PKS)/NRPS hybrid enzyme involved in the induced systemic resistance response in maize.Microbiology 158, 155–165 10.1099/mic.0.052159-0

53. Muramoto N., Tanaka T., Shimamura T., Mitsukawa N., Hori E., Koda K., et al. . (2012). Transgenic sweet potato expressing thionin from barley gives resistance to black rot disease caused by *Ceratocystis fimbriata* in leaves and storage roots. Plant Cell Rep. 31, 987–997. 10.1007/s00299-011-1217-5

54. Nadal A., Montero M., Company N., Badosa E., Messeguer J., Montesinos L., et al. . (2012). Constitutive expression of transgenes encoding derivatives of the synthetic antimicrobial peptide BP100: impact on rice host plant fitness. BMC Plant Biol. 12:159. 10.1186/1471-2229-12-159

55. Narváez-Vásquez J., Pearce G., Ryan C. A. (2005). The plant cell wall matrix harbors a precursor of defence signaling peptides.Proc. Natl. Acad. Sci. U.S.A. 102, 12974–12977. 10.1073/pnas.0505248102

56. Navarro-García F., Eisman B., Fiuza S. M., Nombela C., Pla J. (2005). The MAP kinase Mkc1p is activated under different stress conditions in *Candida albicans*. Microbiology 151, 2737–2749. 10.1099/mic.0.28038-0

57. Nguyen K. T., He X., Alexander D. C., Li C., Gu J. Q., Mascio C., et al. . (2010). Genetically engineered lipopeptide antibiotics related to A54145 and daptomycin with improved properties.Antimicrob. Agents Chemother. 54, 1404–1413. 10.1128/AAC.01307-09

58. Oard S. V., Enright F. M. (2006). Expression of the antimicrobial peptides in plants to control phytopathogenic bacteria and fungi.Plant Cell Rep. 25, 561–572. 10.1007/s00299-005-0102-5

59. Ongena M., Jacques P., Touré Y., Destain J., Jabrane A., Thonart P. (2005). Involvement of fengycin-type lipopeptides in the multifaceted biocontrol potential of *Bacillus subtilis*. Appl. Microbiol. Biotechnol. 69, 29–38. 10.1007/s00253-005-1940-3

60. Patel S., Ahmed S., Eswari J. S. (2015). Therapeutic cyclic lipopeptides mining from microbes: latest strides and hurdles.World J. Microbiol. Biotechnol. 31, 1177–1193. 10.1007/s11274-015-1880-8

61. Pearce G., Bhattacharya R., Chen Y.-C., Barona G., Yamaguchi Y., Ryan C. A. (2009). Isolation and characterization of hydroxyproline-rich glycopeptide signals in black nightshade leaves. Plant Physiol. 150, 1422–1433. 10.1104/pp.109.138669

62. Pearce G., Moura D. S., Stratmann J., Ryan C. A. (2001).Production of multiple plant hormones from a single polyprotein precursor. Nature 411, 817–820. 10.1038/35081107

63. Pearce G., Ryan C. A. (2003). Systemic signaling in tomato plants for defence against herbivores. Isolation and characterization of three novel defence-signaling glycopeptide hormones coded in a single precursor gene. J. Biol. Chem. 278, 30044–30050. 10.1074/jbc.M304159200

64. Pearce G., Siems W. F., Bhattacharya R., Chen Y.-C., Ryan C. A. (2007). Three hydroxyproline-rich glycopeptides derived from a single petunia polyprotein precursor activate defensin I, a pathogen defence response gene. J. Biol. Chem. 282, 17777–17784. 10.1074/jbc.M701543200

65. Pearce G., Strydom D., Johnson S., Ryan C. A. (1991). A polypeptide from tomato leaves induces wound-inducible proteinase inhibitor proteins. Science 253, 895–897. 10.1126/science.253.5022.895

66. Pearce G., Yamaguchi Y., Barona G., Ryan C. A. (2010). A subtilisin-like protein from soybean contains an embedded, cryptic signal that activates defence-related genes. Proc. Natl. Acad. Sci. U.S.A. 107, 14921–14925. 10.1073/pnas.1007568107

67. Ponti D., Mangoni M. L., Mignogna G., Simmaco M., Barra D. (2003). An amphibian antimicrobial peptide variant expressed in*Nicotiana tabacum* confers resistance to phytopathogens.Biochem. J. 370, 121–127. 10.1042/bj20021444

68. Ponti D., Mignogna G., Mangoni M. L., De Biase D., Simmaco M., Barra D. (1999). Expression and activity of cyclic and linear analogues of esculentin-1, an anti-microbial peptide from amphibian skin. Eur. J. Biochem. 263, 921–927 10.1046/j.1432-1327.1999.00597.x

69. Poon I. K. H., Baxter A. A., Lay F. T., Mills G. D., Adda C. G., Payne J. A., et al. . (2014). Phosphoinositide-mediated oligomerization of a defensin induces cell lysis. Elife 3:e01808. 10.7554/eLife.01808

70. Portieles R., Ayra C., Gonzalez E., Gallo A., Rodriguez R., Chacón O., et al. . (2010). NmDef02, a novel antimicrobial gene isolated from *Nicotiana megalosiphon* confers high-level pathogen resistance under greenhouse and field conditions. Plant Biotech. J. 8, 678–690. 10.1111/j.1467-7652.2010.00501.x

71. Rahnamaeian M., Langen G., Imani J., Khalifa W., Altincicek B., von Wettstein D., et al. . (2009). Insect peptide metchnikowin confers on barley a selective capacity for resistance to fungal ascomycetes pathogens. J. Exp. Bot. 60, 4105–4114. 10.1093/jxb/erp240

72. Rahnamaeian M., Vilcinskas A. (2012). Defence gene expression is potentiated in transgenic barley expressing antifungal peptide Metchnikowin throughout powdery mildew challenge. J. Plant Res. 125, 115–124. 10.1007/s10265-011-0420-3

73. Ramadevi R., Rao K. V., Reddy V. D. (2011). Antimicrobial peptides and production of disease resistant transgenic plants, inPests and Pathogens: Management Strategies, eds Vudem D. R., Poduri N. R., Khareedu V. R., editors. (CRC Press;), 379–452.

74. Ramadevi R., Rao K. V., Reddy V. D. (2014). *Agrobacterium tumefaciens*-mediated genetic transformation and production of stable transgenic pearl millet (*Pennisetum glaucum* [L.] R. Br.).In Vitro Cell. Dev. Biol. Plant. 50, 392–400. 10.1007/s11627-013-9592-y

75. Ramamoorthy V., Zhao X., Snyder A. K., Xu J.-R., Shah D. M. (2007). Two mitogen-activated protein kinase signalling cascades mediate basal resistance to antifungal plant defensins in*Fusarium graminearum*. Cell. Microbiol. 9, 1491–1506. 10.1111/j.1462-5822.2006.00887.x

76. Rebollar A., Marcos J. F., López-García B. (2014). Screening of a synthetic peptide combinatorial library to identify inhibitors of the appressorium formation in *Magnaporthe oryzae*. Biochem. Biophys. Res. Commun. 454, 1–6. 10.1016/j.bbrc.2014.09.145

77. Ryan C. A., Pearce G. (1998). Systemin: a polypeptide signal for plant defensive genes. Annu. Rev. Cell Dev. Biol. 14, 1–17. 10.1146/annurev.cellbio.14.1.1

78. Ryan C. A., Pearce G. (2003). Systemins: a functionally defined family of peptide signals that regulate defensive genes in*Solanaceae* species. Proc. Natl. Acad. Sci. U.S.A. 100(Suppl.), 14577–14580. 10.1073/pnas.1934788100

79. Sagaram U. S., El-Mounadi K., Buchko G. W., Berg H. R., Kaur J., Pandurangi R. S., et al. . (2013). Structural and functional studies of a phosphatidic acid-binding antifungal plant defensin MtDef4: identification of an RGFRRR motif governing fungal cell entry. PLoS ONE 8:e82485. 10.1371/journal.pone.0082485

80. Sammer U. F., Reiher K., Spiteller D., Wensing A., Völksch B. (2012). Assessment of the relevance of the antibiotic 2-amino-3-(oxirane-2,3-dicarboxamido)-propanoyl-valine from *Pantoea agglomerans* biological control strains against bacterial plant pathogens. Microbiologyopen 1, 438–449. 10.1002/mbo3.43

81. Sammer U. F., Völksch B., Möllmann U., Schmidtke M., Spiteller P., Spiteller M., et al. . (2009). 2-amino-3-(oxirane-2,3-dicarboxamido)-propanoyl-valine, an effective peptide antibiotic from the epiphyte *Pantoea agglomerans* 48b/90. Appl. Environ. Microbiol. 75, 7710–7717. 10.1128/AEM.01244-09

82. Schaller A., Oecking C. (1999). Modulation of plasma membrane H+-ATPase activity differentially activates wound and pathogen defence responses in tomato plants. Plant Cell. 11, 263–272.

83. Scheer J. M., Pearce G., Ryan C. A. (2003). Generation of systemin signaling in tobacco by transformation with the tomato systemin receptor kinase gene. Proc. Natl. Acad. Sci. U.S.A. 100, 10114–10117. 10.1073/pnas.1432910100

84. Scheer J. M., Ryan C. A. (2002). The systemin receptor SR160 from *Lycopersicon peruvianum* is a member of the LRR receptor kinase family. Proc. Natl. Acad. Sci. U.S.A. 99, 9585–9590. 10.1073/pnas.132266499

85. Schmelz E. A., Carroll M. J., LeClere S., Phipps S. M., Meredith J., Chourey P. S., et al. . (2006). Fragments of ATP synthase mediate plant perception of insect attack. Proc. Natl. Acad. Sci. U.S.A. 103, 8894–8899. 10.1073/pnas.0602328103

86. Schmelz E. A., LeClere S., Carroll M. J., Alborn H. T., Teal P. E. A. (2007). Cowpea chloroplastic ATP synthase is the source of multiple plant defence elicitors during insect herbivory. Plant Physiol. 144, 793–805. 10.1104/pp.107.097154

87. Schubert M., Houdelet M., Kogel K.-H., Fischer R., Schillberg S., Nölke G. (2015). Thanatin confers partial resistance against aflatoxigenic fungi in maize (*Zea mays*). Transgenic Res. 24, 885–895. 10.1007/s11248-015-9888-2

88. Shi M., Chen L., Wang X. W., Zhang T., Zhao P. B., Song X. Y., et al. (2012). Antimicrobial peptaibols from *Trichoderma pseudokoningii* induce programmed cell death in plant fungal pathogens. Microbiology 158(Pt 1), 166–175. 10.1099/mic.0.052670-0

89. Shiu S. H., Bleecker A. B. (2003). Expansion of the receptor-like kinase/Pelle gene family and receptor-like proteins in*Arabidopsis*. Plant Physiol. 132, 530–543. 10.1104/pp.103.021964

90. Stotz H. U., Spence B., Wang Y. (2009). A defensin from tomato with dual function in defence and development. Plant Mol. Biol.71, 131–143. 10.1007/s11103-009-9512-z

91. Stratmann J. W. (2003). Long distance run in the wound response–jasmonic acid is pulling ahead. Trends Plant Sci. 8, 247–250. 10.1016/S1360-1385(03)00106-7

92. Terras F. R., Schoofs H. M., De Bolle M. F., Van Leuven F., Rees S. B., Vanderleyden J., et al. . (1992). Analysis of two novel classes of plant antifungal proteins from radish (*Raphanus sativus L.*) seeds. J. Biol. Chem. 267, 15301–15309.

93. Thevissen K., Warnecke D. C., François I. E. J. A., Leipelt M., Heinz E., Ott C., et al. . (2004). Defensins from insects and plants interact with fungal glucosylceramides. J. Biol. Chem. 279, 3900–3905. 10.1074/jbc.M311165200

94. Trivilin A. P., Hartke S., Moraes M. G. (2014). Components of different signalling pathways regulated by a new orthologue of AtPROPEP1 in tomato following infection by pathogens. Plant Pathol. 63, 1110–1118. 10.1111/ppa.12190

95. van der Weerden N. L., Bleackley M. R., Anderson M. A. (2013).Properties and mechanisms of action of naturally occurring antifungal peptides. Cell. Mol. Life Sci. 70, 3545–3570. 10.1007/s00018-013-1260-1

96. van der Weerden N. L., Lay F. T., Anderson M. A. (2008). The plant defensin, NaD1, enters the cytoplasm of *Fusarium oxysporum* hyphae. J. Biol. Chem. 283, 14445–14452. 10.1074/jbc.M709867200

97. Verma S. S., Yajima W. R., Rahman M. H., Shah S., Liu J.-J., Ekramoddoullah A. K. M., et al. . (2012). A cysteine-rich antimicrobial peptide from *Pinus monticola* (PmAMP1) confers resistance to multiple fungal pathogens in canola (*Brassica napus*). Plant Mol. Biol. 79, 61–74. 10.1007/s11103-012-9895-0

98. Vincent D., Bedon F. (2013). Secretomics of plant-fungus associations: more secrets to unravel. J. Plant Biochem. Physiol.1:e117 10.4172/2329-9029.1000e117

99. Vriens K., Cammue B. P. A., Thevissen K. (2014). Antifungal plant defensins: mechanisms of action and production. Molecules19, 12280–12303. 10.3390/molecules190812280

100. Wang G., Mishra B., Lau K., Lushnikova T., Golla R., Wang X. (2015). Antimicrobial peptides in 2014. Pharmaceuticals 8, 123–150. 10.3390/ph8010123

101. Wang J., Wong E. S., Whitley J. C., Li J., Stringer J. M., Short K. R., et al. . (2011). Ancient antimicrobial peptides kill antibiotic-resistant pathogens: Australian mammals provide new options.PLoS ONE 6:e24030. 10.1371/journal.pone.0024030

102. Wäspi U., Blanc D., Winkler T., Ruedi P., Dudler R. (1998).Syringolin A, novel peptide elicitor from *Pseudomonas syringae*pv. syringae that induces resistance to *Pyricularia oryzae* in rice. Mol. Plant Microbe Interact. 11, 727–733. 10.1094/MPMI.1998.11.8.727

103. Wäspi U., Schweizer P., Dudler R. (2001). Syringolin reprograms wheat to undergo hypersensitive cell death in a compatible interaction with powdery mildew. Plant Cell 13, 153–161. 10.1105/tpc.13.1.153

104. Yamaguchi Y., Barona G., Ryan C. A., Pearce G. (2011). GmPep914, an eight-amino acid peptide isolated from soybean leaves, activates defence-related genes. Plant Physiol. 156, 932–942. 10.1104/pp.111.173096

105. Yamaguchi Y., Huffaker A. (2011). Endogenous peptide elicitors in higher plants. Curr. Opin. Plant Biol. 14, 351–357. 10.1016/j.pbi.2011.05.001

106. Yamaguchi Y., Huffaker A., Bryan A. C., Tax F. E., Ryan C. A. (2010). PEPR2 is a second receptor for the Pep1 and Pep2 peptides and contributes to defence responses in *Arabidopsis*.Plant Cell. 22, 508–522. 10.1105/tpc.109.068874

107. Yamaguchi Y., Pearce G., Ryan C. A. (2006). The cell surface leucine-rich repeat receptor for AtPep1, an endogenous peptide elicitor in *Arabidopsis*, is functional in transgenic tobacco cells.Proc. Natl. Acad. Sci. U.S.A. 103, 10104–10109. 10.1073/pnas.0603729103

108. Yuan J., Li B., Zhang N., Waseem R., Shen Q., Huang Q. (2012).Production of bacillomycin- and macrolactin-type antibiotics by*Bacillus amyloliquefaciens* NJN-6 for suppressing soilborne plant pathogens. J. Agric. Food Chem. 60, 2976–2981. 10.1021/jf204868z

109. Yuan J., Raza W., Huang Q., Shen Q. (2011). Quantification of the antifungal lipopeptide iturin A by high performance liquid chromatography coupled with aqueous two-phase extraction. J. Chromatogr. B Anal. Technol. Biomed. Life Sci. 879, 2746-2750. 10.1016/j.jchromb.2011.07.041

110. Zasloff M. (1987). Magainins, a class of antimicrobial peptides from *Xenopus* skin: isolation, characterization of two active forms, and partial cDNA sequence of a precursor. Proc. Natl. Acad. Sci. U.S.A. 84, 5449–5453. 10.1073/pnas.84.15.5449

111. Zeitler B., Herrera Diaz A., Dangel A., Thellmann M., Meyer H., Sattler M., et al. . (2013). *De-novo* design of antimicrobial peptides for plant protection. PLoS ONE 8:e71687. 10.1371/journal.pone.0071687

112. Zhang X., Li B., Wang Y., Guo Q., Lu X., Li S., et al. . (2013).Lipopeptides, a novel protein, and volatile compounds contribute to the antifungal activity of the biocontrol agent *Bacillus atrophaeus* CAB-1. Appl. Microbiol. Biotechnol. 97, 9525–9534. 10.1007/s00253-013-5198-x

113. Zhao P., Quan C., Wang Y., Wang J., Fan S. (2014). *Bacillus amyloliquefaciens* Q-426 as a potential biocontrol agent against*Fusarium oxysporum* f. sp. spinaciae. J. Basic Microbiol. 54, 448–456. 10.1002/jobm.201200414

114. Zhou M., Hu Q., Li Z., Li D., Chen C. F., Luo H. (2011). Expression of a novel antimicrobial peptide Penaeidin4-1 in creeping bentgrass (*Agrostis stolonifera* L.) enhances plant fungal disease resistance. PLoS ONE 6:e24677. 10.1371/journal.pone.0024677

115. Zhu X., Li Z., Xu H., Zhou M., Du L., Zhang Z. (2012).Overexpression of wheat lipid transfer protein gene TaLTP5 increases resistances to *Cochliobolus sativus* and *Fusarium graminearum* in transgenic wheat. Funct. Integr. Genomics 12, 481–488. 10.1007/s10142-012-0286-z

CHAPTER 17

TRANSGENIC APPROACHES TO ALTERING CARBON AND NITROGEN PARTITIONING IN WHOLE PLANTS: ASSESSING THE POTENTIAL TO IMPROVE CROP YIELDS AND NUTRITIONAL QUALITY

Umesh P. Yadav,[1] Brian G. Ayre,[1] and Daniel R. Bush[2]

[1] Department of Biological Sciences, University of North Texas, Denton, TX, USA

[2] Department of Biology, Colorado State University, Fort Collins, CO, USA

ABSTRACT

The principal components of plant productivity and nutritional value, from the standpoint of modern agriculture, are the acquisition and partitioning of organic carbon (C) and nitrogen (N) compounds among the various organs of the plant. The flow of essential organic nutrients among the plant organ systems is mediated by its complex vascular system, and is driven by a series of transport steps including export from sites of primary assimilation, transport into and out of the phloem and xylem, and transport into the various import-dependent organs. Manipulating C and N partitioning to enhance yield of harvested organs is evident in the earliest crop domestication events and continues to be a goal for modern plant biology. Research on the biochemistry, molecular and cellular biology, and physiology of C and N partitioning has now matured to an extent that strategic manipulation of these transport systems through biotechnology are being attempted to improve movement from source to sink tissues in general, but also to target partitioning to specific organs. These nascent efforts are demonstrating the potential of applied biomass targeting but are also identifying interactions between essential nutrients that require further basic research. In this review, we summarize the key transport steps involved in C and N partitioning, and discuss various transgenic approaches for directly manipulating key C and N transporters involved. In addition, we propose several experiments that could enhance biomass accumulation in targeted organs while simultaneously testing current partitioning models.

INTRODUCTION

The principal components of plant productivity and nutritional value, from the standpoint of modern agriculture, are the acquisition and partitioning of organic carbon (C) and nitrogen (N) compounds among the various organs of the plant. The initial step of C acquisition is the fixation of atmospheric C dioxide by photosynthesis followed by sugar partitioning to the non-photosynthetic tissues. N is primarily absorbed as inorganic salts from the soil solution and incorporated into amino acids in the root or shoot for subsequent transport to import-dependent tissues. These "sinks" are the primary consumers of newly assimilated sugars and amino acids, and include many essential organ systems such as expanding leaves, roots, seeds and fruits, storage tissue and organs, and secondary growth. As much as 80% of the C assimilated in mature leaves, and comparable amounts of amino acids, are exported in the phloem to satisfy the metabolic needs of the heterotrophic tissues (Kalttorres et al., 1987). Understanding the mechanisms and regulation of C and N partitioning to

Citation: Umesh P. Yadav, Brian G. Ayre and Daniel R. Bush, Transgenic approaches to altering carbon and nitrogen partitioning in whole plants: assessing the potential to improve crop yields and nutritional quality, doi: 10.3389/fpls.2015.00275.

these heterotrophic tissues has always been a central theme to improving crop productivity. Moreover, human manipulation of C and N partitioning is evident from the earliest examples of crop domestication. A good example is the selective concentration of carbohydrate and storage proteins in the larger kernels and inflorescences of maize during domestication from ancestral teosinte as much as 10,000 years ago (Doebley, 2004). Likewise, the dramatic yield increases of the green revolution were, among other advances, based on the selection of smaller stature plants that invested more biomass in reproductive tissue versus vegetative growth. Indeed, yield trials of early twentieth century winter wheat versus late twentieth century cultivars grown in a common garden experiment showed that C assimilation per acre was the same for both varieties, yet the yields of the modern cultivars outperformed the early varieties by 40% because more C was directed into reproductive growth (seeds) versus vegetative tissues (Gifford et al., 1984). Yield enhancement of harvested organs continues to be a primary objective, whether the intent is edible organs for feed/food consumption, fiber for textiles and composites, wood for pulp, paper and building, or general biomass for biofuels. Nutritional and other quality enhancements of feed/food crops are also primary objectives that are in part dependent on C and N partitioning.

The last several decades have produced substantial biochemical and molecular advances in our understanding of cell-to-cell and long-distance partitioning of sugars and amino acids in plants. These advances have created novel opportunities to manipulate partitioning pathways to enhance yield and/or nutritional quality. The basic premise of these approaches is that altering the expression of the transporters for a desired sugar or amino acid will result in the preferential accumulation of that compound (or a derivative thereof) in a target organ or tissue. The success of these efforts are predicated on accurate knowledge of the physiological and biochemical processes involved to predict the outcome. The goal of this review is to briefly summarize current understanding of the functional contributions of sugar and amino acid transport systems to plant growth and then focus on the limited number of studies where ectopic expression has been attempted. In many of these, the desired outcome has been at least partially achieved, and it is apparent that identifying the limitation—that is, why the full desired effect was not obtained—is necessary to achieve greater gains. Thus, the application of existing knowledge for biotechnology gains has in many cases identified new avenues of biological discovery in the complex interaction of nutrient use and growth. In addition, this review ventures experimental approaches for improving and targeting biomass partitioning to specific organs.

Carbohydrates: Transport Mechanisms and Efforts to Manipulate Partitioning by Engineering Transport

Sucrose (Suc) is the primary form of reduced C transported long-distance in the vascular system of plants, and consequently has garnered the most attention. Suc is produced in the cytoplasm, either directly from the products of photosynthesis or from storage reserves. Once produced, it may remain in the cytoplasm to participate in other aspects of cellular metabolism or it may be compartmentalized, principally in the vacuole; it may pass through the symplasm—the collective cytoplasm of the plant—via plasmodesmata (PD); or it may undergo efflux to the apoplasm—the collective cell-wall space of the plant—and from there it may be taken up into the cytoplasm of an adjacent cell. As a relatively large polar solute, Suc requires protein transporters for efficient movement across membranes. Where the prevailing concentration gradients mandate energized transport, Suc/H^+symporters (Suc uptake transporters, SUT; or Suc uptake carriers, SUC; the SUT designation is used here, unless referring to a specific gene) that utilize the proton motive force for Suc/H^+ symport are well characterized biochemically, genetically, and physiologically (Bush, 1993; Braun and Slewinski, 2009; Ayre, 2011). Those SUTs that are best characterized function at the plasma membrane to pump Suc into cells from the apoplasm, but members of one SUT sub-family localizes to the tonoplast to use the prevailing proton gradient to move Suc from inside the vacuole out to the cytoplasm (Endler et al., 2006; Reinders et al., 2008; Eom et al., 2011). Suc/H^+ antiport to load Suc into vacuoles is supported by physiological studies and patch clamping with mutant*Arabidopsis* lines indicate a role for TMT1 and TMT2 in this process (Schulz et al., 2011). Under conditions where passive transport would suffice, a new family of sugar transporters, SWEETs, were recently identified that have mechanisms characteristic of facilitated diffusion (Chen et al., 2010, 2012; Chen, 2014).

The primary fate of photoassimilated C is long-distance transport from photoautotrophic source leaves to heterotrophic organs. C partitioning via long-distance transport of sugars in the phloem has been extensively and recently reviewed (Turgeon and Wolf, 2009; Ayre, 2011; Braun et al., 2014; Chen, 2014) and is thus only surveyed here. Among mesophyll cells, Suc appears to move cell to cell relatively freely through the PD of the symplasm, and then enters the minor-vein phloem for long-distance transport through the sieve tube elements by bulk flow (Figure). Bulk flow occurs when a hydrostatic pressure difference between source and sink tissues is large enough to drive flux through the sieve elements, with the pressure difference primarily energized by solute accumulation in source-leaf phloem. Establishing a sufficient solute concentration in source phloem is generically referred to as phloem loading, and three mechanisms are proposed: (1) In apoplasmic phloem loading, Suc exits the mesophyll symplasm in the vicinity of the phloem via SWEET proteins located principally on the plasma membrane of presumptive phloem parenchyma cells. Suc is then accumulated against a concentration gradient from the apoplasm into the companion cell/sieve element complex of the phloem by SUTs (Figure). Because uptake is energized by the proton motive force, Suc can accumulate to high levels such that the total solute concentration in source phloem can readily exceed 1 Osm. Sugar alcohols (polyols) are prominent transport sugars in some species, and appear to be loaded into the phloem from the apoplasm by proton symporters in a mechanism equivalent to Suc loading from the apoplasm (Noiraud et al., 2001; Gao et al., 2003; Reidel et al., 2009). (2) In the polymer trap mechanism, Suc diffuses from the mesophyll symplasm into the phloem through specialized, highly-branched PD, and a portion is then converted to raffinose family oligosaccharides (RFO), primarily the trisaccharide raffinose and tetrasaccharide stachyose, but also, in some species, longer chain oligosaccharides verbascose and ajugose. These highly branched PD, located between bundle sheath and intermediary cells (specialized companion cells), are proposed to have a precise size exclusion limit that allows diffusion of Suc but not larger RFOs (Figure). RFO synthesis in the intermediary cells thus maintains a diffusion-friendly Suc concentration while generating the high osmolarities necessary for the hydrostatic pressure that drives long-distance transport through the sieve tubes (Turgeon, 1996). (3) In passive loading, source leaf mesophyll cells accumulate high concentrations of Suc, which enters the phloem passively through regular PD (Figure). In this mechanism, there is not an energized step for concentrating solute into the companion cell/sieve element complex, and the high turgor required for bulk flow through the sieve elements is maintained throughout the leaf (Rennie and Turgeon, 2009; Slewinski and Braun, 2010). It should be noted that some species can employ multiple phloem loading pathways in the same vascular bundle (Voitsekhovskaja et al., 2009) and that different mechanisms can be favored depending on the plant's specific needs (Gil et al., 2011).

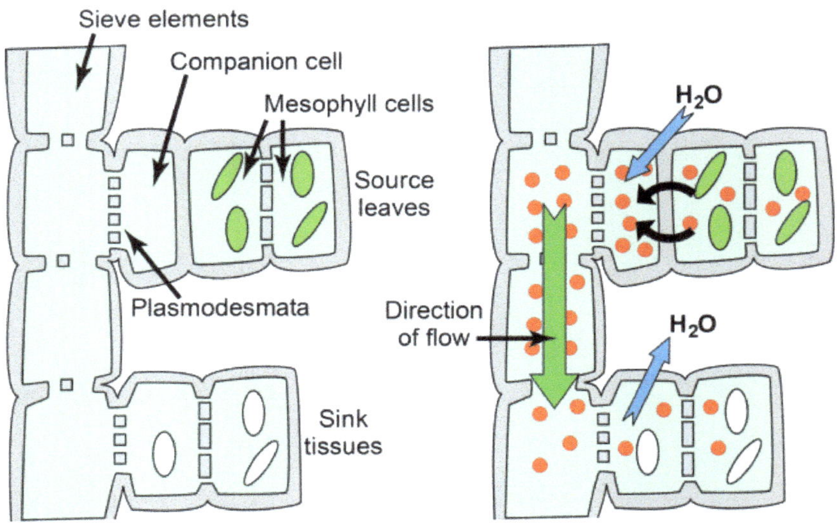

Figure 1: Phloem anatomy and osmotically generated-pressure flow. *Left*: Schematic of phloem connecting source leaves (represented as mesophyll cells with green chloroplasts) and heterotrophic sink tissues (represented as cells with white amyloplasts). Sieve elements .

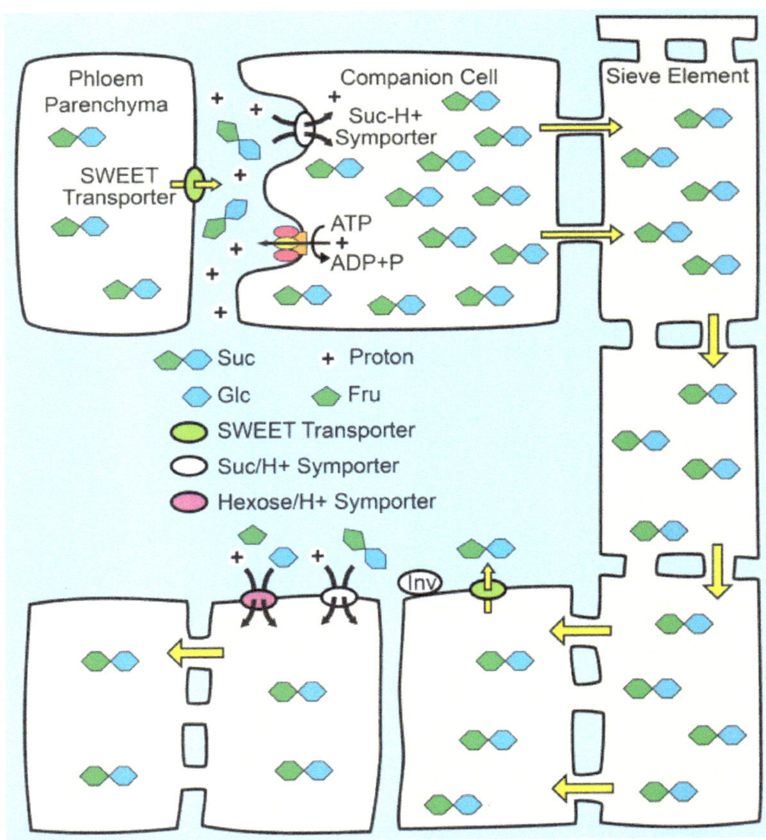

Figure 2: Phloem loading from the apoplasm. *Top*: Suc from mesophyll cells—principally phloem parenchyma cells—enters the apoplasm in the vicinity of the companion cell/sieve element complex via SWEET transporters. Suc, and in some species, polyols, .

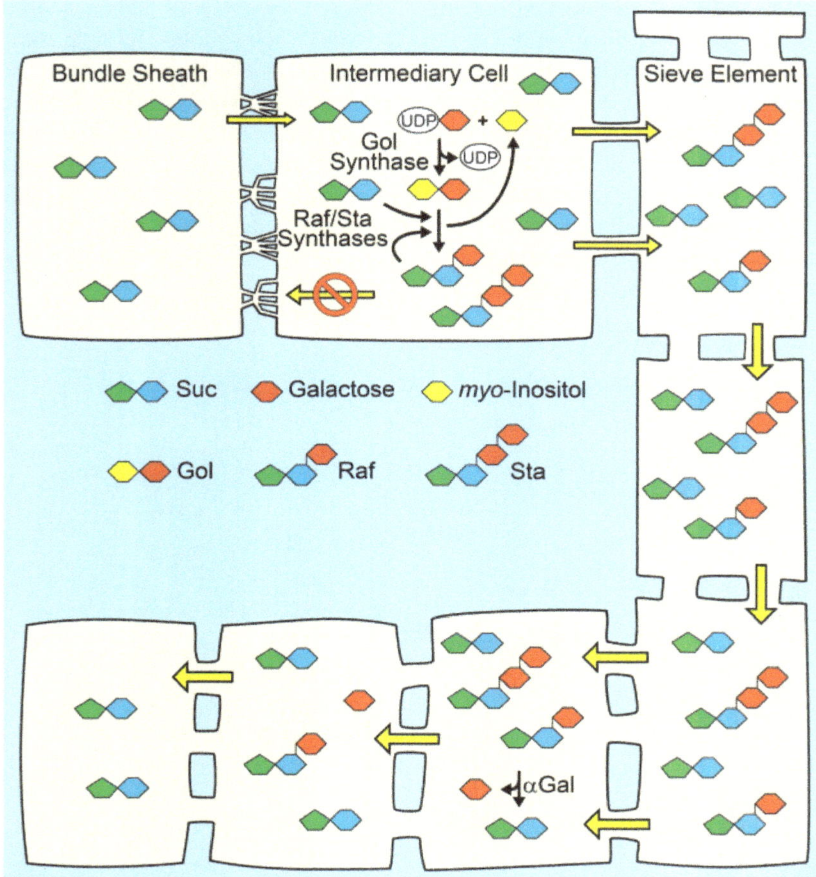

Figure 3: Polymer-trap model for loading through the symplasm. *Top*: Suc from mesophyll diffuses into intermediary cells (specialized companion cells) through highly branched plasmodesmata situated between the intermediary cells and bundle sheath cells, where a .

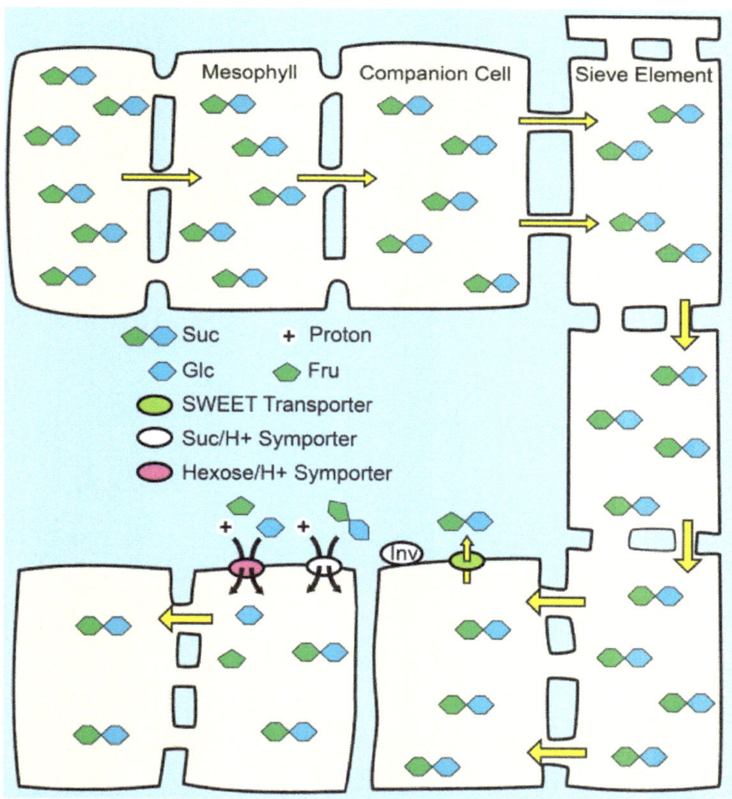

Figure 4: Passive loading through the symplasm. *Top*: Source leaves of species that use passive loading have regular plasmodesmata connections from the mesophyll cells through to the sieve elements and have high solute concentrations throughout the leaf. Consequently, .

Phloem unloading occurs all along the transport path to nourish lateral tissues and storage reserves, and occurs extensively in the sink organs. Unloading may be by efflux across plasma membranes to the apoplasm, or through the symplasm via PD, or both routes may be used in combination. Within the recipient tissues, symplasmic continuity with the CC/SE complex can be extensive for efficient post-phloem partitioning via PD, or isolated domains may force apoplasmic transport (Patrick, 1997). Generalizing, growing vegetative tissues, such as young leaves and root tips, tend to have extensive symplasmic unloading (Oparka et al., 1999; Stadler et al., 2005b). This contrasts with developing seeds, where symplasmic isolation between maternal and filial tissues forces apoplasmic unloading: short distance symplasmic unloading occurs from the phloem strands entering the integument/testa tissue, from where nutrients enter the apoplasm for subsequent uptake into the embryo and endosperm (Wang and Fisher, 1994; Patrick and Offler, 2001; Stadler et al., 2005a). In addition, the principal route of unloading, and the extent of the symplasmic and apoplasmic domains in the post-phloem pathway, may change during organ development, as exemplified in maturing fruit and developing embryos (Ruan and Patrick, 1995; Ruan et al., 2001; Kim et al., 2005). Once released to the apoplasm, Suc may be retrieved back into the phloem or into adjacent cells by SUTs, or may be hydrolyzed by cell wall invertases, and recovered by hexose/H^+ transporters.

The complexity of the loading and unloading pathways involved in C partitioning, and their importance to plant physiology, are in part reflected in the size of the gene families involved. *Arabidopsis* has nine SUTs arranged in three sub-families, and monocots have an additional one or two families not represented in dicots (Sauer, 2007; Braun and Slewinski, 2009; Reinders et al., 2012). The monosaccharide transporter-like (*MST*) gene family of *Arabidopsis* has 53 members, arranged in seven sub-families. The best characterized of these sub-families, and the one with members that localize to the plasma membrane and are thus most likely involved in intercellular transport, is the sugar transporter (STP) family, which contains 14 members (Büttner, 2007, 2010). The SWEET family of mono- and disaccharide transporters contains 17 genes in *Arabidopsis* and 21 in rice, arranged in four clades, with clade I mediating mainly Glc import and export, and clade III transporting mainly Suc (Chen et al., 2010, 2012). Plants have three types of invertases, localizing to the cell wall/apoplasm, cytoplasm, and vacuole, respectively, with six members in the cell wall invertase subfamily of *Arabidopsis* (Roitsch and Gonzalez, 2004). This list only includes the major candidate participants in membrane-mediated C partitioning, and not those involved in polymer synthesis or PD function.

Engineering Partitioning by Manipulating Apoplasmic Loading

Strategic manipulation of C partitioning may lead to more productive plants and enable specific organs to be targeted for enhanced yield. Ectopic expression of *SUTs* from constitutive or tissue-specific promoters, in either sink or source organs, has been attempted in various plants. These studies demonstrate a clear potential, but also emphasize that the natural patterns of Suc flux need to be considered. In potato, constitutive over-expression of spinach *SoSUT1* from the CaMV 35S promoter was used in an effort to increase transport to tubers (Leggewie et al., 2003). This gave higher levels of starch and lower levels of sugar in leaves, but had little impact on tubers yields. These effects were likely caused by futile cycling, in which Suc released to the apoplasm was recovered by mesophyll cells because the promoter drove expression in the mesophyll as well as in phloem. Ectopic expression of potato *StSUT1* in excised pea cotyledon storage parenchyma enhanced Suc influx in a *StSUT1*-dependent fashion (roughly twofold). However, Suc uptake into intact cotyledons was only ~23%, and when biomass accumulation was measured, a similar difference between excised and intact cotyledons was observed (Rosche et al., 2002). This study showed that ectopic *SUT* expression can be used for targeted yield enhancements, but the authors also argue that since most Suc is absorbed at the cotyledon surface by epidermal transfer cells, the full effect of overexpressing *StSUT1* in storage parenchyma was not realized in intact seedlings since the "extra" transporters were not on the transfer cells exposed to apoplasmic Suc (Rosche et al., 2002). As another example of manipulating SUT activity for biotechnology, overexpression of a barley SUT (*HvSUT1*) using an endosperm-specific *Hordein B1* promoter in wheat grains increased levels of storage protein, showing that enhanced Suc transport has positive impacts beyond carbohydrate alone (Weichert et al., 2010). Finally, over-expression of rice *OsSUT5Z* in potato using a tuber-specific, class-I patatin promoter was reported to enhance tuber yield by increasing tuber numbers rather than tuber size. This was not the anticipated outcome, and the authors suggest that altering Suc flux stimulated development of more stolons (Sun et al., 2011).

The above studies emphasize constitutive or sink-specific manipulation, and attempt to "pull" more Suc into the target tissue. Enhancing phloem transport by manipulating *SUTs* involved in phloem loading has been put forward as a "push" mechanism to improve plant productivity. For example, increasing source to sink transport was proposed as a means to enhance crop productivity since more carbohydrate would be sent to sink organs for growth and/or storage, and there would be less Suc-mediated product inhibition on photosynthesis (Ainsworth and Bush, 2011). In addition, Suc loading is modulated, both up and down, in response to the physiological and environmental needs (Vaughn et al., 2002), and it was proposed that heterologous promoters that are uncoupled from the natural regulation may be useful to keep loading rates constantly high (Srivastava et al., 2009).

Despite this, there are few published attempts to enhance phloem transport by over-expressing SUTs in the phloem. The cDNA of*Arabidopsis AtSUC1* (Wippel and Sauer, 2012) and barley *HvSUT1*(Reinders et al., 2012) were cloned downstream of the companion cell-specific *AtSUC2* promoter, and both restored phloem loading in*Arabidopsis Atsuc2 –/–* mutants. These studies showed foreign transporters could restore WT growth to the mutant, but detailed analysis of C transport and impact on growth was not pursued. In a different approach, *AtSUC2* cDNA was expressed from a companion cell-specific promoter derived from Commelina yellow mottle virus (*CoYMVp*) in a homozygous *Atsuc2-4* background (Srivastava et al., 2009). *CoYMVp* is as strong as the natural *AtSUC2* promoter, but is subject to different regulatory cascades: while SUT promoters naturally involved in phloem loading are repressed by Suc accumulation (Vaughn et al., 2002; Usadel et al., 2008), *CoYMVp* is activated by Suc (Dasgupta et al., 2014). This over-expression construct rescued *Atsuc2-4* mutants, showing that foreign promoters can maintain high levels of SUT activity in conditions where expression might normally be repressed (Srivastava et al., 2009).

A more recent study fused *CoYMVp* to alternative *SUTs*, with the aim of keeping phloem loading levels high by uncoupling the natural transcriptional control via the foreign promoter and potential post-translational control with different proteins. Of the *SUTs* tested,*AtSUC1*, *AtSUC2*, and *ZmSUT1* (a *Zea mays* gene from a different *SUT*subfamily) rescued phloem loading in the *Atsuc2-4* mutant; several other *SUT* genes implicated in high affinity Suc uptake were unable to restore efficient phloem transport (Dasgupta et al., 2014). When *SUTs*that did rescue *Atsuc2-4* were overexpressed in WT plants, enhanced phloem loading and transport to heterotrophic organs was evident, but improved growth and primary productivity was not observed. Rather, the plants were stunted. Separate research suggested a link between sugar transport and phosphate requirements (Lei et al., 2011), and this was tested in lines overexpressing *SUTs* from *CoYMVp*. The growth inhibition was accompanied by increased expression of phosphate-starvation induced genes, and was reversed by providing a higher supply of external phosphate (Dasgupta et al., 2014). These findings argue that "hyperloading" the phloem by overexpressing *SUTs* have the detrimental effect of disrupting C/phosphate homeostasis. The implications for agriculture are that efforts to enhance photosynthesis and growth by increasing phloem transport may be imperiled by the plant's perception that it needs more P, unless the links between C and P homeostasis are better understood and uncoupled.

What future experiments could be done to enhance partitioning via apoplasmic loading? Molecular characterization of SWEET and SUT transporters provides, in principle, a complete framework to test hypotheses on the strategic release of Suc from one cell and loading into the adjacent cell. An obvious experiment is to overexpress *SWEETs*in phloem parenchyma cells, particularly *Arabidopsis SWEET11 and SWEET12* (Chen et al., 2012), for enhanced Suc efflux to the apoplasm with simultaneous *SUT* overexpression in companion cells for increased uptake, and then test the impact on growth and whole plant C partitioning. Precise cell-specific deployment will be critical for accurate interpretation of the results, and a caveat to the success of these experiments is the resolution with which the outcome is measured (see below). Simply expecting larger plants in general, or larger target organs, is naïve, even though this may be the desired outcome from an applied standpoint. Instead, these experiments will more likely help identify rate-limiting steps in the overall process of whole-plant partitioning, and thus lead to new prospects (targets) for manipulation, and uncover unrecognized interactions of C homeostasis and other aspects of physiology. A corollary to these experiments aiming to enhance productivity are those that disrupt partitioning to test prevailing and alternative models of C allocation.

Manipulation of source-leaf phloem loading would energize the entire pathway, but by itself does not target resources to desired (i.e., harvested) organs. Targeted sink-specific manipulations may therefore be more fruitful for enhancing yield of specific organs, as evident from overexpression of *StSUT1* in storage parenchyma of pea cotyledons and*HvSUT1* in wheat endosperm, described above. But here too, the complete pathway of Suc unloading, from the sieve elements to the final recipient cells need to be considered (Patrick, 1997). For example, developing embryos and endosperm are symplasmically isolated from maternal tissues, and all nutrient transfer has two obligate membrane transport steps: efflux from maternal cells and influx into filial cells (Zhou et al., 2009). Tissue-specific *SWEET* over expression in integuments/testa of developing seeds may enhance sugar availability at the interface between maternal and filial tissues, but optimally positioned uptake carriers may be required to take advantage of the available resources.

Similarly, Suc unloading to the apoplasm in sink organs is associated with hydrolysis by cell wall invertase and uptake of the resulting hexose by hexose transporters. This is particularly prevalent in fruits undergoing rapid expansion and ripening (Ruan and Patrick, 1995;Patrick, 1997; Zhang et al., 2006, 2007; Jin et al., 2009). As the fruit ripens and accumulates more solutes (simple sugars, etc.), apoplasmic unloading prevents symplasmic "backflow." In addition, hydrolysis of Suc to Glc and Fru doubles the osmolarity to benefit hydrostatic expansion; enhances the Suc gradient between symplasm and apoplasm to favor further passive unloading; and also prevents the retrieval of unloaded Suc back into the phloem by SUTs, which are expressed all along the phloem path.

Although the potential for manipulating the "pull" capacity of sinks is clear, little is known on how manipulation of hexose transporters could be advantageous to increase plant yield. Over-expression of sugar transporter *AtSTP13* in *Arabidopsis* seedlings led to increases in Glc uptake, and also led to increases in the sugars and biomass throughout the plants (Schofield et al., 2009). In tomato, RNAi-mediated knockdown of the high affinity hexose transporter gene *LeHT* led to a 55% decrease in fruit hexose accumulation, implicating *LeHT* in driving accumulation of hexoses into storage parenchyma cells during tomato fruit ripening (McCurdy et al., 2010). Among those hexose transporters that have been characterized genetically, many of the knock out mutants have no visible phenotype. The expression patterns of *Arabidopsis STP* genes, as determined by classical hybridization and PCR approaches, as well as analysis of genome-wide transcriptome data, overwhelmingly place *STP* expression in various sink organs (Büttner, 2010) and this speaks strongly to their potential for targeting biomass to specific sinks by transgenic approaches. Genetic manipulation of additional sugar transporter genes using constitutive and tissue-specific promoters is needed to test this potential. Since manipulating one aspect of the transport system may have limited impact, coupling the targeted expression of genes encoding SWEETs, invertases, and hexose transporters may be more fruitful to capture the full potential of this strategy.

However, manipulating the transporters involved in apoplasmic phloem loading and unloading carries the inherent risk of increasing the apoplasmic concentrations of Suc and other nutrients used by pathogens. Pathogens have been shown to enhance SWEET expression to make reduced C more available (Chen et al., 2010, 2012; Chen, 2014). Several lines of genetic and metabolic evidence implicate sugar transporters in plant defense responses during plant-pathogen interactions. This includes reprogramming carbohydrate metabolism at the site of infection for reduced Suc export, increased Suc hydrolysis by cell wall invertase, and enhanced import of hexose by hexose transporters (Proels and Hückelhoven, 2014).

Engineering Polymer Trapping Metabolism

Other efforts to enhance phloem loading and long-distant transport have relied on the fact that there are different phloem loading mechanisms, and have attempted to superimpose an alternative mechanism on top of the natural mechanism of the host plant. Specifically, polymer trapping biochemistry was introduced to two species, potato (Hannah et al., 2006) and *Arabidopsis* (Cao et al., 2013), both of which load from the apoplasm with SUTs. The aims of these experiments were twofold: (1) to assess the efficiency of RFO synthesis in the phloem and (2) to gage the efficiency with which RFO from the companion cells enters the translocation stream for long-distance transport. In addition, growth and development of the engineered plants were monitored. In both studies metabolic engineering was used to produce galactinol and raffinose in companion cells. In *Arabidopsis*, stachyose was also engineered. Both studies used companion cell-specific promoters to express genes encoding galactinol synthase and raffinose synthase, and in*Arabidopsis*, stachyose synthase. Thus, all the RFO biosynthesis occurred in the phloem after the SUT-mediated phloem loading step. Despite the high concentrations of Suc available in these cells, RFO synthesis and transport in both studies was low. When *Arabidopsis* was photosynthetically labeled with $^{14}CO_2$ for 20 min followed by a 10 min chase period, only 2% of the label was incorporated into RFO, while Suc contained up to 80% and up to 30% was in Glc and Fru. This outcome is surprising and in sharp contrast to plants that transport RFO naturally, such as coleus (Turgeon and Gowan, 1992), cucurbits (Beebe and Turgeon, 1992), and catalpa (Otto, 1968; Turgeon and Medville, 2004), in which RFOs become quickly labeled to high specific activity.

Based on current models, this result is puzzling, and neither study had a satisfying explanation for the low rates of RFO synthesis other than to suggest that RFO biochemistry, especially in relation to phloem transport, was not straightforward. RFO synthesis should be efficient since apoplasmic loaders have ample reduced C in the companion-cell cytoplasm and the engineered proteins are thought to localize to the cytoplasm (Keller and Pharr, 1996); transport should be efficient because the PD-pore units between companion cells and sieve elements are open to diffusion of 10 kDa dextrans (Kempers and Van Bel, 1997;Knoblauch and Van Bel, 1998) and 67 kDa proteins (Stadler et al., 2005b). Further work on the biochemistry in companion cells of apoplasmic loaders and intermediary cells of polymer trap loaders is required to resolve why apoplasmic loaders do not effectively produce and transport RFOs. The precursors for galactinol, UDP-Gal and myo-inositol, or flux through the pathways leading to these, may be insufficient for higher level production.

Alternatively, the inability to produce high levels of RFO in companion cells may be a cell biology problem rather than a biochemical problem. In addition to many highly branched PD, intermediary cells of plants employing the polymer trap mechanism have many small vacuoles and extensive endomembrane systems, and are larger than companion cells found in similarly-sized leaf veins of plants that load from the apoplasm (Turgeon et al., 1993, 2001). A potential function for the extensive internal membranes and vacuoles in intermediary cells has not been put forward, but low rates of RFO synthesis in companion cells suggests a role for these membranes in RFO synthesis. Enzyme localization, stability, and/or interaction with cellular co-factors in companion cells relative to intermediary cells could therefore be a problem. It is worth noting that true intermediary cells and polymer trapping appear to have evolved independently multiple times, arguing that the internal structures are essential for function, and not species-specific characteristics (Turgeon et al., 2001).

If RFO synthesis in companion cells could be engineered to levels that contribute to hydrostatic pressure in the source leaf, then efficient RFO metabolism/utilization could be engineered in specific sinks for targeted partitioning (Figure). Expression of an alkaline α-galactosidase gene specifically in desired sink organs (Carmi et al., 2003), for example, would convert transported RFOs to Suc and galactose, which could then be converted to Glc (Barber et al., 2006; Dai et al., 2006) to enter "regular" pathways of primary metabolism. Since RFOs are not abundant transport sugars in apoplasmic loading species, competing, off-target sinks not engineered for efficient RFO catabolism may accumulate the RFO sugars (Figure). The principle here is to alter the hydrostatic pressure gradients that drive phloem transport throughout the plant: RFO utilization in target organs will result in low hydrostatic pressures and efficient nutrient import, whereas RFO accumulation in competing sinks that cannot metabolize the RFOs will have higher hydrostatic pressure that will decrease bulk flow to that tissue. With the strength of competing sinks reduced, more nutrients may be available for biomass accumulation in target sinks.

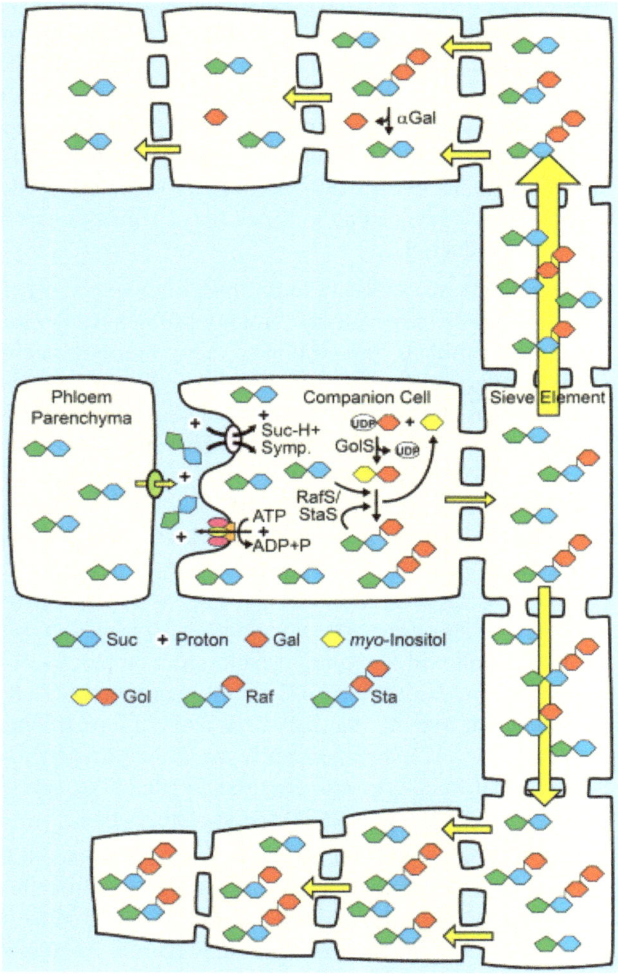

Figure 5: A proposal for combining apoplasmic loading with polymer trap biochemistry and sink-specific digestion of RFO sugars. *Middle*: Companion cells of source leaves loading from the apoplasm with Suc/H$^+$symporters are engineered to convert a portion of Suc .

As a corollary to superimposing polymer trap chemistry on apolasmic loaders, superimposing apoplasmic loading onto intermediary cells may enhance partitioning in species that load by the polymer trap mechanism. Species that phloem load by polymer trapping do not have an obligate need for SUTs or Suc in the apoplasm (McCaskill and Turgeon, 2007). Notwithstanding, Suc is present in the apoplasm and many species with intermediary cells also have "regular" companion cells that express genes encoding SUTs (McCaskill and Turgeon, 2007;Voitsekhovskaja et al., 2009). Cucurbits, and probably other polymer trap species, appear to be able to switch between apoplasmic and polymer trap loading under certain conditions, particularly virus infection (Gil et al., 2011). Engineering *SWEET* expression in phloem parenchyma and *SUT* expression in the intermediary cells may increase Suc availability for RFO synthesis and consequently improve phloem loading and transport. It would also test predictions of the polymer trap model. Polymer trapping proposes that the branched PD allow Suc to diffuse down its concentration gradient. If this is correct, it follows that Suc brought into intermediary cells by SUT mediated loading should be able to diffuse back out. If on the other hand, Suc accumulates in the intermediary cells, it would speak against the polymer trap model.

Engineering Passive Loading

Passive loading species are proposed to have the highest solute concentration in the mesophyll cells, and have PD with relatively large size-exclusion limits from the mesophyll all the way to the sieve tubes. Sugars and other nutrients are thus free to move along concentration gradients from the mesophyll into the sieve elements (Turgeon and Medville, 1998; Rennie and Turgeon, 2009). There is no energized concentrating step, and thus, speaking literally, there is no actual "loading" step. If this prevailing model is correct, specifically that PD with large size-exclusion limits connect the mesophyll and phloem cells to facilitate Suc movement without an apolasmic step nor a polymer trapping step, then it predicts that neither engineering a SWEET/SUT pairing to promote apoplasmic transport, nor polymer engineering to trap sugars in the phloem would improve loading and transport: The model predicts that any efforts to accumulate solute in the phloem would be thwarted by escape through the PD back to the mesophyll. Ironically, testing the passive-loading premise is exactly why these experiments should be conducted: If solutes do accumulate in the phloem after these manipulations, it would argue that the model is incorrect.

Nitrogen: Transport Mechanisms and Efforts to Manipulate Partitioning by Engineering Transport

Although nitrate and ammonia are the primary forms of N acquired from the soil solution by plants, "amino acids are the currency of N exchange" in the physiology of plant growth (Bush, 1999). Inorganic N taken up from the soil is either assimilated into amino acids in the root, or they are transported to photosynthetically active leaf tissue in the xylem via the transpiration stream. Amino acids synthesized from inorganic N in the root are used for basic metabolic needs of root cells, but a substantial amount of those amino acids are transported into the xylem where they make their way to the photosynthetically active leaves in the transpiration stream. In the leaf, amino acids are used for basic metabolism and protein synthesis. However, as generally mature organs, these leaves do not need a large percentage of those amino acids for metabolism. Thus, amino acids arriving from the roots are transported into the leaf phloem where they move to the many heterotrophic tissues of the plant, that include young leaves just building the photosynthetic machinery, roots, stem tissues, storage organs, and developing seeds. When inorganic N from the soil moves directly to the leaves in the xylem, it is assimilated into amino acids in the mesophyll and then, much like the amino acids arriving from the root, they are transported into the phloem for systemic distribution to import-dependent sinks. Taken together, it is clear that amino acids are the primary form of transported N supporting the growth and development of multicellular plants (Bush, 1999).

The systemic distribution of amino acids requires the combined activity of many amino acid transport proteins. Example transport steps include export from xylem parenchyma in the root into the xylem transpiration stream, transport into mesophyll cells, export from mesophyll cells, transport into and out of the phloem, and transport into the heterotrophic cells of all the import-dependent sinks. Remarkable advances in our understanding of the transport properties and identity of plant amino acid transporters in the past 25 years was initially enabled by detailed biochemical descriptions using purified plasma membrane vesicles (Bush, 1993) followed by the successful cloning of a wide array of amino acid transporter gene families (Fischer et al., 1998; Ortiz-Lopez et al., 2000; Tegeder, 2012). To date, at least six families of amino acid transporters have been identified in plants with more than 60 genes encoding putative amino acid transporters. These transporters are functionally differentiated by their transport properties (such as substrate specificity and transport mechanism) and expression patterns that are regulated by both developmental and environmental cues (Liu and Bush, 2006; Tegeder, 2012, 2014). The majority of the transporters described to date are involved with active uptake into the cell, and only two reports describe bidirectional transporters (BAT1 and SIAR1) that may be involved in amino acid export (Dundar and Bush, 2009; Ladwig et al., 2012). BAT1's putative role in amino acid export from the cell is based on its activity as a facilitated carrier and on a vascular tissue-localized expression pattern of a GUS expressing gene-trap inserted in the *BAT1* gene (Dundar and Bush, 2009). However, subsequent localization of the BAT1 protein using a BAT1::GFP reporter localized the carrier to the mitochondria, thereby suggesting a primary role in intracellular amino acid metabolism (Michaeli et al., 2011). In contrast, SIAR1 was localized to the plasma membrane of vascular tissue, thus supporting its role in amino acid export (Ladwig et al., 2012).

Given the complexity of amino acid circulation in the plant, a reasonable hypothesis suggests that altering the pattern and/or timing of the expression of one or more transporters could have a positive impact on plant growth and nutritional quality. For example, Rolletschek et al. (2005) documented a 20% increase in seed N content and seed size as a result of the ectopic expression of the *Vicia faba VfAAP1* amino acid transporter gene in the developing seeds of pea and *Vicia narbonensis*using the seed specific *legumin B4* promoter. These results suggest that amino acid transport into the storage parenchyma of the developing seed is rate limiting. Interestingly, there was also an overall increase in plant biomass, suggesting targeted changes in N allocation in one tissue can have a broader impact on plant growth. In a subsequent report, the transgenic peas were grown in field trials over two seasons (Weigelt et al., 2008). Seed N content was increased as before, but there were compensatory decreases in seed starch content and seed size. Transcript and metabolite profiling indicated the transgenic seeds were experiencing C limitations as more amino acids were being synthesized. Several pathways associated with coordinated C and N metabolism localized to the mitochondria were up-regulated, suggesting metabolic adjustments in response to a changing N:C ratio (Weigelt et al., 2008).

Tegeder and colleagues took a different approach in attempting to increase the sulfur content of legume seeds (Tan et al., 2010). In these experiments, the yeast *S*-methylmethionine permease 1 gene (*MMP1*) was expressed in transgenic pea under the control of the *Arabidopsis AAP1* promoter which directs expression throughout the phloem and in the seeds. *S*-methylmethionine (SMM) transport was targeted because it is identified as a major form of phloem mobile S in plants (Bourgis et al., 1999). *MMP1* expressing plants averaged a 33% increase in biomass, a 19% increase in seed number, a 31% increase in total seed N and a 19% increase in total seed S. There was also an average 67% increase in xylem and 60% increase in leaf SMM. Individual seed S content was unaltered. Analysis of the expression levels of genes involved in SMM biosynthesis suggested increased SMM synthesis in the root. That suggests SMM is transported to the leaf in the xylem, which was supported by the observed increased SMM xylem content. The increase in seed N content was driven by a 33% increase in phloem amino acid content. The surprising result in this experiment is the impact of altered SMM transport on N metabolism and yield. While S content per seed was unaltered, seed N content was increased, suggesting SMM abundance may have a regulatory impact on plant N metabolism. Indeed, SMM represents only 0.2% of xylem and phloem amino acids, suggesting it could be playing a role as a metabolic signal reporting on the S:N ratio in the plant. It is particularly interesting that two experiments that manipulate amino acid distribution in the plant (Weigelt et al., 2008; Tan et al., 2010) have had significant impacts on metabolic pathways that are associated with balancing the relative abundance of three important essential elements (N, C, and S). These observations are perhaps not surprising given the complex interactions between the functional roles of these central nutrients in plant metabolism.

Other approaches for manipulating N distribution in the plant might include ectopic expression of a desired amino acid transporter in the leaf phloem using a promoter that drives companion cell/sieve element specific expression. Under the expectation that anything loaded in the phloem moves with mass flow to actively growing sinks, this could increase the N content of harvested organs, such as mature seeds. Ectopic expression in the leaf phloem might be coupled to increased expression of a bidirectional transporter in the mesophyll that could increase the efflux of the desired amino acid into the apoplast. Likewise, one could simultaneously use a transgenic approach to increase the biosynthesis of a desired amino acid in the mesophyll, thereby increasing the pool size available for export. These approaches focus on the "push" side of the allocation pathway by attempting to increase the flux of an amino acid(s) to the sink tissue of the plant. Alternatively, one can increase the "pull" side of allocation by enhancing the uptake capacity of a desired sink, as Rolletschek et al. (2005) did with the ectopic expression of the *VfAAP1* gene.

In a combined strategy, the Tegeder lab (Zhang et al., 2014) has applied both a push and pull approach by ectopically expressing the pea*PsAAP1* gene in transgenic pea using the *Arabidopsis AtAAP1*promoter. In pea, the *Arabidopsis* promoter drove expression in the leaf companion cell/sieve element complex as well as the epidermal cells of the seed cotyledons, thereby generating enhanced amino acid transport capacity in both the source and sink tissues (Figure). The transgenic peas had an average 24% increase in biomass, compared to controls, a 228% increase in leaf phloem amino acids, a 35% increase in seed yield and a 6% increase in seed N. As previously observed, the activity of the transgenes had a pleotropic effect on overall plant N metabolism with greater N acquisition and significant increases in both leaf and root N content.

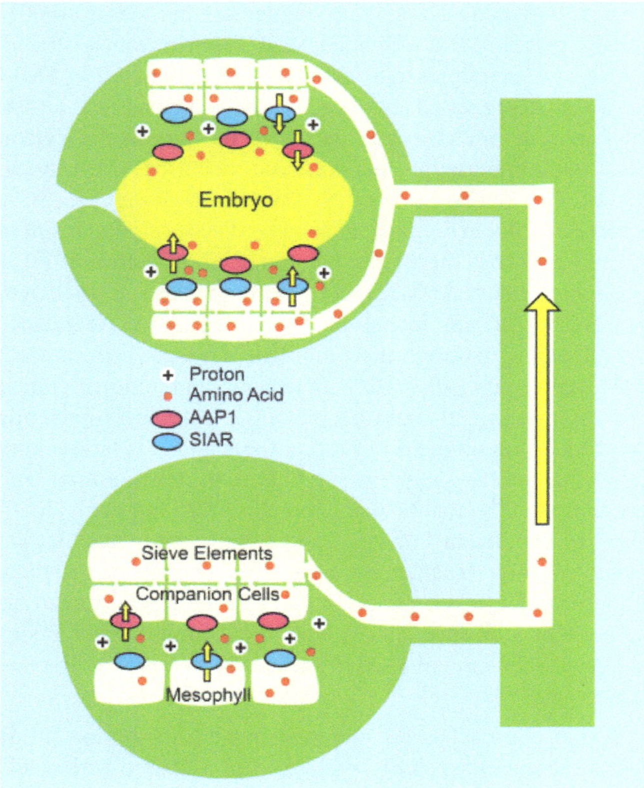

Figure 6: Combined "push" and "pull" strategy for enhancing N transport to developing pea seedling (Zhang et al., 2014). Pea plants (*Pisum sativum*) were engineered to overexpress *PsAAP1*(encoding Amino Acid Permease 1, magenta ovals) .

Transgenic manipulation of amino acid transport in plants has a real potential to increase the nutritional value of harvested tissue by increasing the overall N content and/or by targeted increases of specific amino acids, such as cysteine and methionine. However, it is clear from early transgenic experiments that altering amino acid distribution in the plant has a major impact on both N and C metabolism, which will require additional research to better understand the complex interactions between these macronutrients.

Concluding Remarks

From the few published reports on efforts to manipulate C and N partitioning, it is evident that there is potential for agricultural improvements but also unexpected pleiotropic effects which limit the desired gains, and in some cases, have unpredicted effects on plant vigor. Of particular interest from both an applied and a basic science perspective are those reporting an influence on another nutrient, such as*HvSUT1* overexpression in endosperm increasing storage protein content (and consequently more N; Weichert et al., 2010); and conversely, *VfAAP1* expression in seeds increasing seed biomass (and consequently more C; Weigelt et al., 2008); *SUT* overexpression in companion cells impacting P requirements (Dasgupta et al., 2014); and MMP1-mediated transport of SMM in phloem enhancing seed N content (Tan et al., 2010). These studies point to the complex cross talk between nutrients in specific organs and throughout the entire plant. Also in relation to nutrient cross talk, it would be very interesting to combine manipulated C, N and other nutrient transport systems to look for synergistic effects.

Studying the phloem, phloem transport, and partitioning of nutrients directly is notoriously difficult (Turgeon and Wolf, 2009), and this limitation hampers interpretation of overexpression studies. Analysis of gross morphology (organ size, etc.) is an indirect metric. However, labeling experiments to monitor nutrient transport pathways, either directly with radiolabeled or stable isotopes, or indirectly with dyes or fluorescent proteins, have been, and continue to be, informative and essential. Key technological breakthroughs would be those that allow high-resolution analysis of C and N metabolite concentrations in intact tissues, preferably in real time. Nuclear magnetic resonance imaging has been applied to whole plants, mostly in the context of water relations, but has limited resolution and utility for low-concentration metabolites (Van As et al., 2009). Fluorescence-activated cell sorting and laser capture microdissection allow cell-specific analysis of transcripts and proteins, but have limited utility for small metabolites because of their high solubility and rapid turnover. Imaging mass spectrometry using matrix-assisted laser desorption/ionization (MALDI-IMS) is an emerging technology for spatial metabolomics in tissue sections (Zaima et al., 2010; Horn et al., 2012; Horn and Chapman, 2014). MALDI-IMS resolution is currently at the multiple-cell level and could, in principle, resolve differences across loading/unloading zones—for example, between integuments and cotyledons of the embryo—but the tissue needs to be suitably fixed to overcome the problems of diffusion and metabolism. FRET (Förster resonance energy transfer) nanosensors are engineered proteins that change their fluorescence emission wavelength when a specific metabolite is bound, and have been used to monitor changes in Glc and Suc levels in living plant tissue in real time (Deuschle et al., 2006;Chaudhuri et al., 2008; Chen et al., 2012). Nanosensors also exist for various amino acids (Okumoto et al., 2005; Gruenwald et al., 2012;Zhang et al., 2013). Their continued improvement and deployment to plant apoplasmic and symplasmic compartments in plant tissues may help improve our ability to monitor metabolite levels after manipulating partitioning systems.

Although advances in our understanding of C and N partitioning systems allow us to test hypotheses by manipulating transporter streams from the "push," "pull" and combined sides, it is also clear that our current understanding is limited when looking at the bigger picture of source/sink relationships. Moreover, complex metabolic interactions between assimilation and utilization of essential nutrients confounds attempts to engineer C and N partitioning. A systems biology approach that compares whole plant nutrient partitioning in high yielding domesticated plants and less productive ancestral varieties may shed light on the labyrinth of interconnected networks that were modified during crop domestication and improvement. Nested association mapping populations, genome wide association studies, and whole-genome re-sequencing are some of the tools that could identify genetic changes in key genes and quantitative trait loci (Meyer and Purugganan, 2013; Olsen and Wendel, 2013; Wallace et al., 2014). In addition, coordinated "omic" approaches could identify key changes in transcripts, proteins, and metabolites. In spite of the many advances in our understanding of the pathways of assimilate partitioning, we still lack fundamental insights into the control and regulation of this complex, system wide, physiological process in plants. Notwithstanding, the potential of developing targeted biomass strategies to optimize partitioning for practical yield gains in crops is an applied goal that supports and justifies continued basic science research in the fundamentals of C and N allocation.

ACKNOWLEDGMENTS

BA and DB acknowledge the experimental and intellectual contributions of their respective laboratory personnel. Research on C partitioning in the laboratory of BA is currently supported by the National Science Foundation (grants 0922546, 1121819). Research in the laboratory of DB is supported by Department of Energy grant (DE-FG02-08ER64629) and USDA-AFRI grant (2011-67013-30056).

REFERENCES

1. Ainsworth E. A., Bush D. R. (2011). Carbohydrate export from the leaf: a highly regulated process and target to enhance photosynthesis and productivity. Plant Physiol. 155, 64–69. 10.1104/pp.110.167684
2. Ayre B. G. (2011). Membrane-transport systems for sucrose in relation to whole-plant carbon partitioning. Mol. Plant 4, 377–394. 10.1093/mp/ssr014
3. Barber C., Rosti J., Rawat A., Findlay K., Roberts K., Seifert G. J. (2006). Distinct properties of the five UDP-D-glucose/UDP-D-galactose 4-epimerase isoforms of *Arabidopsis thaliana*. J. Biol. Chem. 281, 17276–17285. 10.1074/jbc.M512727200
4. Beebe D. U., Turgeon R. (1992). Localization of galactinol, raffinose, and stachyose synthesis in *Cucurbita pepo* leaves.Planta 188, 354–361. 10.1007/BF00192802
5. Bourgis F., Roje S., Nuccio M. L., Fisher D. B., Tarczynski M. C., Li C. J., et al. (1999). *S*-methylmethionine plays a major role in phloem sulfur transport and is synthesized by a novel type of methyltransferase. Plant Cell 11, 1485–1497. 10.1105/tpc.11.8.1485
6. Braun D. M., Slewinski T. L. (2009). Genetic control of carbon partitioning in grasses: roles of sucrose transporters and TIE-DYED loci in phloem loading. Plant Physiol. 149, 71–81. 10.1104/pp.108.129049
7. Braun D. M., Wang L., Ruan Y.-L. (2014). Understanding and manipulating sucrose phloem loading, unloading, metabolism, and signalling to enhance crop yield and food security. J. Exp. Bot. 65, 1713–1735. 10.1093/jxb/ert416
8. Bush D. R. (1993). Proton-coupled sugar and amino acid transporters in plants. Annu. Rev. Plant Physiol. Plant Mol. Biol.44, 513–542 10.1146/annurev.pp.44.060193.002501
9. Bush D. R. (1999). “Amino acid transport,” in Plant Amino Acids: Biochemistry and Biotechnology, ed. Singh B. K., editor. (New York, NY: Marcel Dekker;), 305–318.
10. Büttner M. (2007). The monosaccharide transporter(-like) gene family in *Arabidopsis*. FEBS Lett. 581, 2318–2324. 10.1016/j.febslet.2007.03.016
11. Büttner M. (2010). The *Arabidopsis* sugar transporter (AtSTP) family: an update. Plant Biol. (Stuttg.) 12, 35–41. 10.1111/j.1438-8677.2010.00383.x
12. Cao T., Lahiri I., Singh V., Louis J., Shah J., Ayre B. G. (2013).Metabolic engineering of raffinose-family oligosaccharides in the phloem reveals alterations in carbon partitioning and enhances resistance to green peach aphid. Front. Plant Sci. 4:263. 10.3389/fpls.2013.00263
13. Carmi N., Zhang G. F., Petreikov M., Gao Z. F., Eyal Y., Granot D., et al. (2003). Cloning and functional expression of alkaline α-galactosidase from melon fruit: similarity to plant SIP proteins uncovers a novel family of plant glycosyl hydrolases. Plant J. 33, 97–106. 10.1046/j.1365-313X.2003.01609.x
14. Chaudhuri B., Hormann F., Lalonde S., Brady S. M., Orlando D. A., Benfey P., et al. (2008). Protonophore- and pH-insensitive glucose and sucrose accumulation detected by FRET nanosensors in *Arabidopsis* root tips. Plant J. 56, 948–962. 10.1111/j.1365-313X.2008.03652.x
15. Chen L. Q. (2014). SWEET sugar transporters for phloem transport and pathogen nutrition. New Phytol. 201, 1150–1155. 10.1111/nph.12445
16. Chen L. Q., Hou B. H., Lalonde S., Takanaga H., Hartung M. L., Qu X. Q., et al. (2010). Sugar transporters for intercellular exchange and nutrition of pathogens. Nature 468, 527–532. 10.1038/nature09606
17. Chen L.-Q., Qu X.-Q., Hou B.-H., Sosso D., Osorio S., Fernie A. R., et al. (2012). Sucrose efflux mediated by SWEET proteins as a key step for phloem transport. Science 335, 207–211. 10.1126/science.1213351
18. Dai N., Petreikov M., Portnoy V., Katzir N., Pharr D. M., Schaffer A. A. (2006). Cloning and expression analysis of a UDP-galactose/glucose pyrophosphorylase from melon fruit provides evidence for the major metabolic pathway of galactose metabolism in raffinose oligosaccharide metabolizing plants.Plant Physiol. 142, 294–304. 10.1104/pp.106.083634
19. Dasgupta K., Khadilkar A. S., Sulpice R., Pant B., Scheible W.-R., Fisahn J., et al. (2014). Expression of sucrose transporter cDNAs specifically in companion cells enhances phloem loading and long-distance transport of sucrose but leads to an inhibition of growth and the perception of a phosphate limitation. Plant Physiol. 165, 715–731. 10.1104/pp.114.238410
20. Deuschle K., Chaudhuri B., Okumoto S., Lager I., Lalonde S., Frommer W. B. (2006). Rapid Metabolism of glucose detected with FRET glucose nanosensors in epidermal cells and intact roots of *Arabidopsis* RNA-silencing mutants. Plant Cell 18, 2314–2325. 10.1105/tpc.106.044073
21. Doebley J. (2004). The genetics of maize evolution. Annu. Rev. Genet. 38, 37–59. 10.1146/annurev.genet.38.072902.092425
22. Dundar E., Bush D. R. (2009). BAT1, a bidirectional amino acid transporter in *Arabidopsis*. Planta 229, 1047–1056. 10.1007/s00425-009-0892-8

23. Endler A., Meyer S., Schelbert S., Schneider T., Weschke W., Peters S. W., et al. (2006). Identification of a vacuolar sucrose transporter in barley and *Arabidopsis* mesophyll cells by a tonoplast proteomic approach. Plant Physiol. 141, 196–207. 10.1104/pp.106.079533

24. Eom J. S., Cho J. I., Reinders A., Lee S. W., Yoo Y., Tuan P. Q., et al. (2011). Impaired function of the tonoplast-localized sucrose transporter in rice, OsSUT2, limits the transport of vacuolar reserve sucrose and affects plant growth. Plant Physiol. 157, 109–119. 10.1104/pp.111.176982

25. Fischer W. N., Andre B., Rentsch D., Krolkiewicz S., Tegeder M., Breitkreuz K., et al. (1998). Amino acid transport in plants.Trends Plant Sci. 3, 188–195 10.1016/s1360-1385(98)01231-x

26. Gao Z., Maurousset L., Lemoine R., Yoo S.-D., Van Nocker S., Loescher W. (2003). Cloning, expression, and characterization of sorbitol transporters from developing sour cherry fruit and leaf sink tissues. Plant Physiol. 131, 1566–1575. 10.1104/pp.102.016725

27. Gifford R. M., Thorne J. H., Hitz W. D., Giaquinta R. T. (1984).Crop productivity and photoassimilate partitioning. Science 225, 801–808. 10.1126/science.225.4664.801

28. Gil L., Yaron I., Shalitin D., Sauer N., Turgeon R., Wolf S. (2011). Sucrose transporter plays a role in phloem loading in CMV-infected melon plants that are defined as symplastic loaders. Plant J. 66, 366–374. 10.1111/j.1365-313X.2011.04498.x

29. Gruenwald K., Holland J. T., Stromberg V., Ahmad A., Watcharakichkorn D., Okumoto S. (2012). Visualization of glutamine transporter activities in living cells using genetically encoded glutamine sensors. PLoS ONE 7:e38591. 10.1371/journal.pone.0038591

30. Hannah M. A., Zuther E., Buchel K., Heyer A. G. (2006).Transport and metabolism of raffinose family oligosaccharides in transgenic potato. J. Exp. Bot. 57, 3801–3811. 10.1093/jxb/erl152

31. Horn P. J., Chapman K. D. (2014). Lipidomics in situ: insights into plant lipid metabolism from high resolution spatial maps of metabolites. Prog. Lipid. Res. 54, 32–52. 10.1016/j.plipres.2014.01.003

32. Horn P. J., Korte A. R., Neogi P. B., Love E., Fuchs J., Strupat K., et al. (2012). Spatial mapping of lipids at cellular resolution in embryos of cotton. Plant Cell 24, 622–636. 10.1105/tpc.111.094581

33. Jin Y., Ni D.-A., Ruan Y.-L. (2009). Posttranslational elevation of cell wall invertase activity by silencing its inhibitor in tomato delays leaf senescence and increases seed weight and fruit hexose level. Plant Cell 21, 2072–2089. 10.1105/tpc.108.063719

34. Keller F., Pharr D. M. (1996). “Metabolism of carbohydrates in sinks and sources: galactosyl-sucrose oligosaccharides,” inPhotoassimilate Distribution in Plants and Crops: Source-Sink Relationships, eds Zamski E., Schaffer A. A., editors. (New York: Marcel Dekker;), 157–183.

35. Kalttorres W., Kerr P. S., Usuda H., Huber S. C. (1987). Diurnal changes in maize leaf photosynthesis. 1. Carbon exchange-rate, assimilate export rate, and enzyme-activities. Plant Physiol. 83, 283–288. 10.1104/pp.83.2.283

36. Kempers R., Van Bel A. J. E. (1997). Symplasmic connections between sieve element and companion cell in the stem phloem of Vicia faba L have a molecular exclusion limit of at least 10 kDa.Planta 201, 195–201.

37. Kim I., Cho E., Crawford K., Hempel F. D., Zambryski P. C. (2005). Cell-to-cell movement of GFP during embryogenesis and early seedling development in *Arabidopsis*. Proc. Natl. Acad. Sci. U.S.A. 102, 2227–2231. 10.1073/pnas.0409193102

38. Knoblauch M., Van Bel A. J. E. (1998). Sieve tubes in action.Plant Cell 10, 35–50 10.1105/tpc.10.1.35.

39. Ladwig F., Stahl M., Ludewig U., Hirner A. A., Hammes U. Z., Stadler R., et al. (2012). Siliques are Red1 from *Arabidopsis* acts as a bidirectional amino acid transporter that is crucial for the amino acid homeostasis of siliques. Plant Physiol. 158, 1643–1655. 10.1104/pp.111.192583

40. Leggewie G., Kolbe A., Lemoine R., Roessner U., Lytovchenko A., Zuther E., et al. (2003). Overexpression of the sucrose transporter SoSUT1 in potato results in alterations in leaf carbon partitioning and in tuber metabolism but has little impact on tuber morphology. Planta 217, 158–167. 10.1007/s00425-003-0975-x

41. Lei M. G., Liu Y. D., Zhang B. C., Zhao Y. T., Wang X. J., Zhou Y. H., et al. (2011). Genetic and genomic evidence that sucrose is a global regulator of plant responses to phosphate starvation in*Arabidopsis*. Plant Physiol. 156, 1116–1130. 10.1104/pp.110.171736

42. Liu X., Bush D. R. (2006). Expression and transcriptional regulation of amino acid transporters in plants. Amino Acids 30, 113–120. 10.1007/s00726-005-0248-z

43. McCaskill A., Turgeon R. (2007). Phloem loading in *Verbascum phoeniceum* L. depends on the synthesis of raffinose-family oligosaccharides. Proc. Natl. Acad. Sci. U.S.A. 104, 19619–19624. 10.1073/pnas.0707368104

44. McCurdy D. W., Dibley S., Cahyanegara R., Martin A., Patrick J. W. (2010). Functional characterization and RNAi-mediated suppression reveals roles for hexose transporters in sugar accumulation by tomato fruit. Mol. Plant 3, 1049–1063. 10.1093/mp/ssq050

45. Meyer R. S., Purugganan M. D. (2013). Evolution of crop species: genetics of domestication and diversification. Nat. Rev. Genet. 14, 840–852. 10.1038/nrg3605

46. Michaeli S., Fait A., Lagor K., Nunes-Nesi A., Grillich N., Yellin A., et al. (2011). A mitochondrial GABA permease connects the GABA shunt and the TCA cycle, and is essential for normal carbon metabolism. Plant J. 67, 485–498. 10.1111/j.1365-313X.2011.04612.x

47. Noiraud N., Maurousset L., Lemoine R. (2001). Identification of a mannitol transporter, AgMaT1, in celery phloem. Plant Cell 13, 695–705. 10.1105/tpc.13.3.695

48. Okumoto S., Looger L. L., Micheva K. D., Reimer R. J., Smith S. J., Frommer W. B. (2005). Detection of glutamate release from neurons by genetically encoded surface-displayed FRET nanosensors. Proc. Natl. Acad. Sci. U.S.A. 102, 8740–8745. 10.1073/pnas.0503274102
49. Olsen K. M., Wendel J. F. (2013). A bountiful harvest: genomic insights into crop domestication phenotypes. Annu. Rev. Plant. Biol. 64, 47–70. 10.1146/annurev-arplant-050312-120048
50. Oparka K. J., Roberts A. G., Boevink P., Santa Cruz S., Roberts L., Pradel K. S., et al. (1999). Simple, but not branched, plasmodesmata allow the nonspecific trafficking of proteins in developing tobacco leaves. Cell 97, 743–754 10.1016/S0092-8674(00)80786-2
51. Ortiz-Lopez A., Chang H. C., Bush D. R. (2000). Amino acid transporters in plants. Biochim. Biophys. Acta 1465, 275–280 10.1016/s0005-2736(00)00144-9
52. Otto K. (1968). "Carbon metabolism: nature and formation of end products," in Harvesting the Sun—Photosynthesis in Plant Life, ed. San Pietro A., editor. (New York, NY: Academic Press, Incorporated;), 131–152.
53. Patrick J. W. (1997). Phloem unloading: sieve element unloading and post-sieve element transport. Annu. Rev. Plant Physiol. Plant Mol. Biol. 48, 191–222. 10.1146/annurev.arplant.48.1.191
54. Patrick J. W., Offler C. E. (2001). Compartmentation of transport and transfer events in developing seeds. J. Exp. Bot. 52, 551–564. 10.1093/jexbot/52.356.551
55. Proels R. K., Hückelhoven R. (2014). Cell-wall invertases, key enzymes in the modulation of plant metabolism during defence responses. Mol. Plant Pathol. 15, 858–864. 10.1111/mpp.12139
56. Reidel E. J., Rennie E. A., Amiard V., Cheng L., Turgeon R. (2009). Phloem loading strategies in three plant species that transport sugar alcohols. Plant Physiol. 149, 1601–1608. 10.1104/pp.108.134791
57. Reinders A., Sivitz A. B., Starker C. G., Gantt J. S., Ward J. M. (2008). Functional analysis of LjSUT4, a vacuolar sucrose transporter from *Lotus japonicus*. Plant Mol. Biol. 68, 289–299. 10.1007/s11103-008-9370-0
58. Reinders A., Sivitz A. B., Ward J. M. (2012). Evolution of plant sucrose uptake transporters (SUTs). Front. Plant Sci. 3:22. 10.3389/fpls.2012.00022
59. Rennie E. A., Turgeon R. (2009). A comprehensive picture of phloem loading strategies. Proc. Natl. Acad. Sci. U.S.A. 106, 14162–14167. 10.1073/pnas.0902279106
60. Roitsch T., Gonzalez M. C. (2004). Function and regulation of plant invertases: sweet sensations. Trends Plant Sci. 9, 606–613. 10.1016/j.tplants.2004.10.009
61. Rolletschek H., Hosein F., Miranda M., Heim U., Gotz K. P., Schlereth A., et al. (2005). Ectopic expression of an amino acid transporter (VfAAP1) in seeds of *Vicia narbonensis* and pea increases storage proteins. Plant Physiol. 137, 1236–1249. 10.1104/pp.104.056523
62. Rosche E., Blackmore D., Tegeder M., Richardson T., Schroeder H., Higgins T. J. V., et al. (2002). Seed-specific overexpression of a potato sucrose transporter increases sucrose uptake and growth rates of developing pea cotyledons. Plant J. 30, 165–175. 10.1046/j.1365-313X.2002.01282.x
63. Ruan Y. L., Llewellyn D. J., Furbank R. T. (2001). The control of single-celled cotton fiber elongation by developmentally reversible gating of plasmodesmata and coordinated expression of sucrose and K^+ transporters and expansin. Plant Cell 13, 47–60. 10.1105/tpc.13.1.47
64. Ruan Y. L., Patrick J. W. (1995). The cellular pathway of post-phloem sugar-transport in developing tomato fruit. Planta 196, 434–444 10.1007/BF00203641
65. Sauer N. (2007). Molecular physiology of higher plant sucrose transporters. FEBS Lett. 581, 2309–2317. 10.1016/j.febslet.2007.03.048
66. Schofield R. A., Bi Y. M., Kant S., Rothstein S. J. (2009). Over-expression of STP13, a hexose transporter, improves plant growth and nitrogen use in *Arabidopsis thaliana* seedlings. Plant Cell Environ. 32, 271–285. 10.1111/j.1365-3040.2008.01919.x
67. Schulz A., Beyhl D., Marten I., Wormit A., Neuhaus E., Poschet G., et al. (2011). Proton-driven sucrose symport and antiport are provided by the vacuolar transporters SUC4 and TMT1/2. Plant J. 68, 129–136. 10.1111/j.1365-313X.2011.04672.x
68. Slewinski T. L., Braun D. M. (2010). Current perspectives on the regulation of whole-plant carbohydrate partitioning. Plant Sci.178, 341–349 10.1016/j.plantsci.2010.01.010
69. Srivastava A. C., Ganesan S., Ismail I. O., Ayre B. G. (2009).Effective carbon partitioning driven by exotic phloem-specific regulatory elements fused to the *Arabidopsis thaliana* AtSUC2 sucrose-proton symporter gene. BMC Plant Biol. 9:7. 10.1186/1471-2229-9-7
70. Stadler R., Lauterbach C., Sauer N. (2005a). Cell-to-cell movement of green fluorescent protein reveals post-phloem transport in the outer integument and identifies symplastic domains in *Arabidopsis* seeds and embryos. Plant Physiol. 139, 701–712. 10.1104/pp.105.065607
71. Stadler R., Wright K. M., Lauterbach C., Amon G., Gahrtz M., Feuerstein A., et al. (2005b). Expression of GFP-fusions in*Arabidopsis* companion cells reveals non-specific protein trafficking into sieve elements and identifies a novel post-phloem domain in roots. Plant J. 41, 319–331. 10.1111/j.1365-313X.2004.02298.x
72. Sun A., Dai Y., Zhang X., Li C., Meng K., Xu H., et al. (2011). A transgenic study on affecting potato tuber yield by expressing the rice sucrose transporter genes OsSUT5Z and OsSUT2M. J. Integr. Plant Biol. 53, 586–595. 10.1111/j.1744-7909.2011.01063.x
73. Tan Q., Zhang L., Grant J., Cooper P., Tegeder M. (2010). Increased phloem transport of *S*-methylmethionine positively affects sulfur and nitrogen metabolism and seed development in pea plants. Plant Physiol. 154, 1886–1896. 10.1104/pp.110.166389
74. Tegeder M. (2012). Transporters for amino acids in plant cells: some functions and many unknowns. Curr. Opin. Plant Biol 15, 315–321. 10.1016/j.pbi.2012.02.001

75. Tegeder M. (2014). Transporters involved in source to sink partitioning of amino acids and ureides: opportunities for crop improvement. J. Exp. Bot. 65, 1865–1878. 10.1093/jxb/eru012

76. Turgeon R. (1996). Phloem loading and plasmodesmata. Trends Plant Sci. 1, 418–423 10.1016/S1360-1385(96)10045-5

77. Turgeon R., Beebe D. U., Gowan E. (1993). The intermediary cell—minor vein anatomy and raffinose oligosaccharide synthesis in the Scrophulariaceae. Planta 191, 446–456 10.1007/BF00195746

78. Turgeon R., Gowan E. (1992). Sugar synthesis and phloem loading in *Coleus blumei* leaves. Planta 187, 388–394. 10.1007/BF00195663

79. Turgeon R., Medville R. (1998). The absence of phloem loading in willow leaves. Proc. Natl. Acad. Sci. U.S.A. 95, 12055–12060. 10.1073/pnas.95.20.12055

80. Turgeon R., Medville R. (2004). Phloem loading. A reevaluation of the relationship between plasmodesmatal frequencies and loading strategies. Plant Physiol. 136, 3795–3803. 10.1104/pp.104.042036

81. Turgeon R., Medville R., Nixon K. C. (2001). The evolution of minor-vein phloem and phloem loading. Am. J. Bot. 88, 1331–1339. 10.2307/3558441

82. Turgeon R., Wolf S. (2009). Phloem transport: cellular pathways and molecular trafficking. Annu. Rev. Plant Biol. 60, 207–221. 10.1146/annurev.arplant.043008.092045

83. Usadel B., Bläsing O. E., Gibon Y., Retzlaff K., Hoehne M., Günther M., et al. (2008). Global transcript levels respond to small changes of the carbon status during progressive exhaustion of carbohydrates in *Arabidopsis* rosettes. Plant Physiol. 146, 1834–1861. 10.1104/pp.107.115592

84. Van As H., Scheenen T., Vergeldt F. J. (2009). MRI of intact plants. Photosyn. Res. 102, 213–222. 10.1007/s11120-009-9486-3

85. Vaughn M. W., Harrington G. N., Bush D. R. (2002). Sucrose-mediated transcriptional regulation of sucrose symporter activity in the phloem. Proc. Natl. Acad. Sci. U.S.A. 99, 10876–10880. 10.1073/pnas.172198599

86. Voitsekhovskaja O. V., Rudashevskaya E. L., Demchenko K. N., Pakhomova M. V., Batashev D. R., Gamalei Y. V., et al. (2009).Evidence for functional heterogeneity of sieve element-companion cell complexes in minor vein phloem of *Alonsoa meridionalis*. J. Exp. Bot. 60, 1873–1883. 10.1093/jxb/erp074

87. Wallace J. G., Larsson S. J., Buckler E. S. (2014). Entering the second century of maize quantitative genetics. Heredity 112, 30–38. 10.1038/hdy.2013.6

88. Wang N., Fisher D. B. (1994). The use of fluorescent tracers to characterize the post-phloem transport pathway in maternal tissues of developing wheat grains. Plant Physiol. 104, 17–27. 10.1104/pp.104.1.17

89. Weichert N., Saalbach I., Weichert H., Kohl S., Erban A., Kopka J., et al. (2010). Increasing sucrose uptake capacity of wheat grains stimulates storage protein synthesis. Plant Physiol. 152, 698–710. 10.1104/pp.109.150854

90. Weigelt K., Kuster H., Radchuk R., Muller M., Weichert H., Fait A., et al. (2008). Increasing amino acid supply in pea embryos reveals specific interactions of N and C metabolism, and highlights the importance of mitochondrial metabolism. Plant J.55, 909–926. 10.1111/j.1365-313X.2008.03560.x

91. Wippel K., Sauer N. (2012). *Arabidopsis* SUC1 loads the phloem in suc2 mutants when expressed from the SUC2 promoter. J. Exp. Bot. 63, 669–679. 10.1093/jxb/err255

92. Zaima N., Goto-Inoue N., Hayasaka T., Setou M. (2010).Application of imaging mass spectrometry for the analysis of*Oryza sativa* rice. Rapid Commun. Mass Spectrom. 24, 2723–2729. 10.1002/rcm.4693

93. Zhang C., Wei Z. H., Ye B. C. (2013). Imaging and tracing of intracellular metabolites utilizing genetically encoded fluorescent biosensors. Biotechnol. J. 8, 1280–1291. 10.1002/biot.201300001

94. Zhang L., Garneau M. G., Majumdar R., Grant J., Tegeder M. (2014). Improvement of pea biomass and seed productivity by simultaneous increase of phloem and embryo loading with amino acids. Plant J. 81, 134–146. 10.1111/tpj.12716

95. Zhang W.-H., Zhou Y., Dibley K. E., Tyerman S. D., Furbank R. T., Patrick J. W. (2007). Nutrient loading of developing seeds.Funct. Plant Biol. 34, 314–331 10.1071/FP06271

96. Zhang X.-Y., Wang X.-L., Wang X.-F., Xia G.-H., Pan Q.-H., Fan R.-C., et al. (2006). A shift of phloem unloading from symplasmic to apoplasmic pathway is involved in developmental onset of ripening in grape berry. Plant Physiol. 142, 220–232. 10.1104/pp.106.081430

97. Zhou Y. C., Chan K., Wang T. L., Hedley C. L., Offler C. E., Patrick J. W. (2009). Intracellular sucrose communicates metabolic demand to sucrose transporters in developing pea cotyledons. J. Exp. Bot. 60, 71–85. 10.1093/jxb/ern254

INDEX

I

L

M

N

O

P

Q

R

S

T

U

V

W

X